The Friendly Orange Glow

The Friendly Orange Glow

The Untold Story of the PLATO System
and the Dawn of Cyberculture

BRIAN DEAR

Pantheon Books
New York

Library of Congress Cataloging-in-Publication Data
Name: Dear, Brian, [date] author.
Title: The friendly orange glow : the untold story of the PLATO
system and the dawn of cyberculture / Brian Dear.
Description: First edition. New York : Pantheon Books
[2017]. Includes bibliographical references and index.
Identifiers: LCCN 2017013007. ISBN 9781101871553
(hardcover : alk. paper). ISBN 9781101871560 (e-book)
Subjects: LCSH: PLATO (Electronic computer
system)—History. Cyberspace—History.
Classification: LCC QA76.8.P53 D43 2017.
DDC 303.48/34--dc23.
LC record available at lccn.loc.gov/2017013007

www.pantheonbooks.com

Jacket design by Oliver Munday

Printed in the United States of America
First Edition

2 4 6 8 9 7 5 3 1

For Patricia

Contents

Preface

Imagine discovering that a small group of people had invented a fully functioning jet airplane capable of flying long distances at hundreds of miles per hour, *decades before* the Wright brothers cast their fragile craft into the wind for twelve seconds over North Carolina sand dunes in 1903. Imagine how such a discovery would disrupt our common understanding of history. The story of PLATO, a computer system so far ahead of its time and perhaps the least known major twentieth-century technology project, may strike you as just as impossible as a nineteenth-century jet, but PLATO really happened. It is a story of inventors, mavericks, hackers, geniuses, visionaries, scientists, and educators who came together not so long ago in the very heart of the American Midwest, a story that so disrupts the conventional view of twentieth-century technology history that it may make you wonder, as it has made this author wonder, *How could this have happened?* Where are the books, the magazine articles, the documentaries, and the museum displays that should have covered this story? Why has this story gone untold? Why are we only finding out about this now?

Having begun using PLATO in 1979, as an undergraduate at the University of Delaware, I developed an insatiable curiosity about who these people were. I wanted to understand where they came from. I kept asking myself, Why is the world ignoring this incredible phenomenon? After a number of years, I decided I was not going to wait any longer for someone to come along and write the history documented in this book, so I reluctantly set out to do it myself.

And here we are. What follows in these pages is an attempt to provide a glimpse into the largely unknown world of PLATO. As you read the following chapters, pay attention to dates. Considering what innovations are being described—many of them things we now use every day of our lives—it is hard to believe how many years ago these things came about, long before personal computing, social media, and the Internet existed. A book about the history of the PLATO system turns out not to be just about PLATO. It also sheds new light on the origins of diverse fields, from what we now call e-learning or personalized instruction, to flat-panel displays and touch-sensitive screens; from time-sharing to cloud computing; from MUDs and other multiplayer games to online communities and social networking. To absorb the history of PLATO leads to a better understanding and deeper appreciation of all these areas of innovation, and of the nature of innovation itself.

In the American study of economics there are two prevailing schools of thought, one called the "saltwater" view and the other called the "freshwater" view. The economic details of each view are unimportant here. What's notable is that the saltwater view refers to the predominant economics mind-set at universities and think tanks on the East and West Coasts of the United States. The freshwater view belongs primarily to Midwestern economists closer to Lake Superior and Lake Michigan, mainly at the University of Chicago.

Unlike in economics, in the popular history of computing there's one prevailing view, and it's a saltwater view: of, by, and for the two coasts, with its centers being Silicon Valley and Cambridge, Massachusetts. Whether the view was taught in school, broadcast on television, depicted in movies, printed in books or magazines, or published on the Internet, the view holds that most of the great American computer innovations of the twentieth century share one thing in common.: they happened on the coasts. The history of the last sixty years of computing, from mainframes to the rise of the Internet, from silicon chips to the revolution brought forth by the personal computer, to, in more recent years, the marvels of smartphones, portable digital music players, touch-based tablet computers, search engines, billion-user social networks, online stores—these are breakthroughs that came about on the coasts.

So used to saltwater stories of the Next Big Thing in technology are we that we have largely missed what went on between the coasts. To this day, many of these freshwater accomplishments have gone unnoticed.

We celebrate the rise of ARPANET, the Internet, and the World Wide Web. We celebrate the accomplishments of Xerox's Palo Alto Research Center, which in turn inspired Apple's Lisa and Macintosh computers and much of the personal computing environments we still use today. We celebrate the innovations produced by legions of start-up companies and placed into the hands of millions, now billions of people: Innovations that have changed the world. Innovations people cannot *imagine* living without. We are living in the very "shocking future" Alvin Toffler wrote about—warned us about—forty-five years ago. And the history of how we reached this future has been researched, deciphered, studied, analyzed, organized, and disseminated far and wide for long enough that the story has become legend, set in stone. Nerds, geeks, and hackers are no longer outcasts and ridiculed; they're now sought-after "thought leaders," many counted among the tens of thousands of recent millionaires and hundreds of billionaires. The list of heroes' names in the "computer revolution" is long. But there is an equally long list of unknown computer pioneers, the people whose stories fill the pages of this book.

To be in the great state of Illinois is to be hours away by jet, days by car or rail, from the West or East Coasts, each a thousand miles away. To be in Illinois is, instead, to be in the heart of the fruited plain, that vast prairie with soil so rich you could jam a broomstick into it and leaves would sprout. Some 80 percent of the state's nearly 58,000 square and famously flat miles are devoted to farming, much of it corn and, in more recent years, soy. Endless farmland surrounds most Illinois cities and towns, which pop up like islands in a sea of green.

Out in the middle of that fruited Illinois plain there's a place where a lot of the future we take for granted today got started, long ago. The town is Urbana, and the place is one of the largest universities in the United States: the University of Illinois at Urbana-Champaign. On UI's campus there's one particular building that is at the very heart of our story. Most of the events chronicled in the pages that follow took place in this particular building, or if they took place elsewhere (sometimes clear across the world), they did so *only because* of prior things that had taken place in this building. What went on in this building changed lives. It may have changed yours and you don't even realize it.

What happened in this building changed the world. And unlike every other major technology story of the twentieth century, very few have heard of it: the PLATO system was born in this building.

The building's PLATO presence is long gone, the names and numbers on office doors scraped off long ago. The mechanical engineering department took over the building years ago, filling remodeled offices with researchers working on everything from improving air conditioners to "nanoscale chemical-electrical-mechanical manufacturing systems."

The Mechanical Engineering Laboratory
and Power House, circa 1920

The brick-walled building, four narrow stories, stands alongside Mathews Avenue, a quiet tree-lined street running north and south through the campus. The building went up during a noisier, sootier era, when railroad tracks ran through the streets, upon which rolled trolley cars, horse-drawn delivery trucks, Model Ts, and steam locomotives. Outnumbering the trees, stark utility poles supported countless power and phone lines in every direction. One track along the north side of the building crossed the street and led into a long-gone locomotive testing plant, complete with roundhouse, where engineering students worked on real trains and engine designs.

For more than half of the building's hundred-plus years a tower of one sort or another loomed alongside it. In 1910, as part of the original construction as the campus power plant, there stood a 175-foot-tall brick smokestack along the south side, toward the rear of the building. The stack could be seen from miles away belching out thick, black billows of smoke from the coal furnace, boilers, and steam generators inside. But in time, the campus outgrew the power plant's capacity, so a bigger plant was built farther outside town. The smokestack was torn down, the building's length more than doubled, and soon another tower, for radar research, was constructed on its north side.

Doric columns on either side of the entranceway's double doors support a simple stone entablature above, engraved into which are the words "POWER HOUSE." It's now the only hint of the building's distant power plant past. If it wasn't for a small historical marker right outside the doors, there would be no hint at all that this was also the birthplace of PLATO.

The doors open to a small landing and a few steps leading up to another set of double doors. The steps, and the walls halfway up to the handrails on either side, are covered in little stone tile squares varying in shades of faded orange and earth-clay brown. Up the stairs, through the doors, down the hall, there are offices and labs. In those rooms not so long ago, atop tables and desks rested computer terminals sporting impossibly futuristic, high-resolution, touch-sensitive, flat panel gas plasma displays from which emanated a light that was at the time commonly called the Friendly Orange Glow.

The level to which PLATO, its people, and its history have been ignored is extraordinary given not only how seminal the innovations were and how early its online community flourished, but also how recently it all happened. PLATO was a computer system, but more important, it was a culture, both physical and online, a community that formed on its own, with its own jargon, customs, and idioms; its own cast of thousands, a world familiar to us yet subtly foreign, an entire era that clashes with the accepted, canonical history of computing, social media, online communities, online games, and online education. It's as if an advanced civilization had once thrived on earth, dwelled among us, built a wondrous technology, but then disappeared as quietly as they had arrived, leaving behind scraps of legend and artifacts that only few noticed. This book is the result of an effort to capture the history of this lost culture of innovation before it vanishes completely, by someone who had the great fortune to come of age, to "become digital," as it were, within that very culture, thereby having a chance to get to know some of the people there, their stories, their visions, and their amazing technologies—technologies we all now recognize and use.

This book is as much the *biography of a vision* as it is the story of the people behind PLATO. Every technology story, whether it's about the steam engine, textile loom, light bulb, telephone, airplane, Model T, or, more recently, Macintosh, Google search engine, or Tesla electric car, has at its core a vision. It is the unalterable fate of technology visions that they run full life cycles from conception to obsolescence. Tech visions typically start with one person with an idea. Early on that visionary faces skepticism and rejection. But if the visionary has sufficient technical proficiency and the support of colleagues, he or she may hang on long enough to get a prototype working, an accomplishment that attracts others, who soon come to share in the vision, and thus a technology project snowballs, attracting more bright people, more funding, and, hopefully, widespread adoption. In time the vision gets stale, and the visionary grows stubborn, as others dream up new visions that challenge then replace the old. It's how we got Facebook, the answer to MySpace, which was the answer to Friendster. It's how we got Google, which was the answer to AltaVista, Lycos, and Infoseek. It's how we got the iPhone after the Palm Treo, Apple Newton, and flip phones.

PLATO's story is no different. *But what a story.*

The Friendly Orange Glow is divided in three parts. Part One, "The Automatic Teacher," begins before PLATO with the story of the 1950s work of the psychologist B. F. Skinner. In subsequent chapters we'll see how a confluence of people, ideas, national emergencies, and government mandates led to the creation of the PLATO system. While the project's budget and staff were small at the beginning, the scale of its ambition was not. From this project would come major inventions, including the flat panel gas plasma display, the source of the Friendly Orange Glow that many thousands of people would come to know in the 1970s and after. PLATO was a project designed to see if a computer could teach a student as well if not better than a human could. Over the years, many lessons would be learned—not just by the students.

Part Two, "The Fun They Had," is named after Isaac Asimov's short story in which a brother and sister in the far future complain about being "schooled" via boring mechanical robot "teachers" in their bedrooms. The siblings discuss how in the distant past kids met in a physical classroom with other kids, and a real live human teacher taught them. The little girl imagines the fun they must have had. In Part Two we discover that PLATO, a real-life automatic teacher, was nowhere near as boring to work with as Asimov predicted. In fact, even the creators of PLATO had no idea what would happen in the 1970s: young people would in many ways take over the system and turn it into exactly the kind of exciting, endlessly stimulating, and, for many, addictive, online environment that we all live with today. Only they did it forty-five years ago—when nobody in the saltwater states was paying attention. The result was the dawn of cyberculture in all the ways we know it today.

Part Three, "Getting to Scale," chronicles the effort to transform the original dream of building the PLATO system into a new dream to build a gigantic worldwide computer network, not just for education but for online community, business, research, entertainment, and games, and by selling the service to schools, homes, and businesses, hopefully making a fortune in the process. Fortunes were indeed made, though how they were made, who made them, and who didn't—well, therein lies a tale.

Welcome to PLATO. Turn the page to begin.

Part I

THE AUTOMATIC TEACHER

Children seem to be such remarkable learners on their own, but then they enter school.

—Seymour Papert

There will always be naysayers. They're the same people who go from saying it's impossible to saying it's inevitable.

—Elon Musk

Whenever anything is being accomplished, it is being done, I have learned, by a monomaniac with a mission.

—Peter Drucker

Praeceptor Ex Machina

They sat in little wooden chairs in front of little wooden desks. Each desk had a hinged lid that opened to reveal a place to stash books, pencils, and papers. The left and right sides of the classroom were lined with great big windows filling the room with daylight and, on warm-enough days, fresh air. At the moment the room was quiet. The teacher, Mary Eliot, went to the blackboard and wrote some arithmetic problems upon it. Then the twenty boys and girls in the class began to work on the problems. Miss Eliot strolled up and down the aisles between the rows of desks, now and then stopping to answer a child's question or point out a mistake, before continuing her stroll. Suddenly, she scuffed the sole of her shoe against the hardwood floor, right next to a slouching child's outstretched foot.

November 11, 1953, Armistice Day around the nation, happened to also be Father's Day at the Shady Hill School in Cambridge, Massachusetts. Shady Hill was a private school nestled in a leafy suburb just past a big bend in the Charles River about a mile west of Harvard University. Among the Harvard professors who sent their children to the school was the behavioral psychologist B. F. Skinner, whose two daughters, Deborah and Julie, attended. On this particular day, fathers had been invited to come to the school, spend time in their children's classes, and observe the lessons. Skinner sat in on his daughter Deborah's fourth-grade arithmetic class and watched how Miss Eliot taught the children.

He could not believe what he saw.

Decades later, Skinner could still recall Eliot's scuffing her shoe to correct a child's posture, with his own dramatic shoe-scuffing demonstration. But something *else* he saw hit him much harder that day, making an impact that would change not only his own life, but the lives of people all over the world.

Burrhus Frederic Skinner was born in 1904 in Susquehanna, Pennsylvania. He attended the town's small public school for twelve years. As a teenager he wrote for the local paper, and when he went off to Hamilton

College he majored in English with dreams of becoming a novelist. After graduating, writer's block made the novelist's life look less promising. He spent time working in a New York bookstore, where he came across books by the psychologists John Broadus Watson and Ivan Pavlov, introducing him to something called "behaviorism."

Behaviorism arose as a movement in psychology shortly before the outbreak of World War I, largely due to the work of Watson at the University of Chicago. Wat-

B. F. Skinner in 1960

son wanted psychology to be taken seriously as one of the "hard" sciences, based on *facts*—observable, verifiable, repeatable phenomena seen and recorded through experimentation. To Watson, any study of psychology that dealt with notions like "feelings" and "consciousness" had no place in science. The historian Edna Heidbreder once described Watson's rigid views this way: "Consciousness is only another name for the soul of theology, and the attempts of the older psychology to make it seem anything else are utterly futile. To admit the mental into science is to open the door to the enemies of science—to subjectivism, supernaturalism, and tender-mindedness generally. With the simplicity and finality of the Last Judgment, behaviorism divides the sheep from the goats. On the right hand side are behaviorism and science and all its works; on the left are souls and superstition and a mistaken tradition; and the line of demarcation is clear and unmistakable."

By 1924, Skinner had given up becoming a novelist and, fascinated with behaviorism, enrolled in the psychology graduate program at Harvard headed by William Crozier, who became a mentor, shaping his view that the best way to understand an organism, even a human, was by understanding its behavior. *Observable* behavior was the key. No need to check under the hood—no need to guess what's going on in the brain—rather, just observe and undertake an experimental analysis of the organism's external behavior. Over the coming decades, Skinner would become the father of "operant behaviorism," the school of thought that posits that all animals, including humans, are influenced by the consequences of their own behavior. Whenever a given animal acts, Skinner would argue, it follows that it experiences the consequences of its act. Skinner believed that the nature of those consequences could reliably determine how an animal will act in the future.

Skinner eventually joined the Harvard faculty and became one of the most famous, and certainly one of the most controversial, experimental psychologists of the twentieth century. He called himself a "radical behaviorist." Much of Skinner's early experimental research centered on analyzing the behavior of rats and pigeons. His Operant Conditioning Chamber, known widely as the "Skinner Box," was a device in which he placed an animal, often a rat, which after wandering around in the box would eventually bump into or otherwise cause a lever to be pressed, an action that released a pellet of food for the rat to eat. The next time the rat hit the lever, the same thing happened. Pretty soon, the rat figured out how to get the food: press the lever. Over time Skinner would "schedule" the delivery of the food: it might be immediately upon hitting the lever, or it might be only after the rat hit the lever several times. The delivery of food served, in his view, as a reinforcement designed to encourage a repeat of the lever-pressing behavior. Each time the rat pressed the lever the desired number of times, another pellet dropped into the box, thus reinforcing the consequences of the rat's behavior. No surprise: rats caught on quickly, and the pellets flowed. The Skinner Box owed much to a predecessor device called the "Puzzle Box," built by the Columbia University psychologist Edward L. Thorndike. In the Puzzle Box, a cat discovered a piece of salmon that was just out of its reach. Only by accidentally hitting a lever that opened a door on the side of the box was the cat able to get to the food. Over time, the cat, like Skinner's rat later on, learned to skip struggling to reach for the food and instead go for the lever. Thorndike called this "the law of effect," wherein an organism increases over time the frequency of a behavior that yields satisfactory outcomes arising from that behavior, and reduces over time the frequency of behaviors yielding unsatisfactory outcomes.

Another Skinner creation, which he initially called a "baby tender" and later dubbed the "Air Crib," was introduced to mainstream America in a sensational article he had submitted to *Ladies' Home Journal* in 1945. The editors renamed Skinner's original title for the article with the more provocative "Baby in a Box," and published it in the October issue. The article featured a jarring full-page picture that at first glance a reader might think showed a year-old baby girl trapped inside a glass box, holding her hands up to the glass while she looked out at the world.

"When we decided to have another child," Skinner wrote, "my wife and I felt that it was time to apply a little labor-saving invention and design to the problems of the nursery. We began by going over the disheartening schedule of the young mother, step by step. We asked only one question: Is this practice important for the physical and psychological health of the baby? When it was not, we marked it for elimination. Then the 'gadgeteering' began."

The resulting "gadget" was a compartment, about the size of a conventional crib, closed on five sides with a pair of hinged doors made of safety glass in front, which the parents could open so their baby Deborah could be tended to. Heat and humidity were controlled automatically within the compartment ("We keep the temperature at 78 degrees, humidity 50. She is never too hot, or too cold, but just right"), and it was not necessary to use blankets or clothing on the baby other than a diaper. "The human species evolved in a tropical climate and certainly without the benefit of clothing," Skinner wrote. Mother and father were pleased with the freedom of movement the baby enjoyed inside: "When awake, she exercises almost constantly and often with surprising violence. Her leg, stomach, and back muscles are especially active and have become strong and hard. It is necessary to watch this performance for only a few minutes to realize how severely constrained the average baby is, and how much energy must be diverted into the only remaining channel, crying."

World War II was over, and while America was ready to enjoy the fruits of consumerism, it was not quite prepared to embrace the Air Crib, and only a few ever sold commercially. Over the years an urban legend spread that the Air Crib was simply a glorified Skinner Box, this time redesigned for experimenting on a human baby instead of a rat, and that Skinner's daughter Deborah had suffered greatly under the hands of her mad-professor father, to the point, so the story goes, that she grew up to be psychotic and ultimately shot herself in a bowling alley in Billings, Montana. It was all a wild rumor and nothing but, possibly put out there to discredit behaviorism in general and Skinner's name in particular. Many decades later, Deborah Skinner, very much alive, still finds she has to occasionally pen a letter to a newspaper editor, pointing out the blatant inaccuracy of the rumor and how healthy and normal her childhood actually was.

The same year that the *Ladies' Home Journal* article appeared, the future Shady Hill schoolteacher Mary Eliot graduated from Vassar College. She'd heard about Skinner's Air Crib, probably, she says, from that famous article. Only, she and her friends didn't call it an Air Crib; they, like many people then, referred to it as "the box" or as the "Skinner Box," with all of the rat-experiment connotations that that name brought with it. "We were fascinated that anybody would want to raise a child in a box," Eliot recalls. "Of course, the child was very small at this time. But it seemed very strange to me." So when, in 1953, Deborah Skinner, the actual Baby in the Box from the *Ladies' Home Journal* article, entered the fourth grade at Shady Hill School, "I was very much interested to see what she was like," says Eliot. "It turned out she was very much like most children. I didn't see any great effects from being raised in a box."

When asked about his own childhood experience at the Susquehanna School, B. F. Skinner told this author, "I *loved* school! The first and second grades were in one room, the third and fourth were in another, the fifth and sixth in another, seventh and eighth in another, and the rest in high school. One teacher had charge of two classes. We learned a great deal, we were well disciplined; we sat in seats with desks built on them and so on, that kind of a thing, and we had outdoor toilets, big, you know, no plumbing, it was that kind of place. But I *loved* it, I tried to go there as early as I could in the day." He viewed his teachers as generally excellent, and in the first volume of his autobiography, *Particulars of My Life*, he devoted many pages to his favorite, an English teacher, Miss Mary Graves. She was a "very important person in my life," he explained. "She was someone who listened to me, answered my questions, and almost always had something interesting to say or a suggestion of something interesting to do."

If his own schooling had been so idyllic, his desire each morning to race to school so palpable, his own teachers so wonderful, with one, Mary Graves, being a such profound influence on his entire life, why, beginning in the 1950s, did he think education and learning urgently needed improving? What triggered this desire to get in there and try to fix education?

"It was my *daughters'* school that got me going," he said.

As Miss Eliot walked up and down the aisles between the children's desks in his daughter's fourth-grade Shady Hill arithmetic class that November day in 1953, something else besides the screech of Miss Eliot's shoe against that hardwood floor had caught Skinner's eyes and ears. He noticed that some students finished their math problems early and then sat there with nothing to do; others lagged until the end of the period, at which time she collected the papers, which—one hoped—she would grade that night and return the next day.

"I suddenly realized," Skinner would cry out years later, reliving that Shady Hill moment with a slap of his hand against his formidable forehead, *"God! This violates everything we know about the learning process."*

Whether Eliot knew it or not, Skinner felt, she was breaking not one but *two* sacred rules, rules this staunch behaviorist believed were bedrock-foundational principles of learning: *Self-Pacing* and *Immediate Feedback*. If those two principles were not practiced in a classroom setting, the teacher was impeding learning, plain and simple. "The students," Skinner felt, "were not being told *at once* whether their work was right or wrong." A paper, quiz, or exam marked up with corrections and a grade, but not seen by a student until some twenty-four hours later (or longer), could not *possibly* act as effective reinforcement. Furthermore, "they were all moving at the same pace regardless of preparation or ability." If it takes Johnny ten minutes to get through some math problems that Sally whips through in five, why should Sally then have to wait around for Johnny to catch up? And yet that is what Skinner saw: "A few students soon finished and were impatiently idle. Others, with growing frustration, strained."

Skinner hadn't invented those two principles, Self-Pacing and Immediate Feedback, but he was fully invested in them. Others had developed the concepts years earlier, starting with the early behaviorists in the first two decades of the twentieth century. Then, in the 1920s, an Ohio State University professor of psychology named Sidney L. Pressey took the ideas a step further by building a series of "automatic teacher" devices to test his theories. One of his machines issued a piece of candy whenever a student got the right answer. The results of the research made a believer out of Pressey. He observed that when teachers grade student papers but fail to return them to students until some later time, much of the potential *impact* on the learning that was supposed to be taking place is lost. For behaviorists, this slow-poke grading approach epitomized one of the weaknesses of education, a way that seemed designed to be as bureaucratically—not to mention behaviorally—inefficient as possible. "When an examination is corrected and returned after a delay of many hours or days," Pressey wrote, "the student's behavior is not appreciably modified." In behaviorist terms, if the organism had no way of determining what the consequences of its behavior were, right away, then it was unlikely one could predict how the organism would act later as a result of that previous behavior. This would not do.

He not only published papers and presented his findings at conferences, but also attempted to build and sell automatic teaching and testing devices in the hopes that they would receive widespread adoption in schools. Maybe he might strike it rich along the way. Unfortunately, he went a little overboard, as true-believer technology start-up founders often do. Seeing great potential in his products, he "worked himself sick," says one historian, "to contrive and commercialize machines, remedial materials, and tests." The timing of Pressey's attempts at commercializing his machines couldn't have been worse. He introduced his most advanced device right around the time of the catastrophic 1929 stock market crash and the dawn of the Great Depression. Pressey's ideas failed to catch on, the nation's priorities shifted, and for several decades not much happened. In the 1940s, he attempted to rekindle interest in his work, publishing and presenting a few more times, but it wasn't until Skinner saw what was going on in his daughter's arithmetic class in 1953 that things started to heat up again, rekindling behaviorist interest in the two principles to the point that they would quickly begin to propagate throughout Skinner's thinking and his published work—and, soon, throughout the thinking and the work of many others.

Later the very same day Skinner had sat in on his daughter's fourth-grade class, he went to work wondering about what he could do to facilitate more effective learning. Was there a way for a teacher to successfully educate an entire class of students while *not* violating those two fundamental learning principles? Could a teacher, Miss Eliot or any teacher for that matter, teach a class, all the while providing Immediate Feedback and Self-Pacing at appropriate levels for each individual student? Somewhere around this point Skinner made the same leap that Pressey had made all those years before. Maybe this was the wrong question to ask. Maybe the right question to ask was, forget the teacher for a moment—was there simply a better way for a child to learn?

It did not seem humanly possible for Miss Eliot or any other teacher to teach a class of kids *and* do so while providing Self-Pacing and Immediate Feedback. How could *one* teacher provide the proper reinforcement, as Skinner would have termed it, at just the *right* time for *each* of the twenty students, without boring the other students, so they were all effectively learning *at their own pace*? Perhaps if there was a way to make twenty clones of Miss Eliot, or hire enough teachers at the school so there was always a 1:1 student-teacher ratio . . . but neither of those scenarios was realistic then, and they're still unrealistic today.

But perhaps there was a way to solve this conundrum. Especially if the focus were more on helping the child *learn*, and less on helping the teacher *teach*. He realized from his own laboratory work that the answer was "instrumentation," which for Skinner meant more "gadgeteering." If you can't clone the teacher, then why not turn the teacher into a *machine*, and *clone the machine*. A machine, some sort of an apparatus—Skinner would have to figure out what exactly—that not only helped students learn but actually *taught*. For Skinner the question was quickly evolving to one like this: perhaps if the twenty boys and girls seated in their little chairs in front of their little desks, each had little *machines* on top of those little desks, and each child was interacting, at their own comfortable pace, *privately* with their own little machine, with Miss Eliot strolling up and down the aisles of the classroom to mind the children's progress and stop here and there to help when and as needed, we might get somewhere?

For days Skinner gadgeteered, trying to think this machine through. His Harvard colleagues reacted with confusion as he sliced manila folders into pieces using scissors, then took the pieces to create a crude box. Not a box for rats this time, not for babies, but a box for teaching, to help students learn. He then built a sturdier version of the box out of plywood pieces. It was his first "teaching machine" and required no power, operated by manually moving a simple slider mechanism. While it was primitive, it was a start of something that moved closer to his ideals of offering learners Self-Pacing and Immediate Feedback.

Teaching machines became Skinner's new passion, almost an obsession. Over the next few months, he improved the machine's design and ruggedness, and in March 1954 he packed up the latest version and took it to a psychology conference at the University of Pittsburgh, where he demonstrated it publicly for the first time. He submitted a conference paper, "The Science of Learning and the Art of Teaching," that accompanied his demonstration. It caused a sensation.

Skinner's 1954 machine

Skinner's machine is a plywood box some fifteen inches high, wide, and deep. The hinged top opens so that a scroll of pleated paper tape can be loaded inside. Typewritten onto the paper are arithmetic problems to be solved by the student; the answers to each problem are encoded as small holes, like in a player piano, punched in very specific places to denote a corresponding value. When the box top is shut, you can read the math problems, one at a time, through a small opening in the top of the box, showing a small, exposed area of the paper. You might see a math problem, such as "3 + 2," through the viewing window on the top side of the box. To answer the problem, you manipulate a series of what Skinner called "sliders" that can be moved up or down through slits in the wood, serving as number scales. This particular math problem is looking for a one-digit answer, so only one slider need be moved into place. The correct slider to move is the one that causes numbers to appear in the hole under the "3 + 2." Pull the slider toward yourself and the number increases up to 9. Push it away and it drops to zero. Once you've formed your answer you can attempt to turn a big black knob on the front side of the box. If the answer is correct, the knob turns freely and the scroll advances to the next problem. This is how you know your answer is correct. If the knob is locked and the scroll won't advance, you know you're wrong.

Unlike a written test, which requires a teacher to laboriously correct and grade afterward, in theory this device freed her from that labor, at least the labor of grading—she still had to compose the questions. When a student finished an entire scroll of problems, the teacher could assume the student finally got all the answers right.

Skinner's accompanying paper at the conference was subsequently reprinted in the *Harvard Educational Review* and elsewhere, and quickly became one of the most widely cited papers in the field for many years. Like the provocative "Baby in a Box" article from 1945, it sparked discussion and controversy, but most of all great interest. In 1968, he finally extended the ideas born out of that paper into a book-length work entitled *The Technology of Teaching.* He dedicated the book to Mary Graves.

Pressey read Skinner's 1954 paper and contacted him. "He wrote me and asked if we could get together," Skinner told the author. "So we had breakfast and . . . he brought me his papers. Later he sent me one of his machines. He was very cordial. And he saw the point of it all."

But Skinner never considered Pressey's machine to be a *teaching* machine: instead, he felt Pressey's device was more of a *testing* machine, not something that taught something new. Still, Pressey's thinking clearly influenced Skinner, and the eventual outcomes Pressey hoped for in the long term were arguably things Skinner would have agreed with: an "industrial revolution" in education, in which, Pressey had once written, "educational science and the ingenuity of educational technology combine to modernize the grossly inefficient and clumsy procedures of conventional education. Work in the schools of the future will be marvelously though simply organized, so as to adjust almost automatically to individual differences and the characteristics of the learning process. There will be many labor-saving schemes and devices, and even machines—not at all for the mechanizing of education, but for the freeing of teacher and pupil from educational drudgery and incompetence."

As Skinner's ideas spread, and the field of teaching machines and "programmed instruction" exploded over the next few years, a long-retired Pressey began to have a change of heart. "He became unhappy," Skinner told the author. "He felt that I was wrong, that a teaching machine should be a testing machine. You should instruct, and then use the machine. Not use the machine to instruct." Which was exactly the pivot Pressey hoped Skinner and the teaching machine practitioners would undertake in their strategy.

Pressey's concerns were largely ignored.

The Shady Hill School had been founded out of necessity by parents and educators in the Harvard community who had been sending their children to the Agassiz public school until 1914, when the building was condemned. Concerned parents, realizing that a new school building was not likely, took matters into their own hands, starting their own Cooperative Open Air School, initially situated on the back porch of Agnes and William Ernest Hocking, who along with other Harvard faculty taught the students. William Ernest Hocking was a noted philosophy professor at Harvard whom later Skinner met during graduate school. In a few years the Hockings found a more permanent location in the nearby Shady Hill neighborhood, and the school took that name. Students arrived in the newly constructed buildings in 1917. It has grown and expanded ever since, and is a highly regarded private school today.

The school was founded on progressive educational principles, and the faculty were encouraged to study the work of William James, John Dewey, Francis Wayland Parker, Alfred North Whitehead, and Jean Piaget. Katharine Taylor, Shady Hill's founding director, who had been handpicked from the Francis Parker School in Chicago, would write, in 1937, regarding teaching, "The more you think of teaching the more you realize that it can never be classified as science. It is nearer to being an art. It makes use of the findings of science, but the ability to teach can never be fully acquired through scientific pursuit alone, and the procedure of teaching can never be organized purely as a science."

Unlike B. F. Skinner's own childhood school experience that he looked upon with affection and gratitude, such was not the case for daughter Deborah, the proverbial Baby in the Box, who by the time she was eight and attending Shady Hill School was not happy being there at all. Shady Hill itself was apparently not the problem. She simply felt she wasn't prepared. "I hated it," she said, "because I was a poor student and not popular. The reason, I like to think, is that most of the other kids had been taught to read in the womb, but my parents thought school would be the best place to be taught. When I was eight, I was still a poor reader and the teacher complained to my parents!"

"She wasn't the most popular of kids," admits Mary Eliot, "but she wasn't really disliked." She was, however, behind a bit in her skills. Eliot would learn that Deborah's father was not just unhappy with what he saw during that Father's Day classroom sit-in. There was more. "He was upset about the teaching of arithmetic, math, and later he was very much worried about how spelling was taught."

Skinner wanted his daughter to get some extra help. He believed that the breakthroughs he was making with teaching machines and "programmed instruction" could help Deborah. She could be a guinea pig for his ideas. "The school was asked," says Eliot, "to provide a special time for her to go in a closet, a special little room, and work on spelling, and it was just about programmed learning." Eliot remained skeptical. She felt Skinner's approach was too "controlling," and in the ensuing years she would never warm to his idea of applying behaviorist principles to learning. "[Deborah] had things that he had sketched out for her to learn to spell, and I wasn't very much impressed by that, because I found that if children wrote and read and had their compositions corrected, they learned to spell just about as fast as if you had spelling lists."

A contemporary of John B. Watson's, the psychologist E. L. Thorndike, in his book *Education*, wrote,

> If, by a miracle of mechanical ingenuity, a book could be so arranged that only to him who had done what was directed on page one would page two become visible, and so on, much that now requires personal instruction could be managed by print. Books to be given out in loose sheets, a page or so at a time, and books arranged so that the student only suffers if he misuses them, should be worked in many subjects.

Thorndike's vision was remarkable given that he wrote these words in 1912. Hypertext, which serves as the basis of today's World Wide Web thanks to its HTML language, might arguably claim to be one application of Thorndike's vision, the notion extended far beyond simply education to all knowledge. It's also notable that Thorndike's vision was published more than forty years before Skinner built his first machine, and yet Skinner's machine lacked that "miracle of mechanical ingenuity," limiting his machines to the same linear presentation of problems and questions for all students who sat down to use the machine. The idea of a machine that can achieve that miracle, breaking from its linear program to address the truly individual needs of each particular student, was an educational Holy Grail that eluded Skinner's work.

Over the course of his career, Skinner would extend this notion of reinforcement in shaping behavior from rats and pigeons to learning in animals in general, including human learning. But Skinner's theories rubbed some scientists, educators, and psychologists the wrong way. As the decades wore on enterprising researchers realized it might be possible to not only think about brains as computers but even try to model—using a computer—how brains worked. The process of collecting observational data to support "how" the brain worked would have made Watson blow a fuse. Behaviorism's dismissal of "looking under the hood" to see what was going on in the brain—how consciousness worked, how learning worked—actually spurred entirely new and different "cognitive" theories, unafraid to consider human learning that involved brain functions for which external, observable behavior alone could not account.

Controversy would follow Skinner and his ideas for decades. In academic corners, his views became less and less popular in the 1960s and 1970s as, coincidentally, computers became more and more available. Noam Chomsky, an MIT professor of linguistics, was one of Skinner's many rivals and one of the leaders of the cognitive revolution aimed to combat behaviorism with what cognitivists believed was a better explanation of what was going on inside the brain. For years, Chomsky would attack Skinner's entire thesis, whether it be the behaviorist model or Skinner's later treatises on broader themes like freedom and dignity.

Some researchers with a passion for understanding learning and improving education found Skinner a perplexing enigma. Seymour Papert was one. Papert, who was greatly influenced early on in his career by the work of Jean Piaget, helped create and was a major proponent of the LOGO educational computing movement in the late 1960s and onward, and also a major proponent of the "cognitive science" revolution, an entirely different view of the mind and an approach that vigorously attacked, some might even add ridiculed, behaviorism. Instead of a machine teaching a student, Papert was in favor of children teaching machines, and in so doing, learning about mathematical concepts, not to mention gaining skills in computer programming.

"I find Skinner somewhat of a contradiction," Papert once confessed to this author, "because as a person he's intellectually very rich and multi-sided and very literate and likes poetry and I think is a great person. When he thinks about children and education, there's a lot of richness. The form in which it takes when it gets out into the world is extremely," he said, pausing for a moment as if to choose the next word carefully, "*pernicious*. He has a very pernicious doctrine. The pernicious doctrine being that you can break up knowledge into fragments and guide children toward acquiring the knowledge like you might involve the behavior of a rat or a pigeon. I find that a contradiction. I find when you think of Skinner as a whole person, he's so far away from this kind of thinking and practice of education, well, I'm full of wonderment that he isn't the main critic of the way that his ideas are being used in the world."

In the years following the 1954 debut of Skinner's teaching machine, he iterated repeatedly on the design, producing newer, more complex devices. His ideas spread around the world, and many others took up the task of tinkering with mechanical devices that could, in theory, teach.

One of these more advanced machines, heavier, sleeker, enclosed in sheet metal, Skinner designed for use in his Natural Science 114 classes at Harvard University during the middle to late 1950s.

To operate, an instructor had to first load it with what Skinner would call the "program," by opening the top and placing a cardboard disk, about the size of an LP record, onto a platter. The disk would have room for about thirty questions, each typed inside a very thin "pizza slice" of the disk. Each slice provided enough room for maybe two or three lines of text. After placing the disk in the machine, the instructor would shut the top and lock it. With the top of the box shut, the student could only read one question frame at a time through a rectangular viewing hole exposing one of those thirty slices of the disk and could only remove the disk once all questions had been answered correctly. The upper-right-hand corner of the little window had a shield,

Skinner's 1958 teaching machine

a kind of metallic eyelid, hiding part of the text underneath. This is where the question's answer was printed.

Located next to the rectangular window, another area existed for writing your answer on an exposed section of a paper tape. The machine's innards revealed a fantastic assortment of clocklike gears, pulleys, levers, and other mechanisms working together in a harmonious logic that, through a careful choreography of mechanical motions, permitted the student to advance to the next frame by pulling a lever to rotate the disk slightly.

As teachers, these machines were not mechanical equivalents of Mary Graves, to be sure. The machine's elaborate gears and pulleys and knobs, though impressive, had no means of "knowing" whether a student's response to a question was really correct or not. Nor, for that matter, did it know if a response was inappropriate. If asked for the name of the first president, the student could just as easily write "Mickey Mouse" as he could write "George Washington" and move the lever horizontally, telling the machine that he'd answered correctly. The machine had no way to verify the student's answer. Skinner's design placed the responsibility of judging the correct response entirely on the student. Here, the limitations of mechanical engineering forced Skinner to fall back on the safety net of Harvard's Honor Code prohibiting cheating. Another weakness of this machine, in fact all of the machines Skinner would design over the years, is that the "program" is *linear*. Each student gets the same questions in the same order; everything is uniform. There is no possibility of branching; no possibility for the machine to think, *This here is one bright student, how about we skip to some harder questions.*

Skinner also designed machines that would be manufactured by IBM. One was based on the 1954 "slider" design, built to teach spelling or arithmetic. The "spelling" version of this machine employed a dozen sliders with the alphabet imprinted upon them. Thus, one might be asked to spell the word "manufacture" by positioning eleven sliders with the correct letters until the word is formed.

A set of six frames is shown in the illustration below. This example shows what at the time was called the "tiny steps" approach Skinner took, based on his experience with training rats and pigeons. The instructional "program" guided the student, through incremental steps, how to "behave" accordingly—in this case, spell the word "manufacture" correctly and understand what it means. The student must correctly spell the first word before being presented with the second, and so on:

---

1. **Manufacture** means to make or build. *Chair factories manufacture chairs.* Copy the word here:

   ☐ ☐ ☐ ☐ ☐ ☐ ☐ ☐ ☐ ☐

2. Part of the word is like part of the word **factory**. Both parts come from an old word meaning *make* or *build*.

   m a n u ☐ ☐ ☐ ☐ ☐ ☐ ☐

3. Part of the word is like part of the word **manual**. Both parts come from an old word for *hand*. Many things used to be made by hand.

   ☐ ☐ ☐ ☐ **f a c t u r e**

4. The same letter goes in both spaces:

   m ☐ **n u f** ☐ **c t u r e**

5. The same letter goes in both spaces:

   m a n ☐ **f a c t** ☐ **r e**

6. Chair factories ☐ ☐ ☐ ☐ ☐ ☐ ☐ ☐ ☐ ☐ ☐ chairs.

A set of frames designed to teach a third- or fourth-
grade pupil to spell the word "manufacture"

Skinner actively pursued relationships with commercial firms, from IBM to, improbably, a company called Rheem, a boiler and air conditioner maker, which viewed the teaching machine market as a new craze attractive enough for them to try to cash in on. But after a few frustrating years he learned that American companies simply could not deliver on what they promised. "It was really a terrible thing," Skinner told this author. "It was shocking. Rheem was the worst one. They obviously took this on, gave me some kind of an arrangement, [but] I don't remember getting much money out of it. . . . [They] kept me around for several years, never did what they said they were going to do at all. It was a shock."

IBM seemed a little bit more hopeful as a partner but in the end disappointed Skinner as well. "They built this machine, which was very good, it was doing what they said it would do, and as a matter of fact they then redesigned it and got a patent based on my patent. But then they stopped it. And one of the engineers there who was dying to see it developed tried hard to get [IBM president] Thomas Watson's children to try my machine. They had given it to the typewriter division because they had contacts with high schools and he simply said, 'I can use this money to stay ahead of my competition in typewriters' and that was that. I don't think there's anybody within IBM who knows that that machine is now in the Smithsonian."

In a classic example of the right hand not knowing what the left hand was doing, another corporate group within IBM enthusiastically contacted Skinner a few years later inquiring about his machines and how they might be commercialized. They were unaware of IBM's previous dealings with Skinner. Skinner did not pursue the new opportunity.

Interest in teaching machines and what was by now called "programmed instruction" was growing. Across the country, in corporate, university, government, and military settings, a variety of mechanical devices were starting to be utilized to teach and train people on numerous subjects.

Beyond the already cited weaknesses of Skinner's machines, some experts became concerned with another perceived drawback of Skinner's design, that students were not given an opportunity to learn *why* their answers were wrong. The design emphasis of his devices focused on reinforcing and rewarding correct answers. But in the real world, students often get things wrong, and while it is useful to know when one has gotten something wrong, it's even more useful to know *why*, so one can learn to get things right in the future.

As more money and people and resources were brought to bear, the ideas began to develop. One improvement sought to help students know *why* they got something wrong. Immediate Feedback was a bed-rock principle, sure, but how about making that feedback meaningful? Perhaps there was a way to do that. Perhaps using a machine wasn't the right way to do it. Was there another, simpler way?

Harking back to Edward Thorndike's 1912 ideas, researchers believed there was, with textbooks. Not just ordinary textbooks, but "scrambled" textbooks. Make it so a page out of an instructional book—call it a "programmed text"—posed a question to the student, and offered some multiple-choice answers. Tell the student if you chose answer "a," turn to, say, page 50. If you choose "b," go to page 51, and so on. One of those multiple-choice answers was right, but in this framework, the authors of the text could provide meaningful—and instant—feedback to the student dependent on which answer the student chose.

One proponent of programmed texts was Norman Crowder, who held a variety of psychology and training positions in the U.S. Air Force. Crowder believed a scrambled book was all that was needed to make the feedback meaningful. Unlike Skinner's devices, if you got the wrong answer you weren't just locked from moving to the next problem or shown the right answer and forced to move on. You could stop and learn *why* your answer was wrong. For instance, in the 1960 book *Adventures in Algebra*, written by Crowder and Grace C. Martin, and marketed by Doubleday as a "Tutor Text," the reader is presented with concepts and questions of steadily increasing complexity. Each question offers a number of multiple-choice answers, and the reader is told what page to turn to depending upon which answer is chosen. Wrong answers are given detailed explanations, and, occasionally, the authors scold the reader for choosing a particularly boneheaded answer, directing them back to the beginning of the chapter since it appears they didn't understand the concepts at all.

Crowder seems to have realized that while Tutor Texts were powerful, they still relied on this honor system, where all of the material, all of the answers, were right there in plain view. It was easier than ever to cheat. In time he began to think about mechanical devices that took this "branching" concept to the next level—devices that finally fulfilled Thorndike's 1912 vision. He would team up with a company to build the AutoTutor, which presented instructional material as frames in a film, and could be controlled by a computer program on the back end that knew which frame of the film to jump to depending upon a student's answer.

—

In December 1958, all of the major figures in the automated teaching field, including Skinner, Pressey, Crowder, and the rest, met at a conference, "The Art and Science of the Automatic Teaching of Verbal and Symbolic Skills," in Philadelphia. One notable paper, by Gustave Rath, Nancy S. Anderson, and R. C. Brainerd of IBM, stood out from the others. It was the only presentation that explored the idea of using a *computer* to teach. They described how they had built a program on an IBM 650 to "simulate a teaching machine," specifically one that could teach binary arithmetic. The student interacted with the system via an IBM 650 Inquiry Station typewriter. It marked the crude beginning of what would become, as computers became more affordable and more widely available, the obvious direction to pursue with teaching machines. Computers afforded a level of flexibility impossible to achieve in mechanical devices. The key was *software:* computer software was unquestionably more flexible for programming instruction than the painstaking labeling of disks and paper tapes or twiddling with knobs, sliders, and fragile pulleys and gears.

One notable finding in the IBM project was the acknowledgment that a computer spends most of its time waiting for the student to do something. In "computer time," that wait time amounts to what would be thousands of millennia to humans, with brief interruptions when a student types a letter at the keyboard. All computers are this way, even today. Your fancy laptop computer is primarily sitting there idle, checking every tiny fraction of a nanosecond to see if there's anything to do yet, and from the perspective of the computer processor, usually there isn't. But the insight that the IBM researchers came away with would prove to be key to all future computers: "Since the computer spends most of its time waiting for the student, suggestions for the utilization of this time are as follows: by multiplexing, the computer could present and score problems for several students who sat at different 'Inquiry Stations.'" Soon the general name for this powerful notion would be called "time-sharing."

Skinner had finally achieved his dream of becoming a novelist with the 1948 release of *Walden Two*, in which he depicts a behaviorist utopian commune. A few years later, John Gilpin, a math major in Dickinson College, found the book in the college library. Intrigued with *Walden Two*'s notion of a "designed community," and believing that a *Walden Two* community actually existed, Gilpin wrote to Skinner asking for more information. He would learn that there was a group attempting to live out the *Walden Two* vision, so he hopped on a bus and headed for Boston to join the group, but nothing came of it. Gilpin asked Skinner for a job but was told nothing was available. Gilpin told him he'd even work for free—he just wanted a chance to work at the lab. Skinner relented, offering him a one-eighth part-time job, at the lowest possible pay scale, working with Skinner's research colleagues Jim Holland and Susan Meyer to glean information from the teaching machine data gathered from Skinner's Natural Science 114. They spent the summer sorting through the pile of data produced from the fifteen or so machines in the lab. "Every time a student completed one of these disks," Gilpin recalls, "he produced a paper tape, and so they had a paper tape for every disk. . . . They had this huge pile of paper tapes. The task was to go through and see to what extent the students got the right answers. Because, of course, part of the ideology was the student should never be wrong."

Gilpin, fresh out of college, was now immersed in what he believed was an "extremely cutting-edge development," yet something wasn't right. "I found that every time I went down [to the lab]," says Gilpin, "in less than half an hour, I was fighting sleep." Using Skinner's teaching machine, if a student felt she answered a question correctly, she was supposed to move a lever to the right (causing the machine to record a little hole in a paper tape), and not move the lever if she felt she got the answer wrong. Problem was, Gilpin observed, since the programmed instruction had been written in such a way that it was fairly easy to get most of the answers right, a student would move the lever to the right whether or not she thought it was correct or not. "I never quite had the guts to challenge Holland or Skinner about it," he says.

Gilpin eventually moved on from Skinner's lab to Bell Labs, but that didn't last long, and he later accepted a 1959 offer to work on programmed instruction at Earlham College in Indiana. At Earlham he worked with a team developing scrambled textbooks. One day, Max Beberman, who had founded the University of Illinois Committee on School Mathematics (UICSM) and would become widely known for leading the "New Math" movement of the 1960s, came to speak at Earlham. Gilpin was impressed, and later wrote to Beberman to inquire if there were jobs at UICSM. There were, and in 1962 Gilpin moved to the University of Illinois and joined UICSM, where he developed programmed instructional materials.

UICSM was located next door to University High School, which happened to be located near an odd-looking four-story brick building with an equally odd-looking radar tower adjacent to it. This was the Coordinated Science Laboratory (the new name for Control Systems Laboratory). Inside, among CSL's many projects, was one centered around "automatic teaching" via computer. The project was two years old, gaining speed, and had a weird name: PLATO.

2

An Educational Emergency

Seventy thousand fans turned out on the first Wednesday of October 1957 to watch game one of the World Series as the New York Yankees, on their home turf, beat the Milwaukee Braves. The teams met again the next day, but this time the Braves prevailed, 4–2. Friday being a travel day, game three wouldn't be played until Saturday in Milwaukee. Instead of baseball that Friday, Americans had another notable event to consume their attention: the CBS television network airing of the premiere episode of a new situation comedy series called *Leave It to Beaver*. No doubt CBS was pleased that there was no World Series game that day to distract potential viewers from its new show, but it's even less doubtful that anyone at CBS, or anywhere else in America for that matter, could have anticipated something very different happening: an event, a phenomenon, a *thing* that would distract the nation a great deal that evening, and then even more the next day, and then even more the rest of the weekend, and then kept distracting everyone for months and years to come.

That Friday, the future came barging in without so much as a knock, and it had a name: Sputnik. While Casey Stengel and the Yankees entourage chugged along on their sixteen-hour, 1,100-mile railroad journey to Milwaukee, the Soviet Union had successfully launched, at 2:28 p.m. New York time, a rocket that carried a 184-pound, twenty-two-inch shiny metal sphere into space, becoming the first man-made artifact to orbit the earth.

Early the next morning, twenty-five minutes before the sun would rise over Cambridge, Massachusetts, B. F. Skinner, his wife, Eve, and daughters Deborah and Julie were already up and outside, met by a gentle breeze of chilly morning air on the terrace of their Ellsworth Avenue home. The overnight clouds had cleared by the time they went outside, and at around 6:20 a.m., they looked up and saw a tiny dot move across the northeast sky, passing by the Big Dipper and dissolving into the predawn light growing along the edge of the horizon. A delighted Skinner jotted down in his journal, "Newton's theory of celestial mechanics has been experimentally confirmed!"

That same weekend, on a farm near the airport just outside Champaign-Urbana, in east-central Illinois, a group of electrical engineer friends gathered in the dark to hoist up a large antenna pole. The group included Jay Gooch, Dominic Skaperdas, and, at twenty-three, the youngest of the bunch, Donald Bitzer, known as "Bitz" to his friends. Bitzer, with his short haircut, can-do spirit, and "tell ya what I'm gonna do" Midwestern car-salesman affability, could have walked right off the set of the brand-new *Leave It to Beaver* show. They all worked at the Control Systems Laboratory, a classified military research lab on the nearby University of Illinois campus, from which they had "borrowed" a few pieces of equipment. They wired the pole up to some radio equipment, connected the radio equipment to a tape recorder, tuned the radio to the right frequency, and then settled down to wait for Sputnik to streak across the sky. They were hoping to capture and record a certain *beep-beep-beep* signal that Sputnik was said to be broadcasting. Sputnik did not disappoint.

There were two kinds of reactions to Sputnik. Only the first kind of reaction got widely reported in the news: that it was something to be feared.

"A shocker like Pearl Harbor, waking America up and making it buckle down," is how *The New York Times* described the events of October 4, 1957. The Soviets had beaten the Americans to space, and Americans were in a state of disbelief. Sputnik—a common Russian word meaning "fellow traveler"—was worldwide news. On Saturday one Frenchman quipped, "The Russians put up Sputnik yesterday, and it goes around the world saying *beep-beep-beep*. Then, when it gets to the U.S., it says *ha-ha-ha*." In those first few days after October 4, American newspapers struggled to describe Sputnik, many choosing to refer to it as "ARTIFICIAL MOON," or, just as likely, "RED MOON."

For many Americans, Sputnik might as well have been a hammer and sickle orbiting the earth. The exceptional Sputnik defied American exceptionalism: a shiny metal ball made by communists, hurled into the heavens by communists, tumbling across a sky now owned by communists, at 18,000 miles per hour, soaring over America every ninety minutes, and there wasn't a damn thing anyone could do about it. America, busy still with its baby boom a dozen years after the victories of World War II, was suddenly an embarrassed nation.

How could leading scientists and technologists have let the country down so badly? Did we not have enough of them? Were the Reds smarter? How does America get back in the lead? These were questions the politicians and the media were asking. The White House publicly played down the launch but privately worried about the immediate aftermath: Soviet claims of scientific and technological superiority were gaining wide acceptance amid public concern that the balance of military power was shifting to the USSR. One senator made a billion-dollar Manhattan Project–scale proposal to compete with the Soviets in the Space Race.

"The more Americans were told by the men in Washington not to worry," one historian wrote, "the more they panicked. For all its simplicity, small size, and inability to do more than orbit the Earth and transmit meaningless radio blips, Sputnik's impact on America and the world was enormous and totally unanticipated. To the man or woman in the street, it was vastly confusing and most threatening."

"A colossal panic was underway," wrote Tom Wolfe in *The Right Stuff*, "with congressmen and newspapermen leading a huge pack that was baying at the sky." To them Sputnik "had become the second momentous event of the Cold War. The first had been the Soviet development of the atomic bomb in 1953." Says Wolfe,

> From a purely strategic standpoint, the fact that the Soviets had the rocket power to launch [Sputnik] meant that they now also had the capacity to deliver the bomb on an intercontinental ballistic missile. The panic reached far beyond the relatively sane concern for tactical weaponry, however. [Sputnik] took on a magical dimension—among highly placed persons especially, judging by opinion surveys. It seemed to dredge up primordial superstitions about the influence of heavenly bodies. It gave birth to a modern, i.e., technological, astrology. Nothing less than *control of the heavens* was at stake. It was Armageddon, the final and decisive battle of the forces of good and evil. Lyndon Johnson, who was the Senate majority leader, said that whoever controlled "the high ground" of space would control the world.

To achieve and hold on to that "high ground" would require not just a lot of hard work and heavy lifting but also a lot of smart people. Sputnik suggested that America did not have enough smart people. The West's hysteria was not expected by the Soviets, but they were quick observers, and almost immediately they began to pump out propaganda to heighten it and sow confusion even further. The result was an escalating cycle of fear, mobilization, and worry for the future. Given that Sputnik appeared right in the middle of the Cold War—in the six months prior to the launch the U.S. and USSR combined conducted some thirty-nine atomic bomb tests, and fourteen in the six months after the launch—for one superpower to surprise another with the first-mover advantage in space raised legitimate concerns about the Soviet Union's intentions and America's ability to get to space as well.

Among other things, Sputnik fueled calls for immediate and sweeping reform of American education. Getting to space was rocket science, and rocket science required math and science skills, and clearly the American education system had let the country down in those areas. It had to be fixed, and fast. The top priorities? Improving scientific inquiry and mathematical problem solving. How could America have a chance at winning the new Space Race if it didn't have top-notch scientists and mathematicians?

One month after the launch of Sputnik came another launch, the much heavier Sputnik 2, a larger craft that carried inside it a once stray Moscow dog named Laika, instantly dubbed "Muttnik" by American media. Unlike the relatively small silver ball lobbed into orbit the month before, this time the Soviets had managed to hurl a thirteen-foot-high, 1,100-pound dog-carrying metallic cone into space. It did not require much effort to imagine a nuclear warhead in that cone, instead of a dog.

In February 1958, as a direct consequence of the Sputnik launches, the U.S. Department of Defense created ARPA, the Advanced Research Projects Agency, with a mission to embark on technological research and development projects of national importance. Congress scrambled as well. That same year they passed two major pieces of legislation, the National Aeronautics and Space Act, creating the NASA space agency, and the National Defense Education Act (NDEA), which President Dwight Eisenhower signed into law in July and September, respectively.

"The Congress finds that an educational emergency exists," the

NDEA bill declared, "and requires action by the federal government. Assistance will come from Washington to help develop as rapidly as possible those skills essential to the national defense." Title I of the act went on, "The security of the Nation requires the fullest development of the mental resources and technical skills of its young men and women. The present emergency demands that additional and more adequate educational opportunities be made available. The defense of this Nation depends upon mastery of modern techniques developed from complex scientific principles. It depends as well upon the discovery and development of new principles, new techniques, and new knowledge."

In simple terms, this was a cause the nation could rally behind, at a time when Washington had money and was willing to spread a whole lot of it around. Businesses, universities, schools, think tanks, and state governments didn't hesitate to put out their hands. The timing boded well for creators of teaching machines. Back in 1929, Sidney Pressey's timing had been the stuff of tragic comedies: his feverish efforts to drum up interest in his devices collided with the stock market crash and the Great Depression. But now, thanks to the birth of ARPA, NASA, and the NDEA, prospects were looking bright for B. F. Skinner's and Norm Crowder's work, as well as the countless academic research and business ventures that rose up seemingly overnight to popularize and profit from their work. The repeated usage of the word "emergency" in the NDEA legislation helped drive people to pitch in and help the nation's young men and women achieve "the fullest development of the mental resources and technical skills."

There was another kind of reaction to Sputnik, which went largely unnoticed due to the fear-mongering cacophony fueled by the media and political establishment. Another perspective on Sputnik existed, one that did not involve politicians or the media. In this alternate view, the Sputnik event was met with curiosity, excitement, and fascination. This was the scientific reaction.

For example, B. F. Skinner's reaction to Sputnik was that of an intrigued, if not jubilant, scientist. In his diary Skinner expressed a reaction free of politics, free of media hype, and nerdy to its core: he was excited that somebody had proven Newton's theory of celestial mechanics by lobbing a big ball into orbit, and guess what, it worked.

Likewise, engineers and scientists around the world acknowledged that the launch was a historical, technical milestone. What's more, it proved if you threw an object into the emptiness of space and had it send out radio signals, those signals could reach anyone on the ground.

It just so happened that 1957 had been a booming year for science. The year marked the beginning of the International Geophysical Year (IGY), involving sixty-seven countries including the U.S. and Soviet Union, undertaking scientific experiments, measurements, and expeditions all over the world. In fact the project was so major, the "year" of the IGY, which had officially started in July 1957, went on for eighteen months, ending on the last day of 1958. Within the IGY's mission was also a goal, stated as early as 1955 by both the U.S. and the Soviets, to launch small orbital satellites during the IGY project time frame. So things like Sputnik were known to be coming. It was expected within the scientific community that rockets would soon launch little "moons" that would reach orbit and tumble across the heavens overhead. Many discoveries were made during the IGY, including the Van Allen radiation belts, the undersea mountain ranges in the Atlantic and Pacific (bolstering the theory of plate tectonics), and numerous discoveries and advances in the Antarctic region. The Sputnik launch fueled concerns, of course, about space being a new "platform" for missiles and bombs. But as with any technology there are good uses and bad uses. Scientists believed there would be good uses for space. And it was a new frontier, begging to be explored.

This was the other side of the Sputnik coin: possibilities now opening up with space, with implications both scientific and technological. A Space Age that was no longer science fiction but here. *Now.* Perhaps a few months sooner than expected, but here and now nevertheless. It was expedient for the media and the politicians to drum up fear as a result of the launch. America has witnessed such reactions and manipulations many times since. That very fear, uncertainty, and doubt was what motivated legislators and the president to enact laws that formed NASA, ARPA, and the NDEA and justified pouring millions of dollars into institutions, organizations, and businesses around the country. But money alone was not going to make things happen. There was more to it. For science and technology to advance, there needed to be a catalyst. Sputnik became the convenient catalyst.

—

How Sputnik, ARPA, NASA, and the NDEA would change priorities at academic institutions can be seen in the story of the Control Systems Laboratory at the University of Illinois.

The story starts back during World War II. Universities around the country lost top faculty to the Manhattan Project to develop the atomic bomb. The University of Illinois was not immune to the "brain drain," losing a number of its top scientists and engineers to the war effort. Many had moved to MIT to secretly work on radar at that university's storied Radiation Lab, but when the war ended, those Illinois brains didn't return home.

"Wheeler Loomis was head of the physics department at that time," recalls Illinois physicist Richard M. Brown. "He had been in charge of recruitment for the MIT Radiation Lab during World War II, and had succeeded in raiding a number of the physics departments of their very good people to come to that laboratory for work on radar and other things, and saw what it did to those departments including his own and had vowed that if another war started of this kind, he would start a laboratory of similar form here at Illinois in Urbana. So when the Korean War started, he proceeded immediately to do that."

So Illinois created CSL, to undertake new projects including air defense, radar, and missile guidance, all augmented through the use of brand-new digital computers. One notable recruit in 1951 was the physicist and electronics engineer John Bardeen, who along with Walter Brattain had invented the point transistor at Bell Labs in 1947 (and, together with William Shockley, would share the 1956 Nobel Prize in physics). In short order, CSL would claim roughly a third of the UI physics faculty.

Loomis hired Frederick Seitz to head up the lab, and Jack Desmond to be its business manager. Seitz, who resembled President Eisenhower, was a renowned physicist with degrees from Stanford and Princeton. In 1940 he wrote the book on solid-state physics, and during World War II had applied his expertise and interest in metallurgy to ballistics and how to penetrate armor. He also worked on radar and was a member of the Manhattan Project. He was heading an atomic energy program at the Oak Ridge National Laboratory in Tennessee when Loomis invited him to join and direct CSL.

CSL was what was called a "locked lab." The engineers and scientists put their documents in locked safes at the end of the day, and every month the safes' combinations were changed. Not only was nothing allowed to leave the lab without permission, but the academics working in the lab could neither discuss what they were working on nor share their experimental results. In the "publish or perish" world of academia, the inability of professors to publish or even talk about what they were doing could hurt their academic career, but Wheeler Loomis had taken that into consideration, relaxing academic requirements for people who took a chance on their career by joining the lab.

Funding for the lab came from Army, Navy, and Air Force grants. The lab's cozy arrangement with the military was such that it received over $1 million of funding each year. The lab was then turned loose to work on projects it thought were important to the Defense Department. At the end of the year the results were reviewed by a technical advisory council, and depending on how the review went, the lab got the same, more, or less money for the subsequent year.

In 1957, Loomis retired and Frederick Seitz was picked to take over his position. That left CSL looking for a new director. While Loomis conducted a search, he also would run the lab until they found someone new. That someone was Daniel Alpert, another respected physicist with global stature. Loomis offered Alpert a deal whereby he would be given a half-time professorship in the physics department, plus he would start as CSL's new technical director. If all went well, in a year or two he'd assume full directorship of the lab.

Then, right out of the blue, Sputnik happened.

Daniel Alpert was born on April 10, 1917, in Hartford, Connecticut, to parents who originated from Russia and Lithuania. Originally a blacksmith, his father became a tinsmith once he arrived in the United States. His father's tin business did well, the family prospered, and he retired around 1927. "I remember vividly," Alpert recalls, "his taking me around when I was about ten and saying, 'Daniel, this apartment building that I just bought is going to send you to Yale.'" It no doubt would have, but the stock market crashed and the Great Depression hit. His father lost most of his property. The life the Alperts had known up to that point suddenly changed. The Depression forced the family to move to a farm, where Daniel would spend the rest of his childhood. He graduated from high school as class valedictorian, still dreaming of attending Yale, but his parents could not afford it, so he went to the less expensive Trinity College in Hartford. At Trinity he discovered a love and a talent for physics. He did well and wound up pursuing a graduate degree at Stanford University, falling in love with the Palo Alto, California, area in the process. He finished the oral examinations for his PhD degree in physics on Saturday, December 6, 1941. The next morning, Pearl Harbor was attacked and the United States entered World War II.

Alpert took a postdoctoral fellowship at Westinghouse Research Lab in Pittsburgh. In 1942 he married Natalie Boyle, a landscape architect. In October 1945, they read B. F. Skinner's article in *Ladies' Home Journal* about the Air Crib for babies. Inspired by the article, Alpert designed and built his own Air Crib, and over the next few years the Alperts raised their two infant daughters in homemade boxes based on Skinner's designs.

During World War II, Alpert worked on a number of projects to help in the war effort, including airborne radar components and gas switching devices. He also worked with Ernest Lawrence on the isotope separation of uranium for the Manhattan Project. For a while he worked out of the Lawrence Livermore Lab near the University of California in Berkeley. He loved California and tried to get a job there so he could stay. "I'd done my graduate student days there," he recalls. "I loved the area, it was beautiful at that time." From his office in the Radiation Lab at the top of Berkeley, he had a spectacular view of the Golden Gate Bridge in the far distance across the San Francisco Bay. "I wanted to stay. I looked around to see who's there. Lockheed offered me a job at 55 cents an hour, not as a physicist but as an engineer. Fifty-five cents an hour! I was doing better at Westinghouse."

So he stayed at Westinghouse and in 1946 began a fruitful decade of basic research at Westinghouse's Pittsburgh lab, where, Alpert says, "I would do whatever the hell I wanted." He pioneered work in a field that he himself named "Ultra-High Vacuum Physics." As part of that work he co-invented a device called the Bayard-Alpert gauge, still in use in laboratories to this day. "I became known in the world of physical science, all over the world," Alpert says. "In 1953, for example, my wife and I took a trip to Europe and I was invited to give lectures at every major industrial lab. Philips, GE, and so on. I went around for a month giving ultra-high-vacuum lectures and where we stood, and the physics that was involved, and wherever I went I got the red-carpet treatment."

By 1957 he already had been promoted to manage the entire physics department at the Westinghouse lab and learned he was on the short list of candidates for the top role of the lab's director. He didn't get the job. "I wasn't shocked, because at age forty to be the director of a whole laboratory for a company with fifty-one divisions—fifty-one divisions!—it was kind of a big job. My friends at GE and AT&T who were my professional colleagues, they weren't forty, they were fifty, they were in their fifties before they could aspire to be director of a lab."

He wasn't happy, though, about being passed over, and the memory still stung nearly fifty years later. He had been one of three candidates for the prestigious position, a major deal to all three candidates. So major, that, as Alpert bluntly put it, "the other guy

Daniel Alpert (*l*) and Wheeler Loomis (*r*), c. 1957

my age committed suicide after he wasn't selected." Instead of selecting Alpert the powers-that-be at Westinghouse went for the third candidate, Clarence Zener, who had made a name for himself by inventing what became known as the Zener diode. Alpert felt he had more relevant and meaningful experience than Zener. "Including factory experience," he said. "Including the making of the first devices that we had, of getting into the factory and seeing what the factory was like, that it didn't take intellectual labor on my part, it was *intuitive* on my part. . . . They picked a guy who was ten years older than me, better known as a physicist than I was, but who was a disastrous choice as director." He told his wife, with whom he was raising a family that he'd realized was now growing deep roots in Pittsburgh, that he didn't think he'd be able to last more than five years under Zener. "Why don't you wait it out, trust your judgment," Natalie told him. "I don't want to be a bitter person," he told her. "I could easily be a bitter person in five years." Alpert wasn't threatened with being fired, but he was unhappy. "I had a very excellent secretary, and I was associate director of the lab. And I was still manager of the physics department in name, although Clarence would come around and tell my people what to do without ever even bringing me into the room. It was just asinine."

Alpert was right about not lasting five years under Zener. In 1957 he made up his mind to leave Westinghouse. But where to? He turned to science colleagues, distinguished people who sat on the boards of

IBM and Texas Instruments and elsewhere. He tried to get another industrial job. "It was the only work I'd ever done. Instead of offering me advice for another job, they offered me jobs in universities with which they were associated." One job was at Berkeley; one was at Illinois. This would not be the last time he would be faced with having to choose between living in the San Francisco Bay Area and a small Illinois town surrounded by cornfields.

The Control Systems Laboratory was situated on Mathews Avenue in the second, third, and fourth floors of the old Power House building next to the Mechanical Engineering Laboratory. The University of Illinois had struck a deal with the government that in exchange for designing and building the Ordnance Discrete Variable Automatic Computer (ORDVAC), they'd get funded at the same time for a second computer of their own. This second computer, the Illinois Automatic Computer (ILLIAC), became operational in 1952 and would stay in use for more than ten years. Within two days of the launch of Sputnik, a team of Illinois scientists had used the ILLIAC to calculate the exact orbit of the spacecraft.

Two of the creators of the ILLIAC standing
by the machine in 1952

An effort to build a large, vacuum-tube-based digital computer, the ORDVAC, was large in every sense: row after row of vacuum tubes mounted in floor-to-ceiling steel racks, with miles of wiring running throughout. Once completed, ORDVAC was shipped off to the Army's Aberdeen Proving Ground.

One unusual aspect of ORDVAC and the ILLIAC is that UI's electrical engineering department was not leading the project—the physics department was in charge. Says Gene Leichner, an engineer who worked at CSL and later became a professor: "The head of CSL was a physicist. The head of the computer lab was a physicist. And that's a funny thing. Those guys were running the show, not the double-E's. The double-E department had said no to this endeavor, when the Army wanted the computer built. . . . They didn't want to have anything to do with the computer."

To test out the numerous radar projects, CSL had built a tall steel tower right next to the north wall of the building. On top of the tower they mounted a large radar device with a rotating antenna. After the tower was built, someone got the idea that the space inside the tower's four legs could be filled up and used as additional office space. So on the second, third, and fourth floors of the Power House building, holes in the exterior brick of the north wall were knocked out, doorways put in, and three small, square offices were added, one on top of the other, inside the legs of the tower. Alpert took the tower's second-floor office.

Richard M. Brown headed up the new project, named Project Cornfield "to emphasize the remarkable fact that for the Navy, this was being done in the middle of an Illinois cornfield." The project arose out of the realization that radars were providing a lot of information on the battlefield, but it was difficult to analyze all that data quickly and effectively. Perhaps the newly emerging field of computers could help? The ILLIAC analyzed incoming radar data, taking advantage of the principle of the Doppler effect, the same as when one hears a change in pitch when a train goes by. "If you could measure," says Brown, "how much the phase differs as a function of the motion of the object being observed, you can tell if it is coming towards you or away from you and at approximately what speed."

By 1959, Alpert had become full director of the lab. He was asked to put together a plan that laid out the future of the lab. The Korean War was long over and CSL was either going to have to adapt or probably be shut down. So he came up with a plan.

The first thing Alpert recommended was that they declassify CSL, as well as change its name to reflect better integration with the rest of the university. "I came to the conclusion that CSL could not survive in the size that it was and on the campus," he says. While he was all for declassifying the work the lab did, he was happy to continue getting funding from the various branches of the military. But it would take a year to convince the military that declassifying was a good idea. Cleverly, he changed the lab's name but not its initials, from Control Systems Lab to Coordinated Science Lab. That way, the Army, Navy, and Air Force would still recognize the "CSL" acronym that for years had such a good reputation in Washington circles, and, more important, would keep sending those annual checks.

"I had to look for some projects that would be viewed as valuable to the military but would not be classified," Alpert says. "And there were several. One was air traffic control. One was the electric vacuum gyro, a new type of gyro for navigation, which was useful for the military but not in a military classified sense."

He spent a lot of time wandering around the lab learning about everyone's projects. He was proud of the CSL organization, which had assembled by the late 1950s one of the most formidable groups of brilliant scientific and engineering minds anywhere in the world, including Nobel Prize winner John Bardeen. Even on the campus of the University of Illinois, it was CSL that was recognized as dominating electronics, computing, and physics research and not the respective departments. In addition, since CSL and the Digital Computer Lab had built the ILLIAC, not the electrical engineering department, CSL always had first priority on its use, though it was a resource open to any department of the university.

One CSL physicist with whom Alpert had a long chat was Chalmers Sherwin. Known as a visionary who loved to ask probing, provocative questions, Sherwin began discussing the idea of using computers to help people learn. At one point Sherwin wondered, "There hasn't been a great invention in the field of education since the book. Why couldn't computers be applied to education?" By the late 1950s computers were becoming a lot more reliable, even the ILLIAC, which might run without a hitch for as many as forty hours straight—considered a very impressive feat at the time. Computers were going to get a lot faster, more reliable, and cheaper in coming years. Given all of the post-Sputnik cries to improve American education, especially in math and science, and given the flurry of activity by universities and businesses to create teaching machines and programmed instruction books, perhaps computers, which would be much more flexible than a mechanical machine or a printed programmed text, could be harnessed to help in the effort? In time, perhaps even on a mass scale? Perhaps a computer could finally deliver Thorndike's "miracle of mechanical ingenuity"? For a technically savvy university physicist thinking about this problem in the very late 1950s and early 1960s, there really was only one answer. From a technologist's perspective, if you believed in the principles of Self-Pacing and Immediate Feedback—and during this era, nearly everyone did—and you wanted to improve education at a mass level, it's easy to see how computers were going to be the only answer. If you wanted to secure the "high ground" of education's future in America, you had to figure out how to harness the computer.

If CSL was going to declassify itself and become more integrated into the mainstream university, yet still expect funding from military agencies, using computers to advance the university's mission was the way to go. Sherwin described his idea as a computerized "book with feedback," not unlike what Thorndike had envisioned back in 1912. By 1959, articles on the wonders of teaching machines were commonplace in the nation's magazines and newspapers. Academics convened conferences to discuss the latest thinking and advances in the field. Skinner was busy attempting to commercialize his latest machines with IBM and Rheem. Norman Crowder had chosen a different path: his "scrambled textbooks" and random-access film-based AutoTutor devices merely hinted at potential applications if you had been able to take the content of the book and program it into a computer. If you were working on applications of digital computers in an engineering laboratory in 1959, and you heard about teaching machines and programmed instruction, it was not hard to put two and two together and see the potential that lay in educational applications of digital computing.

Sherwin also took his idea to William Everitt, dean of the Engineering School. Everitt and Alpert were both intrigued. "Chal's idea appealed to me very much," Alpert says. It appealed so much so that he wanted Sherwin to lead a project to pursue answers to the great questions he was asking. "I twisted Chal's arm real hard. I said, 'Look, I got money, you take this on,' and he said, 'Well, I'll think about it.'" But Sherwin was more of a visionary than a project director and in the end was not keen on heading up a project. "I don't want to do that kind of thing, I want to have ideas, I don't want to do it," Alpert recalls Sherwin telling him.

Sherwin dropping out didn't diminish Alpert's enthusiasm for the notion of a computer-powered "Book with Feedback." But like all great ideas, it would never go anywhere without leadership and execution. He had to figure out how to turn this into a real project at CSL and get someone to lead a team to do it. He next approached CSL physicist Richard M. Brown, whom Alpert would later describe as "a strong person in the use of computers for naval applications," asking him to form an interdisciplinary committee to explore Sherwin's idea in detail, and then make recommendations on how such an idea could be turned into a successful project for CSL. Brown put out feelers across the university to find people for his committee. He recruited Larry Stolurow, a professor from the psychology department, as Stolurow was interested in programmed instruction and teaching machines, and had been present at the 1958 conference in Philadelphia along with B. F. Skinner and Sidney Pressey. Other recruits included Max Beberman from UICSM, and Rupert Evans, head of the education department. Chalmers Sherwin sat in on the meetings.

Meanwhile, Donald Bitzer submitted his dissertation, a crisp, ninety-eight-page affair entitled "Signal Amplitude Limiting and Phase Quantization in Antenna Systems," in December 1959 and took a vacation in Cuba over the Christmas holiday. On January 1, he turned twenty-six. Later that month he was awarded his PhD degree. Bitzer was still considered relatively junior among the CSL staff and was not invited to participate in the committee, despite his being known as a very sharp, competent engineer. Years later he would recall seeing the assembled group meeting from time to time. He remembers occasional shouting and arguments coming from the room.

The committee began meeting in the first few months of 1960. "We met usually with bag lunches," recalls Stolurow. "It was somewhat informal, in talking about what might be done. I would leave the laboratory in the psych department and go over and join the group, and one of the people I developed a relationship with, most closely related to, was Beberman." The committee met roughly every week for a few months. Alpert sat in on the last few meetings, which for many years since he described in the exactly the same way: *"A worse Tower of Babel I have never heard in my LIFE!"* He found that the psychologists in the room looked at the problem from only a psychology perspective. The engineers viewed it from only an engineering perspective. The educators viewed it from only an education perspective. Nobody could agree on anything. They were not getting anywhere. Says Alpert, "Each guy had the position that if it was going to teach math it had to teach *his* grade of math, if it was going to teach science it had to be *his* field of science, if it was going to be engineering then the *engineers* would have control. Then there was an education person who thought *he* was the key person, and if it was going to succeed then they had to have *him* as the leader of the project. Well, it wasn't long before I realized with a bunch of prima donnas who had a specialty—who knew their field of math in the eighth grade or math in the fifth grade, or engineering—*it wasn't going to go.*"

Brown wrote a five-page letter to Alpert on May 3. Whatever enthusiasm for the project Brown conveyed in the letter was overshadowed by the letter's grim conclusion that delicately questioned whether the project would ever succeed. "Many technical questions and disagreements arose," Brown wrote. "It is apparent that this field is in a considerable state of flux and is subject to most of the conflicts now afflicting the field of education in general."

The committee kept close to Sherwin's original Book with Feedback vision. "The immediate and long range objective of this research should be to produce an educationally (but not necessarily economically) practical mechanization of the textbook," Brown wrote, more prophetically than he or anyone else could possibly have known at the time.

Brown went on to recommend that the initial project focus on the delivery of education to one student only. "The burden of proof," he wrote, "regarding suitability for parallel use by many students or use by groups of students should rest with outside critics. This is not to ignore the fact that parallel use is of prime economic importance, but to be saddled with that requirement at the outset is to ignore the probable benefits of such developmental work now under way and to unduly restrict the initial efforts."

The committee had difficulty recommending what subject the computer should initially be programmed to teach. Physics? Medicine? Math? French? Biology? Electrical engineering? So instead of identifying a specific subject, Brown's committee did something that committees from time immemorial excel at: they punted on the issue. "The question of the subject matter to be programmed is a difficult one and reflects the present ignorance of the nature of educational processes as well as an understandable uncertainty as to the optimal use of a teaching machine." In other words, *let someone else figure it out.*

Brown continued, listing problems that telegraphed to Alpert that the committee not only couldn't agree on what subject the system should teach first, but also that they didn't know where to begin or how to go about building the machine. "It is believed that the two most critical problems in the teaching machine area are: 1. the detailed design and construction of the educational program; and 2. the technical realization of a method for storage and retrieval of the educational materials used in the program." He went on to list a random assortment of suggested devices, including "closed T.V. recorders," and various audiovisual systems built by Kodak and Magnavox. But the committee was stumped as to what was the best way to build a random-access system that could fetch educational content, be it for display or for audio playback.

In terms of personnel, the situation was equally fraught. Brown recommended that the project start with "two distinct groups," one for programming and one for engineering. Brown's committee was especially at a loss to identify an overall project leader for these groups, and suggested that if such an individual existed at all, he might have to have superhuman abilities.

> The most critical personnel problem remains to be discussed: that of overall direction of the project. This is a truly interdisciplinary project and it is clear that considerable effort is necessary to keep the efforts in the disciplines of education, psychology, and engineering well correlated. It is essential that the project be headed by a person capable of understanding the elements of each discipline entering into the project. Many decisions will have to be made on an intuitive basis which affect all parts of the program. Without an appreciation of all the components in the program on the part of the leader, the project will be crippled if not doomed.

Brown's conclusion to Alpert was that it all came down to people. Find the right people, and the project might have a chance. Hand over the project to the wrong people, and you might as well forget it:

> It is believed that the program outlined above is both natural to and desirable for CSL to pursue. However, at the present

time, two critical elements are lacking: 1. a senior staff person to assume responsibility for the programming effort, and 2. a suitable project supervisor. If these cannot be found within this University or obtained from outside, it is believed unwise for CSL to proceed further.

Alpert tried to "volunteer" Brown to take the project on himself. "He asked me if I would be the head," Brown remembers, "and I was burning with a gemlike flame to do some computer ideas of mine. I wasn't ready. I'd just come out of that Cornfield project and I wasn't ready for what was obviously going to be a long-term, big-haul, long-haul project. And I said I didn't care for it, didn't want to do it at this time."

So Alpert went to Stolurow. "I declined," Stolurow recalled, "because I felt I didn't have a strong enough engineering background. I knew what the system had to do, at least in my perception at that time, but I didn't feel I knew enough about the engineering aspects for implementing a design."

Sherwin, Brown, and Stolurow: all no. On May 24, Alpert began drafting a letter to Dean Everitt, summarizing the Brown Committee's conclusions. Like Brown's assessment, Alpert's focused in on project staffing, and in particular, who would lead the project, as the key to success. Alpert believed that the project leader needed to be a generalist capable of understanding the overlaps from all the conflicting constituencies. "In view of their interrelatedness," Alpert wrote, "it is most unlikely that four specialists, in say, a given subject, computer technology, communications media, etc., could make these decisions in committee."

He never finished his original marked-up three-page draft. He had a trip scheduled to Washington to meet with military funding agencies, so he set the letter aside. On his return from Washington, Alpert had an epiphany. "Flying back . . . I suddenly realized that, if there wasn't at least one person who was not only good and interested in teaching but could teach a subject and could design a computer-based system . . . it was obvious that you needed a person who was interested in *teaching how to use a computer*. Because if you're going to have the talent to design it, that's the obvious subject, right? I mean, it was clear. It had never occurred to anybody on that committee.

"What we wanted was an *inventor*," Alpert would recall. "From what I'd seen in the past of invention, of creativity, I knew the project would have to go to someone with both the ability to design a very different kind of machine and the motivation to try it." Someone who had a genuine passion for teaching people how to use a computer. Someone who was just crazy enough to take the project on despite its slim chance of success.

Alpert started looking around CSL, wondering if there was "a person in captivity who has the qualifications to design a computer-based system."

He believed there was such a person.

3

The Super-Achiever

Whoever swung the bat must have had slippery hands. The bat flew into the air, and as it flew it twirled, and as it twirled it smashed into Don Bitzer's head, right along an eyebrow. Bitzer and his Beta Theta Pi fraternity brothers were at a picnic that some of the university fraternities and sororities had organized. On this occasion everyone had decided to play softball. Now Bitzer was bleeding. "A big slash," recalls Jim Dutcher, a classmate and fraternity brother of Bitzer's. "We took him to the hospital and had to stitch him all up."

Three or four days later, Dutcher, Bitzer, and some other Betas were hanging out, shooting the breeze, when the subject of physical achievements came up. That got Bitzer and Dutcher talking about sit-ups. All of a sudden, Bitzer declared that he could do a thousand sit-ups without stopping. The Beta brothers were skeptical. "He was not *out* of shape," Dutcher says, "but he wasn't *in* great shape. To look at him, he was kind of heavy, and we all pooh-poohed it." But Bitzer was serious, and pretty soon it developed into a bet. "We put up a dollar apiece, probably fifteen guys betting against it, so it was big money," Dutcher says. Bitzer, stitches still fresh in his forehead, lay down on the floor and started doing sit-ups. He did ten. He did twenty. He did fifty. He did one hundred. Minutes rolled by. Still, he kept going, first to the astonishment of and soon to the concern of his Beta brothers. "He gets to three hundred, four hundred, and we're *begging* him to stop," says Dutcher. *"STOP! We'll pay you the money."*

Bitzer did not stop. He did all one thousand.

"You talk about grim determination," says Dutcher. "I thought he was going to split his stitches open and we'd have to take him back to the hospital . . . 'cuz he's getting red in the face and all, but he just went and did it."

Donald Lester Bitzer was a New Year's baby, born at the stroke of midnight on January 1, 1934, in East St. Louis, Illinois. He grew up in Collinsville, "the Horseradish Capital of the World," a small southern Illinois town situated a few miles east of the Mississippi River, on the west side of which sprawled the St. Louis metropolis.

Hailing from Germany, the Bitzer clan made their way to Pennsylvania as farmers, eventually settling in Illinois. Donald Bitzer's great-grandfather arrived in Collinsville in 1905 to open a feed store and livery stable. In 1917, Walter Bitzer, the oldest of the sons, went to St. Louis to see about getting a Ford automobile dealership. "In those days, car dealers were called 'garages' when they first came out," recalled Earle Bitzer, a cousin of Don's, "and they were in the alley like the blacksmith, they weren't out front with the showrooms and the glass." Much to his surprise, the Ford people in St. Louis told Walter there already was a dealer in Collinsville. Walter hadn't realized it because, sure enough, the Ford garage was hidden in some back alley. The Ford people suggested Walter talk to the Dodge people instead, which he did. There, he found an opening, and Bitzer & Co. was launched.

Donald's father, Jesse, born in 1902, and Jesse's three brothers kept the Bitzer automobile business going for decades. By the time Don was in high school in the late 1940s, the Bitzer auto business was booming. The postwar American economy was growing quickly and everyone wanted—and was starting to expect—a great job, a home of their own, and, perhaps most important, a shiny new car. Business was so good that the Bitzer clan would over the coming years open up four separate dealerships in towns across southern Illinois.

Bitzer was not just brainy. In high school he played on every sort of sports team, and at home played on a lighted basketball court in the backyard. "We'd play basketball until eleven or twelve every night," he says. "I didn't do a whole lot of studying, I didn't really have to. . . . Generally speaking, school was not very good. If you came from our high school, only maybe two or three people would go to college. It wasn't considered a great school, but the mathematics, the geometry teacher I had was Polish and he was just terrific. Instead of turning me off at geometry, like many people, it really turned me on. So, generally speaking, the math part was fine, the rest of the education was pretty poor, but the University of Illinois took care of that really fast."

A testament to the level of Bitzer's high school visibility and popularity is the frequency with which his name and photo appear through-

Donald Bitzer, c. 1964, in PLATO III classroom

out his high school yearbooks. Like Dan Alpert years before him, in senior year he would be named class valedictorian. That same year he was vice president of the Honor Society, vice president of the Monogram Club, president of the Math Club, president of the Pioneers Club, co-captain of the football team (alas, it finished with a 0-8 season), and a member of the Senior Class Council. He was also the escort to the Homecoming Queen. His senior class voted him one of the five top students in the class, in his case, "outstanding student." The school yearbook had this to say: "Don not only chose the hardest courses but excelled in all subjects. Science, Math, and Electricity were his best."

Science and math certainly (he was an avid reader of *Popular Mechanics* and *Popular Science*), but there was something about Bitzer's intimacy with electricity that brought to mind inventors and innovators like Nikola Tesla. During the World War II years he developed an interest in building radio kits with an uncle who was also a hobbyist. He'd tune in to London and other far-flung places.

His father, Jess, built the family house, and Don was encouraged to help out. "Laying the wire, put the heating in and things, and I did that," Bitzer says. "Worked until about midnight, and at about midnight when things were done, I'd sit down and do my homework, geometry or whatever, and get up at 6:30 the next day and go to school."

One summer, around 1949 or 1950, Bitzer was sent to the Carter Carburetor School in St. Louis on behalf of the family automobile agency in Collinsville. The Carter Carburetor Company, founded in 1909, became the premier manufacturer of carburetors in the United

States. By the early 1950s if one purchased a Ford, Buick, Oldsmobile, Nash, or Pontiac, the car probably had a Carter carburetor under the hood. Carter's school was the Harvard of carburetor academies. Dealerships and repair shops across the United States would send their people to Carter to get a diploma, the bestowal of which then often triggered a newspaper article back home with a headline along the lines of "Bob Phelps Ready to Take On Your Carburetor." These proud local shops would also run ads with messages like *Our mechanics are Carter graduates. Are you sure you can trust your car in the hands of those other shops in town? They don't have any Carter diplomas on their wall. We do. Call us today!*

Bitzer discovered that he was too young to attend, as the school's insurance policy would not cover someone younger than seventeen. "I think I was sixteen, or under sixteen," says Bitzer. The Bitzer automobile agency told him, "Don't worry about that, just tell them you're older."

"It was the only time in my life I remember lying," he would recall years later, "because I was told to." He filled out the forms, and supplied his correct birth date. But that date now conflicted with what he had verbally represented to the school.

The Carter official, "a really sharp cookie," according to Bitzer, looked at the birth date as written on the form, then looked at Bitzer, and said, "Which is it?"

"The birth date is right," Bitzer said. "I was told to tell you this other. I'm sorry, I should have never done that."

"That's all right, we understand," Bitzer remembers him saying. "It's an insurance problem, we have to get special insurance for you, we just need to know." He added, "You're in."

Bitzer loved his time at the school. But the experience of telling the lie ingrained in him a determination to, in his words, "never listen to anybody again if they told me to say even a little lie. I shouldn't have done that. It bothers me to this day that I was talked into that."

Beyond learning about carburetors, the Carter experience was also where Bitzer first learned what good teaching was really like. "The short period of time that this guy taught," he recalls, "he could put across ideas better than any teacher I'd had at school. And I'll never forget the time after the first hour lecture, he turned to everybody there, most everybody there was grown-ups, thirty or forty years old, turned to everybody and he said, 'If I've been doing a good job teaching, you won't be able to tell me what I'm wearing.' And he moved behind a screen, and said, 'Am I wearing a tie? What am I wearing today?' Nobody in the class had paid any attention to what he was wearing. Not a word. They could only remember what he said. That shows you when you're being effective."

Bitzer gained a great deal of knowledge about carburetors, but what made the biggest impression on him was the quality of the instruction. "I learned an awful lot about education in that carburetor class," he says.

The Bitzer family assumed that Don would make a career in the auto business, but his passion lay elsewhere. From around when he was just six years old, he'd always wanted to be an engineer. "I just loved putting things together and taking them apart," Bitzer says. "Figuring out how they worked and the like, and I was good at it."

His stepmother, Ruth, let him and his father deal with the issue of Don's future. "I tried to stay out of that," she says, "because I felt that Don and Jess should make the decision themselves. The only thing that Jess ever said to me was, 'I believe I could make a real salesman out of Don for the business, but,' he said, 'Don should make his own decision about it.'" He did. He would be going to the University of Illinois.

One day during Bitzer's senior year at high school, he asked a friend to pass a note to a girl in his study hall, which by some whim of scheduling combined students from the senior class with students from the sophomore class. *Donald Bitzer would like to meet you*, the note said. The recipient was Maryann Drost, a sophomore. She was delighted to meet Don. They dated, and four years later they married.

Don's competitiveness, the same competitiveness that would drive him to dare to do a thousand sit-ups a few years later, was already apparent during his high school years. "He always liked to outdo somebody at something," says Ruth. Bitzer had been that rare breed of student: the all-around super-achiever. He was the kind of kid who never seemed to study but always got straight A's. After acing high school and entering the University of Illinois, he not only got straight A's again, but had such a stellar, perfect grade-point average that by graduation four years later, his name would be etched into his 1955 graduating class's coveted Bronze Tablet, the U of I equivalent of *summa cum laude*.

Bitzer arrived at the U of I during the Truman era, the beginning of the Cold War. Atomic bomb tests in the Pacific were frequent. The

Korean War had already been raging for over a year, and thousands of Americans were being sent overseas to fight. College students were excluded from the draft as long as their grades were good. "They put a lot of pressure on the grades," says Jim Dutcher. "If you let your grades slide, then you were in the draft and off to Korea *fast*. They were killing people pretty quickly over there."

Bitzer and Dutcher became pledges at the Beta Theta Pi fraternity. With fifty-six fraternity houses, the University of Illinois had the largest and strongest "Greek" system in the country. For decades, leading up to and ending with the Vietnam War, the Betas beat out all the other frat houses, ruling the campus in terms of scholarship. "There was a rich tradition back in those days, the Betas just didn't rush anybody who was not in the top 10 percent of their high school class," says Dutcher. "So this was a brainy group of guys. People thought they were a bunch of nerds, which I guarantee they were not; it was a very, very exciting environment to be around seventy or eighty guys and all pretty darn well accomplished."

Dutcher remembers Bitzer being a bit of a slob. "He'd drive these guys nuts, because he wouldn't keep the room clean. I can still recall walking in there and seeing him study, and he would lie on his back, and he would have his calves up on his chair, his study chair, and he'd look like he's asleep, and the rest of the guys, accomplished, semi-cocky if you will, and everybody's saying, 'How the hell is he ever . . . he's not going to get any kind of grades at all. He's kind of nutty the way he studies.' He was not typical at all. Of course, midterms come out, and he's 5.0. It just really upset the upperclassman he was rooming with. They were getting ready to lower the boom on him, tell him to shape up, and hell, he was getting better grades than *anybody*." Bill Forsyth, for a while his roommate, believed Bitzer had a photographic memory, which would help explain his straight-A performance, and felt he was destined for the engineering life. "He knew more than the professors," says Forsyth. "He would work out a formula that was different than what was in the book, but came up with the right answer."

Bitzer took electives on a wide range of subjects, like English literature. "The son of a gun would take some lofty-to-tough course as an elective," Dutcher recalls. Everybody would be griping about the course, how tough the professor was—the usual litany of college student complaint—but Bitzer still got an A. He finished his undergraduate electrical engineering degree in 1955, got married to his high school sweetheart Maryann that summer, and came right back in the fall to begin work on a master's degree. His family would have been happier if he'd gone back to Collinsville to join the family auto business, but Bitzer was determined to continue his engineering education. Soon he was working at the Control Systems Laboratory on some of their classified projects, using the ILLIAC to process real-time radar data.

Five years later, at twenty-six, his master's and PhD degrees complete, Bitzer found himself being invited by Dan Alpert to take the lead on something entirely new: building a computer that could teach.

The very earliest days of the PLATO system's creation are fuzzy and details are lacking. There's almost no printed record, and little remains of lab notes, photographs, film, prototypes, the day-to-day documentation of progress. Had Donald Bitzer and his collaborators been meticulous documenters of every detail during those early days (and especially nights), instead of focusing on the actual design, building, and testing a real system, the project very might well have lost its momentum, and wound up abandoned. They were too busy building to be documenting. Bitzer was finally being given a shot to be what he'd dreamed of being since early childhood: an electrical engineer. History would just have to stay out of the way.

"I dislike history," Don Bitzer has said on more than one occasion. "I generally don't read books," he's said as well. He's too busy trying, testing, inventing, and building new things to be bogged down by someone's subjective and incomplete accounting of the past. Too many instances of those subjective and incomplete accounts have left him skeptical. He's admitted that what's written about him is often inaccurate to the point of being, in his words, "bogus."

While Bitzer was known to his colleagues in PLATO's early years as an excellent writer of code (the ILLIAC was a formidable beast, and only a few elite "code whisperers" knew the right incantations to make it do its magic), in the fifty years of PLATO he was hardly ever known to be a writer of words. (Alpert in interviews was quick to point out that it was *Alpert* who wrote anything published that listed Bitzer and Alpert as coauthors.) Unlike the conventional "publish or perish" pressure most professors are under during their careers, especially early on when they're just starting out, Bitzer somehow managed to play the academia game without having to write much at all. Just like in high school and throughout his undergraduate years, he was able to sail through the politics and pressures of academic life through a combination of luck, magic, salesmanship, and chutzpah—he was the son of a car dealer, after all—and a mission-focus bullheadedness that made others stand back and wonder, *How does he do it?*

"Publish or perish" would not be the Donald Bitzer Way. With him, the mantra became "PLATO or perish." His approach fit more with the present-day Silicon Valley mind-set of build something, anything, get it up and running, and show it to other people as soon as possible. Silicon Valley calls it *demo or die.* Stop yakking and build the damn thing and prove it does what you say it is supposed to do.

Compared to Bitzer's bemused but detached view of the long history of PLATO, Alpert decades later seemed bitter about the story's inaccuracies and could get downright cranky about the details of who did what, when, and why. "Bitzer remembers it flat out wrong!" was usually Alpert's first utterance in response to the words "PLATO" and "Bitzer." One possible explanation for Alpert's frustration is that, as the fates would have it, he would play the role of "champion" in this story, and Bitzer would play the role of "inventor." Inventors are the builders, the ones who usually get the recognition. Champions are the sponsors, the adult supervision, the mentorship. They're believers in the inventor's vision, a vision they share, and they're usually the ones with the money or at least the *access* to the ones with the money—just as important. The champions are the ones who block and tackle anything that might get in the way of the inventor being successful. Champions are essential. You might say Alpert was like a venture capitalist and Bitzer was the whiz-kid start-up founder.

Good technology champions help create a favorable environment for inventors. Daniel Alpert was, for PLATO's first dozen years, that dedicated champion who fought behind the scenes to make PLATO happen and make sure PLATO kept happening. He kept Bitzer and PLATO under his wing, and when bureaucratic forces attempted to pull them apart, or when Bitzer would occasionally disappear to pursue other interests, Alpert was the one who kept the funding flowing and the momentum going.

—

It would not have been unusual for someone in Alpert's position to take the findings of Richard M. Brown and his committee, charged with investigating the feasibility of CSL starting a computer-based education project, and tell his boss, Dean Everitt, that he agreed with the committee's findings. He could have simply said, "It's a great idea, but unless we find some superhuman with deep technical, educational, psychological, and subject matter knowledge, my recommendation is that we should not pursue this." He might not have even brought it up, telling the dean instead that his lab is still exploring ideas for new projects and *we'll get back to you*. But Alpert was not an ordinary laboratory director, and CSL was no ordinary laboratory. They relished hard problems. Impossible problems. Problems nobody else would or could take on. Building a computerized teacher was a hard problem. Finding someone to lead such a project was not entirely impossible. Genuinely intrigued and enthusiastic about Sherwin's original computerized Book with Feedback idea, he wasn't about to let a few minor details like the lack of a key senior staff person to oversee a brand-new, high-risk project get in the way. Still, it remains remarkable that Alpert chose to disregard the general gloom and doom of Brown's assessment. Even more remarkable is that Alpert came to reverse his own thinking, in the period of a few days, between May 24 and June 3, 1960.

It was during this time that he took his quick trip to Washington and had his lightning-bolt epiphany on the flight back home. Now that he had decided that "teaching people how to use the computer" should be the subject matter for the initial project, it was a "simple matter" of finding someone who had the drive, enthusiasm, interest, and expertise to not only design a computer system that could teach, but design one that could teach people how to use a computer.

Also remarkable is his choice of a junior engineer at CSL, someone with a noticeable lack of major accomplishments, lack of publications, and lack of teaching experience. Bitzer himself once confessed, "Perhaps the main reason I got the chance was that most of the people with more experience didn't think the project had a prayer."

If it were up to the university bureaucracy, Bitzer would have been deemed unsuited for leading such a project: he was low on the academic totem pole, and low in the pecking order within CSL, only twenty-six, and a freshly minted PhD with little teaching experience. He wasn't a tenured professor. CSL knew Bitzer was brilliant; everyone knew it. To Alpert, Bitzer was still as green as a month-old stalk of corn, "a graduate student who had never written anything in his life, and had never taught anything in his life. I didn't trust him as far as I could throw him," he said. Jack Desmond, CSL's business manager, remembers the meeting "in which Alpert told Bitzer that yes, they would be forming a new group within CSL to pursue this kind of research and development, and that yes, Bitzer would be the titular head of that group. I wasn't in that meeting, but I was in the anteroom just outside of where Bitzer and Alpert met, and Bitzer came running out, saying, 'Jack, what in the hell does *titular* mean?'"

And yet, offering him the project shows that Alpert *did* trust him, though with a short leash. There was something about Bitzer that gave Alpert confidence that he was the guy. Perhaps Bitzer was in some small way like the son he never had. Perhaps Alpert saw in Bitzer some of that grim determination Jim Dutcher had witnessed on the porch outside Beta House when Don Bitzer bet his Beta brothers he could do a thousand sit-ups, and then went and did just that. Perhaps it was that Alpert saw Bitzer as a builder, not just a talker. He was a dedicated engineer who had the technical ability, the tenacity to stick with a problem until it was solved, and a catchy enthusiasm and a vision that attracted respect, admiration, followers—and believers.

What Bitzer had going for him was a genius for engineering, particularly electrical engineering, that others recognized. You could throw down a challenge to him and he would not walk away, he would figure out how to build a viable solution. He had an ebullient personality that reflected a passion and joy for his work. It was infectious. Everyone saw it. Many were drawn to it. Everyone also saw the never-say-no car-dealer side to his personality (countless colleagues would describe this side of Bitzer as "snake-oil salesman"). Alpert was not blind to this, but nevertheless he had enough confidence in this twenty-six-year-old wunderkind—Bitzer's confidence no doubt fed Alpert's confidence—to turn him loose and give him a shot at running the project.

By now Alpert wasn't pitching the idea as just a Book with Feedback. He pitched it to Bitzer with a lot more focus. He described it as a "teaching system," and the first thing it needed to do was teach people how to *use* a computer, *via* the computer.

Bitzer, never one to say no, told him he'd think about it.

Alpert insists that Bitzer wasn't interested right at the start. "I had to twist his arm to take it, and two or three weeks later he said, 'Okay, I'll do it.'"

Bitzer's version of the story says he told Alpert, "First of all, hold the letter, because certainly that's not the case, we can do better than that. If you want a plan, give me a couple of days, and I'll come up with one, how to approach it, and then I'll see what we can do to implement it." Later he added, "After a few days I came back, I had an idea about how we might do it using some equipment that was already around, got permission to order some other stuff that we didn't have, got myself a programmer by the name of Peter Braunfeld, who was in the laboratory, and one technician . . . and that was how we started."

The technician was an accomplished engineer named Wayne Lichtenberger, who had not only also graduated from the University of Illinois in 1955, but was, like Bitzer, a winner of that year's Bronze Tablet award. Lichtenberger had become involved with digital computing at CSL in 1957 while he was in graduate school, and had known Bitzer since freshman year.

"The theory was," Braunfeld recalls, "a) here we are at a university, so by definition we know how to educate, and b) we know how to do what was in those days called real-time data processing. Which I guess today there isn't even a word for that, because it's all real-time. But the notion of real-time data processing was a big deal in those days, because computers typically had as their inputs cards or paper tape, and it came in and it did its thing and it came back out. The notion of actually getting stuff in at random times was rather rare."

Having all come fresh from the Cornfield project, these CSL engineers immediately saw an analogy between Cornfield's need to track the real-time comings and goings of multiple airplanes and naval vessels, identifying which was friend, which was foe, while they moved in different directions at different speeds, and the new project's need to track the real-time interactions of multiple students each sitting at a terminal, each interacting with the computer at their own pace, each student having essentially an ongoing, private, live conversation with their digital tutor. The big breakthrough that gave Bitzer and his colleagues a surge of confidence was that they could see at an abstract level the similarities between these two very different problem spaces—air traffic control and student data. Step back far enough, and from a purely architectural point of view, naval destroyers' radar data could be seen as data from student terminals. To the computer it was all just zeroes and ones. Students would be interacting with the system in real time, so the data would be pouring in from multiple sources in real time. No different, to the computer, from multiple aircraft radar pings being fed back to the central machine in real time. The notion of a central computer handling lots of random incoming data, processing it as fast as possible, and spewing it back out to graphical terminals? Just think of the students as aircraft and it all started making sense.

On June 3, 1960, Alpert finally sent a letter to Dean Everitt announcing his decision to launch a new project to build an automated, computerized teaching system, having found someone with the requisite enthusiasm, expertise, and energy to lead it. He described the project this way:

> As suggested by Professor Sherwin some time ago, the advent of the high speed computer makes possible a new approach to education in which the principle of feedback (from the student) may be applied to such traditional educational tools as the book, visual aids, etc. While this idea is in principle applicable to the teaching research objective of any course of study at various levels of sophistication, the first research objective will be toward the design of an "automatic teaching machine" to teach students at various levels how to use a high speed computer. This particular objective is especially suited to the talent and motivation of the people who would have to design such a machine and represents a unique educational need in the coming decade.

Alpert explained why teaching how to use the computer itself was the ideal first subject for the automatic teaching machine:

1. The design of the first machine requires decisions as to the philosophy of the approach, the design and use of a computer complex (the "machine"), and the programming of a course, all of which are intimately related. This relatedness means that the decisions must be made by people who not only know computers but also the proposed course of study. By selecting a subject which is (by definition) understood by any computer expert we have an obvious advantage in staffing a creative program.

2. One of the important aspects in learning to use a computer is to learn its language. It seems an advantage rather than a disadvantage to omit the human teacher and thrust the prospective student into contact with a machine.

3. This teaching objective is finite and specific. The effectiveness of the teaching program will be measurable in an objective way.

4. The research program will be uniquely suited to studying man-machine relationships. Both the language of the machine and new methods of input-output will necessarily be related areas of investigation.

5. While it is expected that the techniques developed can be applied to other subjects, it is evident that teaching the use of the computer to people in all walks of life is of itself a broad educational objective with far-reaching implications. It is particularly suited to an educational institution.

Such was Alpert's confidence in his laboratory's expertise and intellectual power that they would propose to embark on such a project that had no precedent and no certainty that it would work, and that indeed in all likeliness would fail. It is also noteworthy that the project would not directly involve educational psychologists, learning theorists, and others from the social sciences. Alpert seems to have become so fed up with them that when the idea of using Bitzer came to him, it must have looked like the light at the end of the tunnel. With Bitzer, Alpert had found a way to keep an eye on the project and keep it entirely within CSL. Over the next five decades, this decision would fuel criticisms that PLATO was too engineering centric: built by electrical engineers and physicists with little input or direction from real educators, when, considering PLATO's mission, it should have been led, designed, and run by educators.

Peter Braunfeld did his undergraduate work at the University of Chicago, and decided in 1949 to pursue a graduate degree in mathematics at the University of Illinois. He became a teacher's assistant in the math department, but needed a job for the summer. "In those days they had guys going around with sticks with spikes in them to pick up pieces of paper in the Quad," says Braunfeld. "I thought that it would be a great job. It would be outdoors, it would be mindless, and it would bring in a little bit of income." He went to the student employment office to apply for the trash-picker's job, but was surprised when they called him to say, "We want you to go over to a place called the Control Systems Laboratory because they're looking for people." A physicist named Lloyd Fosdick hired him "essentially as a gofer," Braunfeld recalls.

Braunfeld had a reputation for being an excellent mathematician. "He was an extremely good mathematician, extremely good," says Jeannine Leichner, one of only a few women who worked at the lab. "Peter was a really nice guy and very, very bright." But it took time for the lab to appreciate how good Braunfeld was.

Fosdick envisioned CSL building a computer that could track many airplanes in the sky simultaneously, in real time. A Navy aircraft carrier, escorted by destroyers, cruisers, supply craft, and other fleet vessels, would collect the radar data from many of these vessels, which would be scanning the skies for airplanes. The radar data would then be fed into a central computer aboard the carrier, which would sort out the data, identify directions, speeds, and altitudes of aircraft, evaluate threats, and, when foes were found, provide targets to eliminate them. This was the secret project Cornfield developed on the fourth floor of CSL.

"As time went on," Braunfeld recalls, "I got more and more interested in this stuff and I guess I got to be pretty good at it, and literally, kind of a Horatio Alger story, from just doing dog-work, they kept giving me more and more difficult things to do, and ultimately I was involved in doing a lot of programming for them."

Eventually, CSL tried not only to track airplanes but also to shoot them down. "There was a big project," says Braunfeld, "that involved the interactions of humans at screens and what functions they could perform best and what functions could the computer. Did the computer really help at all?"

Cornfield was beginning to wind down by the time of Sputnik, "and it was clear," says Braunfeld, "the military wasn't so inclined to support these things anymore." While the benefits of a Cornfield-type system appealed to the Navy, the actual implementation of Cornfield turned out to be not viable.

A quick overview of the design of the system may provide some insights into the early thinking of PLATO. When one considers a 1950s-era digital computer, of which there were only a few in the world, it is important to remember that they were hulking machines full of vacuum tubes requiring enormous electrical power to operate and considerable physical space in a climate-controlled building. Cornfield's designers felt that installing such enormous computers on every ship in the fleet, or at least the destroyers that escorted an aircraft carrier, was simply impractical. Instead they proposed that the aircraft carrier serve as a "mainframe," or data center, a central processing facility that could crunch through the data fed from the various destroyers' radar systems and then feed back the results to each ship. CSL's engineers proved that such a system could work, but the requirements for ILLIAC-style digital computers were so extreme that the Navy concluded the Cornfield concept was not suitable for operational use. The Navy saw the centralization of the data processing as one of Cornfield's less attractive design attributes: in engineering-speak, it's what's known as a SPOF, or "single point of failure." They would have preferred that each ship have its own computer and be able to assess its own radar data rather than depend on the aircraft carrier's computer. (Perhaps the Navy's memory of World War II and the heavy losses of ships accounted for their reluctance to put all their computing eggs in one basket.) In the 1960s, as the cost and electrical and physical space requirements of digital computers became more reasonable with each successive year, thanks in large part to the tiny, more efficient semiconductors replacing big, bulky vacuum tubes, some aspects of the Cornfield concept did eventually go to sea, in the form of the Naval Tactical Data System, which featured a more distributed model with computers on multiple ships in a carrier fleet.

"There was a feeling," Braunfeld says, "that the people who were involved in this Cornfield system were about the best bunch of people that had ever been assembled in any kind of work. . . . We were an incredible, elite group." This feeling may have contributed to Bitzer's confidence in accepting the challenge from Alpert to organize and lead a team to build a teaching computer. Bitzer and Lichtenberger would handle the engineering for PLATO, and Braunfeld would handle the software. Their shared experience on Cornfield guided their thinking for the new PLATO project. But it didn't take long for them to see that computer-based education was a wholly different kind of challenge.

In June 1960, coming up with a name for their new computer project "took the first week of research," Bitzer recalls. "We had a meeting the first week. Peter Braunfeld and Chal Sherwin and I met and . . . we each came with names. I came in with the name 'PLATO,' and I said 'Programmed Logic for Automated Teaching Operations.'" Bitzer recalls Braunfeld saying, "No, no, we can't say that, 'Automated' means self-mated, we don't want to give that image, we'll call it 'Automatic.' So we made that change, and that was it, and then we got down to work."

4

The Diagram

Bitzer was frugal. "Don never spent a nickel if a penny would do," says Garrie Burr, a longtime technician on the PLATO project. Consider Bitzer's legendary attire—the man was widely believed to own just one suit and that one suit was widely believed to have been a hand-me-down bought at a Goodwill store. "Clothes meant very little to Don," says his stepmother, Ruth. Even back in his undergraduate years, Don's clothes were a thing. Says one Beta Theta Pi brother, "His clothes, he'd just drop them on the floor, didn't care if they were rumpled up or what, and would hop back in 'em the next morning and you frequently wondered if they ever got washed, because he just didn't care." Judy Sherwood, who would work on PLATO a few years later, vividly remembers, "He would go out jogging at lunch time. And he had as usual these raggedy awful-looking jogging clothes."

One time, Bitzer threw his white T-shirt and white shorts into the washer with some red towels, and they came out pink. "My dad gets pretty attached to just one item of clothing," says his son, David. "He went right ahead and wore them, and jogged in this pink shirt and these pink shorts, and I think this was probably at the time that the Peter Sellers movies were out and therefore he got the nickname, jogging through campus, of the 'Pink Panther.'"

He wore the pink running outfit for a long while afterward, and at least one time it became an issue. After a run, he arrived back at the office and found his secretary still out at lunch. There was a man sitting in the outer office. The conversation unfolded like this:

BITZER: [in glorious raggedy pink running outfit] Hello, can I help you?
GUEST: I'm Professor _____ and I have an appointment with Don Bitzer. And I'm just waiting. He's at lunch and I'm just

waiting. People in the hallway told me I could sit here, he'd be
back in a little while.

BITZER: Well, I'm Bitzer, can I help you? Why don't you come
on into the office? And I'm glad you got here.

GUEST: You're *not* Don Bitzer.

BITZER: Well, I am.

GUEST: No, no, no, you're not Don Bitzer.

BITZER: How do you know I'm not Don Bitzer?

GUEST: Well, I've never met Bitzer and I don't know what he
looks like, but he doesn't look like *you*.

"That was pretty typical of Bitzer's attitude and approach towards
people and his garb," Sherwood says. "He goes into the office and he
finds somebody there and he didn't know him from anybody, and he
welcomes them heartily, you know, let me show you around, in his raggedy pink jogging shorts, I mean, somebody with more sense of style
would have stopped in the john or gym or someplace and changed his
clothes before he went back to the office. But not Bitzer. He probably
had his clothes in his own office and he'd shut the door and he'd change
clothes. Oh, he was a character."

He was frugal with cars as well. There was the Volkswagen so past
its prime, so decrepit, that the floor ("*What floor?*" quipped one relative) rusted out and you could see the ground while riding—hopefully
you did not ride over a puddle, as the splash came right up into the car.
Later he got a Dodge Dart (from Bitzer Motors, naturally), for which
he was reluctant to get power steering, considering that an unneces-
sary luxury. He ran that Dart for many years, again long past its prime.
"Every car Don had was into the well-after's," says a relative.

That same frugality came in handy with PLATO as it was getting
started—and it no doubt pleased Alpert that he'd picked someone to
lead the project who was not going to burn a lot of CSL's budget while
building the system. PLATO had to get by on whatever spare change
CSL was willing to toss at it, plus whatever Bitzer could scrounge
together. CSL had priority access to the ILLIAC. That machine might
be ugly, might be slow, might be severely limited in capability, and
was, in 1960, already eight years old and aging fast—recent advances
in integrated circuits and solid-state electronics went far beyond the

ILLIAC's short-lived vacuum tubes, and would soon become the standard for new computers—but that's all they had to start with.

Bitzer, Lichtenberger, and Braunfeld already knew how to write code for it, if you can call it that: it was an act requiring long hours of thinking and planning, large amounts of stamina, determination, and, above all, paper: one didn't simply march over to the machine and type the code into it and watch it run. Instead one tested out ideas on paper, wrote and rewrote the routines, met with others, and commented on them, and only after a thorough vetting would one march over to the machine—assuming you had booked some time beforehand—and type in the magic incantations that then generated punched holes in a special paper tape.

To make the ILLIAC do anything required first getting yourself into a special place mentally, getting into a "zone," a place that put you, in the author Ellen Ullman's words, "close to the machine." It required the equivalent of a Swiss watchmaker's intimacy with tiny gears, wheels, springs, rollers, and escapements, only in this case the intimacy was with logic circuits, vacuum tubes, and a lot of binary and hexadecimal mental gymnastics. It required *thinking* like the machine, one that was extremely limited in memory and speed, and understanding that a machine only ever did what it was told. If you gave it erroneous instructions, it faithfully went and made errors, because as far as it was concerned, that's what you told it to do.

Modern computers take advantage of what are called "interrupts," signals sent either by the program running at the moment or by some component of the hardware, to the central processor to tell it to stop what it's currently doing, set aside its current task and the memory associated with that task, and go deal with another task using *that* task's memory until told otherwise (that is, until it receives another interrupt). Once the higher-priority task has been dealt with, usually in a tiny fraction of a second, the processor can resume working on its original task. Every computer today uses interrupts. The ILLIAC's 1952 vintage afforded no interrupts whatsoever. "You had to be very, very careful how you wrote the code so that ILLIAC could respond suitably," Lichtenberger says. "I remember fooling around with that for quite a while until we worked out appropriate techniques."

Despite the ILLIAC's numerous quirks and limitations, the PLATO staffers were confident they could deal with them. They had experience connecting peripheral devices to the ILLIAC, back during the

Cornfield project. It was already wired up, ready to go. So the ILLIAC was going to be the engine that powered PLATO, whatever PLATO turned out to be. There were other pieces of the PLATO system that they would have to build from scratch. And the ILLIAC knew nothing about teaching, or computer science; all kinds of software would have to be written. The amount was in fact absurdly ambitious, and the chances of a tiny team getting it all done were hardly confidence-inducing.

The ILLIAC had only about five thousand bytes of computer memory, not enough even to store the text of this chapter of this book. To a CSL engineer, it was just another constraint. Engineers tend to relish constraints. With the ILLIAC the PLATO team had much to relish.

Over the summer of 1960, they built a crude prototype, based on a diagram they'd sketched that would wind up printed in one of CSL's Quarterly Progress Reports a few months later. This would mark the first time the PLATO acronym was mentioned in print as well.

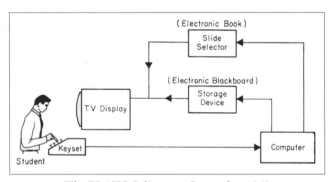

The PLATO I diagram, September 1960

The diagram, labeled the "Equipment Diagram for PLATO," contains the core design DNA of the PLATO system that would set the course for the next four decades. The diagram is as simple as can be—practically primitive. It's as if Bitzer and Braunfeld said, "Let's see, we'll need this and that and this and that and this and that," drew a quick sketch on the back of an envelope to reflect all those things, then later had an artist tidy it up and make a more presentable illustration for the CSL report. And yet, despite its utter simplicity, the diagram speaks volumes. It suggested what the system could do. It suggested what it could not do. It reflected the thinking of its creators, not only in terms of electrical engineering, but in terms of education.

There are three general components to this diagram, and while they're not spelled out, they can be summarized as Input and Process and Output. What gets *put in*, either by a human or by some device—either way, it has to get converted into electronic signals—gets sent to the computer to be evaluated and acted upon. Depending on how the computer's evaluation of the input goes, there may or may not be anything to *put out*, those electronic signals sent onward to those parts of the system that blink lights, print something out, display something on the screen, or play some media.

The diagram shows five boxes, labeled "Keyset," for student input; "Computer," for processing; and not one but three components driving output: "Storage Device," "Slide Selector," and "TV Display." But there's a sixth part of the diagram, the most important part: the student. Fitting with the era in which the diagram was created, the well groomed student has his shirt tucked in and is wearing a tie. In a few years he'd more likely have long hair and a big peace symbol dangling from his neck, but nevertheless, there he is, the student, drawn into the diagram as if to emphasize the importance of the human being to this system—as if to remind everyone building the system to remember what this project is all about.

Arrows run between each of the five boxes, indicating how information flows. If you step back and view the entire diagram as a whole, there are two clear ways to interpret it in terms of the model it is representing. You can see the diagram representing a *student-centric* model, where everything starts with the student and ends with the student. Or you could view it as a *computer-centric* model, where everything starts and ends with the machine, and the student is just along for the ride.

Viewing the diagram in the student-centric way, we see the student reading material on the screen and then responding to what he sees by reaching out to touch a key on the keyboard—which Bitzer and company called the "keyset." The information loop starts with the student and ends back at the student. The student is in fact the bridge between the keyboard and the "TV display." Two lines are missing but are implied: information from the TV display reaches the student's eyes, where they're processed in the student's brain, and signals are then sent to the student's arm and finger and through a typed character on the keyboard, the input traveling to the computer triggering another round of the loop.

This wasn't some Air Force corporal sitting at a console watching little blips on a radar display, identifying which were friendly jets and which were foes. This was going to be a student sitting at a terminal. Someone who, you could be sure, had never used a computer before or even seen one. The computer was the teacher, and needed to exhibit some characteristics that all good teachers exhibit: patience, leadership, and subject matter proficiency. The teacher had to know his or her stuff. In 1960 the computer was not going to "know" its stuff. It was, however, the "miracle of mechanical ingenuity" that had escaped Edward Thorndike in 1912: a machine capable of arranging instruction in such a way that "only to him who had done what was directed on page one would page two become visible, and so on." But it did something neither Thorndike's vision nor even Skinner's gadgeteering could pull off: answer judging. The ability of the computer to not only evaluate the input of the student, but also figure out its meaning, and not only determine if it was correct or not, but also figure out if it was partially correct or partially incorrect, and provide the most meaningful feedback to the student no matter what had been input.

It's notable that Bitzer positioned the "Computer" box in the diagram as far away as possible from "Student" and "Keyset." The student and the student's keyset and screen could be physically distant from the computer, hopefully as far away as possible: this would remain one of the core characteristics of the PLATO system, and time-shared computers in general, in all of its manifestations for decades to come. It didn't matter if the user was sitting in the same room as the computer or sitting twelve thousand miles away: a loop is a loop, and this is the way PLATO worked. In 1960, it was the only way it was going to work: computers were too expensive and too large for each student to have his or her own. Time-sharing was the answer, with terminals connected remotely all feeding into the same shared machine. Time-sharing would be the beginning of what we refer to today as "the cloud."

The growing body of late-1950s academic research and commercial projects inspired by the behaviorist work of Pressey, Skinner, and Crowder had raised awareness of the advantages of teaching machines and programmed instruction, among those being Self-Pacing and Immediate Feedback. Self-Pacing was a given, just as it had been with the earlier pre-computer boxes. The PLATO team took particular interest in the Immediate Feedback concept, imbuing the system with a need for responsiveness right down deep into the very core of the hardware. Another influence no doubt was the team having experienced, in prior CSL projects like Cornfield, the advantages of fast and responsive signals in the military radar and air traffic control systems: when in combat, it helps to know right away whether that blip on the screen is friend or foe.

In order for there to be Immediate Feedback, PLATO in its role as teacher was going to need to maintain an illusion of being very responsive to the student. Real teachers expect to encounter inattentive students all the time, but a student encountering an inattentive teacher is a surefire way of losing that student's focus. PLATO was going to need to be so fast that it appeared to work *at the speed of thought*—certainly at the speed of typing. As soon as the student pressed a key on the keyset, the system needed to recognize that a key had been pressed and quickly echo a representation of that key onto the display to be seen by the student. There was a kind of Skinnerian cause-and-effect satisfaction to achieving this simple feat: so subtle the student might not even notice what's going on. Tap on the bar and the pigeon gets a pellet of food. Tap on a key and the student gets something new appearing on the screen, providing Skinner's much designed "immediate knowledge of results." So far so good.

This was what we might call the "Fast Round Trip," the loop that consisted of the input followed by the processing followed by the output. The PLATO team decided that a Fast Round Trip would happen so very fast that the student sitting at the terminal would not even think about the magic that happened when he or she typed a letter and then immediately saw it on the screen. It just worked.

But behind the scenes, electrons embarked on a journey down tiny wire paths, every time a student tapped a key on the keyset. The journey began the moment the key was pressed, triggering an electrical connection inside the keyset, which launched an electronic signal out of the wire coming out of the back of the keyset, and then down through whatever length of wire might separate the keyset from the computer, be it ten feet or, in theory, ten miles or ten thousand miles, the bits of information that announced that "the user just pressed the '3' key on the keyset" would fly down that wire and in just a few hundredths of a second would reach the computer, traveling across more wires and through layers of circuitry and logic and relays and vacuum tubes until the information was absorbed and considered by the deepest reptilian layers of the computer's brain and recognized for what it was, "ah, a keypress," and then passed through whatever machine-code logic had been written to decide what to *do* with this signal that way out there somewhere in the realm of humans the "3" key had been pressed, and then either do something with the "3" or just send a signal out to display a "3" on the screen—so the human would see visual confirmation of the pressing of the "3." And all of that had to be very fast: from start to finish in under a quarter of a second. Any slower and the person sitting at the terminal would start to notice an annoying time lag between typing on the keyset and seeing the typed characters appear on the screen. Much of computing even today is about maintaining this elaborate illusion, just to keep the human's attention on-task and not on the computer itself.

One aspect of PLATO that one cannot glean from a glance at the diagram is the fact that the connecting lines drawn between the various components do not reflect how much information is passing through the lines at any given point. In fact there's an asymmetry to the information being sent out to the students' display versus what's being sent to the machine. If the diagram had been drawn to reflect the asymmetry, the lines coming out of the "Storage Device" and "Slide Selector" boxes heading toward the "TV Display" would be massively thick compared to the hairline connecting the "Keyset" to the "Computer." The "pipes" leading out to the student were big, while the pipes leading from the student to the system were little. It's the difference between a fire hose and a straw. There were all sorts of reasons why this was a practical approach. The speed of the little pipe was "good enough" for the Fast Round Trip and to enable working at the speed of thought. PLATO was like a building where the computer's processor was on the top floor, the students worked on the ground floor, and data going down was able to hop on an express elevator, but data going up required taking the stairs. To this day, what most Internet service providers provide is an asymmetric Internet service to the home: they give you a big pipe down (so you can watch streaming videos, listen to streaming music, download big files, etc.) and a smaller pipe up. Of course, today, that pipe up is still pretty big. At the start of PLATO, that pipe only allowed a tiny trickle of data back to the computer, and that decision would have serious repercussions in years to come.

The box labeled "TV Display" in the diagram was obvious enough, but what about the "Slide Selector" and "Storage Device"? The label didn't say "Slide Projector," it said "Slide Selector." What the PLATO team decided to do was use photographic slides to show pictures, diagrams, and other complex illustrations that would simply be too difficult or require too much memory or power to be generated by the computer itself. The other thing about slides: people—especially instructors, educators, and people in the audiovisual trade—knew how to design and produce slides for presentations, lectures, and classes. They wouldn't have to learn machine programming language to create the displays, even if they could, which they couldn't due to the limitations of the ILLIAC. What the ILLIAC could control was a digitally addressable slide selector, which fed its output to the "Storage Device," which Bitzer and company decided would take the form of a Raytheon QK-685 storage tube from which a video signal could be generated and sent to the "TV Display." The storage tube, which was a type of cathode ray tube that could store a computer-generated display image, was the key to making this work. CSL just so happened to have some Raytheon storage tubes lying around. They had been used in the Cornfield radar projects, so a computer could "draw" the radar images of approaching and departing aircraft and assist the operator to identify "friend" and "foe" in the sky. The drawn "image" was then read from the storage tube and displayed on a video screen. Couldn't the same principle be used to, for instance, draw a triangle, label the sides A, B, and C, provide the lengths of A and B, and ask the student to solve for C? A TV display could show the image that had been "written" into the electrostatic memory of the storage tube.

The storage tube had four operating modes: write, read, erase, and selective erase. Under computer control, the tube could be told which mode to switch into. Bitzer and team described it at the time this way, reinforcing their notion that the storage tube was an "electronic blackboard":

> When the tube is in the write position, letters and diagrams can be written on the storage surface point by point. In the read position the information on the storage surface is read out onto a dis-

play for presentation to the student. When the tube is switched to the erase mode, the entire surface is erased, while in the selective erase mode any individual point on the surface can be erased.

Raytheon storage tubes were expensive and didn't last very long, burning out after about one thousand hours of use. They also had the annoying habit of "forgetting" the contents of their memory after a while. A student sitting in front of the TV display being fed from a storage tube would see this "forgetting" as a gradual fade-out on the screen. Soon the entire picture would fade away, like an increasingly overexposed photograph. For this reason, for as long as the PLATO system made use of storage tubes—and it would, for a long time to come, the first dozen years—the keysets had a special "REFRESH" button on them, which when pressed told the computer to redraw the image into the storage tube's memory, which would in an instant then appear as a fresh, crisp picture on the screen. Until it started to fade again after a few minutes . . .

For slides, the team originally considered having the ILLIAC control a device that would feed from a set of everyday 35mm slide magazines and automatically select any random slide for presentation to the student. In PLATO's earliest incarnation, as the student reached a certain point in a lesson that required a slide to be selected, a technician in another room picked out the appropriate slide and scanned it with a camera, merging the image into the video from the storage tube. ("He got pretty fast at it," says Bitzer.) But these manual approaches didn't last long. Nor did the idea of some sort of ILLIAC-controlled 35mm slide carousel or slide magazine reader. Such devices would be too slow and unreliable: too many moving parts, too many things could break down. They needed something heavy-duty, and, always, something *fast*.

That something didn't exist. So they designed something that they could build. It was a machine that would hold three thin, disk-shaped platters, each about a foot wide, and upon each, slides would be printed as annular bands. Next to each slide a special binary code would be printed, so the computer could tell the machine exactly what slide to select, and the machine could read the codes at lightning speed and jump right to the correct slide. They envisioned fitting 224 slides on each disk, which should be plenty for any individual lesson.

The nearly nonexistent budget for PLATO in its first few months meant that Bitzer, Braunfeld, and Lichtenberger needed to be as creative in finding resources as they were in designing the system. For some parts, that meant the occasional "borrowing" of something lying around the lab or home. For the "TV Display," Bitzer found someone selling a used TV for a few dollars. The TV's channel tuner was broken. Not a problem for an intrepid electrical engineer like Bitzer. They didn't need the tuner working; they'd rig the TV to "tune" directly into the closed-circuit video signal coming over the wire from the storage tube.

The crude "Keyset" in the diagram was an apt choice of term, for this little box was far from being a keyboard. It contained sixteen keys for inputting answers to questions, as well as some additional keys such as "Continue" for moving ahead and "Reverse" for moving back through the "slides" presented in the lesson. The lack of a full keyboard limited the range of possible student answers, as well as the range of possible teacher questions. That would be solved in time, but PLATO I was a prototype, a proof-of-concept.

PLATO I terminal

Among other function keys were HELP and AHA, both notable and worth a moment of consideration, as they represent, like the diagram itself, an insight into PLATO's original DNA. If while taking a lesson you got stuck or confused trying to answer a question, you could always press HELP to see one or more pages of material providing more detail on the concept you were stuck on. Once you were ready to go back to the problem, you would press AHA to indicate to PLATO that you now understood and wanted to go back to where you were in the lesson before you pressed HELP. The AHA key would fall out of use in a few years as PLATO evolved, but it became an unspoken rule that the HELP key should always provide you with contextual help, whether you were a student in a lesson or an author doing something else on the system. This tradition would become so ingrained in the PLATO community that even games, as we'll see in Part Two of this book, had extensive HELP sequences. If you wrote a program on PLATO, it was assumed that you would spend as much time as necessary to write a thorough collection of context-sensitive help pages to go along with the program. On PLATO, when you pressed the HELP button anywhere, no matter what you were doing, the help page that appeared was always about the very thing you were doing at that moment. (Reliable, fast, context-sensitive help would become the norm on PLATO over the coming decades—something that the personal computing, smartphone, and Web industry has never embraced as vigorously.)

Each time a student entered an answer to a question, the system immediately responded with OK, NO, or a customized message, depending on the correctness of the answer.

Another aspect of PLATO visible right from the beginning is what today's Internet industry calls "analytics." PLATO collected student data while students worked on lessons, and that data was then available for analysis by the instructor. A late 1960 progress report on PLATO had this to say:

> The machine keeps an accurate record of each "move" the student makes. Thus at the end of an instruction period, the experimenter has at his disposal a print-out of how long the student spent on each page, what sequence of right and wrong answers were given, how long a problem required solution, as well as at what point help was requested, etc.

The advantages of this kind of built-in data collection and immediate analysis over Skinner's teaching machines and Crowder's scrambled textbooks were abundant and obvious. Whereas PLATO and non-computer-based teaching machines both offered Self-Pacing and Immediate Feedback, a system like PLATO extended Immediate Feedback *to the teacher* as well, enabling her to quickly see how a student was doing. In 1960, this was a giant leap forward.

During the fall of 1960, the PLATO team managed to create several small lessons to begin experimenting with the system, in elementary number theory, German grammar, and computer programming. Alpert had green-lit the new PLATO project with the idea that Bitzer would initially develop interactive lessons teaching people how to use the ILLIAC itself. Bitzer and Braunfeld did develop some binary arithmetic lessons that were key to understanding ILLIAC programming, but they did not hesitate to venture beyond that subject matter, knowing that PLATO would need acceptance and interest from a broad range of faculty disciplines if the system were to be viable long term.

The PLATO team knew that sitting a single student down at a single "Keyset" and "TV Display" connected to the gigantic ILLIAC machine was not going to be a compelling demonstration of the economic viability of computer-based education. The only way CBE was ever going to be economical was if multiple simultaneous students each could sit down at their own terminal and undertake their own lessons at their own pace, regardless of what else was going on on the system at the moment. This meant that PLATO had to become a time-sharing system. By December 1960 Bitzer and company were already working on it, mentioning the concept for the first time in another CSL progress report, describing a plan for "construction of a machine with the same teaching logic . . . which will be able to handle more than one student concurrently. In this version of the machine, two or more students will be able to work completely independently." This would become PLATO II.

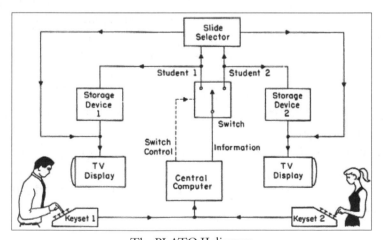

The PLATO II diagram

Bitzer tried to get the University of Illinois to file a patent for their implementation on the ILLIAC of "time-sharing," but the university botched the patent application. "We filed for a patent application right away," says Bitzer. "The person who supported PLATO dearly throughout the whole life was the president of the university, David Dodds Henry; he was the president who also managed to get the patent lost in the university. Not because he wanted to, but in those days, universities and patents, you know, it was a curiosity piece. It wasn't his fault. He wasn't trying to disturb PLATO—he loved PLATO. It's just that the university was not set up to handle things like this. You're a pioneer in many ways."

A few years later the time-sharing patent application was found, stashed away in some archive. It was re-filed, but it was too late. "As a result," says Bitzer, "we lost time-sharing. And that's something we would have had as well."

Crack open any computer history book or encyclopedia article today and you'll learn that MIT is credited with the first implementation of time-sharing, and usually there's no mention of PLATO or Illinois. "We lost a lot of clout," says Bitzer. It was a tough lesson that he took to heart. He would be ready the next time.

Al Avner, an educational researcher at the University of Illinois who would in time join the PLATO project, once described Don Bitzer himself as the original "time-share system." Those wishing to program on the ILLIAC I, Avner would recall, "would usually be told by Don that whatever needed to be done was trivial and to simply drop by Don's house that evening to see how. Upon arrival at seven or so, one would find a group with similar interests gathered around the Bitzer dining room table. Don would work on each person's problem for about five minutes in rotation until three or four in the morning and the 'trivial' problem would be solved in a few mind-numbing evenings (everyone was also working normal hours). . . . Anyone who wanted to use PLATO so much that they were willing to go through the agony of this process was deemed to have a strong enough interest in [computer-based education] that their inputs were of value in designing the system."

With the arrival of 1961, the PLATO project was gaining speed.

The PLATO II terminal

The effort to build PLATO II was under way, and Bitzer and Braunfeld were giving more demos all the time. In fact, "the demo" was to become the main thing Bitzer seemed to be doing: showing the system off to whatever interested party would sit down to take a look. In March, the interested party was David Dodds Henry, the president of the university. He put on a conference, called "Improving Our Educational Aims in the Sixties," at Allerton Park, a secluded retreat among tall pines some thirty miles to the west of the campus. It would mark a milestone in PLATO's history: the first time the system had been demonstrated remotely, dialing in from a distance.

Bitzer and Braunfeld recruited a young grad student named Rick Blomme to play the role of "student" during the demonstration. "The engineers thought Bitzer was crazy," Blomme says. The idea was to use a regular phone line and a primitive, hacked-together modem to send the student "keyset" data back to the ILLIAC. But you couldn't use a phone line for the signal coming to the student's "TV display"—that would have to be an actual video signal. Cable television did not exist yet, and they certainly were not going to run a fresh coaxial cable from CSL all the way out to Allerton. So they came up with a "kludge," in keeping with Bitzer's frugal, pragmatic style. The university ran a television station in town, WILL-TV. It did not have any broadcasting in the morning. But it did that day. Bitzer rigged up the closed-circuit video output from the storage tube, and ran it over to WILL-TV, which broadcast it over its antenna. Had anyone turned on their TV at home that day, they would have seen the same output that President David Dodds Henry and the assembled dignitaries saw on a television screen out at Allerton: the output of a PLATO lesson.

Bitzer gave the main talk about PLATO and then Braunfeld introduced Blomme to the assembled audience as the live, remotely dialed-in demo began. "Here's a student, now, he's going to study this little lesson, and push keys." As Braunfeld lectured the audience about the workings of the PLATO system and the design of the lesson, Blomme would wait to get a signal from Braunfeld to do something. "Now here's what the student would do, he doesn't know what to do so he would press the HELP key," which was the signal for Blomme to press HELP, after which Braunfeld would describe how HELP and AHA worked.

Blomme may have been portrayed at this conference as the innocent student, but in reality he was no slouch at computers. "I started in chemistry and went into engineering physics," he says, "went to grad school, did mathematics, got a master's, and that was about as far as I went. They were getting tired of me, and I had gotten involved with computers. . . . My understanding was, in those days there were no computer programmers. People were electrical engineers or chemists and physicists and they programmed the computer to do whatever they needed to do." The idea that someone would be a programmer—as in, that was what they did, *program*, full-time—was looked down upon in 1961. Blomme was one of the first to break the mold, and proudly identify as being a programmer.

Blomme would soon be recruited by Larry Stolurow, who had also presented at the Dodds conference at Allerton, to work on a competing computer-based automatic teaching system at UI. He would be Stolurow's programmer.

5

Soldering Irons, Not Switchblades

The orchestra played "Arrivederci Roma" as passengers drank, danced, and celebrated the last evening of their westbound transatlantic journey that had begun nine days earlier at a dock in Genoa. The captain had scheduled the ship to pass through the narrows between Brooklyn and Staten Island at sunrise, offering everyone a dramatic view of the Statue of Liberty and the New York skyline as they steamed into the harbor. Twelve-year-old Andy Hanson, his ten-year-old brother, Donnie, his seven-year-old sister, Ardith, and his six-months-pregnant mother, Elizabeth, were looking forward to coming home. They'd packed and gone to bed early so they'd be up and on deck in time to witness the exciting end to their voyage.

Andy's father, Alfred, a nuclear physicist, wasn't aboard. After working on the Manhattan Project in Los Alamos, New Mexico, where Andy had been born only months before the first atomic bomb was tested, Alfred packed up the family, moved to the University of Illinois, and became an assistant professor in the physics department and eventually project director for the Betatron, a four-hundred-ton particle accelerator. He began a long, prolific period of research studying the structure of protons and neutrons, research that helped lay the groundwork for the multibillion-dollar, miles-wide particle accelerators and colliders built in more recent times. In 1955, Dr. Hanson was awarded a Fulbright scholarship to help physicists at the University of Turin build their own Betatron. The whole Hanson family would move to Turin for the year. That summer, the family sailed to Europe and began their Italian stay. In just a year, Andy would cultivate a lifelong fluency in Italian and love for Italy. Alfred Hanson was invited to confer with nuclear physicists working on an experiment at the Brookhaven National Laboratory, so he flew back to New York early, and planned to meet his wife and three children when their ship docked in New York Harbor on July 26, 1956.

On the night of July 25 they approached the mainland of the United States, south of Cape Cod. As passengers danced and drank the night away up on the main decks, the four Hansons lay asleep in what Andy describes as "not quite the worst cabins" of tourist class, situated one level above the waterline. At 11:10 p.m., they were jolted awake by a violent crash followed by a chilling, unearthly, grinding groan of metal against metal—*EEEEE-zhurrrrrr*—a deathly sound that Andy Hanson still recalled more than fifty years later as if it had happened yesterday. He bolted upright, looked out the porthole from his bunk, and saw, from what seemed to be just an arm's length away, lights from another ship, a moving wall of steel, pass right by the window. That did not make any sense. He immediately slammed shut the cast iron cover of the porthole and tightened the screws. Then he climbed out of bed.

Within seconds, everyone noticed that their ship, the 701-foot-long *Andrea Doria*, a world-class ocean liner beloved by many celebrities of the time, was listing starboard. Within a minute it was listing 15 degrees. Soon it was listing 20 degrees. Confusion reigned among passengers and crew alike. In the middle of a fog bank, on the final evening of a long cruise from Europe, the *Andrea Doria* had crashed into something, or something had crashed into it, and now it was sinking.

It turned out the collision was with another ocean liner, the eastbound *Stockholm*. It had entered the same fog bank that evening, some forty miles south of Nantucket. A series of miscommunications, misunderstandings, and mistaken indicator readings led the crews of both vessels to commit to fateful wrong turns. The *Stockholm*, sporting a special ice-breaking bow ideal for navigating the ice-filled waterways of Sweden, gouged a massive hole into the *Andrea Doria*'s steel flank, crushing numerous cabins, instantly killing forty-six passengers, and tearing open several fuel tanks, at the time nearly empty and full of air after the long voyage. The ruptured starboard tanks rapidly filled with ocean water. The combination of empty port tanks full of air and the weight of all that onrushing water into the starboard tanks caused an uncorrectable starboard list that quickly went extreme. Within twelve hours the *Andrea Doria* would sink to the bottom of the ocean.

The Hansons were about to find out just how low on the totem pole tourist-class passengers were. Steel gates blocked stairways up to the wealthier classes. "It was very well partitioned off," Andy Hanson says.

He compares the class demarcations to scenes in the 1997 film *Titanic*, where the poorest passengers were kept behind caged-off stairways deep belowdecks. "By the time we got up through our tourist class cabin labyrinth up to the deck, all the lifeboats were gone, there were no rescuers to be seen, so we sat there for a couple hours."

As they waited for help, their ship continued to list further, slowly drifting away from the *Stockholm*, which was now invisible in the dark, foggy distance. The crew of the *Stockholm* would in time realize that despite the fact that the bow of their ship was crushed—there was no longer any bow at all—their ship appeared to be stable and was not going to sink. So they decided to release a couple of lifeboats to rescue *Andrea Doria* passengers. Out of the dark mist down in the water, Andy made out several lifeboats appearing out of nowhere. Elizabeth Hanson was determined that her children were going to be rescued. "We're going for the lifeboats," she declared. The *Doria* was now listing 40 degrees, the edge of the deck now only twenty or thirty feet above the water. "We went down," Andy says, "and there were no ropes, no ladders, no nothing, on the deck we were on, no way to get to the lifeboat except jumping in the water." Andy jumped in first, finding the water warm as bathwater, and swam to the first lifeboat, where he was quickly pulled on board. "My mom says I sat there grinning like a Cheshire Cat," Andy recalls. His sister jumped in next, followed by a reluctant Donnie, who needed a decisive push from his mother. Then she bunched up her knees to protect her unborn child and jumped in next. They all swam to the lifeboat and got on.

Other passengers followed suit. One small Italian family, frantic to save their four-year-old daughter, Norma, "just went completely crazy," Andy recalls. The father tossed the child down to the boat before people in the lifeboat were ready to catch her. Andy watched the child fall and hit her head against the side of the lifeboat, instantly knocking her unconscious. Rescuers had to call for a helicopter, which transported the Italians to a Massachusetts hospital. The child died the following day.

The lifeboat made it back to the *Stockholm*, which, while still afloat, was so heavily damaged that the captain maintained a speed of only two knots. Another ocean liner arrived a few hours later and picked up the rest of the passengers. It would take two more days before the *Stockholm* would limp into New York Harbor and the Hanson family could be reunited with their father.

More than fifty years later, now Professor Emeritus of Computer Science at Indiana University and former head of the department, Andy still thinks about that night. Over the years he's figured out the math. For him it all boils down to a difference of 2.7 seconds. That's all it would have taken. "I had gone back into the records," he says, "and found that the *Andrea Doria* was moving at 21.6 knots. And if you do the conversion and you calculate that, that's thirty-five feet a second. I'm ninety feet away from the collision. Forty-six people died instantly at the collision point. That's 2.7 seconds. I'm either here or I'm not here. The other ship could have been slower, the *Doria* could have been faster. Two-point-seven seconds would have been me instead of those other forty-six people."

From an early age Andy Hanson had an interest in science and technology, no doubt influenced not only by his physicist father but also three physicist uncles. In 1955, before his family left for Italy, he'd built a robot of sorts in his garage. "I did lots of things," he says, "built rockets, blew things up." When he needed to find or make parts for his latest gadget, it helped to have a father who ran the Betatron, in a building with not only tools galore but a complete machine shop. "We'd get stuck with some kind of gear we needed, and we'd go over . . . and machine it at the Betatron."

Hanson enrolled in Urbana Junior High School and, one night in early October 1957, watched from his backyard as a tiny sunlit speck crossed high up in the sky: Sputnik. "I remember huddling around a little old tube AM radio in our basement listening to the beeping," he says. Even though he was just in eighth grade, he took a high school trigonometry class over at Urbana High School. In that class he met a kid named Roger Ebert. Susan Gilbert, another student at Urbana High, and editor of the school yearbook, was Ebert's "archrival," Hanson says. "[She] succeeded in getting his goat by inserting the aphorism 'A self-made man who adores his creator' under Roger's yearbook picture." Decades later, Hanson, who regularly attended the Ebertfest film festival in Champaign, bumped into Ebert, who remembered him not only by name, recalls Hanson, but "as the precocious freshman in his junior year geometry class with Miss Bauer, UHS's legendary geometry teacher."

Urbana High School participated in a program called JETS, the

Junior Engineering Technical Society, a national organization founded in 1950 at Michigan State University with the goal of encouraging high school students to immerse themselves in challenging engineering projects, which, it was hoped, would spark an interest in engineering as a career. Many high schools near Big Ten colleges had affiliations with JETS, and by the late 1950s, thanks in part to Sputnik, some four hundred JETS clubs had been formed. Andy Hanson got involved in JETS at Urbana High, as did several classmates, including Gary Gladding, who would go on to have a long career as a professor of physics at UI.

Don Bitzer wound up being one of the UI faculty sponsors for the Urbana High School JETS program. Just a year into the PLATO project, he was already sponsoring a few kids at University High School (known to everyone as "Uni High" or just "Uni"), located on the other side of Springfield Avenue from CSL. Bitzer had already tasked two Uni students, George Frampton and Steve Singer, to help him with PLATO. According to Hanson, Bitzer was having trouble raising money for the nascent PLATO project, and also had trouble attracting graduate students or postdocs to pitch in and help. "Frampton and Singer were being trained," Hanson says, "as programming assistants for Bitzer to do projects he was unable to get done by other people. I remember him mentioning this to me many times, 'I don't even know why I would bother with college students and postdocs, I've got the smartest high school students in the universe right here, I'm just going to grab 'em!'"

This attitude was shared by Dan Alpert. Says Lou Volpp, who would become involved with PLATO a few years later, "Dan Alpert's idea about getting good people to work on something was to announce that he had a problem, and he wanted some people to help with it. He would never mention that he had money. And then they would come and they would work, and he wouldn't tell them that there was some way to get them research time or anything like that. He would find out who was compelled to work on it by the nature of the problem. And when he found out who those people were, then he'd hire them. And I think that's why he and Don would get along so extraordinarily well."

The "turn 'em loose" recruiting style led to great things. Steve Singer, Hanson says, "wrote an entire compiler from scratch, he was quite amazing." The compiler (a system program that takes a human-readable programming language and converts it into machine-readable form for execution by the central processor) was called CATO, standing for, appropriately enough, "Compiler for Automatic Teaching Operations."

Frampton, like so many Uni kids would do in coming years, had one day wandered over to CSL and marveled at the work being done in the lab. He was particularly impressed with the Cornfield project. "There was a group of four or five people who were doing this, and it was fascinating to watch the screens," Frampton recalls. "You'd see two planes come together, and then they would run all kinds of test runs, and occasionally the planes would hit and then 'oh shit' and back to the drawing board.

"It must have been one of the very first efforts to do automated air traffic control where you have the program calculating quite a bit—and their idea was to put planes in different envelopes many, many hundreds of miles away, create the layer in the sky and give them signals and I can vaguely remember as they got closer and closer the signals would either blip more, be a different color, but it was a fascinating computer programming." Everything was written in machine code, he says. "There was no programming language. . . . I can remember that there was a certain amount of back-and-forth with the PLATO people, who were facing some conceptual issues that were not that different."

Frampton remembers PLATO's early days at CSL as being almost extracurricular in nature compared to other, better-funded, better-staffed CSL projects. "It was a little like the PLATO stuff was, you know, 'going out to have a beer afterwards.' The PLATO stuff was the easy stuff, the fun stuff, this other stuff was . . . hard and tedious and complicated and PLATO was, 'Oh, well, we're actually going to have some fun and do something.'"

Steve Singer hailed from Joliet, Illinois, just outside Chicago and a two-hour drive north of Urbana. In the fall of 1958, he stayed with his grandparents in Urbana and enrolled as a sophomore at Uni High. It was quite a departure from what he'd experienced as a freshman back at Joliet Township High, which dwarfed Uni High with one thousand students in its graduating class. "Uni High occupied two small buildings," Singer says, whereas the Joliet school "occupied three or four city blocks. . . . Except for basketball and track, Uni had no athletic program—not that I missed that. There was also no vocational training; more than half its students had a parent on the faculty of the U of I who expected them to become professionals or academics as well. There was more than a whiff of elitism in the air."

One day Singer discovered he had "somehow offended" an Urbana High football player, someone he didn't even know, and worse, he then found out this football player was "looking" for him. Singer had no interest in a confrontation. Being a Uni High student had its advantages, one being excellent connections with faculty at the university, since so many children at Uni had parents who taught or did research on the campus. It was time to make use of one of those advantages in hopes of steering clear of the Urbana High bully. One Uni classmate's father, a physics professor, heard that Bitzer was looking for help over at CSL and introduced Singer as a possible candidate. He got the job. "I was glad," Singer says, "to have a job close by where I could disappear after school—even at 90 cents per hour. Random choices sometimes lead to a vocation."

Frampton and Singer were both a year older than Andy Hanson, and spent most of their time at CSL. Hanson considered them the "grand old men" of JETS. Hanson became more involved on a different project, with Bitzer's guidance, where the JETS kids would build their own crude computer—or at least functioning components of one; it wasn't clear yet how long it might take or how far they'd get—out of a nearby "ramshackle building." Frampton and Singer were more involved in programming, and in this little building they would gather, sometimes with Hanson, to "program," far away from the computer. The only realistic way to write a program on the ILLIAC computer, the temperamental hulking beast in the CSL building, was to work it all out on paper first, reducing their ideas to numbers and machine code—not an easy task even for adults. Only after it was on paper did one dare make an appointment and take it over to the ILLIAC to be keypunched into the machine and run.

Frampton and Singer programmed a checkers game on the ILLIAC. Writing a game on any computer is an achievement, but writing a game in 1961 on the ILLIAC—one of the first computers in the world, with extreme constraints that would confound most present-day programmers—was not a minor feat. The game was even playable. What made it special, Bitzer remembers, is that it "learned" as the game went on. The high schoolers were especially amused when adults tried their game and found the going tough, the computer winning. What they didn't tell the adults was that the game had a convenient bug that in effect made the computer cheat.

Bitzer's JETS group became affectionately known as "Bitzer's Boy

Scouts," and as each summer came to an end and the high school seniors went off to college, new kids took their place, and the projects continued. But for Andy Hanson, the "Boy Scouts" moniker reminded him of something else: a Disney comic strip from the 1950s, *The Junior Woodchucks of the World*, featuring the antics of Huey, Dewey, and Louie, and their scoutmasters Donald Duck and Launchpad McQuack. For Andy, the local Urbana JETS chapter would always be known as "Junior Woodchucks of the Engineering School."

Another JETS member was Carolyn Leeb (née Leach), a self-admitted "science and math geek" who had been captivated by Sputnik back in 1957 and still holds on to a folder full of yellowed newspaper clippings about the satellite. "I have vivid recollections in third grade," she says, "even before Sputnik, of spending much of my days with a grocery bag over my head with an eye place cut out, as a space explorer." Her dreams of going into space were "shut down," she says, "because at my generation, there were to be no women in space. But thanks to Sputnik there was some money and emphasis on speeding up our math and science education."

By the time she was ready for high school, Carolyn found out she qualified for Uni High, but it cost money to go there, "and there was no way that my parents could have done that." Instead she went to Urbana High School, the quality of which exceeded her expectations. "Many of my high school teachers were women with PhDs, and of course this was in a time when women's opportunities were severely restricted, so these women would get PhDs, their husbands would get the job at the university, where were they going to go? So my chemistry teacher, my math teachers, had PhDs."

"It's amazing how good, stuck off in the middle of nowhere in a corn-field . . . how good things were there," says Nick Altenbernd, a class-mate of Carolyn's and a fellow JETS member. "A lot of the teachers at UHS were graduate students who were finishing up their degrees at the university. . . . They were out of school and writing their dis-sertations, or something like that, and their job to keep them alive was teaching at Urbana High School. They were young and idealistic, and we really profited from that. We had a wonderful education as a result. Our first French teacher, as it happens, was Igor Stravinsky's daughter-in-law. His son Soulima Stravinsky was on the music faculty at Illinois. And his wife, Françoise Stravinsky, who held a law degree from the Sorbonne, taught French at Urbana High School."

It was decided that the JETS project, as mentored by Bitzer, was to build, from scratch, a genuine homemade computer, component by component. "We made everything, coils, metal parts, switches, by hand," says Hanson. For starters they would build a 12-bit adder, which eventually they even got to work. An adder is exactly that: a digital circuit that electronically adds numbers in binary form. By building such a gadget, the JETS kids would learn about electricity, binary numbers, binary arithmetic, combinational logic circuits, AND gates, XOR gates, digital circuitry, and perhaps most important, how to wire the whole thing together without burning the house down. "We had quite detailed plans for how to make the timing controller and program execution but that turned out to be a bigger project than high school students could do," he says.

In Andy's basement, they set up on top of a Ping-Pong table what he described as a "computer component assembly line," with lathes and coils of wire. His parents were not too appreciative of the clutter. The JETS kids then went to work spinning the coils of wire. Seemingly endless wire, countless coils.

"The whole thing was based on reed switches," he says. A reed switch is an electromagnetic device invented by a researcher at Bell Labs in the 1930s. At its simplest, the device consists of a pair of slightly separated ferromagnetic metal "reeds," sealed in an airtight tube to prevent contamination. The far ends of each of the metal reeds are connected to wires forming a circuit. The circuit is "off" or "open" in its default state. But if the gadget is placed close enough to a magnetic field, it could cause the two reeds to come together to the point of touching, allowing electricity to flow through the wires and across the two reeds. A common application of reed switches is in security alarm systems for windows and doors: if the door opened, for example, it might cause the switch to pass by a magnet briefly, and set off an alarm. Reed switches were cheap and relatively easy to make—two criteria that appealed to the always practical, always frugal, Donald Bitzer. Hanson recalls that they were "a relatively clean but bulky way of handling digital logic."

"I remember hours and hours in Andy's basement," Carolyn Leeb recalls, "trying to perfect, then mass-produce, the switches. We listened to comedy and music from the likes of the Smothers Brothers and the Limelighters." She can still recall the words to the Limelighters' song "Vickie Dougan" and the Kingston Trio's "Scotch and Soda."

One time Bitzer was trying to drill a piece of sheet metal but the metal spun and cut his hand and once again he was off to the emergency room for stitches. "I remember Dr. Bitzer arranging for computer time for us in the middle of the night and how cool we thought we were hanging out in the Union or at the Spudnut shop," says Carolyn. (Spudnuts were not just ordinary doughnuts: these were made with potatoes, and devoured by Midwesterners of the era.) The JETS kids discovered camaraderie on the project. Some dated each other. Andy and Carolyn did for a while. Everyone had membership pins to wear on their lapels. The pin? A miniature golden slide rule. Given the fact that *West Side Story* had won Best Picture in the 1961 Oscars, it wasn't surprising that this nerdy ensemble of Urbana kids would joke about their name being shared by one of the gangs in the movie. "We made jokes about it all the time," says Carolyn. "Stuff like, 'soldering irons instead of switchblades.'"

Bitzer's enthusiasm, electrical engineering wizardry, and most of all his positive attitude toward failure impressed Carolyn, who described his approach as, *Well, let's try this material. Oh, that doesn't work, well, let's try this then.* "The notion that failure means you got to try something different," she says. "Failure was never a roadblock. It was just rerouting." Altenbernd views Bitzer as a "first exemplar of what a free-wheeling mind can do." To an optimist like Bitzer, *failure was actually success*—congratulations were in order—in that it meant one could scratch one approach off the list and move on to try a different approach. One fewer thing on the list. "It's an attitude that will serve you well in any kind of discipline," says Altenbernd. "We built and tested lots of failed ideas," says Hanson, "converging towards a working design over a long period of what was really research. For example, I remember inventing the principle for a 'memory' switch sitting on Bitzer's floor one evening fooling around with some metal strips and some magnets—possibly even patentable, but we never pursued it."

Hanson was looking for something to do over the summer of 1961. "I had a really boring summer job the year before and was looking for something interesting," he says, and somehow he heard about an opportunity at the PLATO project. He describes Bitzer's management style as *monkeys and typewriters:* "He'd define a problem only if he needed something specific done. The rest of the time I think he sort of trusted us to come up with things, and many people did just because of the freedom to do so.

"Bitzer at least made me feel as though I were the main designer," says Hanson, "with him collaborating on figuring out what main idea needed work next and getting resources if we needed them. Then I served as the main conduit to the other students, who could grasp what was going on. Mike Walker was one of the few who really turned out to have the right stuff to pick up the ball where I left off."

A friend of Andy's, Mike Walker got interested in the JETS activities and would eventually take over the leadership of the project when Andy graduated from high school. Together they programmed the ILLIAC to play a version of Kalah (known also as Mancala) and didn't understand enough complexity theory to calculate how long it would take to run through the possible outcomes. "One weekend we brought sleeping bags and started our program running to try to 'solve' Kalah by exhaustion, and about four in the morning, a microscopic way through the exhaustive search, a campus policeman found us sleeping behind the computer and kicked us out. I don't know if Bitzer got heat from that or not." It was not the last time the police would kick kids out of the building late at night.

Hanson couldn't believe CSL even allowed him and the other JETS kids in the building, and yet, there they were, sometimes working twenty-four hours straight through. He attributes their unhindered access to Don Bitzer. "Bitzer had this amazing, very explicitly articulated philosophy," says Hanson, "that he didn't care who was doing what, as long as he had the most capable people he could rustle up from any walk of life to do his work for him. And he just basically gave us a lot of freedom and he got a lot of stuff out of us for essentially no money. I mean he was paying a dollar an hour or something." Decades later, Bitzer's involvement in JETS can be viewed at least in part as a recruitment effort as much as one of genuine encouragement and mentoring. He knew there were bright kids out there in local high schools. He knew he could help them discover talents and creativity they didn't even know they had yet. And he knew he needed help on the PLATO project. So it is not surprising to find the cream of the crop from JETS winding up working on PLATO in serious and meaningful ways.

In the spring of 1962, the Urbana JETS presented their computer project at a local JETS science fair. It should have been a wild success. It should have brought great recognition to Andy and the rest of the JETS for accomplishing so much. They did win some sort of statewide award, but it was a bittersweet win. Not because Andy Hanson and his teammates had done anything wrong. On the contrary: they'd done everything right. To their shock they discovered people felt they were *too* good, and judged their work not to be their own.

JETS members, Urbana High, 1962

"There were lots of little pieces of really interesting research," Hanson recalls, "many details of which I had the only real overview, though many of the JETS members knew their 'specialty.' So I was all charged up to try to explain all the concepts and put everything together in one big picture, not only for the judging panel but for the other students." Hanson got halfway into what he thought was a "bang-up job" in his lecture, describing his group's project in detail, when the judges cut him short and walked off in a huff. One of the professors came over, took Hanson aside, and proceeded to, in Hanson's words, chew him out. "I was told by our sponsoring teacher later that they were extremely offended at my 'attitude' of 'pretending' to know it all, and had decided on that basis to downgrade the rating of the project, though not too much, in deference to the fact that the other kids had put so much work into it. I was incredibly offended at the idea that here were actual university professors who were not interested in the concepts, the work, the research, but only in trying to make sure my tone of voice didn't imply that I knew more than they did." For a long time afterward, Hanson would question "whether scientists were really to be trusted when they said they believed in research for its own sake; clearly this group did not." After the disaster at the JETS fair, Hanson's friends told him they wanted to hear the rest of his talk, but Hanson was disillusioned and "never really got up the enthusiasm to do it again after the put-down." Hanson remembers numerous people coming by his team's exhibit booth, but they would take one look, snootily

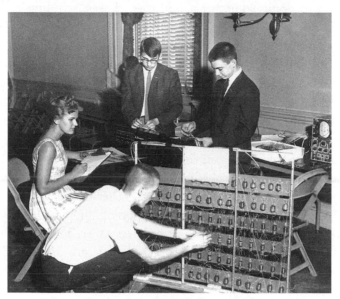

JETS exhibiting REGITIAC, 1963: Carolyn Lcch,
Mike Walker, Maurice Miller, Nick Altenbernd

mutter, "*Oh*, well, they had *help* . . . ," and walk away. "We *never* had that kind of help," Hanson says. "Bitzer made sure that everything we knew we learned the hard way."

Bitzer's house became the new meeting place for JETS. With Hanson graduating from high school, the handmade JETS computer, by now christened REGITIAC (the Urbana High School sports mascot was the tiger, so they reversed the letters and added "IAC" as a homage to the ILLIAC), was no longer welcome on his family's Ping-Pong table in the basement. Mike Walker gathered it all up and moved it to another location—Bitzer's basement—and then took over the reins as the leader of the JETS. "I ended up," says Walker, "basically living in Bitzer's basement for a couple hours every evening after I delivered my papers and before I went home to have dinner."

Walker had been introduced to JETS by his "terrific" high school math teacher, Miss Brannon, but he says it was Bitzer who drew him into the world of electrical engineering and computing. "It was Don

Bitzer's enthusiasm and acceptance of young people and bringing them in and putting them to work. I mean, truly. The guy's got absolutely infectious enthusiasm. He'd bring you in and he would take the time, set you up to doing something, and there I was, winding coils on a lathe, and doing this sort of stuff, and hey, he wasn't too proud to do some of that himself."

Some parents of the JETS kids were growing suspicious of what exactly was going on in Bitzer's basement. "One day, I was in a book club," says Bitzer's wife, Maryann, "and one of the members came and she said to me, 'Maryann, I've got to apologize to you, I just couldn't stand it anymore, but I came to your house, and I was looking in the windows.'" She told Maryann that each day after high school, her son would tell her that he was heading over to the Bitzers' to work on the computer. "I feel terrible admitting this," the woman told Maryann, "because I should trust him, he's always honest." The woman went up to the windows of the Bitzer home and peered in. The basement was mostly at ground level so it wasn't hard to look inside and see what was going on. "I felt so *terrible*," the lady confessed to Maryann, "because *there they all were*. All around this *computer* in your basement."

From the start, Bitzer knew that PLATO was going to need a more powerful computer than the ILLIAC. CSL also realized it needed more computing power for the variety of projects under its roof. In time the lab invited hardware manufacturers in to discuss what machines were available. "It turned out," says Bitzer, "that the Midwest salesman from CDC [Control Data Corporation], by the name of Harold Brooks, visited CSL to show us the CDC computer. Before he joined CDC he used to sell heavy equipment—cranes or the like. What a salesman."

Getting that CDC 1604 was not easy. CSL took competitive bids from a number of vendors, including CDC and IBM. CDC's bid was about $995,000. IBM's, for a 7090 computer, was about $1.9 million. That led to a second round of bidding. CDC came in again at $995,000. IBM this time lowered their price to about $850,000. CSL still wanted the CDC machine. IBM was furious, and contacted the state legislature in Springfield to force CSL to accept the IBM bid. CSL didn't want it. The legislature threw the whole thing out and told them both to rebid. For the third time, CDC came in around $995,000. IBM, for the exact same 7090 computer, came in around $200,000. This time,

CSL was furious. Such were the ways of IBM in those days. After a long fight CSL got their CDC 1604, and IBM's shenanigans would be remembered for many years to come by everyone at CSL, including Don Bitzer.

The first thing they did on the 1604, for which the PLATO project was granted about one to two hours a day, was create a simulator for the ILLIAC. Ten years of algorithms and libraries of subroutines created on the ILLIAC were too valuable to reinvent on the 1604, so it was decided to emulate the ILLIAC on the new CDC machine as a way to jump-start its use and be productive more quickly. Wayne Lichtenberger wrote the program, which they named SIMILE for "Simulate the ILLIAC Exactly." "All of our programs in the laboratory and PLATO were written for the ILLIAC," Bitzer says. "SIMILE was a program that made the 1604 simulate the ILLIAC and the simulation mode ran at roughly the same speed as ILLIAC, which was very slow. Nevertheless, the first thing we did was run PLATO in the SIMILE mode in order to get more clock time than the one or two hours a day on the ILLIAC that was available for the whole laboratory."

Bitzer corralled Andy Hanson and Mike Walker to work on PLATO, continuing the tradition he had begun with Frampton and Singer. Hanson graduated in 1962 from Urbana High, and went on to Harvard. But even as a college student, during the summers he would come home to Urbana and work for Bitzer on PLATO. "What I recall," says Hanson, "is that the work of at least a core group of Bitzer's Boy Scouts in '62–'63 seemed essential to Bitzer's plan to make PLATO into a viable project on the 1604 in the face of zero resources, active opposition from some in CSL, and a power struggle with his collaborator. He needed massive amounts of programming and experimentation and he managed to get what he needed from us for practically nothing, and on top of that built the rest of the empire for which he is justly famous." His "collaborator" was Peter Braunfeld, who was slowly detaching himself from Bitzer and PLATO. Too many cooks in the kitchen, it seemed, and Bitzer had established himself as head chef.

The challenge Bitzer presented to Hanson was extraordinary: write a multiuser, multitasking resident operating system for the 1604 that would support thirty-two simultaneous users. Not even CDC had been able to build such a thing on the 1604, but such was Bitzer's unstoppable optimism that he was confident Hanson, all of nineteen, would be able to pull it off. "My work on the resident program," says

Hanson, "was built around Bitzer's explaining a finite-state machine to me, and then expecting me to figure out the rest." A "finite-state machine" is a concept used in electrical engineering and computer science to describe a "machine" or program that can be in one of several states until something changes causing the state to change. A simple analogy would be a traffic light, with its red, yellow, and green lights. The signal will show one light at a time, starting with red, then going to green after a certain amount of time has elapsed, then yellow, then green after another amount of time has elapsed, then back to red.

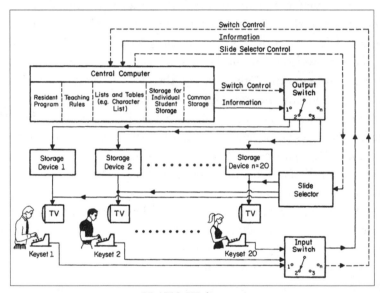

PLATO III diagram

The "resident program" Hanson built was called CATORES, short for "CATO resident." Bitzer was right. Hanson's code was a work of art, and would keep the PLATO III system running until 1973, when it was finally retired to make way for PLATO IV.

Even though Bitzer's Boy Scouts were doing great work, particularly during the summers when he would pay them and give them specific projects, some of the senior staff at CSL expressed consternation and doubt about the value of having a bunch of kids running around in their fancy laboratory. "There were some senior people," says Bitzer, "who objected to these students being paid to come in to do this creative work which they didn't think was productive." Alpert shared Bitzer's views on bringing in bright people regardless of age—his own daughter was one of them. Alpert defended Bitzer's kids but other senior lab people continued to object. "They insisted that these kids give a semi-nar *defending* the work that they were doing," says Bitzer. "These kids would have to defend their work. The kids loved it. They got together and gave this four-hour presentation on the things they had done with computers, and it was impressive. It was *impressive*. Everybody was impressed, and never again did we have a question about our summer students."

Sometime around 1964, while Hanson was an undergrad at Harvard, he heard about an upcoming seminar at MIT. The presenter: Don Bitzer. Hanson attended and witnessed the same sort of skepticism against Bitzer and PLATO that he himself had faced back at the JETS science fair, a disbelief that would haunt the PLATO project for its entire history. Bitzer began his presentation and encountered heckling from the MIT audience. "His trademark enthusiasm was mistaken for 'used car salesman' misrepresentation," says Hanson, "and the elderly— MIT engineering professor?—man sitting literally next to me in the third or fourth row kept muttering out loud about how everything Don said was completely theoretically impossible. He objected in par-ticular to the idea that you could handle as many students at once as Bitzer claimed using a single processor; we all know now that this is easy using task-swapping in multitask operating systems. The 1604 resident program of which I was—or was at least allowed to believe I was—a principal architect accomplished all this by brute force in a very primitive way, but it did work; the fellow next to me paid me no attention when I suggested that not only was our speaker, Bitzer, not lying, but that I had written the code that did it. . . . I was probably not very convincing as a very young-looking twenty-year-old."

6

Gas and Glass

The PLATO project launched just six years after Skinner unveiled his first teaching machine, made out of a wooden box, at a convention of psychologists in 1954. What people thought a teaching machine could be—should be—had evolved at a lightning pace in the ensuing years. First there were the mechanical boxes. Then there were the scrambled textbooks. Then IBM toyed with using a computer to simulate a teaching machine. That was the starting point for PLATO's team, whose stated goal was "to develop an automatic teaching system sufficiently flexible to permit experimental evaluation of a large variety of ideas on automatic instruction." Such a system would need innovations in software and hardware. They were still grappling with late-night paper-coding sessions at Bitzer's dining room table and pecking away in machine language on the ILLIAC and then the 1604. The 1960 PLATO I diagram depicted what components they needed to acquire or build. One implied aspect of their stated goal would remain elusive for years: scale. Bitzer's view was not unlike the mantra of Silicon Valley start-ups decades later: *Go big or go home.* In three years, his fledgling team had built and demonstrated PLATO I and II, not only showing hints of what could be done with automatic teaching, but also attracting more professors to try out the system within their respective subject areas. PLATO II's time-sharing feat demonstrated that two terminals could be plugged into the ILLIAC and used simultaneously by two students: a necessary milestone. PLATO III, now under way, was going to be another important milestone. But the dream lay beyond even PLATO III. It did not stop at two or even thirty-two connected terminals: the dream called for thousands, ultimately *millions* of student terminals scattered all over the country, in schools, colleges, universities, and homes.

Two components of the 1960 diagram, it was becoming clear, needed ditching: the "Storage Device" and the "TV Display." Something had to be done about those things. As word got out about what Bitzer had in mind to replace them, some reacted by saying that he must be insane.

PLATO's early design called for each student to use a makeshift keyboard and a cheap TV display fed by a closed-circuit video signal from an expensive, unreliable Raytheon storage tube. That design had two big problems. First: the video signal itself. Closed-circuit video signals require a lot of communications bandwidth. Every time you added another terminal to the system, that meant another display, which meant you needed that much more video bandwidth. Unlike 1960s broadcast television, where millions of people tuned in to one of a mere handful of channels to see one of a handful of programs, with PLATO every student would be working on something different, working at their own pace, typing their own unique answers to questions, interacting seemingly one-on-one with their automatic teacher. In essence, every PLATO user would be tuned in to their own "show." That was fine on PLATO I and PLATO II, and would be tolerable with PLATO III, but what about a future PLATO with thousands, maybe millions of terminals? Video signals were not a viable solution for that kind of scale.

The second problem had to do with a consequence of the way TV displays worked. Cathode ray tubes (CRT) constantly redraw themselves, thirty times per second, because the CRT has no memory. Instead, every thirtieth of a second, it requires a fresh signal of information describing the complete picture, which the electron beam draws, line by line, over and over. Lose that signal, and the picture instantly goes away. When a TV screen was used as a student's PLATO terminal display, the Raytheon storage tube acted as a temporary "memory" containing any text and graphics that were mixed together in the tube, then blasted over the closed-circuit cable to the TV's CRT. But storage tubes were expensive, unreliable, and required one per student terminal. This approach would not support a very large-scale PLATO system. It was hard enough to support a handful of terminals: sometimes the images would begin to drift while the student was interacting with the material.

Lezlie Fillman had been hired in August 1961 as an educator to observe students working with some of the mathematics lessons, see how they were interacting and responding, and suggest areas for improvement. "We had to put everything that was text on film," Fillman says, "and try to display that to students as the image was drifting across the screen, and the students would put their answers in, and by the time they got their answer in after they'd pondered the problem, the answer had kind of drifted out of the box." The storage tube arrangement was tolerable as a proof of concept, but it was impractical for thousands of terminals in the future. An alternate solution had to be found.

When PLATO began, the notion of Moore's Law was still five years out. Not until 1965 would Gordon Moore, then of Fairchild Semiconductor and later one of the founders of Intel Corporation, write his famous article, "Cramming More Components onto Integrated Circuits" in *Electronics* magazine, from which emerged the principle known as Moore's Law, which stated that the number of transistors and integrated circuits etched into a silicon wafer doubles every two years. As long as this trend continued, Moore argued, computer components would continue to get faster and cheaper. He even believed that this trend would lead to computers cheap enough for home use. Moore only had a few years' worth of actual data upon which to base his prediction, since integrated circuits had only come onto the scene in the late 1950s. But the trend was so clear, so obvious, he believed that there was nothing to stop it, so it would continue for at least ten years. It turns out that trend would continue all the way until today.

It was not yet apparent to the PLATO team in 1960 that Moore's Law was already under way. To appreciate what they were up against, consider how prohibitively expensive random-access memory (RAM) was for a computer then: $2 *per bit*. Take the typical 8 gigabytes of RAM found in a single present-day laptop: at $2 per bit, that much RAM in the early 1960s would cost $68 billion in 1962 dollars, or nearly $500 billion in 2017 dollars. That's for *one* laptop.

Consider video RAM. Instead of using storage tubes, Bitzer could have built a box with some RAM in it that served as the memory of the display fed by the computer and then sent out to the TV screen. There was a simple reason why Bitzer did not go this route. If each PLATO student workstation were to rely on video RAM, each terminal would have cost over half a million dollars (nearly $5 million in 2016 dollars). Video RAM was therefore out of the question. They knew the storage tube route was temporary; something else would have to be done for the long haul. What other technologies were there to consider? There was no existing affordable digital memory and display technology.

Computer graphics displays were already a hot topic by 1963, but they were typically relegated to massive government-backed projects. For example, the Cold War–era SAGE project to detect incoming intercontinental ballistic missiles had gigantic graphics displays but that system cost about $10 billion—more than the Manhattan Project—to design and build. Much was learned from SAGE, including the numerous benefits of interactive graphical displays. CSL itself had worked on some for its air traffic control projects. But neither SAGE nor Cornfield solved how to put twenty interactive graphics terminals in a classroom in a school, and then repeat that many times over in other schools, and do it in a way that didn't bankrupt the school, the town, the state, or the nation.

So, how were they going to scale PLATO but keep the cost of the terminal displays manageable?

As early as 1961, Bitzer began looking around for alternatives. There was one that had promise: gas plasma, where you send a current of electricity strong enough to charge a gas to the point where it glows. If you created a grid of super-thin wires and ran each intersection of wires over tiny "cells" of gas, in theory you could at any one moment charge any cell you wanted in that matrix, with the combined glowing dots forming a picture. The idea had been kicked around for years. But success had so far eluded everyone.

Causing an enclosed area of gas to glow by running a sufficiently high voltage through it was an old idea. Neon signs, invented by Georges Claude in France fifty years earlier, worked on this principle. Just before the turn of the century, Sir William Ramsay of Scotland and Morris W. Travers had discovered the element neon. They injected neon into a tube that had a metal anode (the positive piece where the electrical charge flows in) and cathode (the negative piece that the electrons flow to, across the gas), powered it up, and discovered that the gas glowed the color of fire. By filling the tubes with different mixtures of neon and other gases, and then running voltage through them, you could control which color would glow. But there was something else about neon. In his 1904 Nobel Prize lecture, Ramsay said that neon "showed a spectrum characterized by a brilliant flame-colored light." Travers remarked that "it was a sight to dwell upon and never forget."

They'd discovered the Orange Glow.

—

By the 1950s, Burroughs Corporation brought out a product called the Nixie tube, a small glass tube with a single wire mesh anode and a number of separate cathodes, each ingeniously shaped into numbers, letters, or other symbols. Run a current through the anode and a selected cathode, and suddenly a visible number or letter would blaze to life with an Orange Glow. Over the coming years, Nixie tubes would appear in all kinds of devices, from industrial and technical equipment to indicators in elevators to let you know what floor you're on.

In January 1963, an article entitled "Large Displays: Military Market Now, Civilian Next" would appear in *Electronics* magazine. It described three different engineering designs for large flat, digital displays, and featured an illustration of a large auditorium with seated rows of military men viewing two huge flat wall displays showing maps and trajectories. This was what one company, Lear Siegler, envisioned. They already had a contract with the Navy to build small flat panel displays using two sheets of glass as a "sandwich" around an inner "slice" of transparent ceramic material. One of the two sheets of glass contained horizontal rows of tiny electrode wires (so tiny they were effectively invisible to the human eye) and the other sheet contained vertical columns of electrodes. The inner "slice" of the sandwich was pockmarked with a grid of many tiny 0.05-inch holes drilled into the ceramic and then filled with neon gas. The holes were spaced to sit directly on top of the intersections of the electrodes, which were exposed to the gas like in a Nixie tube. The theory held that if you sent an electrical charge through a particular horizontal electrode and the corresponding vertical electrode, the two of which intersected over a specific "cell" of gas, that cell, and *only* that cell, would illuminate. Repeat quickly enough with different horizontal and vertical electrodes, and you have "pixels" across the display lighting up. Presto: you've got a flat panel display. One problem Lear Siegler continued to run into, however, was that adjacent cells tended to light up as well as the intended cell. That remained a stumbling block for the company, but would later serve as an eye-opener for the PLATO team, who studied Lear Siegler's results carefully.

Lear Siegler's belief that they could build large flat displays for the military was a natural extension of an idea already well understood. With the display Lear Siegler proposed, with its cross-hatched conductors sandwiched between sheets of glass, a grid of thousands of small gas-filled holes wherever the X and Y conductors crossed, the priority seemed to be getting a large display to work, not necessarily figuring out how to solve the digital *memory* that the display would rely on. If the memory wound up costing hundreds of millions of dollars, oh well, that was the military's problem. PLATO didn't have millions of dollars lying around. The memory problem would need to be solved.

For a number of years, two camps of researchers around the country had been exploring ways to use grids of tiny wires across tiny pockets of gas plasma: one camp was interested in creating displays, and the other was interested in creating computer memory, as it was theorized that gas plasma had inherent memory characteristics if you set up the wires, voltage, gas, and glass in just the right way. So far no one, including Lear Siegler, had figured out how to create a single gas plasma device that could function as *both* a display *and* a memory. It was not clear that anyone ever would.

This is where Bitzer and PLATO began poking around. Through their research they found patents filed in the late 1950s that were focused on gas plasma devices. In retrospect, the work is remarkable if for no other reason than that it was undertaken by none other than Douglas Engelbart, who, by the time Bitzer and company came across his patents, had already abandoned the research and moved over to a project that would lead to inventing the mouse and that inspired— perhaps launched—the Silicon Valley personal computing revolution. But back in the 1950s Engelbart was trying to make a name for himself with gas plasma matrices for display and memory use. He was not successful. But his work left clues . . .

Two years younger than Bitzer, Robert Willson was a grad student who had earned a bachelor's in engineering physics and a master's in physics at Illinois. He took a graduate course in circuit theory by Dr. Mac E. Van Valkenburg, and, says Willson, "I was incredibly turned on by what he did. I mean, he was probably my first professor who really got me excited about learning. And I thought, Gee, I want to switch to electrical engineering and do my PhD in circuit theory, something related to that." He switched and found himself with an assistantship at CSL. While he was considering writing his PhD thesis on graph theory or circuit theory, at the same time, Bitzer was beginning to poke around with the idea of a plasma display with inherent memory. "All of a sudden it dawned on me," Willson recalls. He would work with Bitzer to attempt to create a plasma display and base his thesis on the work.

In early 1962 Willson built a crude one-cell model of a plasma display in a glass tube. Inside the tube, emptied of air to the point of being a vacuum and then backfilled with argon—they hadn't begun trying neon yet—was a "sandwich" of super-thin, little square sheets of glass, the kind of glass used to make microscope slides. The outer sheets of glass—think of them as tiny slices of "bread" in this sandwich—each had a thin strip of nickel wire running across it, one serving as anode, and one as cathode. In between the two slices was placed a sheet of glass into which a tiny dot had been drilled. Inside that tiny dot was a small amount of the argon gas. The idea was to see what would happen when they ran an electrical current across the gas inside this sandwich of glass and nickel wires. What they discovered was that this was not going to be easy.

"From my perspective," says Willson, "none of us really knew anything about vacuum physics. We were very naive." After the first three months of work, Willson declared in a CSL Progress Report that he had just spent his time on "familiarization with vacuum techniques" and "preliminary investigations into the characteristics of the one-hole storage tube." He was trying to get consistent data readings while charging a tiny hole of gas inside the glass sandwich, but the readings were far from consistent. Plus, he discovered there was contamination from air leaks in the gas cell and "burning" of the tube's components from the voltage. Over the summer of 1962, Willson fixed the air contamination and burning problems, only to run into a new problem, "sputtering." When he applied voltage to the gas, the excited gas emitted ions, which over time bombarded the surfaces of the nickel anode and cathode, making them less effective. He was going to have to build new tubes and try different gases, pressures, and voltages. So it went: as soon as Willson knocked down one problem, new ones popped up, and the slow-going would continue well into 1963.

In the 1950s, the government of India created the Indian Institutes of Technology, with plans for several campuses around the nation. American universities, including the University of Illinois, partnered with IIT, which would become the MIT of India, to help it grow its own engineering and computer science departments.

In April 1963, UI professor Gilbert Fett proposed that it was time for IIT's Kharagpur campus to get its own IBM 650 digital computer. He suggested that twenty-nine-year-old Donald Bitzer was just the man for the project. At first hesitant to go on such a journey, over some months Bitzer warmed to the idea. PLATO I and II were done; PLATO III was well under way running on the CDC 1604. Even the plasma display project was coming along. Six months, a year: why not?

Dan Alpert was not so happy about Bitzer, the heart and soul of the PLATO project, going off to India, when so much work had yet to be done. The mere mention to Alpert of the word "India" forty years later instantly sparked a response of "Oh shit, just an adventure!" The PLATO project was growing, gaining steam, and its director was going to be eight thousand miles away with unreliable and slow communications with the outside world. "Who was there to run the show?" Alpert said. Despite the concerns, he did not stop Bitzer from going. Bitzer did have high-ranking champions cheering him on. One of them was William Everitt, the dean of the engineering school. "He was fascinated with what Bitzer was doing," says Jack Desmond. "The revolutionary implications of what he was doing were really quite well known to the administration and he had an enormous amount of visibility. Once you see genius like that then you're quite willing to overlook some of the more mundane duties that professors perform, when this guy's presumably making much greater contributions to the educational enterprise through his vision."

As 1963 wore on, Robert Willson was conferring more and more with other CSL people for ideas on the plasma panel project, including Hiram Gene Slottow, known to all as "Gene," a research physicist. Slottow once described the early days of the project this way: "Neither Bitzer nor Willson was an expert on gas discharge phenomena, but Bitzer's style was (and is) to jump into a program with great imagination and energy, make mistakes, and learn as he goes along."

Thirteen years older than Bitzer, Slottow had been an Army engineer during World War II, working at the Aberdeen Proving Ground in Maryland. His wife, Irene, met him there when she was a USO hostess. "I was taking commercial art classes," she says, "and looking for a beautiful profile to draw . . . and there he was, surrounded by USO hostesses. A very charming, very gentle man." Gene inherited a grand piano from a wealthy aunt, and they took it wherever they went from that point on. He was an excellent pianist and often considered going professional. "He really wanted to be a jazz musician," she says. But the world had other plans. Once, during a trip to Chicago in the 1950s, he met a UI physics professor who urged him to consider joining CSL. Gene was ready for a change, and said he would come for a year, just to try it out. He wound up staying, even completing a PhD along the way. "It was really a very interesting group," Irene says, "and he just loved it." If Bitzer was a "wild man" within CSL, Slottow was a quieter one, widely regarded as a scholar and a gentleman, a careful writer and thoughtful presenter. Slottow did not share Bitzer's car-salesman personality. "Don was really one of a kind," says Irene. "He didn't follow any rules . . . [and] he created an awful lot of controversy."

Now Bitzer and Willson were conferring with Slottow on the preliminary work on the plasma displays, still only one-pixel test models plagued with setbacks and surprises as they learned more each day about how unwilling a wild and free Nature was to be harnessed. But Nature would soon learn that it was no match for Bitzer, Willson, and Slottow.

In August 1963, Bitzer learned that IBM had unilaterally changed their mind about donating a 650 computer to ITT, and instead would give a less powerful 1620 machine. By December, IBM was trying to back out of even supplying ITT with any computer at all. Gilbert Fett pushed back, and IBM agreed to stick with the plan. Bitzer, his wife, Maryann, and their toddler son, David, began their westward journey from St. Louis on January 10, 1964, flying Pan Am the whole way. At each stop they rested at a hotel for a day or two before moving on. They arrived in Kharagpur on January 22. They found their bungalow on the outskirts of town at the edge of dense tropical jungle.

Despite the temperature sometimes topping 113 degrees, India did not stop Bitzer from his daily running routine. A British neighbor said that until Bitzer arrived, "only mad dogs and Englishmen went out in the midday sun." The "boiling sun," says Maryann, was not enough to stop him. "He would also race with the rickshaw drivers. He'd be out jogging, and these fellows, they were on bicycle rickshaws—Don doesn't ever meet a stranger without talking, so he would always talk to them or call to them when he went by. It was all in fun, he would race them and they would be pulling their rickshaw with their bicycle and racing him, but he was not looked upon kindly by some of his colleagues for doing this because the caste system was still pretty evident and they didn't like him talking to this lower caste of people." Bitzer, always notorious for his jogging outfits, would later remark, of his time running in India, that "I was the best dressed person, jogging person, in the jungle." Says Maryann, "He was the *only* dressed person in the jungle."

Back in Illinois, the PLATO team in CSL continued their work in hardware and software development. Circuitry was designed and built to enable twenty simultaneous PLATO III terminals to run at the same time. CATO, the Compiler for Automatic Teaching Operations, developed by Masako Secrest and Steve Singer, became operational, as did CATORES, the resident CATO program that Andy Hanson had created. Meanwhile, more and more lessons were beginning to appear. "PROOF" was a lesson to help students with mathematical problem solving. "Alphabat," designed to help young children learn the letters of the alphabet, was a new lesson authored by Amy Alpert (daughter of Dan Alpert), one of the high school kids who like Mike Walker worked on PLATO-related projects with Bitzer as mentor. "Kids who identified the correct letter on the screen were given an M&M," says Mike Walker, "which was ejected by a contraption powered by a washing machine relay. . . . It was a bit too powerful and occasionally obliterated the piece of candy." There were even new lessons designed to help potential lesson authors learn how to create PLATO lessons. This would become a common trait of PLATO over the coming decades: use PLATO itself to teach people how to use PLATO.

William Golden, who had gotten involved with PLATO from Max Beberman's New Math team at UICSM down the street, and was now involved in the project to build lesson PROOF, found Bitzer's India trip ill-timed and "terrible." It "arrested," he says, the forward momentum of the project. Not only that, Bitzer's long absence impacted the special culture he had nurtured. "The culture of PLATO was set by Bitzer," says Golden. "And he was a remarkable administrator. Bitzer didn't care about your education, your background. It was what you could *do*. If . . . an undergraduate member of the group had the best idea, *he* drove the group. It sounds impossible but you could have full professors, researchers, PhDs, graduate students, and undergraduates working together. . . . Whoever had the best ideas drove the projects in most cases."

Bitzer had only been gone a few weeks when concerns grew that other projects within CSL were starting to cannibalize the PLATO project. "Please make sure," Bitzer wrote to Tebby Lyman on February 7, "that none of the PLATO staff or space is obtained by other groups since we will need all the staff and space this coming September." He wrote letters to Robert Willson to instruct him what to do next with his plasma display experiments.

Word eventually reached ARPA, one of the agencies funding CSL and PLATO, that Bitzer had embarked on a plasma display project on top of what he was supposed to be doing with PLATO. "The first money for this came from the behavioral science branch of ARPA, and they didn't want it for this," says Bitzer. "They were very upset with me. I was in India, and I was working on this, but we were also working on advancing PLATO, and they gave the money to essentially find what are the ten most important learning concepts using computers. That was the behavioral science branch of ARPA. The university signed for it to get the money, but when I got back and saw it I said, I would have never signed this because what we want to do, the most important problem is to get a display we can believe in. So I just took the money—it was a small grant, I believe it was $50,000—I took the money and said we're going to use it for people to help develop the display. The behavioral science branch was really upset. They were really going to chew me out."

Back in India, Bitzer found a high-frequency oscilloscope stored away in the electrical engineering department head's office. Two undergraduate IIT students, Brij Arora and Suhas S. Patil, were working on a project at the time and needed that oscilloscope to test their circuits. Bitzer, adept at ways of bypassing bureaucracy so people could stay productive and get things done, showed the two students a way to get access to it. Arora impressed Bitzer and he began attending some of his lectures and discussions on how to use computers and what makes them tick. But he also gave lectures on other things of interest to him, especially his pet project, the plasma panel. "Is it possible to make a flat panel display out of this stuff?" Bitzer would ask the students. Arora was particularly interested in hearing about this plasma display. Little did Bitzer know just how interested Arora was.

Nehru, the prime minister of India, died unexpectedly in late May. The nation of India shut down in mourning. Everything in India stopped, including transport of the IBM 1620. Finally, a mere two weeks before the Bitzers were scheduled to climb aboard Pan Am for the westward journey home through the Middle East and southern Europe, the IBM 1620 arrived at Kharagpur. Then Bitzer received a request from a USAID official at the U.S. embassy in New Delhi asking if he could stay another three months. He responded that would be impossible, due to pressure to get back to the PLATO and plasma projects, not to mention a desire to get his family back home after six months in India. Eight thousand miles away, Alpert went into a brief tizzy over the thought of Bitzer staying longer. He immediately sent a cable: DEAN EVERITT, TEBBY LYMAN, AND DAN ALPERT EXPECT YOU BACK THIS SUMMER. WE KNOW OF NO OTHER PROPOSAL FOR EXTENSION OF YOUR STAY. CARRY OUT ORIGINAL RETURN PLANS PLEASE. Over the two final weeks Bitzer scrambled to work with the IIT students and other faculty to get the 1620 machine up and running.

He was back in Urbana the first week of July.

Sometimes the best way to achieve a breakthrough when struggling with a tough technical problem is to get far away from it for a while, then come back. This may have been the case with the plasma display project, for it was not long—indeed, only days—after Bitzer returned to Illinois that they achieved a major breakthrough. By July 1964, Willson had made good progress on the plasma panel prototype, or at least had a much better understanding of the factors still holding the project back from success. Plus, he had had regular and increasing help from Slottow. Their prototype was still a single pixel, but having Slottow's mind, as well as that of another CSL physicist, Frank Propst, cranking away along with Willson's on how to achieve inherent memory in the display, was paying off. They noticed that when voltage was applied to the gas cell, a charge built up against the surfaces of the cell. They considered whether this charge buildup was useful or a hindrance. Up until this point, Willson had been doing roughly what the Lear Siegler people had been doing: attaching the super-thin electrodes to the *inside surfaces* of the glass sandwich, meaning the electrodes were in contact with the gas. The Lear Siegler team felt that some sort of resistor was necessary to prevent the direct current (DC) voltage from arcing across the gas from the one electrode to the other. Lear Siegler had made some progress with this approach, but had only been able to create a panel with 10 x 10 pixels. But then Bitzer and Slottow had an idea.

One day in July while they were waiting for their wives to pick them up at CSL they got into a discussion. They understood why Lear Siegler was using resistors, but it was expensive and difficult to go that route. In the early 1960s it was hard to find very high-resistance resistors that would do the job anyway. For the team to ultimately build a 512 x 512–pixel display, it would require 262,144 resistors embedded in the glass. It did not seem like the right path to take. "Clearly Engelbart looked at this," says Roger Johnson, who would eventually join the plasma display team as a grad student, but Engelbart "did not quite grok the fact that he could *use* this charge buildup. He saw this charge buildup as a *problem*. 'Cuz he was still thinking DC [direct current]. So he went down that path . . . and that turned out to be possible to demonstrate, virtually impossible to manufacture."

Gene Slottow working on plasma panel prototype, c. 1967

"Then we had a much more novel idea," says Slottow of his afternoon discussion with Bitzer that day in July. "If we simply moved the electrodes from the inner surfaces of the glass to the outer surfaces, we would isolate each cell from the external circuits by a capacitive impedance instead of a resistance." By moving the electrodes outside the sandwich instead of inside, they would not come into contact with the gas at all, which in theory might reduce the undesirable sputtering and arcing effects. They could skip resistors entirely, and instead use the physical sheets of glass as a capacitor. Plus, instead of using DC voltage as Lear Siegler had done, they'd use alternating current (AC). Not ordinary household AC, in which the current alternates sixty times per second. No, they had to crank up the AC for the plasma to nearly 100,000 cycles per second. When the gas glowed, it was actually flickering 100,000 times per second: so much faster than the human eye could notice that the glow seemed stable.

They tasked Willson with creating a new sandwich of glass sheets and electrodes, sealing the sheets of glass once again with Torr-Seal, a powerfully strong epoxy glue commonly used for high vacuum seals, but this time they had Willson place the electrodes on the *outer* surfaces of the glass sandwich. The gas would not come into contact with them, as they would be on the other side of the glass. Then they backfilled the tiny hole in the center sheet of glass with neon, and fired it up.

It glowed blue.

They had a one-pixel cell that seemed to exhibit inherent memory. If you sent a sustaining voltage through the electrodes, then spiked it higher for a brief moment, that spike of electricity excited the ions in the gas enough for them to light up. You could then throttle back to the sustaining voltage and the gas would *still* glow. This was the memory effect they had long sought. You could then dip the voltage down a bit, then back up, and that was enough to calm the furiously excited ions down enough for them to stop emitting photons, so the glass stopped glowing.

On and off. One and zero. Memory. They'd done it.

But: it glowed blue. It should have been orange. The blue indicated contamination. Contamination almost invariably meant one thing: a leak, something they, particularly Willson, were terribly used to. When they had backfilled the neon into the little glass sandwich, something else must have snuck in as well. Most likely it was a tiny amount of ordinary air from the room they were working in.

Knowing that air contains about 78 percent nitrogen, 21 percent oxygen, a tiny bit of argon and water vapor, and tinier bits of other elements, including carbon dioxide (more in today's air than that of 1964, thanks to humanity's impact on the global climate), they assumed that most of the contamination would be nitrogen, since that's what the earth's air is mostly made up of. They confirmed that there was contamination and set out to prevent it. "So we fixed that leak," says Bitzer, then backfilled the glass sandwich with neon again, "and boy, now watch it work." And it didn't work. "The memory disappeared!" he says.

A number of experiments and measurements ensued, revealing hints as to why the memory effect came about when there was contamination. There was something happening because of some small amount of gas mixing in with the pure neon. The scientific term for the effect is the "Penning gas mixture," in which a noble gas like neon is mixed with an extremely small amount of another gas such as argon or nitrogen. The presence of the second gas changed the behavior of the ionization, which in turn changed the color of the glow, but also created a nice "margin" between the sustain voltage and the spike voltage, which in this case helped reveal the memory effect. They didn't have all the answers yet, but by the fall of 1964 they knew enough that it was time to file a provisional patent claim.

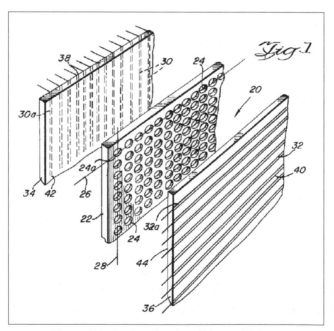

Plasma Display Panel patent illustration, 1966

It would take another year of Willson making endlessly more prototype one-pixel gas and glass sandwiches, but by the fall of 1965 they were able to confidently build the prototypes and explain why things were happening the way they were. Much of it had to do with the fact that during each momentary half cycle of AC voltage ping-ponging across the gas many thousands of times per second, there would be a buildup of charge on the surface of the glass, a so-called wall charge. That tiny bit of buildup helped keep the gas cell "on" and glowing just long enough until the next full AC cycle would come around a tiny fraction of a second later.

With the Los Angeles Dodgers and Minnesota Twins in the 1965 World Series playing on Bitzer's basement TV in the background, Slottow, Bitzer, and Nate Scarpelli, the attorney hired by the university, worked through two big patent applications. One for PLATO, and one for a fuller continuance version of the 1964 provisional patent application. That plasma patent, Scarpelli would say years later, was the longest, most complete, and most detailed patent application he ever filed in his long legal career. The U.S. Patent and Trademark Office took their time reviewing it. It would finally issue in 1971, and subsequent industry licenses would bring millions of dollars into the university, with a fraction doled out to Bitzer, Slottow, and Willson.

In the fall of 1965, Bitzer got a surprise visit, just when the new semester was beginning. Suddenly on his front doorstep at the laboratory appeared Brij Arora, one of the young IIT students he'd met in India in 1964. "He had flown himself over from India," Bitzer says, "and was there looking for a job. He said, 'I want to work on this stuff you've been talking about.'"

Unbeknownst to Bitzer, Arora had applied to the University of Illinois with the goal of enrolling in an engineering master's or PhD program. "I had known Illinois to be a great university to continue further studies," says Arora. Bitzer did not send him away. Instead, he made him a research assistant, and over the next several years Arora received his master's and PhD in electrical engineering with Bitzer as his advisor. "When I joined in fall of 1965, I was not yet assured of a research assistant position. I had some idea that there was something possible in the computer science department. However, someone guided me to the Coordinated Science Laboratory. Yes, Don must have been surprised seeing me suddenly. I must admit it was providential. That moment was a turning point for me."

Willson got to work writing his long-delayed dissertation, and Arora was tasked with attempting to build multiple-pixel plasma display prototypes. Another engineering student, Roger Johnson, had wrapped up his bachelor's degree in 1965 and Bitzer talked him into pursuing graduate school to work on PLATO and the plasma display project, about which Johnson up to that time knew nothing. Bitzer had a way of finding bright talent, and when he spotted it he didn't want to lose it. "He starts throwing me at the laboratory," Johnson says, "and clearly had picked me, in some subtle sense, to join this team that was inventing the plasma panel. He had this Einstein lab up on the fourth floor. He hauled me up there one day—this is how he does stuff, it wasn't like oh, I had to make an appointment to see Don, he was just like, 'Come with me'—he hauled me up there and he's pointing at this stuff which looked like it was right out of *Frankenstein* and I'm trying to figure out what the hell I'm looking at, I was trying to figure out what the hell he was talking about."

Then Johnson saw it. A little orange dot.

They'd mastered one pixel. The next step was a primitive matrix of 2 x 2 pixels. If they could figure out 2 x 2, then they could to go to 4 x 4. At each step of the way, they ran into new challenges. Roger Johnson eventually crafted a 16 x 16 display, a major step forward on the way to 512 x 512.

For the first few years, blue was the predominant color emitted by the charged plasma gas in the cells. Over time they tinkered with the color by adjusting the mixtures and types of gases they used in the cells. Indeed, the idea had not escaped them that if you tinker with the gases and introduce some phosphors into the cells, and then counted one pixel as a *trio* of cells, one red, one blue, and one green, you could in theory build a display consisting of *arrays* of those trios, and what you would have is a full-color display. (Luckily, Nate Scarpelli made brief mention of the idea in the plasma display patent.) For PLATO's purposes, a monochrome display was going to be more practical and affordable. The question was, what color should that mono "chrome" be? They were used to seeing blue, but there was something about the orange color that they kept coming back to, that "sight to dwell upon and never forget."

In time they got so good at understanding what they were doing with the prototypes and the intricate wiring of electrodes and circuitry to drive them and interfaces to computing devices to tell them what to display, it became time to think about how they were going to manufacture these things in quantity and at the crazy-ambitious resolution of 512 x 512 pixels.

By 1967 Alpert and Bitzer had chosen the Owens-Illinois (OI) company, wizards of glassmaking, to be the manufacturer of the displays. OI was a major U.S. manufacturer of so-called glass bottles, which were literally the "tubes" in cathode ray tubes. OI supplied the glass bottles to a number of the major TV manufacturing firms, including Sylvania. The company saw a profitable future in flat panel displays and televisions using the plasma technology. OI were primarily glassmakers, not electrical engineers, but they knew how to etch glass as well as print decorative lines and patterns on glass, so they would approach adding the 512 horizontal electrodes on the one sheet and the 512 vertical electrodes on the other sheet by "printing" tiny strands of gold onto the glass.

One of the problems with the prototype panels that Bitzer, Slottow, and company had built is that they all relied on the "glass sandwich" idea: two outer slices, the outsides of which were lined with the tiny electrode rows and columns, and an inner slice into which had been drilled tiny holes backfilled with neon and other gases. That was all well and good when you were building 1-, 2-, 4-, 8-, and 16-pixel grids of tiny dots. But to build a panel with 512 x 512 dots would mean that the center sheet of glass would have 262,144 holes on it. The chances that some of those holes would be faulty in some way would be high, not to mention how fragile the hole-ridden sheet of glass would be. OI decided to approach the problem in a novel way: get rid of the inner slice altogether, and fill the *entire area* between the outer sheets of glass with gas. Amazingly, it worked. When a horizontal electrode and a vertical electrode were charged up, the gas at the intersection point glowed, but no adjacent dots glowed. For this novel approach to the design of a plasma display, OI filed a patent, called the "Baker patent." "I always considered it one of the major inventions of the plasma display," says Larry Weber, then one of Bitzer's graduate students, "but it went to court, and you never know what the court's going to do, especially in regards to patents. It's almost, you flip a coin to determine what the outcome's going to be, it never seems to rely on logic." In this case, the coin flipped in favor of the Bitzer-Slottow-Willson patent, deep in the text of which was a casual mention that the center sheet of holes is not necessary—you could take the whole center sheet of glass out and fill the space with gas and it would still work—which made the manufacturing process not only simpler and more reliable, but also economically viable.

By removing the center sheet of glass, OI engineers realized that there still had to be something inside that newly empty area in order for the outer glass sheets not to bend inward due not only to the low pressure of the gas relative to outside atmospheric pressure, but also from the fact that humans would be interacting with these screens by touching them, perhaps not always gently. They needed to be durable. Using microscopes, OI painstakingly attached tiny vertical "spacers" here and there within the glass, so when the two sheets were mounted together, the tiny gap remained uniform across the entire area and now had some supporting structure. Naturally, there was a price to pay with this approach: a user sitting in front of the terminal could *see* the spacers as tiny lines that faintly reflected orange light behind the glass. It was an engineering compromise that everyone would have to tolerate.

At first, OI produced a prototype four-inch panel, and secretly showed it to Bitzer and Slottow in 1967. It would take four more years to get to quantity manufacturing of the eight-inch 512 x 512 panels. OI made a strong case for Bitzer or Slottow to move to Toledo to work on-site on the project. Gene Slottow agreed to go. During those four years an enormous amount of work was done. One of the first urgencies was the discovery that the early OI panels were not very reliable, burning out in under one thousand hours of use (less than PLATO III's Raytheon storage tubes, which they were meant to replace). That led to improvements in the type of glass used—Bitzer would call OI's glass magic a "black art"—that led to longer-lived, more reliable panels. Another discovery was that the panels tended to need some coaxing to get going once they were turned on. Oddly, they needed light from whatever room the early prototype terminals were in. The physics explanation was that the presence of photons bouncing around through the gas helped excite it enough that the pixels would light up and glow. This led to the decision to add a border of pixels out at the edges of the panel—beyond the 512 x 512 one would see when using a terminal. Behind the frame, some additional rows and columns of electrodes were printed onto the glass, and those pixels were left on all the time. The photons from the light of those pixels sufficiently excited the rest of the gas that the panel would function correctly. That unseen lit border, behind the bezel when the panel was mounted in a terminal, contributed to the orange glow across the whole screen, particularly in a dark classroom.

PLATO users themselves called the effect "the Friendly Orange Glow." Perhaps the best way to approach the notion is to think of a campfire. It's not the fire itself. It's what the fire does to the smoke and air surrounding it or to the faces of the people gathered near it. Anyone who has ever sat around a summer evening campfire knows that familiar glow—it's in our DNA going back thousands of years. Something similar was going on with the PLATO plasma panel. It wasn't simply the glow emanating from the text and drawings on the screen—that was only part of it. It was also the eerie, ghostlike effect that one saw in the black *background* of the screen. Because of that hidden border of always-on pixels on the four extreme edges of the screen, even the black background of the screen was not quite truly black. It had a faint glow of a campfire. No other computer screen ever had it. It was something that made PLATO special. And friendly.

After much discussion as to the merits of light pens versus touch screens, they made the decision to add touch sensitivity to the plasma screens in the upcoming PLATO IV terminals. Bitzer assigned a team of engineers—Roger Johnson, Fred Ebeling, Richard Goldhor (another of Bitzer's Boy Scouts), and Jim Parry—to invent and build a touch panel. They came up with an array of light-emitting diodes that gave the terminal an invisible grid of 16 x 16 infrared light beams spaced just above the surface of the plasma panel display glass. If a finger were placed somewhere on the grid, the firmware in the terminal could pass back the proper coordinates to the central computer, which would consider it just another form of input from the student. To make room for the grid of beams, they had to set back the plasma panel by about a half inch from the front surface of the terminal, and include little holes around the resulting frame, where the beams of infrared light would be sent out. They created a mold into which they melted plastic pellets so that they would take on the exact shape they needed and fit snugly around the plasma display. Only—the frugality finally catching up to them this time—they used the oven in Maryann Bitzer's kitchen to melt the plastic, stinking up the house and nearly ruining the oven.

7

Two's a Crowd

Larry Stolurow, the UI professor of psychology who had in 1960 participated in the CSL committee considering an automatic teaching project that turned into PLATO, had turned down Alpert's offer to lead the project once it was green-lighted. Nevertheless, he kept an eye on the system as it evolved from PLATO I to II to III. Stolurow liked that PLATO collected data on students, but disliked how it was collected. PLATO associated data with the terminal's ID number rather than the name of the student. If a student stopped halfway through a lesson, then came back later to finish, but on a different terminal, the fledgling system could not yet handle that: there would now be two separate recordings of a single student's work, and merging them was difficult. "That led me to a lot of frustration," Stolurow says. "And the other aspect of my frustration was in the fact that the materials that they were working with on the system were very, oh, should I say, very elementary." Stolurow wanted to collect richer data on student responses and response times, in order to help prove or disprove various theories about student performance. "We didn't have answers to those questions," he says, "and I felt we needed them in order to really design a system that would be effective and do the job that was intended."

Stolurow decided to build his own system. He named it SOCRATES, which stood for "System for Organizing Content to Review and Teach Educational Subjects." A more blunt statement about PLATO is hard to imagine: in ancient Greece, Socrates was Plato's *teacher*. "I used the resources I had in my own laboratory and developed SOCRATES as a result of the frustrations I had with PLATO," Stolurow says. SOCRATES would run on the filmstrip-based AutoTutor device created by programmed-instruction proponent Norm Crowder, but would be connected to an IBM 1620 computer—the same type that

Bitzer helped install at IIT in India—to capture data from the student as well as to determine the sequencing of individual filmstrip frames. AutoTutors were flimsy, "jury-rigged terminals," admits Stolurow, resembling a microfiche machine one sees in libraries. At the project's peak there were only about a dozen terminals in use, each equipped with a primitive "keyboard" that only had about fifteen keys on it.

As a student progressed through a SOCRATES lesson, the Auto-Tutor would jump forward or backward to the appropriate film frame and display it on the built-in rear-projection screen. One could not help but recall Thorndike's 1912 vision regarding an automated textbook, or Crowder's own scrambled textbooks from the late 1950s. (PLATO I or II inherited a bit of the Thorndike and Crowder visions as well: if there was one thing many people agreed on in the early 1960s, it was to use the branching style of "programmed instruction" that Crowder had promulgated as a starting point.) Unlike in a scrambled textbook, all of the AutoTutor's content was recorded on a filmstrip, and the contraption could randomly select any frame of film after noisily zooming ahead or back.

SOCRATES classroom, 1964

The SOCRATES project attracted some government funding, enabling Stolurow to create a Training Research Laboratory and hire some technical people to set up the system, including programming. Among those people were Rick Blomme, who had acted out the role of student at the Allerton House PLATO demonstration in 1961, along with Scott Krueger and Jan Schultz. Together they made SOCRATES work as best they could. "PLATO had so much more going for it in terms of its core technology," says Schultz, who found the 35mm-filmstrip approach cumbersome at times.

A professor of educational psychology, Richard C. Anderson, took a job at the University of Illinois in the fall of 1963, and when he saw PLATO and SOCRATES, at first he found SOCRATES more appealing. But those jury-rigged AutoTutor terminals gave him pause. "It was a technologically impossible system," he says, "trying to run spools of film backward and forward with a mechanical frame counter to keep track of where it was, they beat the *daylights* out of those. The film transport system couldn't take the beating required for random access, and of course if the frame you wanted to go to was a long way away it was hardly 'random access.'" The AutoTutor had other quirks. "This was the noisiest damn thing," says Gerry Faust, a PhD student of Stolurow's. All that noise made for an unintended consequence. The computer would present a question to the student, and the student would select an answer. The computer would evaluate the student's answer and then select what frame of film to next present to the student. To do so, it would have to run through the filmstrip, counting frames as it went, to find that next frame. Skinner would have had a field day with what happened next: everybody in the classroom would know if a student had chosen the right answer, because the AutoTutor would go *PONNK*, then *ZZZZRRRRR* as the filmstrip spun forward counting and looking for the next frame, but then it would stop almost immediately, as the next frame was not that far off. But if the student got the answer *wrong* . . . the AutoTutor would *PONNK* and then *ZZZZ-RRRRR* . . . but the *ZZZZRRRRR* would go on and on and on as the filmstrip kept rolling and the machine kept counting frames and seeking and spinning and *ZZZZRRRR*ing to right near the end of the reel—where the *remedial* material was located. At some point during all that counting and seeking and rolling and *ZZZZRRRRR*ing the student would realize he was wrong before the AutoTutor could deliver the frame *telling* him he was wrong. The AutoTutor might as well have reached out with a robotic arm and placed a dunce cap on the student's head. "I mean, it was *great* reinforcement, it had to be," says Faust.

"Oh, it *clicked* and *clacked*," Anderson says, adding that "no interesting educational software was developed for that system." In time Anderson would switch his own research to PLATO, which he found was turning into a more interesting platform.

One of the undergraduate SOCRATES guinea pigs was Dick But-kus, a superstar linebacker on the Illinois football team who would become, as a player for the Chicago Bears, one of the greatest lineback-ers in NFL history. Students, including Butkus, who signed up to take classes like Psych 100 had a course requirement to volunteer so many hours to be involved in experiments, and it was Butkus's fate to sit in front of AutoTutors pressing buttons and coping with *PONNKs* and *ZZZZRRRRRs*.

The SOCRATES project exemplifies a phenomenon that often occurs when psychologists and educational theorists lead the creation of a computer-based teaching system. The system's design tends to reflect the theories held by the project leaders. It's their baby; it shares their DNA. The design of Skinner's and Crowder's teaching machines reflected their theories. Now it was Stolurow's turn. Stolurow lacked Bitzer's technical abilities and intense engineering ambition, but for SOCRATES it wasn't a matter of *Go big or go home*, but more a matter of *Get good data or go home*. He wanted data, and relied, as much as pos-sible, on off-the-shelf hardware that was just good enough to do the job of collecting data that he could then study.

Bitzer, a pure engineer at heart, held no firm theories of instruction or learning, nor any strong desire to see just one particular theory reflected in the hardware and software. Theories came and went; they changed like the seasons. Who knew what would become the latest hot notions of teaching and learning in the coming years? The only certainty is that they would change. Better to build a flexible system that's capable of any kind of learning activity than build one restricting you in some way. While the earliest lessons on PLATO reflected sim-ple but well-established, behaviorist, programmed-instruction models, underneath all that lay a desire to remain agnostic, detached from any one particular theory. If you wanted to build a lesson on PLATO that reflected your own pet theory, you should be able to go right ahead, as long as other authors could create their own lessons that followed other theories.

Thomas Anderson, then an education professor at UI, says that in the 1960s, UI's education department was strongly behaviorist, includ-ing Stolurow, but he did not see that thinking within Bitzer's PLATO project. "You couldn't really put them under any kind of psychological banner," he says. "They just didn't seem to be driven by psychological underpinnings. They were driven by a more pragmatic approach: you work with students, you work with content, you work with the technology, you put it together in a way that feels good and it will work. Whether it's consistent with somebody's psychology is a quickly irrelevant question."

Lezlie Fillman, who had been with PLATO since 1961, says Bitzer and Stolurow had almost opposite views about how a computer was going to support education. "Larry Stolurow's was much more specific in 'Give me the hardware to do the task I want to do,' and Bitzer's was 'Give me the tasks you want to do, and let's design hardware broad enough to take care of all of it.'"

"People in the education faculty and the Training Research Lab believed that PLATO was all about technology seeking a role in education," says Kenneth Smith, who consulted on SOCRATES. "And they believed that their position was much more secure and firmly rooted: it was education seeking a role for technology. So this tension sort of existed in the air."

UI president David Dodds Henry, who had first seen a demo of PLATO in 1961, was a supporter of Alpert and the PLATO project. In 1965, he promoted Dan Alpert to be dean of the Graduate College, giving him a far greater degree of say in the university administration, as well as direct oversight of research going on in all of the university's departments. It also meant that Alpert no longer was director of CSL. That meant he was no longer Bitzer's direct boss.

Alpert still wanted to have influence over CSL even though it no longer reported to him. The new director of CSL, W. Dale Compton, reported to William Everitt, dean of the Engineering School, just as Alpert had. Alpert's tentacles were still trying to poke around CSL. "That was a source of some of the friction," says Compton. "When he went to the graduate school . . . he wanted to retain his influence and the recognition of PLATO," he says. "I don't think it was quite so much as to control Bitzer but it was to be responsible for PLATO."

Compton was of two minds when it came to Don Bitzer. On the one hand he was impressed enough with Bitzer's boundless inventiveness to write a letter of recommendation to Dean Everitt in January 1966 for the promotion of Bitzer to full professor. "It is not an overstatement," he wrote, "to say that it is because of his enthusiasm, personal dedication, and extremely imaginative mind that the PLATO system has assumed its present eminence. No problem is too great, no situation too complicated, and no suggestion too far-reaching for him to comprehend and wrestle into submission." But Compton was leery of Bitzer's personality. "He grew up as the son of a used-car dealer, had some of the characteristics of a small-town businessperson who could live on the edge of legality. Very, very opinionated and very, very conscious and very vocal about his accomplishments, about what was due him." He also saw that Bitzer was a bit of a wild man. "He was not a good administrator," he says. "He just didn't want to pay any attention to detail." It was important that Bitzer had Gene Slottow and Tebby Lyman on the team, says Compton. They were more organized, more focused on details, making sure many of the more mundane things got done.

"Alpert had been noticed in the administration of the university," says Jack Desmond. "The orientation of Alpert and the orientation of Bitzer became a little bit more pronounced. Bitzer came from a business family and he was often warned by his father as he was embarking upon his graduate career, don't be seduced by the university, because the big money is out here in private enterprise. And Bitzer really did have a business orientation. Alpert came from an industrial laboratory, in which things tended to be more programmed than the individual disciplinary orientation of the university, and I think somewhere along the line, Dan now being in a position of considerably greater power than he had been as director of CSL, certainly didn't want the new director, Dale Compton, to be the new boss of PLATO. It was at that point where he essentially excised PLATO out of CSL and put it under the Graduate College, with presumably a broader charter for purposes of larger amounts of interaction by whatever discipline cared to write lessons for PLATO."

Blomme was eventually snatched from SOCRATES and brought in to work on PLATO. Krueger followed shortly thereafter. One of the first things they realized was how inefficient it still was to author PLATO lessons. They were skeptical of Tebby Lyman's project to create what was called the "GENERAL Logic," a simple fill-in-the-blank kind of authoring tool that enabled a lesson author to create a PLATO lesson in less time than it would have taken previously. It still took way too much time.

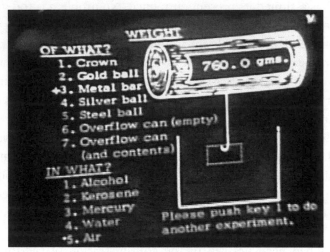

PLATO III lesson example, 1964

For Blomme, the early years of PLATO represent a series of revolutions that were also taking place in computing in general. In just the span of seven years PLATO would evolve from punching in strange hexadecimal codes onto possibly hundreds of feet of oily paper tape, to editing in an authoring language on-screen. "I still remember," he says, "like what 'L5' was, 'clear and add' . . . and that's how you programmed, you put down 'L5,' you typed that in the Flexowriter and punched the tape . . . you didn't say 'add,' and so the idea of writing something to help you produce the program, that was a massive step." CATO and FORTRAN and CATORES had arrived, making programming only a little bit easier. But then Scott Krueger showed up. He took one look at PLATO, and the conversation, according to Blomme, went something like this:

KRUEGER: I want to design some characters.
BLOMME: Oh, okay. Here's a piece of graph paper, with a grid of lines on it, like a 16 x 16 grid, and then you locate all the dots you want on the screen, and then enter a bunch of parenthesis-surrounded pairs of numbers separated by commas into the system.
KRUEGER: (after a long pause, then, whispering) You've got a

graphical display here! Why aren't you using it to design the characters?

Scott Krueger, new to PLATO, fresh from SOCRATES, gazed upon the PLATO III system and saw right away one of its biggest weaknesses—something apparently everyone up to and including Bitzer had not yet noticed: *it didn't use the terminal screen to do the lesson authoring.*

Krueger went to work on a new utility program called CHARPLT (pronounced "char-plot"), an on-screen bitmap editor for creating not only text character sets, but also little graphical images. "That was a revolution," says Blomme. "That changed what happened at the place." We take for granted bitmap editors today. Up until CHARPLT came on the scene, all PLATO III authors were stuck using ugly upper-case letters to display computer-generated text on the screen, including anything the student typed. Suddenly, with CHARPLT, you could create lowercase letters and other text characters in your own designed typeface, and load them into your lesson. PLATO authors immediately saw the advantage of this approach and began using the tool right away. Soon, all kinds of new lowercase text and graphical images began appearing in lessons. In fact, it began to dawn on some authors that maybe they didn't need to prepare fancy 35mm slides any-more, slides that, up until now, had been the basis for the "screens" in many PLATO lessons, with the ugly, bright, computer-generated uppercase text serving as prompts and student input, superimposed over the nicer-looking slides in the background.

There was another unexpected consequence to the rapid adoption of CHARPLT. "Engineers came in one day," says Blomme with a certain amount of glee, "and said, 'What the hell are you guys doing? The storage tubes are overloaded. The voltage! You've got so much crap on the screen that we had to turn up the voltage.' It changed the way the hardware was done, because there were *so many more dots* being put on the screen. We had uppercase, lowercase, special [characters]—and anybody could design this stuff."

Blomme and Krueger weren't done yet. They realized there was another way to dramatically improve author productivity. They cre-ated an enormous, ambitious program, MONSTER, an acronym for "Multiple Online Nifty Speed and Terminal Editing Routine," which enabled people to edit the code to their lessons right on the screen. Until MONSTER, authors endured an onerous process to insert, edit, or delete specific lines of code, by glancing at a printout, finding the line number, then using a teletype machine to type editing commands, the results of which were spat out by the teletype onto oily paper tape, which was then run through a reader to be stored on magnetic tape. Make one mistake, replace line number 751 when you meant line 752, and you could quickly wreak havoc on your whole program. Very often a new print of the code was necessary, since the line numbers kept changing with all the edits. The whole process was wasteful and time-consuming. With MONSTER, you looked at the screen, you saw your code, and you moved around in it and did what needed to be done, not unlike a present-day text editor. "That changed the world too," says Blomme. Says one author who began to use MONSTER at the time, "You didn't have to do these darn edit tapes anymore, and you could go off into the classroom and sit at a student terminal instead of having to be in the noisy computer room."

Recall that Bitzer's original 1960 diagram of PLATO depicted a student in front of the screen and keyset. Halfway into the decade, it seemed like Bitzer, with his hardware orientation, was still thinking that only a student would ever actually sit down in front of a PLATO terminal. But Blomme and Krueger had a software orientation and the time had come for software people to make more of an impact on PLATO. With MONSTER and CHARPLT, they hastened the demise of what might be called the Perseverance Era of authoring lessons on PLATO, ushering in in its stead the Productivity Era.

During this time, entire classrooms of PLATO III terminals were built at locations around Champaign-Urbana, including Parkland College and Mercy Hospital, where Maryann Bitzer had developed a whole series of college-accredited PLATO lessons for nursing students. A program called TEXT TESTER was built in which material from textbooks was reproduced on slides, and then students would answer questions online relating to the material. Twenty lessons in seventh-grade remedial arithmetic were created with TEXT TESTER, as well as twelve lessons in political science. A series of circuit analysis lessons was developed for junior electrical engineering students. Twenty-eight lessons in library science were created. Lessons on FORTRAN programming, arrays of symbols for fourth graders, recursion for high schoolers, braille for blind students, French pronunciation and spelling, geometry, number bases, mathematical proofs, a genetics simulator for high schoolers, mechanical engineering lessons on beam construction, lessons in political science . . . the list kept growing.

Perhaps the core problem with SOCRATES, beyond its flimsy film-strip terminal, was its funding source. Stolurow managed to line up funding from the Office of Naval Research, long one of the major sources of funding for PLATO and a stalwart source for CSL going back years. Alpert had deep ties with the brass at ONR and other federal agencies, and now ONR was pouring money into Stolurow's Training Research Lab? This was not going to end well.

"I gather the people at ONR in Washington had some questions by people around the country as to why they were funding two systems at Illinois," says Stolurow, admitting that the situation was "somewhat of a sore point." The point grew sorer once Alpert became dean of the Graduate College. "Illinois had a plan whereby external research funds were funneled, were divided up, so that the graduate dean's office, the department, and the actual investigator's unit get various percentages of the overhead," says Stolurow. Alpert invited Stolurow to visit his office to talk about SOCRATES and PLATO. "He made it emphatically clear that there would be only one CAI [computer-assisted instruction] system on campus, and any proposal I would submit to an external agency including ONR—who had been funding me—would have to go clear through the graduate dean's office. So I found myself in a strange predicament. Obviously he was going to make PLATO 'it' and see that we didn't continue."

Rupert Evans, who had become dean of the School of Education, believes Alpert and Bitzer were afraid PLATO would lose its funding because the federal government might look at two competing proposals at the same university and fund neither. Evans, even decades later, was still incensed about the SOCRATES fiasco. "This is the only instance of which I am aware in which a small, tightly knit group of people who have a vested interest in one research project were able to keep another, competing project from submitting a proposal to an outside group and thus driving a highly competent professor to leave his employment. My only regret is that I didn't resign in protest. Looking back, I am ashamed that I did not do exactly that.

"Bitzer and Alpert really wanted to run the thing themselves," says

Evans, "and they were not really interested in getting input from any-body else." Compton believes there was another factor at play between the two camps: "It was just plain jealousy. Bitzer had no regard for anybody who was a competitor."

Stolurow resigned and moved to Harvard, a community that at that time lacked the spark, the will, or the vision to undertake building a major computer-based education project rivaling PLATO. He would never regain his stature in the field.

With Alpert kicked upstairs to a high administrative position else-where in the university, PLATO was vulnerable. Its funding was vulnerable. Who was going to write the proposals? Who was shak-ing the hands—who knew which hands to shake—of the Washington civilian and military brass who signed the checks? Who was going to decide how that money was spent? Alpert had overseen Bitzer and the PLATO and plasma panel projects for five years. And every day that went by, PLATO got more sophisticated, more interesting and compelling, and the plasma panel got closer and closer to becoming something that could be installed in terminals. This was not the time to stunt the growth of these projects: this was the time to give them plenty of water and sunshine.

When Compton took over CSL, one of the things he learned was that the team working on the PLATO plasma display was part of CSL—all or part of their pay came from CSL—so they were not exactly tied to the PLATO project per se. That was another source of friction. "There's always friction with Don because he always wants more than his share of the resources. . . . I was trying to encourage a broader display activity that would include other types of displays. He was *violent* about—really upset about—that, because he saw the possi-bility of some resources not going his way." Newspaper and television reporters would visit the lab, see the plasma display prototypes, and immediately envision a future of flat-screen televisions, then go back to their offices and publish stories that talked about wall TVs ("TV of Future: Wire-Brained Looking Glass," blared one *Chicago Tribune* headline of the time). Bitzer would bristle at such notions, insisting that the mission was *PLATO*, that the plasma display was for *education*. The idea that some of his people might be working on plasmas for tele-vision use and not PLATO was not acceptable. "He just did not want anything competing, and of course he had Alpert's backing, and so, it was impossible to do anything," says Compton.

It was not long before the micromanaging by Alpert, now graduate dean, of PLATO and Bitzer within CSL pushed Compton too far. "Dan kept intruding, and I got so fed up with him I said one time, 'Why in the hell don't you transfer it out if you don't like the way I'm doing it?' And so they did." According to Compton, the dean of engineering, William Everitt, to whom CSL reported, was not even consulted about the move. Alpert just went and did it. "That kind of illustrates it," Compton says.

It was October 1966 and PLATO was once again under Alpert's wing. "And then something begins to happen one gets mixed reactions on," says Jack Desmond, "and that is, Bitzer's overwhelming self-confidence and brashness. It excited many people, and it turned a lot of other people off, because it tended to be heaped out in fairly large doses. There was no denying that here was a true believer and someone who was committed and enthusiastic and was indeed showing great progress toward allowing this development to germinate, and it was largely his doing."

It is at this point that a gentleman named Louis Volpp enters the picture, right after he had chosen a path that should have caused him to not enter it at all. Volpp was a professor and associate director in the Graduate School of Business at UI. He had received an offer, around 1965, to take his family abroad to Malta for a year-long university appointment there. "My family and I all had our shots," he says, "and I was sitting in my office signing the letter of acceptance and I had a call from Lyle Lanier, who was provost." Lanier told Volpp to hold off signing the letter and come see him first. When he did, Lanier offered him a promotion to associate provost, which Volpp accepted, canceling his Malta plans.

One of Volpp's new duties was to deal with the State Technical Services Act of 1965, a new federal law described by Congress as "a bill to promote economic growth by supporting state and regional centers to place the findings of science usefully in the hands of American enterprise." The University of Illinois, being the largest such "state center" in Illinois, needed to coordinate with the state government in Springfield on how exactly to comply with this new law. Although still the associate provost, Volpp in effect became the de facto chief of technical services for the state of Illinois. "That brought me in contact with Dan Alpert," he says, "because as dean he was also the administrator in charge of research programs at the university and one of my state functions was to evaluate research proposals and allocate funds under this act to the various researchers in different universities." The two worked together on numerous fronts regarding research proposals, federal and state grants, and statewide community services. In 1966, Alpert recommended to Lyle Lanier that Volpp should be director of the PLATO laboratory that had been wrestled away from CSL.

The point was emphasized to Volpp that CSL had a full slate of physics, engineering, and chemistry projects, and PLATO had become tangential. Alpert helped CSL find a brand-new building to move to, and PLATO took over the entire Power House building.

Bitzer had not expected that Alpert wanted someone else to run the new lab, leading to more friction. Alpert wanted someone with a business orientation to run the organization, and handed Bitzer the role of technical director. For Bitzer this meant that for the first time in PLATO's history there would be a layer between him and Alpert.

Up until the split from CSL, the PLATO project had relied on Alpert and others within CSL to handle its finances. The new lab would now have to manage finances on its own. Volpp set out to do just that, cleaning up what he regarded as a managerial mess. "I wanted . . . to set up some decent internal controls, good financial controls, good cost accounting, some of those things, so that we'd know where we stood because it was, well, sort of *chaotic* beforehand. And so I worked on that and tried to get that straightened out and get people involved in it in that way."

It took a while to name the lab. "We futzed around for the better part of two months for a title," says Volpp, "and we didn't come up with anything, that's why we ended up with that clumsy thing, the 'Computer-based Education Research Lab.' " And with that CERL was born.

Volpp's previous office had been in UI's Business School, "a marvelous old building," he says. Now he found himself in the Power House. It was old as well, but was not marvelous. What he found were spartan digs. "It was higher quality than things built in Russia," he says, but "it didn't have any frills, it didn't have anything that was nice, it didn't have anything that was elegant, it had no charm, it was just a building to get work done in." Volpp's own office was a tiny square room in a hallway full of such rooms. "I remember our furniture was so rickety that I had a desk that had a single pedestal. It was an old oak desk and it was kind of a mess, but it was functional. Somebody came in the lab one day and said, 'You know, you really ought to have a nicer desk than that.' And so that weekend I went over and I painted the darn thing. And that was my nicer desk. People there didn't fuss at all about that kind of thing. It was getting the work done and making PLATO work and teaching with it that really counted."

Volpp got to know people around the lab, including Tebby Lyman. Her real name was Elisabeth, but as a child she liked to spell her first name backward—"Htebasile"—and that evolved and morphed into the shorter form, "Tebby." Her husband, Ernest Lyman, ran the Betatron project that Andrew Hanson's father had worked on. She was also a physicist, but ran into the problem so many women of science ran into in the 1950s and 1960s: universities made it nearly impossible for a woman scientist to have any significant role as a professor—if they could even get that title. So she wound up working at CSL and then at CERL. "She worked closely with Don," Volpp says, and what he and numerous others had found with Tebby was that she provided some adult supervision to the wild-man side of Bitzer. "She was a real smoothing influence on getting the educational components developed. Don was an extraordinarily creative guy, and I can remember the first time he ever got his flat panel display to work, he had four dots that lit up and he came running down the hall saying, 'It works! It works!' and we all came down and looked at his tiny, tiny plasma display."

It took no time for Volpp to see what kind of creative force Bitzer was. "Don also did all the stuff to get the telephone company to put in a proper substation that would handle the remote connections into PLATO," he says. "They were convinced it wouldn't work, and he demonstrated to them that it would." But Volpp would discover that Bitzer's creative thinking extended far beyond PLATO. "He and I attended a seminar on the physiology of the ear, on teaching the deaf to hear," he says. "And the professor who was presenting his findings and his thoughts and his concepts started talking about the ear, the inner ear, and how it works and what was happening and immediately Don lit up and he said, '*Oh. That's how this works.*' And he started explaining to the specialist what was really going on in the propagation of sound into electrical waves for the brain, and I think he had heard of it for the first time right there, but bang-go, he *had* it, and then the two of them did some work together, because they really struck up a kind of technological spark with each other. But it's just the kind of person he was that he would just—I want to use this in the best sense—'fire off.' Something would trigger in him and he would have ideas that would make the experts really impressed. And I really enjoyed working with him and around him and observing that."

One incident at CERL would influence Volpp's thinking about students, learning, and teaching for years afterward. "There was a young lady taking the obstetric nursing course," he says, "who apparently had failed her exam and was going to be failed out of the nursing school. And they said, 'She likes this course so much, let's let her finish this.' So, she kept coming, I think it was four o'clock in the afternoon, when that nursing course was on, and she would always have to be kicked out of the lab at the end of the time because she would run into the next class. She would not give up on it. She just kept going and kept going and kept going. And that was one of the things Tebby had to do was go in and say, 'Now you must leave as the next class has to start.' So Tebby kept kicking her out. And it turned out that this young lady took some national exam after that even though she was going to fail out of nursing school, and she scored very high in the nation. It may be even number one in the nation . . . she had an extraordinarily high score." The nursing instructors and CERL staff could not understand how that could be.

Luckily, PLATO recorded every keystroke, enabling instructors to go back and see how a student had gone through the lessons, keystroke by keystroke, even how much time they took to answer questions and read a page of text, to get insight into the student's thinking and learning process. When they looked at this nursing student's data, they discovered something extraordinary. They found out that this young lady had an entirely different learning style, leading them to theorize that might be why she was not doing well in nursing school. "So they decided to keep her," Volpp says. "They told her what her objectives were and they turned her loose to do it her own way and she did very well. And I often thought about that when I was teaching a regular stand-up course, that somebody didn't catch on, I always wondered, *Is it that their learning style doesn't match my teaching style?* and tried to find new ways, but that always kept haunting me all the rest of my career. That you really better pay attention to that."

While PLATO III experienced a slow, steady increase in the number of lessons being created by authors using CATO, the growth was not yet impressive enough. Authoring was difficult and painstaking work, often requiring programming help. Tebby Lyman tried to relieve author frustrations by creating the GENERAL Logic, but the going was still slow. While the number of student terminals was gradually growing thanks to a few new PLATO III classroom sites installed at Parkland College, Mercy Hospital, and nearby grade schools, there were still impediments to authors working simultaneously online. Time-sharing was great for students, but authors were not so lucky. Even MONSTER, the editor created by Blomme and Krueger, only skimmed the surface of problems authors faced. Only two authors could work simultaneously using MONSTER. Most elusive of all was the fact that authors did not have a real programming language in which to work. One designed by authors for authors.

The GENERAL Logic represented, according to John Gilpin, a quick-and-dirty solution to the problem of authoring. "The PLATO project had always gotten by," he says, "by Don's brilliant ability to make multiple uses of small resources. . . . They didn't have very much in the way of programming resources to devote to anything, so the idea was that they would use what they had to produce the GENERAL Logic and then all sorts of people could come in and write their own lessons by substituting parameters into the GENERAL Logic. And what became clear over the next three or four years was that, well, even though the GENERAL Logic was pretty close to what *most* people wanted, who wanted to do programmed-instruction-type instruction, it was never quite what *anybody* wanted." The result was that authors began tweaking the GENERAL Logic itself to be better attuned to the specific needs of that author's lesson. Then another author would come along, copy the previous author's modified version of GEN-ERAL, go to town, tweak it a little more, and there were now multiple versions of GENERAL. "If anything was discovered that was a flaw in the original," Gilpin says, "it was already propagated to all these offshoots. . . . The whole thing turned out to be kind of self-defeating."

Some authors decided to bypass GENERAL altogether and write their own "logics" or hire programmers to do it for them.

Gilpin had hired a programmer to create some UICSM math lessons for him, but wound up doing the work himself. In time, he was no longer going around looking for help. Instead, people were coming to *him* for help. "Very soon I found myself in the role of being the trainer for PLATO," Gilpin says, offering seminars and showing people how to develop lessons on PLATO. At one seminar Gilpin met a biology graduate student named Paul Tenczar. "He was very straightforward and outspoken," says Gilpin, exclaiming to Gilpin things like *I don't want to do GENERAL*, and *How do I get an arrow on the screen?* There was no easy answer: if there wasn't a way in GENERAL to do something, you had to fall back to CATO and FORTRAN, and even assembly language if you dared, and custom-build it yourself. Tenczar, unhappy with how hard it was to develop on PLATO, nevertheless pressed on, hanging around and asking more questions. "You could check the computer out for three or four hours," says Tenczar. "You might change three or four spelling errors and three or four things in an hour's worth of work." In nine months he had managed to get three interactive pages of material up and running in a lesson. "And I was as top-notch as they could ever expect in a programmer to do this type of thing." One enormous time-burner was compiling a program. Says Jim Payne, who worked on a number of PLATO III lessons, including, over several years, a few foreign-language lessons for Professor Keith Meyer, "I spent several all-nighters there. Just because it took twenty minutes to compile the program. There's only so many edits you can make in the course of the evening."

In the spring of 1967, Bitzer and Max Beberman asked Gilpin to run a National Science Foundation–sponsored Summer Institute, inviting educators from around the country to learn how to develop lessons on PLATO. Gilpin agreed to run it. One night Tenczar stayed up late transforming some of the authoring routines that were available— largely packages of FORTRAN subroutines and procedure libraries that any author could take advantage of so they didn't have to rewrite everything from scratch. An idea came to him that there might be a better way around all this mess. Up until that time the main way to use all these procedures was to call them using the "CALL" statement. For example, to erase the entire storage tube (thus erasing a student's display), you could issue a "CALL ERASE" command. To display one of the prepared 35mm slides for a student terminal, you issued a "CALL SLIDE(N)" statement, where "N" was the number of the slide you wanted to show. If you wanted to plot some text, you issued a "CALL PLOT()" subroutine and passed it a bunch of variables including the X and Y location on the screen along with the text you wanted displayed. But first you had to call all sorts of other subroutines before you could even do the CALL PLOT(). It was all very cumbersome.

Tenczar decided to try something different. His goal was to dramatically simplify what authors had to go through just to put text and graphics on the screen. He created simple commands like "at" and "write," such that if you typed the following:

```
at      1010

write   Hello, world!
```

TUTOR example code

and ran that code, PLATO would display "Hello, world!" ten spaces in from the left of the screen, and ten lines down from the top of the screen. No more CALL commands, no more subroutines. Prior to this point, displaying such a message on the screen might have taken fifty lines of code. Now it took two.

Just as Gilpin's Summer Institute was about to start, Tenczar came in and declared to Gilpin and colleagues, *I know how to do it!* Know how to do what? they asked. *I know how to get an arrow on the screen! I know how to do the language! I know how PLATO ought to run!*

Tenczar called his new programming language TEACHER, but quickly changed that to TUTOR. Before the Summer Institute was even over, TUTOR was far enough along that Gilpin, intrigued, invited some participants to try it out instead of GENERAL, just to see what they'd create and how productive they could be. The results were eye-opening.

To be an author developing a PLATO III lesson in CATO or GENERAL in mid-1967, you had to be your own systems operator. While the project had evolved light-years beyond the ILLIAC days, PLATO III was still cumbersome. It made no sense for these visiting Summer Institute educators to learn how to operate the 1604, so Rick Blomme served as the operator for all of the participants. Meanwhile, Tenczar was working night and day on TUTOR. "Every day he would have a new command or two up," says Gilpin, "and so we would take the stuff into the sessions, and the people would do stuff, and write their stuff, and in fact . . . several people who did that were able to come up with a fair volume of material. Even though they didn't have any computer skills themselves at all."

Rick Blomme had seen firsthand what could happen to productivity when you gave authors decent software development tools. He saw in Tenczar's TUTOR another looming productivity boost. He and Tenczar began to collaborate and build out a fuller set of commands for the language.

Tebby Lyman was not happy with this development. TUTOR was to GENERAL what SOCRATES had been to PLATO: a threat to the officially sanctioned direction of the lab. But unfortunately for Lyman, that wasn't how the lab worked. If there was an officially sanctioned direction at CERL, it was "Whatever works." Lyman thought TUTOR "was a lot of nonsense," says Gilpin. "This was an upstart from somebody who's not one of us, and it's deflecting effort away from further development of the GENERAL Logic, which is where we ought to be going."

Tebby had been the principal programmer for Maryann Bitzer's nursing lessons, all written using the GENERAL Logic. "When Paul Tenczar came on the scene," says Jim Payne, "here's this brash grad student . . . not a lot of people liked him, but he was a real savvy politician. There was a very hot meeting over whether both of these languages were going to coexist. . . . He ruffled an awful lot of feathers in the process."

Don Bitzer, a hardware guy through and through, stepped in, and, says John Gilpin, "declined to make a decision." Which in its own right was a momentous decision, just the kind Bitzer was known to make. "He said," says Gilpin, " 'Both kinds of efforts will continue until one of them works and one of them dies.' "

Around this time, PLATO III got another boost that would seal the fates of TUTOR and GENERAL: disk drives. With some fresh grant money CERL obtained an NCR 160 computer and two big Control Data disk drives. They needed the 160 computer because the 1604 didn't know what disk drives *were*, so the CERL engineers hacked the 1604 and the NCR 160 in such ways as to make all of the hardware happy and functional. The result was a massive boost in productivity across the system.

TUTOR divided chunks of a lesson into what were called "units," using the -unit- command, which was an approximation of a custom function statement in other programming languages. To form a more complete interaction with a student, for example, to ask the student to name the third president of the United States, but also take into account wrong answers, took only a few lines of code:

```
unit    example

at      805

write   Who was the third president of the United States?

arrow   1003

answer  <T,Thomas> Jefferson

write   That's right!

wrong   <J,John> Adams

write   He was the second president.

endarrow
```

TUTOR code with an -arrow-

In this simple example of TUTOR code, the student can type "Jefferson," "T Jefferson," or "Thomas Jefferson" and any of those will be judged correct. If the student types "Adams," "J Adams," or "John Adams," or any other answer, the answer will be judged wrong. Over time, many additional TUTOR commands and enhancements would be added, making answer judging incredibly powerful, including requiring strict or lax spelling, case sensitivity options, stripping out ignore words, and long lists of matching vocabulary and concepts, not to mention powerful numerical expression analysis. In keeping with PLATO's philosophy of being flexible and extensible to meet the needs of any instructor, the language would continue to evolve based largely on the feedback of the authors developing educational lessons.

In order for a particular class of students to take their lesson, the instructor had to make sure that the proper disk pack on the new disk drive was installed and running. Given the communications limitations of PLATO III, whereby only twenty terminals could be active on the network at any time, it was nothing new that the inactive sites were idle. But now, there was another reason to be idle, thanks to the arrival of the disk drive. "All of a sudden everything would shut down at CERL, because they had to put on [a different disk pack]," says Marilyn Beckman. "The nurses over at Parkland also had courseware that they were taking. They would flip some switch, there were terminals over there, and everyone was cut off at CERL. It was very funny."

Despite the ongoing constraints, PLATO III was progressing at two steps forward for every one back. Soon, multiple authors could sit down at student PLATO III terminals and fire up MONSTER and edit their lessons, saved straight to disk instead of the elaborate ordeal of recording edits to paper tape, then magnetic tape. The combination of disk drives and a new authoring language made authors orders of magnitude more productive—and happy.

TUTOR worked. GENERAL died.

The battle between TUTOR and GENERAL left Tebby Lyman a lesser figure in PLATO. "She really faded away from the scene as a result of that," says Jim Payne. "I was in some meetings where it was pretty uncomfortable with all the shouting. . . . It just became, as much as anything, a personality battle here. Each one of them had a large amount of their life invested in, a large part of their ego invested in their respective languages. . . . I do know that Paul ruffled an awful lot of feathers. Tebby had been there a long time, she was one of the original people who with Don started the whole PLATO concept, and she got knocked off her pedestal pretty big-time when Paul came in."

Louis Volpp's reign as director of CERL turned out to be brief. "It was my thought that Bitzer would do better if he had someone else administering and doing the business," says Alpert. "As it turned out Don wanted very much to be the head. We had the kind of relationship where I could say, 'Look, Bitz, I think this is the way it ought to be done,' and my concerns had been that there are a lot of nitty-gritty things that I had always done for the project which I was finding it difficult to do—I had a full-time job. And I appointed Lou Volpp as director and Tebby Lyman as assistant director. It seemed like a reasonable thing. Well, Don practically refused to do any work."

In April 1967 Max Beberman wrote a letter to Dan Alpert expressing concerns about CERL. "I agree with Don that we have been committing ourselves much too rapidly and that the organization of

CERL is becoming too formal." He suggested that Bitzer focus on the development of the upcoming PLATO IV system, Lyman handle production of materials and operations, and Alpert appoint a part-time business manager.

"The only impression I ever had from Dan Alpert toward Don," says Volpp, "was that he was unusual and that he was great and that he needed to be supported but that you can't put him in a harness, that he needed to do it his own way. . . . Dan seemed to me always to express great admiration and love for Don's work and wanted him to do well. From the time I got involved with the program, up until that time, Don had been the centerpiece for that effort in the Coordinated Science Lab, and that didn't change when the new lab was created. Don was really the boss of everything. I didn't ever have any inclination to try to tell Don what to do. . . . My role was to find out what Don wanted to do and make sure we had it supported properly. . . . Get funding, and get people out of the way, and get people in that he could cooperate with and if somebody out of another school wanted to help, then get that dean's nose back out of joint so that person could be released to come and help Don, and work with Don."

But by early July 1967, Volpp was out. His heart was in business and economics—something Bitzer must have noticed. "While I could appreciate a lot of the things that went on in PLATO," Volpp says, "I was not going to be an innovator in it, that was not my strength at all, and so I did what I could, and then it was time to go and do something different." By the fall of 1967 Volpp had moved to Duke University as a professor in, and eventually dean of, its new business school.

8

Knocking on the Same Doors

PLATO was not the only computer-based education project started in the 1960s. Other commercial and academic efforts were under way even as PLATO was just getting started. Government funding from the National Defense Education Act legislation of 1958 was still finding its way to schools, universities, and businesses. From one end of the country to the other, and in spots all over the world, educators, researchers, and engineers were tinkering with computers as the next big wave in education after the realization that teaching machines and programmed instruction were promising but could only go so far. The computer, they believed, could take you the rest of the way.

At Stanford University, two professors, Patrick Suppes and Richard Atkinson, began years of productive research in what they preferred to call "computer-assisted instruction" (CAI). Suppes, who shared with Bitzer the ego, the confidence, the ambition, and, it must be said, the arrogance, embarked with Atkinson on a project to study the psychology of learning around 1960. In 1962 they submitted a proposal to the Carnegie Corporation, which was looking for interesting projects. Says Suppes, "As we wrote it, we realized we could give it an educational learning theory 'twist.' We came up with the idea of computer-based instruction." (Two years had passed since the PLATO project had started at Illinois, but Stanford was apparently unaware of it.) Carnegie awarded the Stanford team $1 million—a huge sum, vastly more than PLATO's budget at the time. In 1963 Suppes and Atkinson launched experiments with elementary school children using teletypes connected to a small computer teaching elementary math and reading.

Suppes and Atkinson eventually teamed up with a group within IBM eager to tap into the education market. Together they developed a simple authoring language called Coursewriter, and eventually commercialized the package as the IBM 1500 system, with a mainframe, simple graphics terminals, random-access audio, and light pens for interacting with the screens. Whereas CSL and then CERL were determined to tackle the display memory problem (including addressing the prohibitive cost of video RAM) with the high-resolution plasma panel, for the 1500 system, IBM punted on the RAM problem by dedicating a disk drive to serve the same function as PLATO III's expensive storage tubes. Up to thirty-two IBM terminals would each have their "screens" stored as files on the disk drive, available for constant display by the terminals. The displays were fairly low-resolution, at 320 x 192 pixels, which only required about 7 kilobytes of disk space per terminal. IBM never made any money from the 1500 system, and only sold a few dozen, but they were nevertheless used in a wide variety of school settings from coast to coast.

In September 1966, Suppes, perhaps even more adept at promotion than Bitzer, wrote an oft-cited article in *Scientific American* entitled "The Uses of Computers in Education," which probably did more to introduce the concept of CAI to laypeople than any other article up until that time. No stranger to audacious goals, Suppes opened his article stating, "One can predict that in a few more years millions of school-children will have access to what Philip of Macedon's son Alexander enjoyed as a royal prerogative: the personal services of a tutor as well-informed and responsive as Aristotle." The article focuses on the work being done at Stanford with IBM, and briefly makes reference to the existence of other centers of CAI work around the country, including Illinois, but never mentions PLATO. Suppes and Atkinson also managed to get major articles in *Time* and *Life* magazines around the same time.

Suppes's emphasis was drill and practice. "In the early days," says Andrew Molnar of the National Science Foundation (NSF), "CAI was not all that popular. You could add two and two but you had to wait about five to ten seconds to get the answer before you could go on to the next frame. Many of the kids became bored. Drill and practice was usually limited to ten minutes, because that was the limit of the student's span of attention when doing drill and practice. While it improved performance, it could be boring." Suppes and Atkinson's computer lessons applied the age-old principles of Self-Pacing and Immediate Feedback, but now, unlike with Skinner's teaching machines, the computer was able to generate custom questions based on a real-time evaluation of how well a given student was doing as she went through the problem set. "Suppes was able to go into very poorly run educational environments and produce results immediately," says Molnar. "The state of California carried out a well-controlled study . . . the arithmetic did work; it improved performance significantly, but the reading was not as successful."

In 1967, the ever-ambitious Suppes founded a start-up company, Computer Curriculum Corporation, which lost money until the 1970s, when it finally became profitable, and went on to be acquired in the 1990s.

Thus, Stanford, through the work of Suppes and Atkinson, established a firm reputation as the research hub on the West Coast for CAI. In addition, they pumped out a number of graduate students who went on to play important roles in various organizations including Xerox's Palo Alto Research Center (PARC). "These guys, Atkinson and Suppes," says Dexter Fletcher, an educational psychology doctoral student of Atkinson's, "were real money machines. That's what it takes to become a professor at Stanford. You attract your research funding. And they were just phenomenally successful at that."

An interesting contrast can be made between the CERL lab and Suppes's IMSSS (Institute for Mathematical Studies in the Social Sciences). CERL was the exception to the rule; CERL's openness was legend, thanks largely to the management style of Bitzer. IMSSS exemplified the more normal university lab. Fletcher described the IMSSS culture as follows: "It's an institute and a laboratory, and you can't just sort of walk in the door, 'Hi, here I am,' to a bunch of busy people and have them turn, drop everything, and [help you land a job there]. So there's some dues to be paid, and I think a lot of people weren't willing to pay them."

Buzzwords have always plagued the broad field of educational technology, long before computers arrived on the scene. "Distance learning" was a fancy way of describing "correspondence courses," and then with the arrival of phonographs and radio and television and tape recordings and film, it meant something else. With Pressey and Skinner came "teaching machines," and later in the 1950s there was "programmed instruction," and then with the arrival of PLATO came "automatic teaching" and "computer-based education," two favorites of Bitzer's in the early days. It can be taken as a rule that if you are a practitioner in this field, and happen to be ambitious, competitive, and want to establish a name for yourself, you first have to talk about what you're doing in a way that is slightly different from everyone else who's doing roughly the same thing. How could Suppes and Atkinson use "computer-based education" as a buzzword if that is what Bitzer and Alpert were using? Instead, the Stanford buzzword was "computer-assisted instruction." In Suppes's perfect world, mere mention of CAI would be equated with Stanford's project. Suppes pushed hard to make real that perfect world.

"Nobody liked the term 'computer-assisted instruction,'" says NSF's Andrew Molnar. "Everybody felt it was a misnomer. But its use in newspapers was so widespread and so many people were familiar with the term 'CAI' it became apparent that there was no way of converting them to anything else. Computer-based instruction, computer-managed instruction were used, but you could not change the popular usage of CAI. . . . Probably no other name is more closely associated with CAI than Suppes. For years they were synonymous."

Bitzer stuck with CBE—computer-based education—as the main buzzword associated with PLATO. He once described its meaning at a conference: "Our definition of 'computer-based education' is that if a human and a computer get together, if either one of them happens to learn something, we consider that our field. So it's very wide indeed." Not surprisingly, no other major project used the "computer-based education" buzzword—CBE meant PLATO.

In the 1970s, a professional society, ADCIS, an acronym for Association for the Development of Computer-based Instructional Systems, pushed the somehow gentler CBI acronym, for computer-based instruction. Because ADCIS was so heavily weighted down with PLATO-affiliated academics and industry people, CBI generally hinted at PLATO's CBE.

Over the years the gurus of educational technology have spun out buzzword after buzzword: computer-aided instruction; computer-assisted learning; electronic learning; e-learning; online learning; digital learning; computer-based training; Web-based training; personalized learning. The game seemed to be, if you can get academia, get industry, and ultimately get the media and the public, to adopt *your* buzzword, you win. (*What* you win at that point is anyone's guess. In reality the constant coinage of new buzzwords probably does no one any good, particularly the confused media and public.)

Thus the situation has continued unabated since the 1960s. Today we have MOOCs (massive open online courses), e-learning, online learning, blended learning, flipped classrooms, learning management systems, and so on. The buzzword game is still in effect, and the media and the public are no better off understanding what's what now than they were fifty years ago.

Corporations like Systems Development Corporation, RCA, General Electric, and IBM all had CAI projects under way by the mid-1960s. University labs continued to pop up around the country and the world experimenting with their own versions of CAI/CBE, either using whatever computer system was around, leasing time on someone else's time-sharing system, or, less frequently, attempting to design and build their own system from scratch.

All these projects needed money, and there was only so much to go around. Corporations usually could rely on their own sources of funding, but academic labs were always looking for money. Behind every successful lab were one or more "money machines," like Suppes and Alpert, well connected with foundations and government agencies.

Then in 1967 it appeared that the federal government was going to hand out some more money. "A report by the National Academy of Sciences, 'Digital Computer Needs in Universities and Colleges,' made a strong case for universities having access to computers for research, but said little about education," says Andrew Molnar, who spent four years at the U.S. Office of Education before joining the National Science Foundation. "In 1967, the President's Science Advisory Committee commissioned a study of computers in higher education. John Pierce from Bell Labs was the chairman and held extensive hearings. They concluded that an undergraduate college education without adequate computing was as deficient as an undergraduate education would be without an adequate library. They also felt there was value in using computers for pre-college education. These recommendations had a significant impact on educators. I think it was a trigger for the involvement of the National Science Foundation. The most significant event occurred when President Lyndon Johnson in his February 28, 1967, speech directed the National Science Foundation to work with the U.S. Office of Education to establish an experimental program to develop the potential of computers in education. In July of 1967, in response to the directive, NSF created the Office of Computing Activities to provide federal leadership in the use of computers for research and education. I joined after that."

Alpert reached out to Molnar. PLATO III was moving along, PLATO IV was on the drawing board, the plasma panel display was progressing well, the TUTOR language had been invented, boosting lesson author productivity by orders of magnitude. PLATO had momentum. It was time for the NSF to pitch in and support the early research and development of Bitzer's enormously ambitious PLATO IV. The commonly bounced around number in the mid- to late 1960s to describe the scale of PLATO IV was 4,096 terminals. PLATO IV would run on some sort of supercomputer mainframe costing millions of dollars, and drive 4,096 terminals in a massive demonstration project that would not only put PLATO on the map but also awaken the nation to the rosy potential of CBE.

NSF duly began doling out funds. CERL got $475,000 in early 1968. Seymour Papert got some to work on the LOGO computing environment for children. John Kemeny and Tom Kurtz at Dartmouth received support for their BASIC programming language. Molnar calls this period from the mid-1960s to the early 1970s "the golden age of education."

NSF was reluctant to fund science education, says Andrew Molnar. But NSF's Office of Computing Activities found a way to do it, by funding the development of computer-based education systems—a sort of trickle-down approach whereby getting computers in the hands of educators would lead to innovative applications related to science, and, inevitably, into the hands and minds of learners in the classroom. The National Science Board, which governed NSF, was not sold on the idea. To help make the sell, at some point around 1970, NSF invited Dan Alpert and Donald Bitzer to come to Washington and make presentations and demo PLATO to the board. This was the standard dog-and-pony show, a live demo using phone lines back to Urbana, with a trusty recorded tape and a tape player at the ready should the phone lines not work (they could plug the tape recorder audio output into the phone setup, and the terminal would be fooled into thinking it was receiving displays from CERL when in fact they were recorded from a previous demo).

CERL engineering technicians Jim Knoke and Ray Trogdon had been tasked with taking one of OI's four-inch prototype plasma panels and incorporating it into a full-size plywood mockup of the PLATO IV terminal for the demo. The result was a big blue plywood box with a tiny square screen front and center, with a lid on the top of the box you could open, but you had to be careful fiddling about inside. "We called it the 'Possum Trap,'" says Knoke. Inside the lid was a crude but functional microfiche slide mechanism, but it was possible to get your hands caught in it if you weren't careful. The technicians carefully set up the big blue clunky Possum Trap terminal on a table, set all the cables, and got the live telephone connection back to CERL up and running. "Everything worked fine during the test, and then just as we were to make the presentation, just before, we had difficulty with the phone line," says Knoke. Then the plasma display didn't light. Moments before the demo was about to start. The electronics inside the terminal were so extremely delicate that Knoke had doubts the Possum Trap would survive the trip out to Washington, but it had made it, and Knoke and the other technicians had taken great care to gently set everything up. And now it was not working, seconds before Bitzer was about to start.

The "Possum Trap" terminal in storage in 2003

Bitzer assessed the situation and then did what any leader of a technology project does when he is about to go on with a demonstration that will determine whether his project gets funding from the United States government or not. He took his hand and slammed the side of the Possum Trap hard. "My heart went into my throat," says Knoke, "when he slammed the side of the box, which, again, was so sensitive. But by some miracle, the whole thing lit up with text on the screen and the phone line came through and we were ready for the demonstration."

Bitzer proceeded with his usual demo to the gathered dignitaries, but he wasn't winning over the board. "Initially the board was extremely skeptical," Molnar says, "since most were unfamiliar with the developments in instructional uses of computers." Luckily for Alpert and Bitzer, they'd brought along in their PLATO entourage chemistry professor Stan Smith, who'd developed a reputation at the University of Illinois as one of the top designers and implementers of creative uses of PLATO. Smith was the ace up their sleeve, and it helped further, for this meeting, that one of the NSF brass had a chemistry background. "Stan Smith," says Molnar, "was an outstanding chemist, a marvelous teacher." Rather than just do a canned demo, he threw down a challenge to the board. "Tell me anything you find hard to teach in chemistry," he said. The board responded with a chemistry problem. Smith duly typed the problem into the PLATO lesson he was running. Chemical symbols came up on the screen and he solved the problem.

The board was astounded. "They did not believe this could be done," says Molnar.

The Stan Smith demo did the job and won the National Science Board over. Bitzer would go on to demo PLATO to members of Congress, which approved the NSF funding. "But [they] did not provide any extra money," Molnar says. "Therefore, we had to take the funds out of existing programs." Luckily, some of the NSDEM money left over from the Johnson administration was available. It certainly was not going to come from the incoming Nixon administration.

There were more NSF demos leading up to the money being released to CERL in 1972. At one, a large CERL entourage, including Bruce Sherwood, Stan Smith, and Paul Tenczar, came out to Washington again, with Bitzer leading the demo not just with a new PLATO IV terminal but also with a new, freestanding random-access audio device attached. Recalls John Risken, who led CERL's project to develop elementary reading lessons, "They all went down and did this demonstration, and of course one of the things they were demonstrating is the random-access audio device, which was pretty impressive really, and so Bitzer was up there onstage, demonstrating the touch panel and the random-access audio device, and the terminal was sitting on a table that had a skirt around it. And the audio device was sitting up there on top of the table along with the terminal, mind you, this is a presentation to a multimillion-dollar funder, and yet, people were comfortable enough with Bitzer and with Bitzer's personality that after Bitzer finished the audio demonstration, and turned around to talk to the people from the stage, Tenczar had set things up so that about five minutes later the audio device would fire off one more message. While Bitzer was talking, suddenly from under the table came this 'Don, can I come out from under the table now?' Throwing the actuality of the whole audio device into doubt. And they knew him, people knew him well enough to know that he would think that was funny rather than become outraged and defensive and everything else."

Erik McWilliams was tall with a Scandinavian look about him and had a background in math and computing at Cornell University, where he'd administered the university's Office of Computing Services toward the end of the 1960s. He'd done similar work at the University of Chicago back in the mid-1960s, where he got his master's degree. After working at Cornell for a while, he took a one-year rotating leave of absence to join the National Science Foundation's Computer Applications in Research program, whose mission was to stimulate the use of computers in scientific disciplines like math, physics, and chemistry. After about six months, Arthur Melmed, who worked at another division in NSF, approached him regarding a program he was setting up to fund two oddly named computer education projects: one called PLATO and one called TICCET, for Time-shared Interactive Computer-Controlled Educational Television.

McWilliams had attended one of the numerous Alpert and Bitzer dog-and-pony shows in Washington. Bitzer demonstrated PLATO III using a live PLATO III terminal connected back to Urbana, as well as the Possum Trap. McWilliams immediately recognized one of the members of the PLATO entourage at the demo: Bruce Sherwood. Sherwood had been working on his degree in physics at the University of Chicago when McWilliams was there, and both of them knew each other from working at the university computer center. Sherwood recognized McWilliams as well. They talked, got reacquainted, and kept in touch afterward.

Meanwhile, Melmed began talking seriously about PLATO and

TICCET, and eventually began including McWilliams in the planning meetings. Ken Stetten, who worked across the Potomac in Northern Virginia at MITRE Corporation, came in for a visit to brief NSF about TICCET, which would eventually be slightly renamed as TICCIT (the second "I" standing for "Information"). "I never worked in the television medium much, so that was my first introduction to that approach," says McWilliams. "And it sounded interesting. I mean, the times were quite different: if you look at it in hindsight, it probably looks a trifle naive to think that you could actually gin something up for ten or twelve million and really get somewhere. But those were different times. Things were much more optimistic, it was more a period of experimentation, particularly in education."

The stories of TICCIT and PLATO intertwine, and how they intertwine sheds useful light on both projects. A good place to start is with M. David Merrill. He attended Brigham Young University (BYU) in Utah as an undergraduate with dreams of becoming an electrical engineer. A Mormon, he began at twenty to serve as a missionary for two years. "I was assigned to work in Ohio, Indiana, and Michigan," he says. "We went from door to door telling folk about our church. I was impressed at this time that our teaching was ineffective. I began to think about better ways that we could teach."

His interest in education eventually made him abandon his pursuit of an engineering degree. In time he would enroll in the University of Illinois' doctoral program in education, studying under Larry Stolurow, creator of SOCRATES. Stolurow's graduate program was designed to expose the students to theories of learning and instruction and hopefully trigger the students to synthesize these ideas into new theories. Merrill completed his PhD at the University of Illinois in 1964, becoming familiar with both PLATO and SOCRATES during his time there.

A year later Merrill had secured a visiting professorship at Stanford University—Suppes and Atkinson territory. "I did not work with Pat Suppes. However, I wanted to and was really frustrated that I didn't get to," says Merrill. Though he got an offer to stay at Stanford, he chose to go to the University of Texas at Austin to work with C. Victor Bunderson, who led a CAI lab. Merrill eventually landed at Brigham Young University. Bunderson wanted Merrill to join his lab, but soon

Merrill wanted, Bunderson to join BYU. Thus began a tug-of-war for several years. Bunderson would continue building up his CAI lab at the University of Texas, acquiring an IBM 1500 system along the way.

At one point Merrill and Bunderson found themselves at a conference on the East Coast. Bunderson told him about his project that NSF was funding. It involved developing an entire remedial English and math curriculum at the junior college level, and described how the system was going to work. The scale of the project was bigger than Bunderson had anticipated, and he was concerned that his Texas lab didn't have enough people to do the work. Merrill offered his BYU team as additional help, and Bunderson accepted. That led them to travel to Washington, to present to NSF and MITRE. Merrill went home with a large amount of the initial contract to do the remedial courseware development, "without ever having written a lick of a proposal," Merrill says. It was something of a coup back at BYU.

Merrill had one more hurdle to overcome. With this new TICCIT project, Bunderson's Texas lab stood to receive substantial federal funding and visibility in the field. Merrill still wanted Bunderson to move himself and his lab up to BYU. Merrill let NSF know that BYU had a standing, lucrative offer to Bunderson for a full professorship at BYU. Then NSF decided to reduce the planned budget for TICCIT such that two university labs were one too many. The solution was to consolidate everything at BYU.

Merrill's team at BYU worked in an abandoned three-story home economics building on the lower campus of the university. They had the entire building to themselves. They used one of the classrooms as a conference room, and met there regularly to discuss plans and share ideas.

There were a few things that Merrill's team at BYU needed badly to make progress. Specifications, for one. They kept asking Bunderson for specs, and specs were slow in coming. "What does it look like? What does the screen look like? What does the keyboard look like?" Merrill had assumed all those things had been decided already. They hadn't been. "I had authors that were writing instructional materials for a system that we don't even know what it looks like. We don't know what it can do, we don't have any of the characteristics."

All Merrill had was an early paper on learner control by Bunderson and Steve Fine, a computer scientist who worked for Bunderson at the Texas lab. Fine had submitted a paper to NSF in May 1972 entitled "Learner Control: Commands for Computer-Assisted Instruction Systems." "Those were very vague ideas, they certainly weren't implementation," says Merrill. "They didn't talk about screen design or any other thing. They talked about 'macros,' they talked about the whole notion of a 'discovery system,' where students would choose from a menu which kind of system and then be kind of locked into that." Merrill and company weren't satisfied that this went far enough. His math and English authors were "going crazy," he says. "We were telling them general instructional design principles and they had no idea what they were designing."

The BYU team's debates on learner control ultimately resulted in taking the TICCIT system on a path the complete opposite of PLATO. Whereas PLATO was open, flexible, and devoid of any particular instructional theory limiting lesson authors to developing lessons that followed a particular design, TICCIT would reflect a single instructional theory burned not only into the software, but into the hardware as well, right in the keyboard with its special keys for RULE, EXAMPLE, PRACTICE.

The culture of the TICCIT lab at BYU could not have been more different than CERL. For one thing, staff meetings were far more formal affairs, always starting with a prayer. Meticulous minutes were kept, typed up in memos, and filed away. BYU was largely instructional designers and educational psychologists, with few engineers, as MITRE was two thousand miles away. This meant there were no impromptu, interdisciplinary hallway conversations allowing cross-fertilization of ideas to take place, whereas at CERL, interdisciplinary cross-fertilization was the norm.

In 1971, just as the TICCIT and PLATO camps were drafting their respective proposals to NSF, Ken Stetten of MITRE Corporation disseminated a position paper comparing TICCIT and PLATO that sent CERL into conniptions. In the paper he argued that PLATO IV's design was fatally flawed and could never work. CERL, he asserted, had not done modeling simulations using Stetten's favored modeling tool, a tool that would have provided insight into how much memory and how many processors would be needed to handle all of the input and output and processing of the PLATO IV system. On the other hand, MITRE, Stetten argued, had done careful analysis and modeling and they were confident their numbers were correct and that TICCIT would perform well. "We felt badly about this," says Bruce

Sherwood, "in part because there had been a *tremendous* amount of actual modeling of the system using PLATO III. It wasn't a simulation, it was actually put together. And reaction times and keystroke rates and output rates and so on had been measured. There had been a *lot* of modeling based on *real* data, and the TICCIT group didn't actually have any data at all."

Bitzer was furious. In a letter to Bunderson, he said, "In reference to the document written by Ken Stetten and sanctioned and circulated by the MITRE Company, our position has always been to stop its circulation. However, Ken Stetten and MITRE evidently circulated enough of the documents before they changed their policy and decided to stop it, so that most people in the field are well acquainted with its contents. In the hard sciences, such documents become property for scrutiny and I suspect that this document will be treated no differently. We have never had, nor do we have now, plans to compare TICCIT and PLATO, but the MITRE document will force the scientific community to make the comparisons that Ken outlined. Ken Stetten and MITRE produced it, and they will have to live up to its contents."

MITRE invited two CERL staffers, Bruce Sherwood and Paul Tucker, to come out to Virginia to take a look at TICCIT firsthand. Before the trip, Sherwood did his own analysis of the TICCIT system architecture, and to his surprise could not figure out how it was possibly going to work. Sherwood believed that MITRE had overlooked some key system design considerations necessary to achieve the equivalent of PLATO's Fast Round Trip: he believed they were going to run into problems just having a user type on the keyboard and see their text echoed on the screen within a reasonable fraction of a second. Tucker and Sherwood flew out to MITRE and got a demo. Sherwood was presented with an electronics lesson asking him to identify a particular component of an example electric circuit. He knew the correct answer was "resistor," so he began typing "r" and then "e" and the rest of the word, but no "r" nor "e" nor the rest of the word appeared on the screen. Sherwood recalls the ensuing conversation going like this:

SHERWOOD: Um, excuse me, why are the characters not echoing on the screen?
MITRE: Oh, well, we haven't implemented that part yet.
SHERWOOD: But how are you going to do that? The way I figure it, there isn't any way to do that.

MITRE: Oh? Why do you think there's a problem?

[SHERWOOD explains what he believes the problem is.]

MITRE: Oh, gee, maybe that *is* a problem. Well, maybe we'll
have to add another whole processor.

"Double the number of mainframes they're going to use!" Sherwood
would recall in horror years later. "And in front of Paul and me,
bemused as we were, these people started brainstorming about how
it might be possible to echo keys onto the screen, because they hadn't
really thought through whether there was a difficulty there. They could
now see there was. I was *shocked* out of my mind. Because as I say they
had actually almost kind of scurrilously attacked PLATO and Bitzer
for not having done a careful analysis that would merit going through
the next stage, whereas they had. And they couldn't even echo keys
on the screen. It was terrible. And it was sort of my first, starry-eyed
introduction to the world of high-powered, Beltway-defense-contract-
vaporware. It was quite an experience."

The TICCIT terminal

The surprises kept coming. BYU and MITRE were starting from
scratch, and had no authoring language to develop lessons on TICCIT,
whereas PLATO had over a decade's experience going, starting out
with machine code, evolving to FORTRAN to CATO to GENERAL
and finally to TUTOR. PLATO had painfully paid its dues when it
came to authoring tools. The TICCIT people had no such experi-

ence to fall back on. For a while they considered using the ALGOL language, but eventually that idea faded away, as it would have been expensive to repurpose a language that had never been designed for CAI. One day, as if to add to CERL's growing apoplexy, BYU and MITRE reached out to Bitzer and Sherwood to mention that TICCIT still did not have an authoring language . . . might they use TUTOR? This resulted in a flurry of back-and-forth letters and phone calls, but in the end, perhaps in part to save face, the TICCIT people decided they would implement their own authoring language. TAL, the TIC-CIT Authoring Language, was the result.

One academic surprised by TICCIT was Michael Allen, who got to know the system and its creators during its early days. Allen, who directed CAI research and development at Ohio State, found that the big issue for Vic Bunderson back then was learner control. Should the student have a lot of control, or a little? For Bunderson, that was a central issue. Allen argued with him about that, telling him, "I don't think there can be too much student control. What you can have is too much student control without enough guidance and feedback, but I don't think you benefit by taking control of the student. If we want to have people who are good learners and going to learn independent of the system and take the good habits forward, then they need to have some discretion and control and then we have to help them practice good habits until they learn them well. To give them no choice doesn't seem to me to be a very good approach."

Allen remembers Bunderson's presentation at an educational computing conference in New York. "I was just amazed at his feeling," he says, "that he had an educational paradigm that was so effective that it was time to implement hardware and software so that you even had a keyboard that specifically was limited to that instructional design."

"I remember my mouth just dropped open," Allen recalls when he first saw TICCIT. "I thought, you can't be serious." The TICCIT keyboard looked like any other keyboard, except that it had three keys that stood out from the others, boldly declaring their take-it-or-leave-it pedagogical model: RULE, EXAMPLE, and PRACTICE. "I thought it was really premature," Allen says, "to implement hardware that was really tailored to that specific approach."

If nothing else, TICCIT had one big thing going for it. Its creators had bought into the REP model so deeply, so thoroughly, so completely, and the system reflected this deep, thorough, complete buy-in so absolutely, as did the courseware written for the system, as did the data that spewed out from the system when students used it, that the research could not help but reflect that model 100 percent.

The National Science Foundation decided in the end to fund both PLATO and TICCIT, two diametrically opposed systems with 180-degree different personalities, philosophies, and system architectures. The fact that the projects were so radically different appealed to NSF. They already loved PLATO and had been familiar with Bitzer and Alpert for years. They recognized that TICCIT represented a wholly different approach, both at the scale (supporting a maximum of 128 color TV terminals running on a minicomputer, versus PLATO IV's much heralded 4,096 terminals running on a supercomputer) and in terms of its instructional design model. "TICCIT was prepared," says Arthur Melmed, "to demonstrate a certain kind of interaction in a relatively efficient way, and I thought that deserved a crack."

In 1972 both systems received roughly $5 million each from Congress.

The race was on.

9

A Fork in the Road

In July 1968, ARPA invited graduate students working on projects it was funding at universities around the country to attend a conference at UI's Allerton House. One of the graduate student attendees was a brilliant computer-philosopher genius, viewed by some as a "wild man," named Alan Kay, then working on his PhD at the University of Utah. He gave a talk on what he called the FLEX computer, which he'd started working on the year before and which formed the basis of his dissertation, "The Reactive Engine," which he would finish in 1969. The presentation did not go over very well. Audience members, largely other computer science grad students, thought Kay's ideas were crazy. The FLEX machine was Kay's idea for a "personal computer," a notion in 1968 (and years afterward) often met with laughter or worse, an idea still foreign to practically everyone involved in computing in the 1960s, including, importantly, the PLATO team. Kay's dissertation would introduce FLEX as "a personal, reactive, mini-computer, which communicates in text and pictures by means of a keyboard, line-drawing CRT, and tablet."

Kay would later describe FLEX as "a small tabletop computer, and it didn't have a very good user interface on it, and only after we had done a lot of it did we realize the user interface was really the critical part. We already had the idea that somehow the novice programmer should be able to program on it and it had an object-oriented language on the first one that I designed but not terribly well-integrated into the user interface." Kay was influenced by the work of Ivan Sutherland, also at the University of Utah, who was a pioneer in computer graphics, as well as Douglas Engelbart, who had long ago abandoned his plasma storage device research to embark on research into user interfaces and online software for "human augmentation," resulting in, among other things, the invention of the mouse. Kay also stumbled upon Wallace

Feurzieg's and Seymour Papert's LOGO work under way in Massachusetts. In Papert he found a kindred soul: Papert was someone who had worked with Jean Piaget and approached education and learning the way Kay did, views that were about as diametrically opposite to Skinner and the behaviorists as you could get.

While attending the ARPA conference, Kay for the first time saw what Bitzer and company were up to with PLATO and the plasma display project at CERL. As for the PLATO III system and CERL's plans for PLATO IV, Kay was not that impressed. "My take on that was very different from Bitzer's," he says. "From my standpoint as a computer scientist, I always thought that the PLATO architecture was stupid. What you need for anything that's going to be interactive is guaranteed cycles. The way you get guaranteed cycles is putting them at the user, not trying to time-share them." Kay and his colleague Adele Goldberg would summarize their distaste for "time-sharing" (the very word was anathema to them) in a famous essay entitled "Personal Dynamic Media," published in 1977:

> Children really needed as much or more computing power than adults were willing to settle for when using a timesharing system. The best that timesharing has to offer is slow control of crude wire-frame green-tinted graphics and square-wave musical tones. The kids, on the other hand, are used to finger-paints, water colors, color television, real musical instruments, and records. If the "medium is the message," then the message of low-bandwidth timesharing is "blah."

PLATO meant time-sharing, and Kay was religious in his dislike of it. But the prototype of PLATO's plasma display, with its tiny, flicker-free pixels from which emanated the Orange Glow, was something else entirely. When Kay laid eyes on that prototype he was astounded. "I saw a one-inch-square lump of glass and neon gas in which individual spots would light up on command," Kay says. "When I saw that flat screen display, I thought, oh boy!" Seeing that panel was an epiphany, something he would later describe as a "big whammy." Kay had assumed that flat screen displays were still science fiction, devices that might come true at some point far off in the distant future, and therefore his FLEX machine would use, as he had stated in his dissertation, "a keyboard, line-drawing CRT, and tablet." But all that changed when he laid eyes on the plasma prototype. He realized that he wasn't going to need a CRT after all. Kay, a believer in Moore's Law, could now see that the future would include affordable chips and flat displays, components that would make personal computing not only possible but inevitable. He spent the rest of the conference calculating when Moore's Law would reach the point where all the processing and memory could fit on the back of the FLEX machine's flat panel display.

"People had talked about flat screen displays since the 1950s as a concept," Kay says, "but I never really thought of it as a *computer* display until I realized that people could actually start building them like that. . . . The thing that hit me on the plasma panel was that people could actually do thin-film deposition now over large areas and get away with it, and so it was only going to be a matter of time before you could actually do some decent display. So yeah, that had a lot to do with it. That immediately gave me a focus for thinking about user interface. And the crux of the thing was that I remembered a saying of [Marshall] McLuhan's, which was, 'I don't know who discovered water, but it wasn't a fish.' And I realized that one of the problems, one of the reasons user interfaces were lousy, was we had adults trying to design for adults. And thinking they knew what they were doing but actually taking so many things for granted that it just wouldn't work out. Somewhere around in there I realized that what Seymour had shown is that the computer is more than just a tool, it could be like a medium and extend into the world. Media are things you want to extend into the world of the child."

As 1968 wore on, Kay visited with Papert and the LOGO project in Massachusetts, and then in early December attended a presentation in Palo Alto, California, given by Douglas Engelbart of the Stanford Research Institute on his "NLS" or "oN-Line System." Known in the history books as "The Mother of All Demos," the event marked a turning point in thinking about computers, what they could be used for, and how they were best designed. Engelbart walked his rapt audience through a demonstration of a working system that would pave the way for every desktop computer in use today. It was an extraordinary, breathtaking demonstration, and convincingly depicted a future very different from the mainframe-based PLATO IV system, which was still four years away. In the minds of people like Kay and Engelbart in December 1968, PLATO was doomed, unless Bitzer and company embraced distributed computing and the idea of "guaranteed cycles," which was then and would be for a long time largely anathema to CERL mainly on economic grounds. Soon, Kay's vision for the FLEX computer, a desktop machine, morphed into a portable tablet, much like present-day iPads. He called this the "Dynabook"—his ultimate dream for a lightweight, portable, multimedia computer that incorporated the flat panel display technology he'd seen at Illinois.

Two approaches, two different missions, two ways of thinking: two visions for computing collided and bounced away from each other in 1968. It was not the last time they would collide or the last time they would bounce away.

That same year, high-flying Xerox Corporation, one of the hottest technology companies of the 1960s, was looking to expand beyond photocopying by venturing into the computer business. Documents were no longer sexy. *Information* was the future. The company recruited Jack Goldman, a veteran head of research at Ford Motor Company, to assume the top research post at Xerox, which was freed up by the departing John Dessauer. Xerox CEO Peter McColough made Goldman an offer he couldn't refuse. "The Xerox job meant money, stature, power," wrote Douglas Smith and Robert Alexander in their Xerox PARC history, *Fumbling the Future*. Goldman was even offered a seat on Xerox's board. He took the job.

Xerox already had a research laboratory, located near its Rochester, New York, headquarters. Goldman toured it and met with the lab's director. He was shocked to find that they did almost nothing with computers and had very little familiarity with digital technology. The lab was focused on the technology behind copying machines, Xerox's cash-cow business. Then Goldman was hit with another surprise: the SDS acquisition.

When Xerox explored the idea of getting into the computer business, it realized it couldn't start from scratch—there was no way it was going to build its own systems. The competition, including IBM, Control Data, Honeywell, Sperry, Digital Equipment Corporation, and a long list of others, was far too far ahead, and Xerox would never catch up. The only way it was going to get into the computer business was to acquire an existing computer business. Control Data and DEC were considered but unobtainable. Then they looked at a hotshot young company called Scientific Data Systems (SDS), makers of a computer called the Sigma, and found it willing to be courted. In 1969, Xerox successfully acquired the company, for nearly $1 billion in stock, a massive sum at the time. Incredibly, Goldman, Xerox's new chief technologist, had not been consulted, nor was he at all involved in the acquisition. He would have fought against it had he known, believing SDS a lackluster business.

After seeing how weak Xerox was with digital computing at the copier research lab, it occurred to Goldman that even with SDS now under Xerox's wing, a new, second research lab, one focused solely on computing, was needed. He wrote up a detailed proposal and presented it to the board, which reacted negatively to it, but he managed to secure McColough's go-ahead to proceed anyway. For recruiting ideas, Goldman turned to fellow Xerox board member Bob Sproull, a well-known physicist, former head of ARPA, and at this time provost of the University of Rochester. "I knew him well from the physics community," says Goldman. "I took him on a plane ride since we were both going to New York and I said, 'You got any suggestions on who I ought to go for?'"

According to the generally accepted historical record, Goldman would pick a physicist named George Pake to be the director of the cutting-edge research lab. Here's what *Fumbling the Future*, which Goldman himself would later recommend as the most accurate history of Xerox's Palo Alto Research Center (PARC), had to say:

> Jack Goldman recruited a long-standing acquaintance of his named George Pake to set up and manage the proposed Xerox research center. While Goldman could, and would, continue to speak at corporate headquarters on behalf of the effort he'd inspired, he had too many responsibilities as the company's chief scientist to operate the facility himself. He had to find someone else for that job, and Pake was his first choice.

This was not exactly true. Goldman and Pake kept a secret from Xerox historians for more than thirty years. The secret was that Pake was not Goldman's first choice. Pake was Goldman's *second* choice.

When Goldman had asked Bob Sproull for names while they flew to New York, there was one name Sproull touted that Goldman recognized right away. "One of the guys he mentioned was Dan Alpert,"

Goldman says. "He mentioned one or two others who were also well regarded and had made their way in the computer community. And since I knew Dan from my Westinghouse days—I had worked at Westinghouse, as my first job in research was at Westinghouse. And Dan was one of the very well-respected Westinghouse fellows." One of the other names Sproull mentioned was Ivan Sutherland, the wizard of interactive computer graphics. Goldman himself had come up with the idea of Pake, someone else he knew, because, it turns out, Goldman, Pake, and Alpert had *all* worked at Westinghouse at the same time, years earlier, when they were all still climbing up the ladder of their careers. Pake had eventually gone on to a PhD and academia, and was now provost at Washington University in St. Louis. Alpert had gone to Illinois, Goldman to Ford. Goldman, four years younger than Alpert, greatly respected Alpert's leadership and abilities. And now on the plane, Sproull was recommending him. "Since I knew Dan, I just picked up the phone and called him," Goldman says.

Sproull had also known Alpert for years. "I first knew Dan Alpert shortly after the war, I believe in connection with the Division of Electron Physics of the American Physical Society. By 1969 I had served with him on a number of committees and at the Institute for Defense Analyses and the Defense Science Board. I had worked at CSL the summer of 1951 and subsequently followed CSL and especially its educational technology. In 1969 I was both a director of Xerox and a consultant and was actively discussing the creation of PARC with Jack. I had a high opinion of Dan as a scientific and engineering leader and administrator, and it would have been natural that I suggested his name to Jack to direct the fledgling PARC."

It was the summer of 1969 when Alpert received the call from Goldman. Goldman was delighted to find out that Alpert was thriving at the University of Illinois, busy running the Graduate College and overseeing Bitzer's lab and its PLATO project, which had just the year before received a large infusion of NSF money to ramp up the planning and design effort for PLATO IV. "I was so happy to see a great guy like Dan Alpert was succeeding in running it properly," he says. PLATO "was strictly for the educational world, and I think Dan's vision was just that, to be an available resource for schools, universities, and whatnot, around the country, around the world, for that matter. And I thought, gee, that is a wonderful idea, anything to take the initiative away from

IBM. It was a nonprofit kind of thing, with government support, and that would have been great considering the state of technology in those days."

Goldman found PLATO intriguing, in fact the whole idea—and market potential—of educational computing was intriguing. Educational computing was something SDS, soon to be renamed Xerox Data Systems, or XDS, was interested in moving into. Xerox already was dabbling in the education market and wanted to dabble more. "The SDS people envisioned the Sigma computer as the ideal educational computer," Goldman says. "So there obviously should have been somewhere in this case a marriage between what we hoped would happen with the Sigma computers, and the educational world, meaning things like PLATO. Now, we began to see flaws, after we took over SDS, we began to see flaws in their lineup of products, their whole approach to it. They were destined for ignominy very quickly because it was a time since they were producing largely for the research community, not for the business community, and in the '69 time frame government funds were beginning to diminish. They were starting to have problems once we took them over. . . . So in thinking about what role this new lab that I was contemplating would take, it was inevitable that I would look at where there are holes in the marketplace, that this company that just took over a computer company, and this lab which is supposed to be a lab backing up the computer business, what holes could it fulfill. And it was inevitable that we would think of the educational market. Because the educational market in those days was a market that Xerox was interested in. . . . We're involved in education, so a computer in the educational world would clearly be of interest to us, and in this respect, PLATO clearly stood out as an option."

It was summer, and Alpert was about to go on a family vacation to Snowmass Village in Colorado, where only the year before he had bought a plot of residential land at 8,500-foot altitude, right down the road from the ski slopes he routinely visited year after year. (He'd tried to talk Bitzer into buying some land nearby, but Bitzer didn't bite.) Goldman had called out of the blue with what Alpert viewed as "a very attractive offer. . . . I would have tripled my salary, plus a bonus." At this point, Xerox had not decided where to locate the lab, though Goldman's preference was New Haven, Connecticut, near Yale University. Santa Barbara was another possibility, as was Palo Alto. Says Alpert, "I could pick whatever place I wanted." He still longed for the rolling hills of Palo Alto and the campus of Stanford University, where he gave the oral defense of his dissertation on the day before Pearl Harbor was attacked, twenty-eight years earlier. The Xerox offer was tantalizing: big money, big stock options, big responsibility, a lot more visibility than he had in his duties at Illinois, and a wide-open future that he could orchestrate. "It was an absolute sweetheart of a deal," says Frank Propst. "It was just an amazing deal. And Don and I kind of were telling Dan, you know, you're kind of crazy not to pick this up. This is a remarkable opportunity and something that could be very, very rewarding for you financially."

Beyond the beckoning riches, the offer would bring him right back where he'd always wanted to be: industrial research. He had only left Westinghouse because he had become disillusioned by the choice of director for that lab (and the fact that he'd been passed over for the position). Academia had not been his first preference, but the CSL job back in 1957 was attractive enough that he had taken it.

And now, opportunity was knocking on his door once again. Dan Alpert, and, in a very concrete way, Don Bitzer, Frank Propst, the CERL staff, and the entire PLATO project, had a massive decision to make. Here is where Alpert's account differs from Propst's, who says he urged Alpert to accept the offer. "Bitzer and Propst," says Alpert, "twisted my arm every way they possibly could, to stay. Because, well, they were living a pretty comfortable life. Hell, they had, in their boss, a guy who understood and cared how PLATO went and was supportive of them as individuals."

Alpert eventually had a long conversation with Peter McColough, CEO of Xerox. But he still wasn't sure what to do. Alpert wasn't very impressed with Xerox's SDS acquisition, which he viewed as "one of the jarring things" about the corporate environment he would be parachuting into. "I could see," Alpert says, "that my interests in computers and my talents were somewhat conflicting with the former president" of SDS, Max Palevsky. "It wasn't much of a company, didn't have any innovation to speak of, just making cheaper machines, making money by being more productive in production sense but not in a design sense. . . . He and I had different perspectives on what the future of computing was going to be about."

There was one thing he was sure about: take some time off, get out to Colorado, and think about it. So he decided to stick to his plans and take his family on the vacation. They loved Colorado and were experienced backpackers around the Rocky Mountains. Out to Colorado they went. But he still could not stop agonizing about the Xerox offer. "I had a hard time making that decision," he says. "I remember vividly looking up at the sky, the beautiful Colorado sky, wondering what the hell I should do."

When he got back to Illinois, he decided to reach out to a trusted friend, Arnold Nordsieck, a theoretical physicist at the university. "I talked to Arnie one time," says Alpert. "I went out to see him and I told him what my dilemma was, and he said, 'Well, Dan, follow your dream. Don't get seduced by industrial firms. Keep on to your dream.'"

Bitzer and others remember Alpert as a big worrier. The Xerox decision was a big worry. Alpert also brought the offer up to Bitzer, though was reluctant to discuss it deeply. "He agonized for some time over that kind of decision. . . . He talked to a lot of people about things, but he kept his opinions of what he should do pretty close to his own chest. Now, we used to play squash almost every day then. And he'd bring it up and talk about it and all, and I'd say, you don't want to do that." Bitzer was used to the steady stream of job offers Alpert received: mostly high-level administrative positions at other universities. Xerox was a different story altogether, and Bitzer realized this was a good deal.

Goldman tried to sweeten the deal. Says Alpert, "Jack Goldman said, 'Well, come on out, Dan, you'll have a budget, you can take the whole damn PLATO project if that's what you want to do, to Palo Alto.'" Alpert vividly recalls this sweetened deal. "They said you can bring Bitzer here. That's the way they put it, that's the way Jack put it. He knew that I was a key figure over at Illinois, he knew they reported to me."

Decades later, when asked about Goldman's sweetening the offer by inviting Alpert to bring Bitzer and PLATO with him, Bitzer paused, surprised. "I don't think he ever told me that," he said. "And he wouldn't say anything like that if he had made up his mind first. . . . I don't recall Dan talking about moving PLATO out to Xerox, and the reason I don't is he probably kept that pretty close to his self, he made up his own mind. But I do remember the occasion when everybody was afraid he might leave and I thought he was crucial and I knew it was a good offer and so you can't say he didn't want to do that because it's stupid. . . . You could see him agonize but he'd play very low-key, and we'd go and play squash and he'd talk a little bit about it."

There was another factor Alpert would have to weigh in his decision. Bitzer and Propst had formed a company called Education and Information Systems, Inc., or EIS. With the huge PLATO IV system coming down the road, it was clear that a lot of hardware was going to be required for PLATO, and like Owens-Illinois, which had taken on the plasma display manufacturing project, some company was going to have to take on the job of designing and manufacturing the peripherals for the interactive audio and microfiche slide mechanisms for the PLATO IV terminals. EIS was formed to do that. If all went well, and PLATO IV exploded, then EIS—and its owners—stood to do well financially. Dan Alpert had been invited to become an owner of the company and had agreed. (He described the arrangement as "the Troika.") He too saw the potential for making some money should PLATO IV scale up the way they all hoped. Particularly if the system ever got commercialized, perhaps via Control Data. The scale might be huge. Given these considerations, EIS was something he could not ignore. "In that period," says Alpert, "I was the guy to whom PLATO reported. . . . I put a tremendous amount of my energy and spirit and life into that project." If Alpert took the Xerox job, none of these great outcomes might happen, or they might happen without his involvement. If Alpert took PLATO *with* him to Xerox, one can easily imagine interesting consequences to follow.

But it did not happen. In the end, he realized he had too much invested in PLATO at Illinois, plus he didn't like Palevsky, wasn't impressed with SDS, might lose out on EIS if he left Illinois, and the Xerox lab was still just a blue-sky notion with zero staff. It all added up to a feeling that it was too risky to move, and so Alpert turned the offer down. It was now November 1969. The PLATO IV system was right around the corner. EIS and its "Troika" might benefit greatly from the PLATO IV rollout. And with Control Data looming in the background, it was anybody's guess how big PLATO could get.

Jack Goldman would keep Alpert's rejected offer a secret for more than thirty years. No doubt the rejection stung, possibly reflecting poorly on his ability to attract top talent to Xerox and on the wisdom of launching this new lab. He decided next to recruit George Pake at Washington University, his second choice. Goldman found out that Pake had recently been offered, and then turned down, his own former research job at Ford. He called him on Thanksgiving Day of 1969. He did not tell Pake about Alpert. But Pake suspected that someone else had already been offered the job and turned it down. Pake, no dummy, knew exactly who to call.

"I got this phone call after I turned down the Xerox offer," says Alpert, "and first he said, 'I wasn't told your name, I wasn't given your name by anybody, but I put two and two together and I figured you're the only person it could've been.' So, I said, 'Yes.' And he said, 'So, why did you turn it down?'" Alpert gave his reasons, but encouraged Pake to take the job.

The timing was better for Pake than it had been for Alpert. Pake was growing tired of the politics at Washington University. He accepted the offer, and then, like Goldman, kept the story of Alpert's prior rejected offer secret, helping Goldman and Xerox, and Pake himself, save a little face. For more than thirty years he would play along with the alternative story that reporters and the historians of Xerox would establish as gospel. Alpert too kept the secret all that time, until asked about it one day thirty-four years later, in a beautiful home he had later built on that 8,500-foot-high plot of Colorado land surrounded by tall pine trees looming over Snowmass Village far in the valley below.

Pake ultimately decided PARC would be located in Palo Alto instead of New Haven, and began hiring. One of his hires was Bob Taylor, who had moved from ARPA to the University of Utah. Taylor eventually hired Alan Kay, and the rest is history. But one minor detail about Kay's early days at PARC is revealing, in that it shows a lasting impact of that "big whammy" moment when he saw the plasma panel display in Illinois in 1968. "He wanted me to convert," says George Pake, "a big chunk of PARC's budget and resources to research on displays, including panel displays, and I looked at this very hard because a group at Stanford Research Institute was doing research on—I can't remember what the display technology was—and it was sort of being cut loose by SRI and we could have hired them." Pake looked around the landscape and saw the work that PLATO and Owens-Illinois were doing on manufacturing plasma displays, and saw that IBM had also begun a major effort to design and manufacture displays, having licensed the Illinois patent. Japanese companies had also begun work on what would become LCD displays, and later, plasma TVs also based on the Illinois patent. Pake realized that there was plenty going on in the industry already. "We couldn't add much to that," he says. "I told Alan, I appreciated the cathode ray tubes were too bulky and power hungry but I said why don't you go ahead and build a prototype . . . of the Dynabook you wish you had, even though it's not portable. He had already started to do that. . . . So he did come to me to want to invest in flat panel displays and I just thought we didn't have the resources to do that. It turned out retrospectively it was a good decision. We couldn't have got very much farther with the displays that were already being done. And we didn't have a lot of money either. Compared to IBM we were a drop in the bucket."

It is worth taking a moment to ponder what might have happened if Dan Alpert had accepted Jack Goldman's offer, packed up and left Illinois, and joined Xerox. Particularly if Alpert had done so only under the proviso that the PLATO project come along with him. This scenario is not that far-fetched: Alpert, after all, agonized over the decision for weeks and very nearly accepted the offer. Had he done so, very likely the Xerox lab would still have wound up in Palo Alto, Alpert's old stomping ground that, despite the distance of time and career circumstance, he missed dearly. Consider the cascade of historical milestones that might—and, likewise, the milestones that might not—have transpired under this scenario. At the top of the list, Alan Kay would never have joined Xerox. He stated this in no uncertain terms to this author when presented with the Alpert hypothetical, such was the level of his dislike for PLATO. Without Alan Kay, there might not have been a leader within PARC pushing for "guaranteed cycles" at the desktop level, and certainly there would have been no visionary pushing for the Dynabook and all that it entailed. As a consequence, the Alto personal computer would probably have not been made, or if it had been made, it would have lacked all the design details that Kay brought to bear. When asked about this what-if scenario, George Pake remarked, "I think that the whole push toward distributed computing would not have been as strong as it was." The "Personal Dynamic Media" article might not have appeared in publication in 1977. Perhaps most crucially, there probably would have been no reason for Steve

Jobs to visit Xerox PARC in 1979, only to be blown away by what he saw, taking ideas back to Apple, then subsequently coming out with the Lisa desktop computer, followed, famously, by the Macintosh, which changed the world.

What would Silicon Valley look like today had Alpert and Bitzer and PLATO moved there in 1970? It is easy to speculate. What is certain is that much of the fabled history of the Valley might not have transpired, or might have transpired in quite different ways. It is also certain that PLATO itself would have evolved quite differently, particularly due to the lack, in the corporate office parks around Palo Alto, of the unforeseen, monumentally transformative influence and impact of the waves of creative high school kids and college undergrads who would soon wander into the CERL building and fall under the spell of the Orange Glow, as we will see in Part Two of this volume. Finally, in this what-if scenario, it is possible that PLATO would have died just as quick and ignominious a death as SDS soon did under Xerox, a company that would become known for fumbling the future.

There is one PARC veteran who, when presented with this what-if scenario, held a more positive view. Unlike Kay's immediate negative reaction to the notion of Alpert running PARC and introducing PLATO to Silicon Valley, Bob Taylor, the man who had recruited Kay to PARC, says he would have warmly welcomed Alpert. "I would have been *a lot happier* if Dan Alpert had taken the job," says Taylor. "I liked Dan a lot, we were friends, and he would have been *great* to work for. I didn't get along with George Pake *at all*. It was *awful.*"

Alpert stayed on as dean of the Graduate College at Illinois. George Pake made a pilgrimage to CERL shortly after he took the reins of Xerox PARC. "I had heard about it, and I wanted to see what I could see. We were already thinking a little bit about distributed computing, and I was frankly not too keen on operating everything as a terminal off a shared mainframe."

CERL and PARC went their separate ways. PARC would go the way of distributed computing, CERL would stick to the path of the centralized mainframe. Alan Kay had once famously said, "The best way to predict the future is to invent it."

Both labs remained determined to invent very different futures.

—

From time to time the labs would check in on each other over the ensuing years.

By 1972 a Xerox PARC team, led by Alan Kay, had built the Alto computer, a prototype of a "personal" computer with a graphical, mouse-driven, windowing display, running in an object-oriented Smalltalk programming environment. Kay and Pat Suppes's former graduate student Adele Goldberg continued to keep an eye on PLATO. CERL and PARC began several years of collegial, open channels, with each lab inviting the other to send people out and see and learn and exchange ideas.

Kay and Goldberg went to CERL and met with a number of staffers there, including Roger Johnson, with whom, over glasses of beer, they had "tremendous battles," Johnson says, debating the advantages and disadvantages of mouse interfaces versus touch screens, as well as plasma displays versus other, cheaper display solutions. CERL had different, more immediate-use cases for PLATO than PARC had for the Alto. PLATO IV terminals had to be workhorses, withstanding many hours of use per day by many different people, some of whom would not be kind to the terminal. A mouse, stylus, light pen, or other handheld device would mean one more thing that could break or suffer abuse at the hands of users. No, CERL had decided that the simplest user interface for a young child—or any PLATO user, for that matter—was the finger. In 2007 Steve Jobs, long a proponent of using mice, would finally echo the exact same rationale CERL had espoused thirty-five years earlier, when he introduced the revolutionary touch screen–based iPhone: "We're going to use the best pointing device in the world. We're going to use the pointing device that we're all born with. We're born with ten of them. We're going to use our fingers." But in 1972, PARC was not convinced of the benefits of touch screens, and they stuck to using a mouse.

Adele Goldberg found in the touch screen an effective interface for children, at least when doing "gross manipulations" such as touching a picture of an animal and dropping it in a picture of a bucket. But the 16 x 16 grid of touch-sensitive regions on a PLATO IV terminal was far too low-resolution for any detailed drawing or other interactions. PARC was less complimentary on PLATO's graphics capabilities, which led to more arguments between PARC people and CERL staffers like Paul Tenczar.

At one point, Tenczar hopped on his motorcycle and rode out to California to visit PARC. David Frankel, CERL's precocious, teenage systems programmer, flew out to join him, and for two weeks they hung out at PARC playing with Alto computers attempting unsuccessfully to use Smalltalk to create the kinds of interactions that they found easy to create in TUTOR. "Smalltalk was very, very pure," says Frankel. "Smalltalk is very, very basic and you have to build up. Now, you can argue that really that's the core of becoming some sort of object-oriented language where you build a bunch of small objects, and then once you have those you can reuse them and other people can use them and so forth. But it was pretty cumbersome to dive in and get started with that language. Whereas I think a language like TUTOR you could much more quickly in ten commands, *boom boom boom*, have a little lesson that would invoke spell checking and answer judging and all sorts of other fancy things that were buried inside those command subroutines."

The trips continued between the labs, at one point Bruce Sherwood and Rick Blomme visiting PARC. While the visits were cordial, it was impossible to ignore the vast difference between the labs' philosophies and missions. They were simply set up for different reasons and the pace and output of each lab's research reflected those differences. PARC found much of what CERL was doing interesting, and some of the software tools that CERL people had created influenced similar tools that would pop up on the Alta, for example PLATO's character set editor that let you design new fonts. But on their respective system architecture designs—CERL's being thousands of dumb terminals connected to a supercomputer time-sharing mainframe, and PARC's being networked desktop computers with all the power at the desktop—neither lab could fathom why the other lab was doing what it was doing.

"We were explaining to him how it was all gonna go," says Kay. "What the future was going to be like, and Paul Tenczar looked at me and said, 'YOU ARE A MADMAN!' It was great. I mean, he really got upset, incredibly upset."

"David Frankel and I went out there and had an enormous amount of difficulty trying to understand the system," says Tenczar. "There's very little documentation. It was a very open atmosphere out there, very nice, Alan Kay very graciously opened up the whole place to us. We tried to write some lesson materials as we would write on PLATO and it was almost impossible. You had to start off by defining what numbers were, in Smalltalk. You know, one plus one did not equal two; you had to define what one plus one equaled. And well, that's great, when you're a mathematician studying your navel. But it has not much practical importance, when we were interested in doing other things. We were interested in doing biology, foreign languages, stuff like that."

Kay and Goldberg also were not impressed that children could not program on PLATO, that everything was programmed for them. This was Seymour Papert's essential argument: kids should program the computers, not the other way around.

As for their respective choices of programming languages, the two labs could never see eye-to-eye. Tenczar despised Smalltalk. Kay despised TUTOR. "It was always a terrible language design," Kay says. "People who did it like Tenczar never understood how to design a language."

"There was a fair amount of religious hostility," says Bruce Sherwood, "between the people working on PLATO and people who were card carrying professional computer scientists. . . . It was the contrast between the pure aesthete and the technician or engineer getting his hands dirty. We undoubtedly suffered from the fact that we did not have anybody in the group who had anything like a real computer science background. Plenty of very good people on the hardware side who were electrical engineers and computer engineers, and there's no question there. . . . People who were working on the software, on the system software, were people who came from the sciences and had used computers in the sciences for scientific purposes, so were self-taught in the use of computers. But did not know any computer science theory to speak of."

TUTOR did not fit into any model of a proper computer language that a computer science purist would find acceptable. Says Paul Koning, who was a systems programmer at CERL while attending UI, "To borrow a term from linguistics, I'd probably have to call it a 'language isolate.' It's the programming analog of Basque. In other words, there really isn't anything like it that I can think of. Syntactically it's faintly like FORTRAN, which makes historic sense. But only faintly."

Dave Liddle, who had worked at Owens-Illinois to develop graphics drivers for the plasma panel and then moved out to Xerox PARC, believes that a clash between PARC and CERL was inevitable. "In my view," he says, "the spirit of TUTOR that I felt was good, was that it was an attempt to match language elements to the pedagogical needs of the people writing the programs. Thus, it was written in a gentle form of educational jargon that wasn't particularly bad. The problem was . . . it jumped over all the hard lessons of computer science. So as a result, it actually did contain a number of things that we don't put into programming languages that are done by people who are computer scientists. . . . The usual problem is that computer scientists per se, sit on their ass, and don't work on the interesting applications, so somebody else who doesn't know these things comes along and does it. For example, HTTP and HTML make all the same mistakes, because they were done by a desperate physicist who couldn't wait any longer, and they weren't done by a computer scientist. So, guess what, was that important, or what? Does everybody like it and use it and has it changed the world? Yes. It still has all these problems in it, because if computer scientists had focused on applications more, somebody would have done this right. And the same thing's true for TUTOR. So you have to look at TUTOR and say, well, at least these people got off their butt, and built this thing, and it was learnable, and so people learned it. As any kind of a language design purist, you have objections that you can raise. But as a practical matter, it did get done. And it worked. And people who learned it didn't know they weren't supposed to dislike it, they just learned it, and it worked fine. . . . Smalltalk didn't have any of that stuff. I mean, Smalltalk was carefully language-designed . . . so it avoided all those things, so it's crystal-clear why you couldn't have a more dialectical combination of Paul Tenczar and say Alan Kay and Adele Goldberg."

Years later, Tenczar would admit that there were aspects that he and David Frankel saw at PARC, like the mice and windowing displays and other desktop graphical interface features—the very things that blew Steve Jobs away when he saw them in 1979—that should have had a lot more impact on them (and in theory on PLATO) but did not. "I think that David and I really missed what was going on at Xerox," he says. "I think we both missed it. Maybe it was more my fault than his, in that I was the senior person in PLATO."

10

Lessons Learned

To make way for the transition from PLATO III to PLATO IV, the aging CDC 1604 would be phased out and replaced by a genuine supercomputer, a Control Data 6400. In addition to the new mainframe, there would be many new disk drives—each in that era the size of washing machines—as well as racks of communications and networking equipment, peripheral processing units, and special air-conditioning equipment to keep everything from overheating. Add to all that the arrival of PLATO IV terminals from Magnavox, sporting the brand-new Owens-Illinois–made plasma displays, plus all the wiring and phone connections. PLATO IV was far beyond PLATO III. CERL was for all intents and purposes starting over, finally deploying at the large, meaningful scale it had long dreamed of. A scale orders of magnitude beyond anything Suppes and Atkinson were doing out at Stanford, or MITRE and BYU planned to do with TICCIT, or IBM hoped to do with the 1500 system. PLATO IV's scale called for 4,096 terminals connected all over the campus, the state, and the country, if not the world. And that was just the CERL system. There were bound to be other PLATO systems in the future.

Monumental efforts were under way not only on the hardware front, but with system software and utility programs, much of which were going to have to be rewritten from their PLATO III counterparts. Countless new programs would also have to be written, including an editor for authors to program their TUTOR lessons that accommodated the larger 512 x 512 plasma display. Initially, much of the system-level programs that ran PLATO were written in COMPASS, CDC's assembly language, making it cumbersome and difficult to make changes. CERL decided to rewrite existing system programs in TUTOR itself, by extending the TUTOR language to have privileged "system commands" that could do dangerous things like read and write directly to memory or disk, log users in or out, etc. It would take months, years even, to move everything over to TUTOR, but the effort was worth it. By creating system programs in TUTOR, operators and systems staff could interact far more productively with PLATO by taking advantage of all of the features TUTOR offered for friendlier displays, listing out information, and drawing lines and graphs, for instance.

All of the functionality of PLATO IV did not appear overnight, however, and the early days tended to be bumpy for users. (CERL would not establish formal "prime-time" hours of service until 1974.) Ruth Chabay, an author of chemistry lessons, tried to run a class of students in room 203b on the second floor of CERL. Room 203b was the "author's room," but was also for a while the only room full of the new PLATO IV terminals. It could be booked by a department for its students to take assigned lessons. "Trying to run a class in room 203b," she says, "was a challenge worthy of an Olympic athlete and a career diplomat. My appearance in the room was invariably greeted with groans from the authors working there, who knew they would be asked to leave. Getting students into the correct lesson meant helping each of them type the file name—there were no student records, routers, and of course no restarts. This meant that when, halfway through the class session, the systems staff decided to reload [reboot the mainframe] to try out a change . . . I'd have to sprint up the stairs to the computer room, and come panting in the door prepared to offer cookies, pizza, or other bribes to persuade them to keep the system up until the class ended in half an hour. Then I'd dash back down to 203 to throw out the persistent young authors who had snuck back into the classroom. Of course, the system invariably did crash."

Chabay had to resort to adding a special code in her chemistry lessons that students could enter using the TERM key, so that should the system crash, when it came back up they could more or less return to where they'd left off. "Between the desperate dashes up the stairs and the negotiations with disgruntled authors, one was glad that classes were usually scheduled for only an hour."

Nineteen seventy-two was a turbulent year on the campus of the University of Illinois. Much of that turbulence was due to a report, written by assistant provost Barry Munitz, which laid out sweeping recommendations for restructuring the administration of the university. The heady days of the 1960s were over. Richard Nixon was in the White House. Belts were being tightened everywhere. Munitz urged the university to modernize its organizational chart, optimizing communications and productivity by pruning wasteful and redundant reporting structures. Munitz had been brought in by John E. Corbally Jr., who had taken over as president of the university when David Dodds Henry, long an enthusiastic supporter of Alpert, Bitzer, and PLATO, stepped down the year before. Unfortunately for Dan Alpert, the Munitz report recommended gutting the Graduate College, its functions and areas of responsibility to be divvied up by existing and new administrative posts, one of which would eventually be manned by Munitz himself. In addition, a senior faculty committee had been formed to consider plans for campus reorganization, and unfortunately for Alpert, members of the committee did not see eye-to-eye with him or his way of running the Graduate College. It was as if all the karma from ripping PLATO out of CSL to form CERL to keep PLATO under Alpert's wing, as well as summarily dismissing Larry Stolurow and his SOCRATES project, were now coming back to haunt him. PLATO IV had just received its millions of dollars from the National Science Foundation, and yet Alpert would not be overseeing Bitzer and CERL any longer.

"Dan was very supportive of what Don Bitzer was doing," says George Russell. "I guess not everybody was, but Dan hung in there, he was a good man and he hung in there, and kept it on track. So Dan is owed a lot of credit for fighting a good battle for Don Bitzer." Despite that long track record, resulting in massive NSF funding for PLATO IV, the university's politics, long tolerating if not openly supportive of the effort, could no longer be counted on. "Dan Alpert's more industrial research orientation," says Jack Desmond, "finally got him into trouble as dean of the Graduate College, because even though he was enamored of the hard science community, he was somewhat critical of other aspects of the university's complexion. He became kind of hard to deal with, irascible, judgmental, outspoken about the quality of university research, and finally, some senior people asked that he be removed because he was not 'one of us.'"

All through the spring and into the summer of 1972, *The Daily Illini* reported new scoops on the Munitz revolution under way on the campus. Alpert, it was reported in May 1972, was out as dean of the Graduate College, and he announced he would resign on September 1, on which date he would start two new posts. George Russell would take over as acting dean of a Graduate College that would soon be dismantled. Alpert was offered a consolation prize of sorts, directorship of the Center for Advanced Study, in no uncertain terms a demotion from which Alpert arguably never recovered. In addition Bitzer offered Alpert, still eager to oversee the success of the PLATO project, the role of associate director of CERL—*reporting to Bitzer*—wherein he would contribute to long-term strategy and oversee the rollout of PLATO IV in community colleges around the Chicago area, one of the key components of the NSF proposal. Alpert accepted. Bitzer was no longer Alpert's protégé. He was finally on his own. "He was clearly," says Desmond, "*clearly*, captain of his own ship." Jack Peltason, then university chancellor, asked Russell to become vice chancellor for research, a role that largely overlapped the work Alpert had done at the Graduate College. Bitzer and CERL from then on reported to Russell.

"It is certainly true," says Desmond of Alpert, "that he had angered many people, but they gave him a very soft landing . . . the directorship of the Center for Advanced Study, which was a nice showplace, but anybody who occupied that position had very little power."

From Richard C. Anderson's vantage point, Alpert and Bitzer had run the PLATO project "in an entrepreneurial way. . . . The university had a tough time covering all the IOUs that were out there for the CERL staff, that Dan Alpert had granted rather indiscriminately in the flush days. Though Dan continued to be an honored member of the university community, they got him the hell out of administration. . . . It was fine when things were flush, or if not fine, they say, 'Well, this is the cost of being the great pioneers in this, we got to do some things that are outside the envelope.' But when we came on harder times, that was a problem for the university."

It had only been three years earlier, in the summer of 1969, that Alpert had been heavily courted by Jack Goldman and Xerox to head up Xerox's new research lab. He'd even been offered to take PLATO with him as an enticement. He'd turned all that down, counting on things continuing to progress in a positive direction at the University of Illinois, in terms of his own career and in terms of PLATO. This was not the outcome Alpert had hoped for. Suddenly, the Xerox deal looked a lot more attractive in hindsight. He would regret not taking that offer for the rest of his life.

—

From the outset, live demonstrations of PLATO had been crucial to the success of the project. By the 1970s, PLATO demos at CERL were so common that they became part of the daily culture of the place. To be at CERL was to be called upon at any moment to give a demo—or sometimes to *be* the demo. Don Bitzer, or one of the other higher-ups, might suddenly appear with someone important by his side, asking you to show his guest a thing or two. You'd be there typing away at your keyboard and you'd suddenly notice people looming over your shoulder. *Don't mind us, we're just going to watch what you're doing there. . . . Now that's interesting, how'd you do that? . . . Can you show us that again? . . .* and you realized only then that you'd been roped into giving a demo. There was no way out. It was the price you paid for the privilege of using PLATO.

At any given moment would appear some entourage from another university, or executives from some American company, or Japanese company (sometimes a literal *busload* of Japanese industrialists), or representatives from government agencies, be they American or from a foreign nation. People would appear, scheduled or unscheduled; if scheduled, expecting a tour, in which they'd stream through the various halls of the frumpy old building, peek in a room or two, stop by one of the more presentable rooms where demos were held, meet a few of the staff, and sometimes make it all the way up to the penthouse on the fifth floor, where many of the "junior systems programmers" stayed up all night, working on some of the more exotic projects, and they'd be asked to give yet more demos. Roger Johnson recalls one time he was suddenly asked by Bitzer to lead a group of Japanese industrialists around the lab. They had shown up with cameras, appeared not to understand a word of English, and, to Johnson's dismay, dispersed to various floors of the building on their own. Johnson found some of them snapping photos of every piece of equipment in the upstairs labs where the plasma display was being created.

"I tell ya, something magical about Don Bitzer was that he had *infinite* cool," says Sherwin Gooch. "He would bring people around to give these demos, but unlike anyplace else I've ever worked, he never told us about them and he never warned us. He never [said], 'This is important because NSF's going to be here' and 'We got to get our next thirty million dollars'—he never, ever, *ever* did anything like that. He just would bring people in and show them what was really going on and so it never interfered with our progress." There was one price the staff had to pay for that "infinite cool," however, says Gooch: "He almost never told us who these visitors were that he brought in." Sometimes there was good reason for not identifying who the visitors were. On some occasions the suits that walked down the hallway and sequestered themselves in the conference room were very shy about having their presence known in the building. Those meetings led to two very secretive groups signing contracts for PLATO terminals to dial into the CERL PLATO system in the mid-1970s. One was the National Security Agency and the other was the Central Intelligence Agency. Even forty years after the fact, the few CERL staffers who even knew about their existence remain hard-pressed to acknowledge that the NSA and CIA were on PLATO. Garrie Burr, one of the CERL field technicians, remembers visiting Fort Meade from time to time. "It was almost impossible to get on the facility," he says, "but once you got on the facility . . . well, you had somebody with you even going to the bathroom, I mean there was somebody *always* with you." Oddly, he says, when you left, nobody checked your car. "I mean, you could have walked off with the store. Nobody checked your car! Only problem was getting in. Once you were in, getting out was no problem." Both agencies had a small number of terminals—the CIA's were believed to be installed at the Warrenton Training Facility, though no one would confirm or deny. It is presumed that their main use was for developing online language training. The general assumption was that by the 1980s, in addition to a number of separate military and nonmilitary government-run PLATO systems, both the NSA and CIA had their own systems, for use in computer-based training. "I had gone to a few of those sites," says Jim Ghesquiere, a CERL staffer during the 1970s who then worked at Control Data, "that are typically off-limits, most people aren't aware of a lot of the government sites. Control Data had many people who were on-site on a government site all the time. And, they would only come maybe once or twice a year to the Control Data building, say for a project kick-off meeting or something like that. And they spent 100 percent of their time working at various government, military, intelligence sites, and they could never talk about what they were doing. I know there was a lot of linguistics work done with the PLATO system at several of the government sites."

Ironically, a CERL technician responsible for maintenance and repair of the NSA's terminals at Fort Meade once admitted that the only time he could ever recall a PLATO terminal being stolen anywhere in the U.S., in all the years he worked at CERL, was the time one was stolen right off the loading dock of the NSA. "I can remember it getting stolen," says Bitzer. "It was stolen from NSA. . . . I thought that was really funny that the terminal disappeared from NSA."

The parade of CERL visitors never stopped. You could wave your hands and draw figures on the board and talk about PLATO to visitors all you wanted, but only when people saw it *firsthand*, caught a glimpse of the Orange Glow coming from a PLATO terminal's screen, watched as they reached out and actually *touched* the screen and saw how the act of physically touching the orange screen caused something to happen, and then invited the visitor to reach out themselves and touch the screen too, and witness the machine reacting to being touched it had a visceral, "holy crap is this the future or what" reaction.

Demos happened all over the world. In the 1960s most of the demos had been domestic. From presentations at major academic conferences to Rotary Club affairs in some small town, it didn't seem to matter: Bitzer made time to be there. As the government funding and the support from Control Data grew, so did the ambition to expose the world to the Friendly Orange Glow. Bitzer, and sometimes a complete entourage, including Maryann and occasionally son David in tow, would hop on a plane and show off PLATO in France, Italy, Romania, Russia, Sweden, South Africa, the United Kingdom, Australia, Belgium, or Holland.

Practitioners from other institutions attempting to show off their own computer creations were often frustrated with PLATO, because the crowds instantly gravitated to the Friendly Orange Glow. John Seely Brown, who in the early 1970s at the University of Michigan was developing an "intelligent tutoring system" called SOPHIE (short for "SOPHisticated Instructional Environment") that used aspects of artificial intelligence and natural-language processing to enhance the dialogue between the electronic tutor and the human student, vividly remembers PLATO stealing SOPHIE's thunder. "I had progressed in the first six months of the SOPHIE contract and I had to go out to Lowry Air Force Base to have a contract review. And, there was a PLATO terminal. And on the one hand we were there giving a demo on some faiirrrrrrly *provocative* uses of intelligence—providing a completely new kind of electronic learning environment—and somebody came in with a PLATO terminal and turned it on and drew a picture of a rose, and I was blown away how people would just gravitate to looking at that rose on this PLATO terminal versus what I thought of course was the world's most significant breakthrough in terms of how to embed intelligence into the learning environment, into the tutoring environment. That was my first exposure to PLATO."

"There was always a standing joke," says CERL technician Garrie Burr, "that Don would demonstrate the system to *anybody*. Even the janitor who was sweeping the floor when we were setting up. He'd give the same kind of demo to the janitor that he'd give to the executives he was going to meet the next day. They all got the same show. Don said more than once, 'You never know where your next nickel's comin' from.'"

Paul Tenczar believes PLATO represented the state of the art for educational computing throughout the 1970s. To wander the halls of CERL during this era gave one the impression that the staff certainly believed it. "Working with Stan Smith," says Ruth Chabay, "I was based in the Chemistry Building, but every day we'd walk over to CERL, and go up to the second floor, where, in the early days, the authors were all working . . . and just walk down the second-floor hall, it was not a very long hall. . . . It could take three hours to walk down the hallway, because you got into so many conversations and arguments and debates and saw so many exciting things and you discussed whether this was a good approach or not—the environment was amazingly intense. And so I think that a lot of that stuff happened informally, there was a tremendous sense of excitement—we learned to seek each other out both online and offline just to see what was going on."

Chabay was a graduate student in chemical physics, used to working with computers as number-crunching devices, not as automatic teachers. If a visitor receiving a PLATO IV demo had had any exposure (outside PLATO III) to computers in education up to the early 1970s, it was typically in the form of teletypes. "Interactivity was using a paper teletype," Chabay says, "and typing characters on a teletype and seeing it slowly chunk back, and the kind of educational use of computing that we all knew about at the time was drill and practice on teletypes where the teletype would print '3 + 5 =' and then the student had to press a key, saying 8." To then get a demo of the PLATO IV system was to experience Toffler's *Future Shock* up close and personal. "To most people at the time," Chabay says, "this stuff seemed like stuff from outer space. When you'd demonstrate this, people sort of didn't believe you some of the time."

Children and teens, on the other hand, had no problem using the system. They took to it like ducks to water. Unlike today, with the proliferation of digital devices everywhere and most people, including kids, having extensive experience with online services, apps, games, productivity tools, social media, and so on, computers were not a part of daily life in the 1970s, particularly for schoolchildren but neither for college students. The constant assumption that a designer of an educational PLATO lesson had to maintain was that the student using the lesson *may have never sat in front of a computer before.* Most likely, this was a first-time experience for the student, regardless of age.

One outcome of all the demonstrations was that a lot of people got to see PLATO. By 1972, PLATO had been a fixture in one form or another on the University of Illinois campus for a dozen years. But its footprint was small. With the arrival of hundreds of PLATO IV terminals with their high-resolution, graphical, orange-glowing screens, there was reason for professors and department heads to check out PLATO all over again. Many did. What they discovered was a state-of-the-art system whose capability seemed endless, and an eager and helpful laboratory in CERL, proud of what they had built and enthusiastic about getting the system out into the world. Professors were invited, encouraged, to design their own lessons for their students, however they wished. The idea, dosed with a heap of wishful thinking, was that professors—without requiring any prior computer experience—could take advantage of a powerful authoring language in TUTOR (with more tools on the way all the time) and design instructional material that fit in with their teaching style, their syllabus, their approach, and their needs. PLATO and TUTOR amounted to a theory of "if we build it, they will come." Luckily for CERL, many professors showed up and gave the system a try.

What follows is a brief overview of some of the ways PLATO was used to teach people from first grade through college:

Elementary Mathematics

When NSF funded PLATO IV in 1972, the money was earmarked not only for hardware and system software, but for lesson development and delivery to live students in K–12 and college settings. One area of concentration was elementary mathematics. For this, CERL managed to attract a nationally known educator, Robert Davis, an MIT math PhD who had moved to Syracuse University to run the Madison Project, a recipient of significant NDEA funding after Sputnik. John E. Corbally, the new UI president, recommended to CERL that Davis head up the elementary math project. Davis brought with him Don Cohen and Jerry Glynn, who had both worked with him on the Madison Project. Davis would have his plate full: in addition to running the elementary math project at CERL, he became an associate director of CERL, joined the education department as a professor, and even signed on as principal over at Uni High. For the elementary math project at CERL he lined up formidable talent, including research assistants Sharon Dugdale, Dave Kibbey, and Barry Cohen, who formed the fractions curriculum team, and Charles Weaver, Esther Steinberg, and Bonnie Seiler (née Anderson) for the whole numbers curriculum team. Cohen and Glynn were new to computers, says Seiler, "but they very much knew his philosophy of *math is a story of the real world*, and so we kind of adapted some of those things to PLATO and some of that worked and didn't work so well. But one thing I remember: [Davis] said, for the first year, just to *try things*. He didn't say, 'We're going to make a curriculum for X grade or X topic,' we just played and we saw what we could do and couldn't do and then after the first year, some decisions were made like, 'We're going to do third grade to sixth.'" Says Kibbey, "We started as programmers, but Bob's approach was more like '*Start creating.*' The atmosphere was, 'Here's this new thing, what can we *do* with it? What's possible to do on this?' Instead of 'Here's a bunch of lessons that we're going to program.'"

They began learning TUTOR on this new thing called PLATO IV, and realized there was a lot that they could do with it. It was the summer of 1972, and many regular staffers disappeared for vacation, leaving plenty of terminals free on the second floor of CERL and a minimum of distracting meetings. They referred to a UICSM math textbook for inspiration, but often created their own lessons in reaction to something they felt was too boring in the textbook. For example, there was a recipe activity, where kids would make cookies using a half cup of this and a quarter cup of that. "Making cookies is not very exciting," says Kibbey. "It's much more interesting to make monsters." That gave them an idea for a Make-a-Monster lesson, using the graphics capabilities of TUTOR, the plasma display, and the touch panel. The textbook also mentioned making and slicing pies, which they transformed into pizzas, which became another lesson: how to slice up enough pieces of pizza so everyone got a piece.

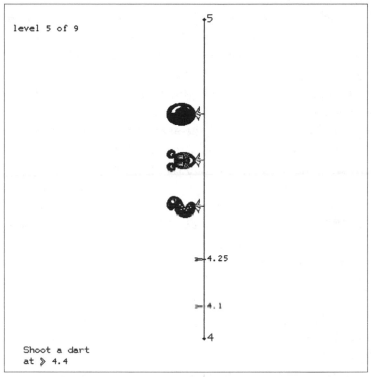

Darts lesson

One of the fractions group's early lessons was Darts, designed by Sharon Dugdale, which would become a perennial kids' favorite for years to come. Darts dealt with fractions. Whole numbers were commonly taught by teachers everywhere and all schools considered them important. Fractions, on the other hand, tended to receive less attention and were ineffectively taught. "Fractions," Dugdale says, "was an area that people seem to consider universally problematic and difficult to teach." When Davis was doing the Madison Project at Syracuse, he had fostered a new approach for teaching fractions using coordinate geometry and number lines. Up until this time teachers around the country were not typically teaching fractions using this approach. Davis had offered the fractions curriculum to Dugdale and Kibbey after finding reluctance from other researchers, who held that fourth graders lacked the "mathematical maturity" to handle fractions, so they were sometimes left until eighth-grade algebra. "That would have set the U.S. way back," says Kibbey. Dugdale was keen to get kids in earlier grades exposed to fractions, so that they were better equipped to enter junior high.

Darts reflected this new thinking. On the screen was a vertical number line acting as the wall onto which PLATO threw darts. At various points up and down the wall were balloons of various shapes. The goal of the game was to tell PLATO where to throw the dart at the various points on the number line in order to precisely hit the balloons. Each level of the game presented a different number line scale (it might be -1 through 1, or 0 through 5) and a different placement, number, and even size, of balloons. As a student got more proficient at the game, the levels got harder and the balloons got smaller and more numerous. The student had to figure out where exactly a balloon was situated on the number line. If it was between, say, 1 and 2, was it "1½"? Was it "1¼"? Thanks to the TUTOR language's flexible answer judging (a powerful system capability that set PLATO above competing computer-based education systems not only at the time, but largely forty years later), students could write "1½" or "1¼" or they could use decimals and write "1.5" or "1.25." They could even enter expressions, like "2+⅔," "½-.03," and "(5+5/6)/2-1/10." The lesson kept track of the student's balloon hits and misses, and moved them up the levels of difficulty as they got better.

Torpedo, another lesson focused on similar fractions problems, presented a situation where a student could play against other students or play against PLATO. The player operated a submarine deep in the ocean, above which swam occasional fish, octopi, and other creatures, and at the surface was a ship. The ocean surface served as the number line, this time horizontal, and the player needed to move their sub backward (by entering the desired negative amount, be it an integer or a fraction) or forward, and then the sub would fire a torpedo upward in an effort to hit the enemy ship. If some creature were in-between, it might get hit by the torpedo instead. The game resembled the popular video arcade game *Space Invaders* that would come out years later, although with *Space Invaders* there was no need to know anything about fractions; players simply moved a joystick left or right and fired away. Some schools had hoped to assign strict, prescribed PLATO time for the kids: six minutes here, twelve minutes there. What they found instead were kids playing Torpedo and Darts for hours on end, being forced to get off the machine by the janitors who were closing up the classrooms for the night.

Over on the whole-numbers strand, Bonnie Seiler developed a series of lessons that exercised a child's ability to creatively mix addition, subtraction, and multiplication operations to arrive at a desired number. Her most famous lesson was West, or How the West Was One + Three X Four: A Race Between the Wells Fargo Stagecoach and the Union Pacific Railroad, in which a stagecoach races a train from Tombstone to Death Valley to Dodge, Dry Gulch, Kansas City, Santa Fe, Urbana, and finally Red-Gulch on a track that zigs and zags its way down the PLATO screen. Seiler had originally designed the game to be a race between a horse and a camel, from one oasis to another in the desert. Then she discovered that Dave Andersen, one of CERL's systems programmers, had created a little train graphic as an alternative character set, which gave her the idea of racing a train versus a stagecoach. Then the idea for the Wild West came to her, and finally, the full name of the game.

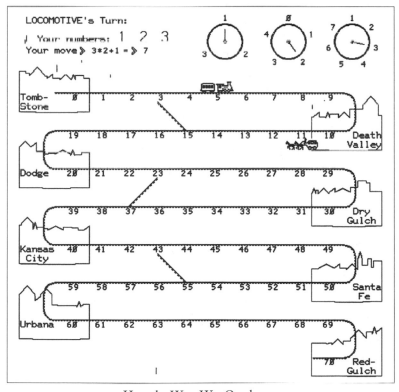

How the West Was One lesson

The child could play against another player or play against PLATO. The lesson had three number wheels, the first offering 1, 2, and 3, the second offering 0, 1, 2, 3, and 4, and the third offering 1, 2, 3, 4, 5, 6, and 7. Each dial in the three wheels would randomly turn to point to the value for that turn, and then the child had to include those three numbers in the best mathematical expression they could think of that would achieve the optimal value, which would then move their piece, be it the train or the stagecoach. For example, if the number wheels gave you 3, 0, and 4 on your first turn, it was probably best to do a simple expression like "3*4+0," which moved you ahead 12 spaces. If PLATO rolled 1, 4, and 6 on the wheels, you could type "6*(4+1)" to get 30, a big move. The child could have typed "6+4*1" but that would only yield 10. Kids needed to understand mathematical orders of operations and the ideal use of parentheses. But even if a child figured out an expression that resulted in a big number, that wasn't always the ideal move. Sometimes it might be better to calculate a smaller value, because there were shortcuts between towns that jumped you ahead past the next town. Also, if you landed right on a town, you were automatically jumped to the next town.

How the West Was One was another perfect example of how the designers working for Robert Davis in the math project took his mantra "Math is a story of the real world" to heart.

Elementary Reading

NSF also funded an elementary reading curriculum, so CERL formed a team to design and develop those lessons. The team included John Risken, who headed up the project, plus Bob Yeager, Priscilla Obertino (Priscilla Corielle), Lezlie Fillman, and TUTOR programmers Tom Schaefges, Brian Shankman, and John David Eisenberg.

Yeager was finishing up a master's degree in 1971 in Chicago, having come from six years of classroom teaching a few years earlier only to grow fascinated by the potential for computers in education. One day a mentor urged him to check out what was happening with PLATO down at the University of Illinois. He went down to Urbana, got a graduate assistantship at the College of Education, found CERL, and got a demo of PLATO III. "My jaw just about dropped," says Yeager. The excitement, ambition, and technical wizardry he saw around CERL, not to mention the immense potential of PLATO IV coming around the bend, had him sold. In short order he changed his career plans, quit his College of Education assistantship, and joined CERL's elementary reading program.

Early on, Bob Davis, who headed up the math program, managed to recruit Seymour Papert from MIT to come out to Uni High and also hang out at CERL to let the elementary math and reading teams pick his brain. Papert ran occasional seminars out of a small house near Uni High. "There'd only be about a dozen of us or so," says Yeager, "we'd sit on the floor and Papert would expound." Papert described the LOGO project and what he'd learned from that so far. But what influenced Yeager the most was hearing Papert's overall philosophy. "Don't try to manipulate the student, let them manipulate the computer. And that's where Papert was at."

Yeager had left one of Papert's seminars inspired but struggling to think of a way to apply his philosophy to the elementary reading project. The idea came to him in the shower the next morning. "I was trying to figure out, how do I do what Papert does, in elementary reading? And I came up with the idea of Sentences."

Sentences was a PLATO lesson that presented a series of sixteen

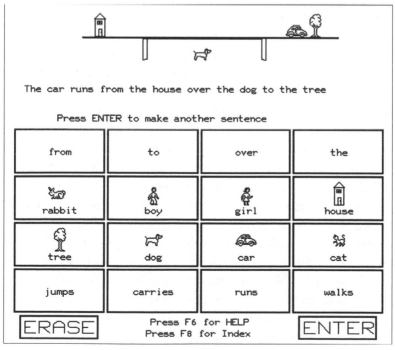

Sentences lesson

words and their corresponding icons inside four rows of touch boxes on the lower half of the screen. There were eight nouns: *tree, car, girl, rabbit, boy, house, dog,* and *cat;* four verbs: *walks, jumps, runs,* and *carries;* and four other useful words: *over, from, to,* and *the.* By touching the words in a certain way, a student could form sentences, which—if they made sense after being evaluated by the TUTOR code—were then animated on the screen. For example, "The boy carries the cat from the tree to the girl." The lesson would evaluate the structure of the sentence, determining if it was a valid sentence or not (if not, the lesson would simply say "Sorry, I do not understand that"), and then drew pictures of the selected nouns "doing" the action indicated in the sentence, in this case, plotting a picture of little tree on the left, with a picture of a boy carrying a cat above himself, slowly walking over to a little picture of a girl on the right. On top of that, the lesson activated an EIS-built random-access audio device to "speak" the words while the animation was going, so that the student could read the text of the just-invented sentence, visualize the action through animation, and hear the sentence spoken back, as a combined multimedia extravaganza providing Immediate Feedback to the student. Things got interesting when students used the word "over" in their sentences. The lesson handled sentences like "the girl walks over the house to the car" not by having the girl literally walk over a house, but rather, by PLATO drawing a road with a *bridge* over a house, then animated a girl moving from the left, across the bridge, over to the right, where she stopped in front of a car.

The elementary reading team made extensive use of multimedia in their lessons. They discovered, as other authoring groups in other disciplines did, that the random-access audio device was woefully inaccurate and prone to playing the wrong audio clip, if it played one at all. The solution, crude but somewhat effective, was to always make sure there were multiple copies of the audio clip adjacent to each other recorded on the big fifteen-inch floppy disk that was placed inside the machine. That way, when a TUTOR command asked for clip number 1234, if the machine gave you 1233 or 1235 by mistake, you greatly reduced the chance of playing the wrong audio to the student.

Yeager had started out as a staunch behaviorist in college, but by the time he got to the UI's College of Education to pursue a PhD, the cognitive science approach was beginning to take hold with some of the professors there, particularly Richard C. Anderson. Anderson's thinking, plus Papert's, led Yeager to "flip," he says, to the "cognitive side." Yeager felt that the elementary reading project team was still largely behaviorist in orientation, and was concerned that they weren't incorporating enough cognitive ideas into their lesson designs. "I was the only person who was moving into the cognitive world." In particular, Yeager was concerned with a strand of lessons covering phonics, which he felt were not up to par with other lessons in the curriculum.

"We had a much more challenging goal," says Risken, "that we had set for ourselves by trying to go to the very beginning of the learning-to-read process, than the elementary math people did, who aimed their stuff at grades 4-5-6 and assumed a certain level of arithmetic skill to start with. They probably achieved their goals much more cleanly and clearly than we achieved ours."

December 1975 was the end of the first semester of the "demonstration year" for NSF and the Educational Testing Service (ETS). Testing revealed that the kids utilizing the elementary reading PLATO lessons were reading at a far lower level than the kids who were in the control classes. "The PLATO lessons were actually having a *negative* impact upon the kids," says Yeager. Shortly afterward, John Risken left to join Control Data in Minneapolis. NSF sent another team out for another project checkup, but Yeager wanted nothing to do with the meeting, so Priscilla Corielle ran it. "They all went away happy that we were going to be in good shape," Yeager says. Yeager, Corielle, and company worked to fix what they believed were deficiencies in the lessons. "We changed things around," says Yeager, "and at the end of the year we sort of broke even."

Chemistry

One day around 1963, Stan Smith, a chemistry professor, stopped by to speak to Bitzer about creating a PLATO lesson in chemistry depicting the rotation of a molecule on-screen. Bitzer sent him to Lezlie Fillman, who made drawings of a molecule in various positions that would each be photographed and loaded as slides in PLATO III's slide selector, to then be played back in rapid sequence to animate the rotation of the molecule for students. She and Smith worked for a long time on making that sequence. It was the first of many lessons Smith would undertake. As the authoring process improved over time, particularly with the arrival of TUTOR, Smith was able to create more lessons for his students. Among the more famous was Titrate, a lesson that gave a student a fairly complete simulation, within all of the limits imposed by the PLATO terminal, of a laboratory experience in which the student would titrate some chemicals (titration being the gradual addition of one chemical solution into a volume of another solution until the mix changes color or otherwise reaches chemical neutralization, in order to determine the concentration of the unknown solution). It is a common college-level lab experiment, and is expensive and messy to roll out to hundreds of undergraduates in large lab settings. Far easier to simulate the entire thing online, and Stan Smith was just the professor to do it. His lessons would be used by thirty-five years' worth of UI undergraduates, on PLATO III through PLATO IV and onward to microcomputers and the Web. Smith epitomized in the late 1960s throughout the 1990s the fearless, enthusiastic academic pioneer who cared deeply about developing compelling interactive simulations and tutorials for students in college settings. His work was often featured around the world in PLATO demos at conferences and other

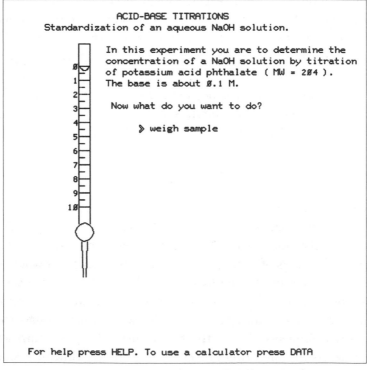

Titrate lesson

venues, and yet for all the recognition he achieved, UI's publish-or-perish chemistry department did not view his work equal to research or publishing peer-reviewed articles, and it made it difficult for him to get tenure. Other departments on campus held similar biases. George Grimes, a professor of veterinary medicine who developed numerous lessons on PLATO, ran into similar problems. He worked day and night for six years on PLATO, but never got faculty support. "I had been promised a promotion," he says, but "those on the committee that decides on promotions had decided this was sort of a mechanical thing that didn't deserve the significance of the research that was going on in veterinary areas."

Bob Yeager has a theory why Stan Smith's work was so good. "Good teachers like Stan Smith used cognitive strategies even if they didn't know it or it wasn't labeled that way. They taught content, but they also taught structure of the content (the mental models that were the earliest of the cognitive contributions). And they probably also taught a passion for their subject matter—something I'm not sure any theory accounts for. When Stan created his PLATO lessons, all that went into them and that's why they were good. They were behavioral, cognitive, constructivist, and things that still don't have a label. They were good teaching."

Biology

Perhaps the most famous lesson ever developed on PLATO was one by David C. Eades, Gary Hyatt, and Paul Tenczar on fruit flies. It had first been done on PLATO III, and was later redone for PLATO IV's higher-quality plasma displays. As Stan Smith had discovered for chemistry, these biologists realized that PLATO was exceptionally efficient at exposing a large population of students to scientific concepts that traditionally required time-consuming, expensive laboratory time but could now be simulated on the computer. In this case, the concepts concerned the genetics of fruit flies and how certain traits are passed down or lost in subsequent generations. By mating certain genetic traits in male and female flies, one could observe the results instantly, whereas it took time in a lab. In fact you could continue crossing different genes over and over again to see the outcomes. But what made the lesson such a spectacle—particularly in demonstrations to funding agencies—were the graphics. Once you had specified the types of traits you wanted to breed in a generation of flies, the lesson then showed you all of the offspring from that mating. Some would have no wings, some might have different eyes or other subtle traits, but students could see them clearly as multiple rows of fruit fly pictures cascaded down the screen.

It is also noteworthy that, as in Stan Smith's chemistry lessons, these simulations did not simply provide a series of choices on the screen at each step in the experiments. Rather, the authors of these lessons simply displayed a statement like "What would you like to do?" with an -arrow- command awaiting user input. The student would be expected to know what to do at each step, such as "see the flies" or "do a cross." Stan Smith had similar prompts in his simulations, asking the student what to do next, without explicitly stating what the choices were.

Music

G. David Peters had received a master's degree in music education from UI in 1965 and then had gone off to join the Houston Symphony and teach music in Houston. In 1968, he received a UI alumni newsletter that had an article in it about the PLATO III system. Peters had been thinking about pursuing a PhD, and when he saw the article and photos, he was intrigued, and decided to pursue his doctoral studies back at Illinois. He received his PhD in music education and computer-based instruction in 1974. He also headed up the PLATO Music Project within the School of Music. Unlike some other UI departments, where just a few professors fully embraced PLATO, learning TUTOR, and programming their own lessons while the department itself tended to look the other way, UI's school of music was very supportive of PLATO and of professors seeking to use it. By 1977 over thirty professors and more than a dozen graduate teaching assistants were using PLATO in their music courses, in which over two thousand students were enrolled. Those courses extended over a vast range of topics including music acoustics, music theory, ear training, tests and measurements, statistics and music, percussion terminology, behavior modification for classroom training, music fundamentals, vocal diction, music methods, trumpet techniques, conducting, composition, electronic music, clarinet performance techniques, and instrumental methods and research. "It was pervasive at that point in time," says Peters.

Despite the broad support for PLATO within the School of Music,

Peters did find that spending a large amount of one's career on the system came at a cost. Like PLATO-using professors in other departments, Peters found that his chances at tenure were in jeopardy if all he did was work on PLATO. "I was racing around doing all this work, and I had some of the senior faculty in music education, they just said, 'Look, this is great, you did some great work, *but*, it's not going to get you tenure. If you want to get tenure, you have to write a book.'" So Peters went and wrote a textbook, twenty music compositions, and other publications. Of the School of Music's tenure committee, Peters says, "I don't know if they didn't value PLATO, they just totally didn't *get* it, they didn't understand it, and they couldn't—you know, a book, you can pick it up, you can weigh it and say, 'Well that's about heavy enough,' but with PLATO, they just couldn't get their mind around whether it was important or not. So I would not have gotten tenure at Illinois if I hadn't written a textbook and some other publications."

Basic Skills for Prison Inmates

The PLATO Corrections Project (PCP) began in 1974 after personnel from the Illinois Department of Corrections came to CERL for a demo. "They were very interested," says Marty Siegel, "in doing a prison project with some federal funds from the Department of Justice. I told Frank Propst that if that project came into being, I was really interested in it, and he said, 'Well, why don't you write a proposal.'" Siegel wrote the proposal, it got funded, and the PLATO Prison Project was begun. Soon it was renamed PLATO Corrections Project (PCP) to better reflect the philosophical change taking place nationwide during the 1970s that prisons were correctional facilities.

PCP developed hundreds of lessons, mainly in basic skills, as well as an early computer-managed instruction (CMI) system, what would be called a "learning management system" today. "An instructor would actually set up a record," says Siegel, "and assign a curriculum sequence to students. Eventually we moved into a kind of mastery-based learning system, where the instruction would be divided into instructional units, bounded with pre-tests and post-tests, and so it was quite a sophisticated system."

The PCP project revealed interesting insights into the use of PLATO by its user population. "You couldn't find," says Siegel, "a more disadvantaged, disenfranchised, turned-off, uneducated group of people than the kind of people you found in prisons. And as you can imagine, survival in a prison means maintaining a kind of tough-guy image. If you are thought of as weak, bad things are likely to happen to you. . . . And so typically what happens is, a lot of people will not even opt for educational classes because that's seen as weak. Or if you're sort of required to sit in an educational class in a prison, you're likely to try to misbehave or be the class clown, or act in some way that shows your disinterest. Not because you really are, but because you can't afford to be wrong in front of your peers. So when the teacher says, 'Where does the comma go in this sentence?' and you don't know the answer to that, you say something that indicates you don't need to—that that's irrelevant, or those aren't the words that someone would say, but it would be the equivalent of that sort of blowing it off. You would blow off the task in front of your fellow classmates. That's the sort of environment that we were stepping into."

Siegel's team had a lot of concern that these valuable PLATO IV terminals and their keyboards and other equipment would be stolen, or disassembled, or otherwise destroyed. "There was great concern about security," says Siegel, "because we knew that movie projectors or slide projectors or any kind of audiovisual equipment was usually taken apart or lost or stolen and broken in some way." The concerns turned out to be unfounded. There was instead what Siegel calls a "most remarkable unexpected outcome." In all the time that PLATO terminals were installed in the prisons, CERL never lost even a single plastic keycap on the keyboard. "Never, ever was there any damage to any of the equipment," says Siegel. "There was more equipment damage to the classrooms at the University of Illinois, with bright college students, than in the prison environment. We heard from the inmates that the word went out really early that this was good stuff, and you don't mess with it."

Jim Knoke, one of CERL's technicians charged with maintaining terminals at remote sites, found that while the hardware was not damaged per se, there was vandalism. Clever vandalism. "We weren't in maximum security prisons, we were in minimum and medium, a little bit scary at first," says Knoke. "We had to be very careful about tools." The PCP lessons did not make use of the microfiche slide selector, so those mechanisms had all been removed from the terminals installed at prisons. What Knoke and company had not removed were the *mirrors*. Mirrors were prized possessions by inmates, not only to help an inmate see down the halls outside their cells, but to provide an inmate with shiv weapons when the mirror was broken into pieces. PLATO IV terminals had a mirror designed to reflect the projected image from a microfiche slide into and through the back of the plasma display, so a student would see the image behind any overlaid text and graphics. "Somehow these guys would get the mirrors out, they would steal them," says Knoke. "We did have to lock down things, padlock them, so they couldn't take them apart."

The inmates discovered not only the advantages of Self-Pacing and Immediate Feedback, but the fact that they were free—in the middle of a prison where there is no freedom or privacy to speak of—to learn, privately, at their own pace, and without fear of ridicule or threats of bodily harm or worse. The computer provided a way to learn that they were not used to. No tough-guy act was required, nor would PLATO have even known how tough a guy you were. You could answer a question and be told you were wrong and why you were wrong, and it was okay. You could answer a question and be told you were right, and that was okay too. PLATO provided a safe space for learning.

Siegel wanted his PCP courseware developers to visit the prisons to better understand who they were creating courseware for. "I actually took them to prison sites," he says, "where they could sit down and see the look on an inmate's face as the system crashed or as a lesson crashed, and to see the student say, 'Hey, I trusted this thing and now I'm not going to trust it anymore.' For some of these people it was the last time they were going to take a risk on learning, on education. It had great impact on the programmers." One of those programmers was Tim Halvorsen. "That experience still stands out in my mind," says Halvorsen. "We'd show up, and I'm talking machine guns on the towers . . . apparently somebody had just stabbed somebody a couple hours before, and there was a lockup. . . . We'd be installing the software and setting up the terminals . . . and I talked to this one person who was in the lab there doing work and he'd say, he'd just blurt out, 'Hi there, my name's so and so, I've killed somebody.' That was his introduction. And I'd just be looking at him like, holy mackerel."

The original idea of PCP was to develop some courseware for inmates and then deploy it at a handful of Illinois prisons. The Department of Justice funding was expected to last only a few years. It had originally come from the administrations of Nixon and Gerald Ford. When Jimmy Carter was elected, the Democrats looked around for funding to cut, and the PCP project was put on the chopping block. "They saw these kinds of projects as pet projects of the Republican administration," says Siegel, "and so there was a freeze on all of these projects. An indiscriminate freeze, I mean they didn't look at 'Is this a good project or a bad project'—it was a Republican project: bad idea. And we were shut down." The state of Illinois sued the government in federal court, but despite testimony that wowed the judge on the success that PCP had made, they lost. Things might have ended right then, but there was another surprise. "The state government was so impressed with what we had done," says Siegel, "and the success that we had with the inmates, the effect that it had, that they decided after they lost the lawsuit . . . they said, 'We will *fully* support the project 100 percent out of state funds.' And from that point on, until the project ended, which was about '87–'88, it was fully funded by the state of Illinois, usually at around two hundred thousand or so dollars a year. So it was quite a bit of investment on the part of the state, and I think a testimony to [its] success."

When asked for an example of a successful use of PLATO with kids during the 1970s, Priscilla Corielle immediately recalled one story she termed a "remarkable success." "It was with a group of kids that were affectionately by their caretakers called 'the weird kids.' This was a state-run residential facility for children with, oh goodness, emotional, cognitive, behavioral, other deep, deep problems, that didn't fall into commonly accepted categories at the time, and we're talking again forty years ago, there might have been autistic kids there—who knows what these poor little kids were dealing with."

The people who ran this facility were interested in PLATO, and after some discussions with CERL, a PLATO IV terminal was installed there. Corielle had developed lessons as part of the elementary reading program, and went out to see how the kids would perform. These were kids who, Corielle says, "had such a terrible time, for a very good reason in many cases, interacting with humans. For example, some of these kids had been abused by their parents in ways that I can't even repeat. So these were really terribly, terribly damaged little people. They were adorable and sweet and lovable, but boy they had been clobbered. So, anyway, there was one little girl that was dealing with the very simple PLATO reading activity. It involves two little girls of, you know, cartoonish characters on the screen: one of the girls is happy, one of the girls is sad, and I think the child was supposed to touch the word 'sad' and that would make the word go up into the sentence and fill it in and it would match the picture. Well, this little girl not only touched it, touched the word correctly, but she said, 'The girl is sad.' She uttered those words."

This little girl had been diagnosed as aphasic, with Down's syndrome. The institution believed she was unable to speak. Doctors had recommended to the family that she be institutionalized. But in front of a PLATO terminal, she spoke. "The family was not a cruel or abandoning family," says Corielle, "they did visit this little person, this little girl, from time to time. But basically the upshot was, she learned the culture of being a Down's syndrome kid, I mean, severely Down's syndrome—she never had models of people talking."

Corielle and her colleagues told the teachers at the residential facility that this little girl was not aphasic. "They said, 'Oh, yeah, she'll never talk,'" says Corielle. "And we showed them that she could talk. It turned out that the child was then tested, it turned out that she had somewhat above-average intelligence and that she did not belong where she was. I'm not sure if the family took her back permanently, I know there were increased visits. But . . . it had an upside in measure potential that was recognized and then cultivated. It had a downside in that acquiring language led her to understand what had happened to her. And that was an emotional blow to the child. I don't know the end of the story. I trust the end of the story is a happy end."

Part II

THE FUN THEY HAD

Now we have the Screen, and it rules.
Our kids are perma-plugged into its promise, admiring all its
  jewels.

—Kate Tempest

Orange dots beckon. They welcome with temptation, power, and
promise. All 262,000 of them reach out with their unearthly
glow.

—P. Gregory Springer

The computer display will be mankind's new home.

—Ted Nelson

11

Impeachment

There is a creek, Boneyard is its name, that runs west to east across the University of Illinois campus, through the Engineering Quad, just south of CERL. It acts as a natural hyphen connecting the two towns of Champaign and Urbana. For years, mere mention of Boneyard to locals triggered strong responses. Hardly a creek, they would tell you: a notorious, no-good, toxic trickle of stinky sewage, more like. Says Marc Andreessen, in the early 1990s a UI undergrad creating the Mosaic web browser, and now a very successful venture capitalist, "I was always worried it was generating invisible fumes that were going to kill me."

The Boneyard, environmentalists insisted, played a key role as an important Champaign County watershed. Past Urbana, it flowed east into one of numerous tributaries that eventually fed into the Wabash River, which in turn dumped into the Ohio River on its way to the mighty Mississippi. That all may be, Boneyard detractors would respond, but on too many days the Boneyard smelled less like a watershed and more like an outhouse. The creek might once perhaps have been pleasant to look at, even safe to drink from, back when Chief Shemauger of the Pottawatomie Indian tribe was born, in the 1700s, so the story goes, under a hickory tree that grew alongside the Boneyard's banks. Then again, less idyllic stories tell of Indians hanging their dead in tree branches directly above the creek, so the bones would eventually fall into the creek and be washed clean and carried downstream. (More recent campus wags say the creek got its name from the bones of engineering majors who flunked out of UI's challenging undergraduate curriculum.)

With the 1820s arrival of white settlers and the construction of cabins and buildings, the creek entered a new era, one marked by a prolonged battle between man and nature. In time, the outposts of

Champaign and Urbana became villages, and with the eventual arrival of the railroad, the villages grew to be intertwined towns. The University of Illinois opened in 1867. More than a century of population growth and the accompanying boom in houses, streets, shops, businesses, parks, and campus buildings gradually obliterated the natural banks of Boneyard Creek. The banks would be replaced by brick and concrete retaining walls, or tunnels, or bridges, or sometimes simply chain-link fences with barbed wire. As the towns grew, so grew the creek's troubles with pollution, odors, rodents, and, in the warmer months, mosquitoes. Boneyard would become an eyesore and an open sewer. For most of the year, the creek never got more than a foot deep, except during those magnificent Midwestern thunderstorms, when the pounding rain drove the creek to rise quickly and flood parts of town, especially around the university campus. For decades environmentalists argued that it was a key component of the Illinois ecology and advocated its restoration, while at the same time others pushed to shut the creek down, fill it up with dirt, and pave it over once and for all. In 2000, some $25 million in public funds were finally put to use: a large section of the creek was cleaned up and diverted underground. In more recent years, additional sections of the creek have been cleaned up and the surrounding landscape beautified, including where it crosses the engineering campus. But this is one ornery creek that never stops finding ways to drive the public, and the city legislators, crazy. Open a newspaper in Champaign-Urbana even today and there's a good chance of finding yet another story about whether or not to continue "improving" the creek.

Boneyard became a hot topic in PLATO circles in 1970. Two University of Illinois graduate students, Stuart Umpleby and Valarie Lamont, were looking for ways to push the PLATO III system beyond mere education. Lamont had arrived at the university in 1968, to pursue a doctorate in political science. Umpleby already had a bachelor's in mechanical engineering and a master's in political science at UI, and was in 1969 pursuing a PhD in communications. During his undergraduate years he'd become something of a protégé of Dr. Charles Osgood's, a psychology professor widely known for his invention of the "semantic differential" (a method of evaluating a person's subjective understanding of the connotative meanings of words, using pairs of bipolar adjectives like "good-bad," "soft-hard," and "valuable-worthless"). Outside of work, Osgood actively opposed the Vietnam

War, as did Umpleby, who managed to push Osgood and other university figures to write letters to the military authorities, citing Umpleby's
valuable and ongoing contributions, and requesting an "occupational
deferment" from the Selective Service System so Umpleby might avoid
being drafted and sent to Vietnam.

Computers intrigued Osgood, and he advocated the discussion and
exploration of alternative futures where war had no place. He thought
there might be ways to use a computer to aid in this exploration. That
led him and Umpleby to PLATO, where they tested out some of their
ideas. One was a game on PLATO III called Delphi. "Delphi enables
individuals to shape—within limits—future social and scientific developments," a press release at the time announced. "The future shaped
by the 'explorer,' or player, is determined partly by his own subjec
tive 'investments,' partly by the relationships between developments
and partly by the conditional probabilities of events happening by the
year 2000. These last two factors are built into the computer program."
The game was designed to get people to think about the future, and
how going down one path to develop a new technology or implement
some social policy might impact other outcomes.

In 1970, Umpleby encouraged Lamont to check out PLATO as well.
They decided to create a program to raise people's awareness about
Boneyard Creek. What would happen, they wondered, if a program
were written on PLATO and presented to local community members—
none of them technical—enabling them to explore multiple possible
scenarios for dealing with Boneyard? The program would present the
scenarios to help the community members in their decision-making
process. Lamont and Umpleby would evaluate how well or poorly the
program did, and draw conclusions regarding the potential for future
computer applications involving community issues and decision making at the local level. It was the kind of project that today gets launched
on the World Wide Web every hour across the world by local community, activist, and political groups, to educate the public, advocate a
certain position, and, if all goes well, get people to change their minds
(and maybe donate money in support). That Umpleby and Lamont
were exploring using a computer network to do this in 1970, decades
before the Web even existed, speaks to their prescience, but as pioneers
in technology have so often found, their naïveté combined with being
first did not translate to success.

Lamont taught herself the rudiments of TUTOR programming and created a lesson called Creek. Creek may be the first case of a "non-lesson lesson" on PLATO, in that it was not a drill and practice, simulation, or interactive tutorial presentation; rather, it was not unlike an online survey or series of informational web pages today. Its intended audience was not *students* at all, but the local public, particularly local politicians, city administrators, and environmental activists. From one perspective, Creek was like an early "Electronic Town Hall," a concept that would itself become a buzzword in another ten, twenty years. Creek resembled a very simple website: throw together a bunch of informational pages online, some with text, some with text and photos, and toss in some questions and branching at the end of the sequence. Behind the scenes, collect some data and run analytics on how people responded and what paths they took through the program. If all went well, the user and the creators of Creek would both learn something (this was Bitzer's "computer-based education," after all). Lamont and Umpleby hoped users would read about the Boneyard and its pollution, see photos of it, get a summary of the environmental issues surrounding it, and then make a decision about the best possible environmental and community outcomes for the creek. Lamont summarized the project in a newsletter she and Umpleby put out during this time:

> One of the most important activities of those concerned with the future has been an attempt to generate public interest and involvement in future-oriented activities. The common methods used have been conferences, travelling lecturers, the promotion of community study groups, and the introduction of future courses in colleges and universities. In addition, there are numerous magazines, newsletters, and journals devoted to articles on the future.
>
> Despite sincere efforts, the vast majority of the population remains outside the realm of future-oriented thinking except possibly as spectators. In order to more actively involve large segments of the population in considering and originating alternatives, it may be necessary to invent new communications media or apply existing technology in innovative ways. There are a number of new communication technologies now coming onto the scene—cable television, communications satellites, video cassettes, and the teaching computer—which offer unique opportunities. The most important of these technologies may well be

those which present the possibility of establishing direct, two-way communication with the populace.

"It was really a different twist on the programs that were being developed from an educational standpoint," says Lamont, "because basically what Stuart and I were saying was that this technology had applications beyond traditionally designed education."

Using CERL stationery and the university's media facilities, Umpleby and Lamont sent out a press release inviting local dignitaries, including the mayor, the city council, state senators, and the local media, to come and try out the program. It wasn't the normal sort of press release the CERL lab would issue, and it's not clear if Umpleby and Lamont first checked with the powers-that-be at CERL for permission. Nevertheless, the press release went out. Not everyone they'd invited showed up, but some did, and the feedback was mixed:

"This is Big Brother!"

"This puts too much control into the hands of the people!"

"This program can't possibly do justice to all the alternatives surrounding this issue."

The turnout was thin. CERL staffers were not amused. Some went so far as to view Creek as a potential PR nightmare for the laboratory. They were aghast that Umpleby and Lamont had the nerve to use such a weak application to demonstrate PLATO—especially to important local government officials. People would get the wrong impression and Creek wasn't even educational. PLATO could do so much more, and didn't those two realize that CERL hadn't received funding from NSF yet and this might put PLATO IV at risk?

It didn't take much to rouse the curmudgeonly systems programmer Rick Blomme, but Creek especially annoyed him. "The Creek lesson was the biggest joke of all time," he said. As far as Blomme was concerned, Creek had no place on PLATO and made PLATO look bad. It should have been a documentary film or slide show shown at City Hall.

"I agree with him," Umpleby would admit of Blomme's harsh critique, many years later. "It could have been done with just an ordinary 35mm slide show. But all I can say is that it was an experiment."

Creek did wind up being used for one unintended educational purpose: Judy Sherwood of the PSO consulting group, CERL's online support team for PLATO authors and instructors, would cite it as one of PLATO's lesser moments. "Umpleby was an unmitigated ass,"

Sherwood would post online in 1977, in a series of heated messages remembering the Boneyard lesson. She regretted that Creek wasn't available on the PLATO IV system, as it would have continued to serve as an example of What Not To Do. Bruce Sherwood was also annoyed with Umpleby's "chasing after big-name people," concerned that it "would have done us all serious damage if any of those invitees had taken him seriously, because Creek was really infantile. It's not a question of someone being a good programmer or not, it is a question of caring about quality and being able to discriminate between good and bad work."

George Carter, who'd been a UI student in 1973, defended the efforts of Umpleby and Lamont. "The invitations to 'big names' was simply youthful exuberance and should be interpreted as such," he wrote in rebuttal. "I am sure anyone of any importance recognizes this. It could not have harmed the PLATO effort." It was not the last time Umpleby would find himself embroiled in controversy and concern about harming the PLATO effort.

In 2006, Lamont and Umpleby (and PLATO) having long gone from Illinois, someone launched a Boneyard website, containing everything you ever wanted to know about the creek. As of 2017 there are now multiple Boneyard websites, run by a number of community organizations and local governments. On these sites, one can see photographs and maps of the creek, read about Boneyard's background and history, its ecological importance as an Illinois watershed, the ongoing work being done to beautify and improve it, and find out ways to support and take further action, including participating in annual Boneyard Creek Community Days: precisely the kinds of uses of computer networks that Umpleby and Lamont had in mind, nearly fifty years ago, before anyone else, even the PLATO people, grasped what was going on.

Despite ruffling many feathers at CERL, Umpleby and Lamont managed, by 1972, to scrounge up a grant of $26,110 from the National Science Foundation to investigate "the use of PLATO," yet another press release announced, "to create a new mass communications medium for the discussion of long range community planning. . . . PLATO terminals will present information on a local issue. Participants in the electronic town meeting will have some control over the flow of information. They will be able to skip sections or have some part of the lesson repeated.

"The purpose of the research," Umpleby explained in the press release, "will be to find the most effective ways of presenting issues on the new medium. We will also try to determine whether patterns of communication and methods of making community decisions change as a result of the availability of the new communications medium."

Umpleby and Lamont giving PLATO demo, circa 1972

By March 1973 Umpleby was penning articles in local papers, advocating for the "community use" of PLATO, fearing that soon the system was going to be swallowed up by corporations. "Try to think how you would like this system to be used," he wrote. "In a few years PLATO is going to be big. But so far the majority of programs in the computer are just normal course material—there doesn't seem to be any movement toward general citizen use. As this new federally funded resource comes into existence citizens should challenge any trend toward exclusive use. It should be a public resource. We must compete for a say in its use and set an early precedent of using it to serve our interests as members of the community."

He soon figured out a new way to set an early precedent.

By the summer of 1973, Umpleby had taken to testing out FORUM, a new, experimental conferencing system created at the Palo Alto–based Institute for the Future. FORUM, a project funded by ARPA and NSF, ran on the new ARPANET network, the precursor to today's Internet, and was designed to work on a variety of computers: from teletype machines connected directly to the network to dial-up terminals connected from virtually anywhere. Earlier in the year he'd started experimenting with a simple conferencing program on PLATO called

Discuss, written by George Carter. Discuss enabled people to post up to ten lines of text as a message, to which others could then read and, if they chose to, respond with their own one- to ten-line messages. If you needed more than ten lines to make your statement, you had to post the first ten lines in a message, then reply to your own message with another, and so on. While other groups of PLATO users were discovering and quickly becoming addicted to live-chat applications (covered in Chapter 14), Umpleby was more interested in message boards. Live chat was *synchronous*—you literally had to be there, participating with the other participants in real time. Message boards, forums, conferencing systems were *asynchronous*—you could post your thoughts at 2 a.m. and not expect, nor care, when others would eventually see them, let alone post replies. Even if it was hours or days later before replies appeared, that was fine. Forums and email work the same today.

ARPANET's FORUM message board content could not be shared on PLATO. Likewise, Discuss content on PLATO could not be shared on FORUM. Thus, participants in the two systems did not see messages posted on the other system. By 1973 the ARPA network and PLATO network reached sites all over the country, but the networks remained isolated from each other, one of the great tragedies in PLATO's history. "They were both coast to coast," recalls Umpleby, "but they didn't overlap very much. The only places they overlapped, as I recall, were Urbana and MIT." "Overlapping," in this context, simply meant that there were PLATO and ARPANET terminals located in the same geographic sites, sometimes in the same room—but again, the connections on these foreign networks were physically separate, ships passing in the night. The closest PLATO and ARPANET would come was inside CERL, oddly enough, which happened to have an ARPANET terminal and a PLATO terminal both sitting along the same wall, like two televisions permanently tuned to different channels.

Umpleby, savvy to PR and maximum coverage, offered to transfer, by hand if necessary, postings from PLATO's Discuss over into FORUM, so ARPA users would see them there. Likewise, he'd retype FORUM postings into PLATO's Discuss. "There was a labor-intensive connection," he says. "It was just an offhand proposal, and it was assuming very low traffic. Back in those days there were many systems and they didn't overlap much, it was not at all like the current [Internet]."

In the summer and fall of 1973, the Watergate scandal was the national preoccupation. New, ever-more-shocking revelations emerged each day. Take, for instance, October 20, 1973, a date that quickly became known in the media as the "Saturday Night Massacre." At 10:58 p.m. that night, Umpleby opened up a new topic in Discuss:

> A news bulletin tonight reported that Nixon had fired Special Prosecutor Cox. Attorney General Richardson resigned. Deputy Attorney General Ruckelshaus then became acting attorney general. Nixon gave Ruckelshaus an order which he refused, so Nixon then fired Ruckelshaus. That made the solicitor general the acting attorney general. These events seem certain to produce a serious move for impeachment in the House with the only reservation being the war in the Middle East.

What started out as a fairly factual opening statement suddenly veered in a different direction. Umpleby continued with a series of paragraphs offering a long set of citations linking the Watergate scandal to the JFK assassination. He finished with a question:

> What do the participants in this discussion think about all of this: the firing of Cox, the chances of impeachment, and the possibility of connections between Watergate and political assassinations?

Umpleby's post was no different, in format if not in content, from any typical online message board posting over the past forty years. Except in 1973, this kind of computer-mediated group conversation was exotic and new. Very few people in the world had any idea what it would mean to have a "discussion" with other people online on a *computer*. Yet note how even at the very dawn of computer conferencing, conspiracy theories were alive and kicking.

A variety of PLATO users, mostly students working at or around CERL, read Umpleby's post and responded into the night and the next day. "The firing of Cox was a bad move," said John David Eisenberg in the first response that night. "Any connection of Watergate and political assassinations is at best very tenuous and, quite frankly, a highly dangerous item of discussion." George Carter responded: "Nixon also ordered the FBI to seal off Cox's offices, presumably the storage location of the evidence Cox and his now defunct eighty-man staff had been collecting on Watergate. Note that this puts Nixon in possession of the evidence against him."

Umpleby was back the next day, asking Eisenberg, "How do you mean dangerous? Surely not knowing what is going on is more dangerous than knowing." And so it went the rest of the day and into the night, with more postings about Nixon and Watergate and impeachment.

Leonard Lurie's book *The Impeachment of Richard M. Nixon* had just come out. Umpleby read it and decided to post a summary of Lurie's arguments on Discuss. But he also decided to see how he could use PLATO and ARPANET not just to debate the issues, but to accelerate the impeachment process by consolidating the political power of geographically separate activist groups. With PLATO and ARPANET, he imagined reaching hundreds, perhaps in time thousands of people. Online was the inevitable future for activism.

"The Undergraduate Student Association on the Urbana campus is engaging in a major effort to lobby Congress in favor of impeachment," Umpleby posted in Discuss on Monday, October 22. "They have been circulating petitions in the student union, dorms, etc. They have set up tables where students can write letters to their congressmen. Tuesday night there will be a meeting in Lincoln Hall to organize further actions—to take the campaign off campus and into shopping areas. What are other campuses doing on impeachment? Would people on other campuses please tell your student government and student press that these programs exist and can be used to coordinate actions if necessary or to pass around bright ideas."

On October 24 on ARPANET, Umpleby posted an update to the FORUM participants:

> Saturday night after the big announcements, we started up a conference on PLATO about impeachment. One lesson space filled up and we are now using a second. People from MIT, Chicago, Ames Iowa, and Ft. Wayne Indiana signed in and we tried to use the program to coordinate campus actions. However, there is the problem that the people with ready access to terminals are not always the student activists. We have been working on conferencing among campuses for almost two and a half months now [and] are disappointed that it is not going better.

He posted again on ARPANET later that day. "Our conferencing activities are proceeding steadily and advancing slowly," he said. "We are trying to interest national activist groups to use the PLATO net for communication among local chapters that have terminals available. We are now working on Common Cause, The Federation of American Scientists, peace researchers, and cyberneticists and general systems researchers. Any other suggestions of likely groups?"

It didn't take long before someone in ARPA's office at the Pentagon came across the FORUM postings on ARPANET—after all, they were funding not only ARPANET itself, but also this experimental FORUM project. It wasn't long after that that someone up the chain of command in the Pentagon read them. And it didn't take long for the brass in the Pentagon to hear about it. Nor did it take long for word to cross the Potomac and reach the people in Nixon's besieged White House. It was right around this time that the White House was making efforts, on Nixon's orders, to cancel FCC applications by *Washington Post*–owned television stations in retaliation for their reporting of Watergate. This was not a good time for Nixon's people to learn of talk of impeachment and political mobilization going on over the wires of some damn Defense Department–funded computer network out in the cornfields of Illinois. Keep in mind that this was an era when "political mobilization" and activism on college campuses was taken very seriously: buildings were often taken over, demonstrations often turned violent, campuses were sometimes completely shut down, and sometimes people were killed, as at Kent State University and Jackson State College just a few years earlier. It is certainly possible to imagine that some of these concerns might have entered the minds of the government officials. If the Nixon administration had no misgivings about abusing its governmental power to go after activists, political mobilization, and even *The Washington Post* for its coverage of Watergate, it does not require a stretch of imagination to see how they'd have no problem shutting down obscure, government-funded computer networks over which anti-Nixon opinions were being expressed. The White House's reaction to this is significant in that this may be the first time in history, and it would certainly not be the last, that a government threatened to shut down people communicating over a computer network because it did not like what they were saying.

Don Bitzer, as usual, was on the road, but it didn't take long for the Feds to find him. It was now November 2. He was about to leave his Philadelphia hotel room and head over to address graduates at the

Moore School of Electrical Engineering at the University of Pennsylvania, but just as he reached for the door, the phone rang. What follows is how he remembers what happened next.

> NSF (National Science Foundation): We just got a call from Nixon's office. The White House essentially said to us that our money would disappear if this goes ahead, and we know you'll disappear and uh . . . we're concerned. What are you going to do about it?
>
> BITZER: Well, I'm going to have a live terminal with me online when I go and give my talk, and I'll look it up and see what's going on. What I'll do is as follows: If this conversation that they're planning is the kind of thing that would be a productive conversation in a classroom, remember you have funded this for education, that would be acceptable as classroom discussion in political science, it will go on. If it is a political rally, which is not what we're supposed to be doing, then I will tell them that they either have to change the topics, change what they're doing, or stop it. And that will be my decision.
>
> NSF: Fine. [click]

Bitzer hung up, but the phone immediately rang again. Now it was ARPA at the Pentagon.

> ARPA: We just got a phone call from the White House. Says that you're one of the sponsors of this. Boy, *you're* in trouble and *we're* in trouble if this goes through.
>
> BITZER: Yes, the NSF and ARPA had poured millions of dollars into PLATO, and, yes, the NSF and ARPA were divisions of the executive branch of the government. Yes, it might seem like there was a PLATO effort to bring down the head of the executive branch of government whose NSF and ARPA divisions were funding PLATO. I haven't seen the actual postings yet, and will check them out and make a decision after reading them.
>
> ARPA: We think what you're doing is right, and we think Nixon's wrong. You go ahead and do it. Unless you think it's wrong, you go ahead and do it.
>
> BITZER: Well, I'm going to look at it. [click]

Bitzer went over to the Moore School, logged into PLATO, and went into Discuss to see what all the fuss was about. He then made a long-distance phone call to Illinois. That same day, as a result of that phone call, word got to Umpleby, who posted a new note in Discuss:

About noon today Bruce Sherwood told me that he had had a phone call from Don Bitzer, Director of CERL. Apparently the Pentagon tracked down Don at a conference he was attending in Philadelphia and asked him about a comment made by me in program FORUM on the ARPA network. The Pentagon, it seems, was concerned that a facility of the executive branch (the ARPA net) was being used to organize an impeachment effort. Don's position is that any discussion that might take place in a classroom will be permitted on PLATO. However, use of the system for political organizing will not be allowed as long as the system is a university facility. When such systems become common carriers, like the telephone lines, then they could be used for any purpose. This sounded like a reasonable view, so I told Bruce we would cooperate.

One of the things Bitzer and Sherwood asked Umpleby to do was add a disclaimer to the opening screen users would see as they entered Discuss. That night a page was added:

PLATO in its present implementation within the University of Illinois is essentially an extension of the classroom. While discussion of current topics is as legitimate on PLATO as in the classroom, it is not permissible in the classroom or on PLATO to organize political mobilization. For this reason, CERL cannot at this time permit the use of the PLATO system for organizing political activities. This note prepared by Bruce Sherwood for Donald Bitzer.

That night, Umpleby was back in Discuss with more:

Following conversations with Bruce in the afternoon and Don this evening, I have deleted those comments which called for other political uses of the system, and I ask that other participants similarly restrain themselves. Comments on this definition of the permissible uses of the PLATO system might be of general interest at this time.

The next day, Saturday, November 3, Umpleby updated the Discuss participants with more news:

> The situation is apparently more serious than I thought yesterday afternoon. The Institute for the Future's program forum on the ARPA network is no longer available. . . . Continuation of their work seems to be endangered. . . . It is hard to believe that a few comments in one program could cause such a reaction. What is also interesting, however, is that apparently on the basis of only one comment, the Pentagon understood the importance of computer-based communications media. Months and even years of talking and attempting to persuade social scientists had produced at best indifference. Such differences in reaction testify far more eloquently than a scientific article why those who are the establishment are there and why social science has been so ineffective.

That same afternoon Don Bitzer, back home from Philadelphia, chimed in online with a posting of his own:

> More discussion is needed in order to decide just what constitutes open discussion with academic freedom and still stays within the extension of the classroom guidelines I've described. Let me try to explain the circumstances which brought on the problem. First the guideline that U of I Plato [sic] is to be used as an open classroom is not new. In fact, that guideline was used, to explain to ARPA—(A generous and enthusiastic supporter of Plato)—that it could only program lesson material also usable in our community college program. Because educational problems in the all-volunteer army overlap significantly with educational problems in community colleges, ARPA has been able to enthusiastically support Plato; even in view of the Mansfield Amendment. Second, when ARPA contacted me on November 2, they were concerned, but brought no pressure. Clearly, not all groups were living under the doctrine that ARPA was given and they were helping support Plato. ARPA's suspicion was not aroused by what

they saw on Plato, but what was contained in a message attributed to Umpleby on the ARPA network. This message in part called for help in mobilizing the impeachment and the use of Plato to aid in such mobilization. I determined that the use of Plato for political mobilization was not appropriate since Plato is part of the U of I educational system, and take sole responsibility for the policy statement at the beginning of each Discuss lesson. I have frequently been informed of the political dangers programs such as Discuss might thrust upon the Plato program. Nevertheless, I have felt that the potential benefits of open discussion on any topic outweigh these dangers and have permitted the use of Plato for these activities even at this early and fragile development stage of Plato. Misuse of Plato will certainly surely set back efforts to make Plato a beneficial tool to aid open discussion. Let me hear from all of you on this subject.—the real dlb.

Participants were quick to thank Bitzer for stepping in and taking a stand. "Professor Bitzer deserves considerable praise for allowing programs like this to exist," said one user. "The response to Pentagon 'suggestions' by other University of Illinois administrators would have almost certainly been to simply delete the program to avoid a hassle, especially one with a funding agency."

Nevertheless, the controversy cost Umpleby his access to ARPANET. "We didn't have an NSF grant, so they couldn't fire me. All they did was change the passwords so I couldn't get on again. I never lost access to PLATO. But a lot of people on the PLATO system became very angry at me, because they thought the Pentagon was going to shut down the PLATO system and we'd all be out of work."

The trouble for Umpleby was not over yet.

"There were two flaps," says Umpleby. "There was the flap when it happened, and then there was the publicity flap. Now, see, I never tried to get any publicity for this during the fall semester when it happened. But then there was a guy named Craig Decker who was at MIT, he was a friend of mine, he was passing through campus, and, you know, he was doing the same thing everybody does, he said, 'Well, hi, what are you up to, what have you been working on,' picked up a few publications, and I said, 'What are you doing,' and he was just about ready to leave and he says, 'Has anything else happened of interest?' And I said, 'Well, we did have a little bit of a flap with the Pentagon a few months ago.' And he says, 'Oh, tell me about it.' So, I told him about it, and he says, 'That's really very interesting, because you know I have a friend at *Businessweek* and I think he might be interested in this story, do you mind if he calls you?' And I said no.

"Now, at the time, I was a believer in talking to the press. The idea of, it's important for the people to know, and the Jeffersonian thing about the value of a free press, and so I said, 'Okay, I'll talk to him.' So this guy from *Businessweek* calls."

An article entitled "No Computer Talk on Impeachment" appeared in *Businessweek* on March 16, 1974:

> The Pentagon is picking a fight with top universities by refus- ing to let its Advanced Research Projects Agency computer net- work be used for a study of impeachment. The network, located at civilian campuses but funded by the Pentagon, is being adapted to provide a nationwide "teleconferencing" system. By using a designated code number, scholars could tap in at any time to con- tribute data and ideas to an on-going conference on almost any subject.
>
> Some subjects apparently are taboo. Political science profes- sor Stuart Umpleby of the University of Illinois says that the contractor for the network, Institute for the Future, turned him down when he proposed a study of impeachment and now is bar- ring him from any access. "They were just scared they'd lose the contract and knuckled under to anything the Pentagon said," he charges.

A writer for *Science* magazine, Dan Greenberg, saw the article. In addi- tion to his work for *Science*, he wrote his own newsletter, from which articles were occasionally reprinted elsewhere in such publications as *Change* or *New Scientist*. Greenberg called Umpleby and they spoke. Soon, an article was published in *Science in Government Report*. "I thought he really missed the point, and I was kind of irritated with it," says Umpleby. "I wrote a letter saying that we weren't using PLATO for political organizing, et cetera."

The story would not die. Soon it appeared in *Change*. Bitzer found out about it when he received a copy of the article from the president of the University of Illinois, with a note on it saying, *What are you guys doing over there?*

Bitzer called Umpleby into his office. Umpleby remembers the exchange this way:

BITZER: Umpleby, this is even worse than the last article. What are you doing? I thought that we had an understanding on this.

UMPLEBY: [Glancing at the *Change* article] This is word for word exactly the same article, I said I'm not doing these things, this is Greenberg's doing.

BITZER: Send off a correction letter.

Umpleby sent a correction. "By that time," he says, "I was persona non grata in the PLATO laboratory. Basically they retracted my office space and I moved out. Because people thought that I was ginning up this publicity program to go over Don's head and generate a groundswell of public opinion about something and that that was even *more* threatening to the laboratory. Well, that's when I became rather unhappy with the press. But at the time I was trying to do my dissertation and so I just retreated into my apartment, wrote my dissertation, and left."

He became a professor at George Washington University, where he's remained for the past forty-plus years.

12

The New Wave

The projects undertaken by Umpleby and Lamont were noteworthy because they veered from the "traditional" use for which PLATO had been designed. No matter how freewheeling CERL's policies were, or how far the "anything goes" spirit of the community extended, in the end there were limits to what was supposed to be done on this precious computer resource, and those limits were generally confined to PLATO's broad educational mission. But then Umpleby and Lamont arrived on the scene, challenging preconceived notions of what could be done with PLATO. One could spin their work as educational in nature, although Umpleby's attempts to use PLATO and ARPANET as platforms for political advocacy stretched to the breaking point the bounds of appropriate use. But maybe PLATO was *more* than education? Maybe PLATO should be viewed as a general-purpose resource? Maybe it *was* or would become a common carrier? What *else* could be done with this thing? CERL was about to find out.

The staff at the lab might have thought the work of Umpleby and Lamont was, if not disruptive, at least offbeat and unusual, but what they did with PLATO was nothing compared to the tidal wave of disruption that was building quickly and would soon hit CERL's shores. There was a new wave coming, and it was made up of young people, teens mostly, coupled with their endless fascination with this computer. Young people had played a key role in PLATO from the start: either as guinea pigs for professors trying out new CAI lessons or pedagogical experiments, or as cheap, nothing-is-impossible labor to help build the system software. It was one thing for a handful of kids like Andrew Hanson and Mike Walker to get involved with the system, but another thing altogether for dozens, scores, hundreds of kids to descend on PLATO as the system expanded in the early 1970s. Many were "hackers," though in the Midwest in the 1960s and 1970s, that term, if used at all, was often viewed as a pejorative for bad golf players and lousy carpenters. The last thing it denoted in the Midwest was skill, passion, endless curiosity, and expertise related to computers.

One early example of this new wave were the PLATO kids of Springfield High School, ninety miles west of Urbana in Illinois' state capital. It was probably not a coincidence that CERL had placed, in 1969, a PLATO III terminal in the heart of a prominent high school in the state capital. "The capitol was just a couple blocks away," recalls one student. "We could see it out the window." The staff at CERL liked giving demos to elected officials in Springfield, so the Springfield terminal was connected to the best, most reliable storage tube back at the lab, yielding the highest-quality displays. "They wanted the legislators to see the best, of course."

At an approximate cost of $18,000 a year (the equivalent of about $110,000 a year in 2017 dollars) that single PLATO III terminal connected via a ninety-mile line between the Springfield PLATO site and CERL did not come cheap. The terminal used a long-distance phone line for the student input from the keyboard back to the lab (around $300 a month), and a leased video line for the display and graphics (around $1,200 a month) from the lab to the terminal. Not that any of that mattered to the high school kids. From their perspective it helped that CERL and the adult supervision that came with CERL were far away. They found that the system was fun to program, and the remote time-sharing aspect—here was a lone terminal connected to a big fancy lab at the University of Illinois—also captured the imagination.

Some of the first students to try out the terminal and see what it could do were David Kopf, Doug Brown, and Mark Rustad. Kopf and Brown were given a lesson file to tinker with, called "dakdwbwk" (named after their initials and the abbreviated word "work"—a naming convention that would become common within the PLATO community in coming years), so they could learn TUTOR and build a program. Bruce Sherwood shared an annotated printout of one of his TUTOR physics lessons with Brown. "I remember being at home . . . leafing through this lesson printout," Brown recalls. "I can still see a little picture of a car that he drew in one place . . . drawing a car and showing the forces of motion and all that stuff. . . . I was learning TUTOR by going through his lesson, his annotations. Figuring it out that way." Brown wound up building a game called Shoot the Q, where you could modify various forces and then launch a small letter "Q" as if it were a cannon ball, seeing it fly across the screen in a ballistic trajectory then dragged down by gravity.

Soon, the Springfield kids discovered there were *people* at the other end of the line over at CERL and it was possible to *communicate* with them. It wasn't easy, it was rather crude, but it was possible. You could type a message into your TUTOR lesson file and someone, often Rick Blomme, would read it and post a reply. Blomme shared Bitzer's style of encouraging—indeed, challenging—kids who came under the spell of PLATO to do something useful on the system, find a way to contribute rather than simply goofing around on it. For a high school kid with hackerlike proclivities, this was like being invited into the secret priesthood of high technology.

One aspect of the CDC 1604 that ran PLATO III was that it emitted sounds. The CERL staff had rigged up a device with a speaker so that the 1604 would play various sequences of beeps and boops depending on what the central processor was doing at the moment. This was a handy way for the system staff to simply hear the occasional "song" of the 1604 to know what was going on inside it. Blomme even rigged the 1604 to play the opening lines of "Pictures at an Exhibition" (a song he was known to play on piano over in the music building) whenever the machine booted up. Then, as keypresses began coming in from students out in the field, the machine would emit a signature *rrrru-uuuuurrreeeeEEEEPPP* sound. (Some sounds indicated imminent trouble: if the system went *REEEEPeeeerruuurrpp* it meant bad things were about to happen. "We came by one time," says Blomme, "and we heard this *REEEEPeeeerruuurrpp* and Paul Tenczar in the other room turns around, yelling, GET OFF! THE SYSTEM'S GONNA CRASH!")

The Springfield kids were soon hooked. And, like Skinner's pigeons, they'd arrive early at school, congregate around the single terminal, and peck away at the keyboard, hoping for a reward—in this case, something, anything, appearing on the TV screen. Any response was a kind of reward. Sometimes there was no response, but they pecked away anyway knowing that eventually there'd be a response. There had to be. Wouldn't there be?

They often got to school before Blomme got to CERL, or if not Blomme, whoever was at CERL that day had gotten around to booting up the 1604. Blomme recalls one incident where he was there in CERL very early, preparing to boot up the system, hearing the trusty

"Pictures at an Exhibition" tune as the system came up. There wasn't a soul around; CERL's own classroom full of PLATO III terminals was empty. But then he heard a distinct *rrrruuuuuurrreeeeEEEEPPP*, indicating someone, somewhere was pressing a key on a PLATO keyboard. But it couldn't be—nobody was around, he thought. Then it dawned on him: it was the kids out at Springfield, patiently pressing NEXT, waiting for something to happen.

Sometimes patiently pressing NEXT got boring, so the Springfield gang resorted to more clever tricks. There was a rotary phone next to the terminal, but a lock had been installed on the dial, as the school didn't want to pay the steep long-distance fees for outbound calls. That did not stop the high schoolers. "We learned to dial the phone by bouncing on the hook to generate dialing pulses," says Rustad. "We came up with the idea of placing a collect call to the PLATO operator's office and ask for 'Springfield PLATO' and if he was out, when would he be back." One time they placed a collect call but Rick Blomme, nobody's fool, picked it up. "'We have a collect call for Mister Put-Us-On,'" Blomme remembers the operator saying. "They got hell for that from me, I said, 'Don't do that again.'"

They discovered that the repeated bouncing of the phone's hook to simulate a rotary dial was not such a good idea. "That mode of dialing turned out to be really hard on the phone equipment," says Rustad, "so we had to stop that. I guess that was when we learned to pick the lock on the phone dial. I guess there was a lot of learning going on."

On most other time-sharing systems around the country, these kinds of high school shenanigans would not have been tolerated. Strict rules would have been put in place, limiting who could get access and what they were allowed to do if granted access. CERL was different. Thanks to Bitzer, Blomme, Tenczar, Sherwood, and other staffers, if you were a kid who showed a fascination with PLATO and a desire to learn how to program, then it was possible to find not only encouragement at CERL but mentorship as well. A big help was finding a sponsor. The two major sponsors were Tenczar and Blomme. Blomme would throw down challenges to the Springfield kids, daring them to debug some code, figure out what the code did and be prepared to explain it, or write some new code that did a particular task. A number of the Springfield kids would pursue this new opportunity and it would change their lives.

—

Mark Rustad had found the PLATO terminal at Springfield while a freshman in 1970, but it wasn't until the following year that he began spending a lot of time hacking PLATO III. Some of his exploits would become the stuff of legend.

It was in 1971 that Rustad created an example of what would be recognized today on the Internet as "phishing": he created a program that *looked* like PLATO's login page, *behaved* like PLATO's login page, but in fact was fake. Rustad had become fascinated with exploring the memory of the 1604 computer. He'd discovered that it was possible to poke around the entire system's memory and look at any bits and bytes he wished: even the memory of programs being used by other users on the system at that moment. PLATO III was an open book, completely unsecure. CERL had not hardened down the system from such unauthorized pokes and peeks. "They didn't perceive the risks of reading [memory]," Rustad says. "I taught them. One of my programs simulated the login page, and had a part that allowed me to put other terminals into my program. . . . I had decoded memory enough to be able to tell what every terminal was doing. So I could select idle terminals and put them into my simulated login page. Then I would wait. Whenever someone logged in, my program would save the password and then return them to the real system so they wouldn't even realize their password had been captured." PLATO III's welcome page sequence was primitive: "Welcome to PLATO," it said, and underneath, "Press NEXT to begin." Secretly, you could type "04891" (a reverse "1984" with a leading zero) to invoke what was called "Author Mode," for authors to edit TUTOR lessons. But students wouldn't know to do that, and simply pressed NEXT. Most students, that is.

Not only had Rustad's hack enabled him to sit back and capture everyone's passwords, including authors, but he had a brilliant stroke of luck right at the outset. "Within fifteen minutes of running this program," he says, "I had captured the high-chief password." This was a super-secret password used only by the systems staff to do privileged things with the system. It didn't take long for Springfield students to find out what Rustad had done, and word eventually got back to CERL. "It was a little trouble," Rustad says, "but they were pretty cool about it. They just asked me to never simulate a system display again. I didn't."

Of course, that didn't mean he didn't hack the system again in *other* ways. He set right to work creating another program. "The next version of my program was able to capture keys from the system input buffer when people were entering their passwords," he says, and this time he didn't have to fake the PLATO welcome page. Instead he essentially wiretapped PLATO III, recording all of the keyboard traffic from any terminal he liked. "I was able to get the high-chief password once more."

One day, Don Bitzer received a letter from a student at Springfield. "'Dear Doctor Bitzer,'" Bitzer remembers it saying. "'Maybe you'd like to know the passwords to everyone on your system.' And there they were." Bitzer was impressed that the Springfield kids didn't do any damage, but instead had reported the security weakness to the "high chief" of PLATO himself.

Rustad's hacks revealed gaping holes in the way programs ran on PLATO. Unless CERL modified the system it was inevitable that some other jokester would come along and pull the same trick on unsuspecting users. For the PLATO IV system, now under way, CERL decided to modify the login process so that users would have to press a new key combination called SHIFT-STOP (hold down the SHIFT and press STOP) in order to proceed. SHIFT-STOP was treated as a special key sequence that would exit you out of any program and was, in theory, the guaranteed way that one could be sure one was not in a fake system program. A similar idea would appear years later at Microsoft when they introduced what would become known across the world as the "three-finger salute": Control-Alt-Del.

"I didn't invent SHIFT-STOP," says Rustad, "but I was the reason for it."

It was one thing to be drawn to a single computer terminal in the corner of a second-floor classroom at Springfield High School. But eventually, if you tinkered with it and the system it was connected to long enough, you'd get hungry for more. And the only way to satiate that hunger was to make the pilgrimage: to hit the road, to head east, to the building where the real action was taking place. CERL.

Doug Brown and Dave Kopf, relative old-timers who graduated high school in 1970, made pilgrimages, just to see what was going on and to meet the people like Blomme they'd interacted with via makeshift online chats and primitive emails on PLATO III. "I was very interested in electronics and computers," says Brown, "so going there was very exciting, seeing all this stuff, all the equipment, the racks of stuff."

It may have been on the first trip or one of the early subsequent ones, Brown no longer remembers, but in their excitement to get to CERL and see the place, they hadn't thought the logistics of their trip fully through. "We came and visited during the day. . . . I don't know how we were getting back home, we hadn't made any provisions for at night, I guess we had just assumed that everybody would be there all night. And, everybody left, and so we tried to go over to Illini Union," which has hotel facilities, but there were no vacancies. "And then we went back to PLATO and we went up to the fourth floor and the room where the PLATO IV computer, the 6400, was going to be was empty, and we slept on the floor up there for a small number of hours." On some weekend trips, they often slept in Blomme's office. "This was a crazy time," Blomme says.

Later, Rustad made the pilgrimage as well. CERL, to its credit, didn't chase them away. It welcomed and encouraged them. It redirected their energy and technical curiosity away from simple hacks and shenanigans into productive work on the system. "You didn't have to be an employee to do great work," says Bill Golden, "you didn't have to be a college student, you certainly didn't have to be a PhD—that was probably a detriment, if you were a PhD. All you had to do was something that one of the people in power—didn't have to be Bitzer—but one of the people in authority said, 'That's clever,' and you were in."

While aging and growing obsolete by the month, PLATO III was still in production and needed help, and the huge PLATO IV project was also under way. CERL could use as many bright minds as they could find, especially with system software. It didn't hurt that these bright teenagers didn't require much pay.

Brown entered the University of Illinois as a freshman in the fall of 1970—and also landed a job at CERL as a programmer. Kopf wound up programming there as well. As did Rustad. "I spent the summer [of 1971] working as a system programmer on PLATO III," says Rustad. "I was kind of a caretaker while most of the staff focused on getting PLATO IV going." It may have not been leading-edge work, but for a high school sophomore, the memory lasted a lifetime. Decades later, Rustad would say of that summer job, "That was probably one of the most important training experiences of my life. I really learned a great deal."

CERL caught on quickly that the Springfield kids were good hackers, and if you wanted to test your system security, letting the Springfield kids loose on the system was the way to go. As the trips to CERL increased, says Rustad, "the staff made security a bit of a game. We were supposed to break in and then tell them how we did it. On one visit, we sent one of our group up to the fourth floor to announce our arrival. While that was happening, the rest of us started hacking. The fellow that had gone up to fourth floor came back saying that they were just going to give us signons because we wouldn't be able to break in. That proved to be wrong. We had already broken in and run batch jobs to put blinking messages on the console announcing our arrival. They had simply forgotten a back door into the old edit logic."

Other Springfield High kids followed in Brown's and Rustad's footsteps. Steve Freyder, Dave Fuller, and Marshall Midden were all a year behind Brown, and when Brown graduated, those three continued the fine tradition of PLATO troublemaking and mischief out at Springfield. "I was one of those attention-deficit kids that wants to get into everything," says Fuller, "and saw this computer and said, *Shit, this is for me.*" Brown had already embarked on pilgrimages to CERL in the past, and that tradition continued, except for Fuller. "There was a bunch that would go to Champaign-Urbana for weekends," he says. "Mom and Dad wouldn't let me." Eventually he was able to travel over and hang out in the Power House, and like so many before him, was given Rick Blomme's now-famous series of computer programming problems. "It was Rick's idea that efficiency is good, and so I pretty much solved them and I kind of went from there. I entered the University of Illinois in 1973, and basically never managed to make a degree there, because I was spending too damn much time on the system."

Word got out that Rick Blomme would help, but only "if you approached him correctly," says Dave Fuller. "If you just wandered in, he was adverse to small talk."

But if you had a good question . . .

"It's like approaching any oracle," Fuller says. "Just don't go there with your mind disorganized. Say, look, I have this problem, I want to solve it. . . . One of the funny stories was, he used to practice piano in his apartment. He used to keep a log of how many bars of piano he practiced that day. Very organized."

Two other high schools served as feeders for the new wave of PLATO kids: Urbana High School and Uni High. Uni High had a particular advantage in this context. It was practically across the street.

"Uni High is an enormous influence on the University of Illinois," says Bill Golden. While it had the reputation for being unstructured, unusual, experimental, compared to the conventional regimen and curriculum of a typical American public high school, Golden says it was in fact quite structured. "Uni High is a five-year high school," he says. "It takes students in as what are called sub-freshmen, they would normally be seventh graders, through twelve. And they should have six years to graduate but they do it in five. The curriculum is very, very structured. There are English classes, and social studies classes, and language classes, math classes, and they're all very rigorous. They're conducted at the college level, basically." About fifty students are taken in each year—after passing a tough entrance exam. The total student body numbered between 225 to 250. Uni High students were bright to begin with, but came out brighter. Many became professors. They went off to the best colleges in the world. "In about 1956 I think it was, Max Beberman gave a lecture at Harvard," says Golden. "And in attendance was the president of Harvard, Conant. Anyway, at the end there was a reception as usual, and the president of Harvard comes, and says, 'You're from the University of Illinois, where is that?' And Beberman says, 'It's in Urbana, in the middle of the state.' And he says, 'Oh, *Urbana*! That's where that high school is. Keep sending us these wonderful students.' And Beberman says, 'Yes, that's where I work, it's my school.' So Uni High was known by the president of Harvard, when the University of Illinois was just, 'Yeah, there must be one.'"

Uni High students were treated as if they were adults, says Golden. "I'd say it's run as though it were a college, based on the responsibility given to students. . . . Unlike most high schools, you must attend classes but the rest of the time is yours, and they have a wonderful library. It's a great school. It's where you want to send your kid if you can. And so it's natural that those with a scientific bent would be lured, where else were you going to play with expensive computers in the 1960s, when you were sixteen years old? Fifteen years old?"

In the late 1960s and early 1970s, a few of Uni High students, including Sherwin Gooch, brothers Phil and Kim Mast, Rich Goldhor, Dave Woolley, and David Frankel, wound up working in one capacity or another at CERL, drawn by the many technological wonders within, plus a welcoming attitude among the staff, as long as you stayed out of trouble. CERL's proximity to Uni High was one of those great fortuitous accidents that wound up benefiting both institutions for years. "Why did the Uni student cross the road?" went the old joke. "To get to CERL."

Woolley's first exposure to PLATO was around 1969, when his Uni High class was tasked with taking a geometry lesson on PLATO III. While the black-and-white TV graphics were crude, and the teletype keyboard required students to clumsily replace some of the keycaps with specially labeled ones designed just for their specific lessons, the interaction and quality of the geometry instruction were intriguing enough for Woolley to want to find out more.

Woolley and Kim Mast would get more exposure to PLATO from Mast's older brother Phil. He had already gotten a job as a junior systems programmer working on PLATO III at CERL, and he let the two try out the system when it wasn't being used by an actual class. Both kept coming back. While still in their senior year at Uni High, they managed to get paying jobs at CERL, creating utilities and simple programs and whatever projects were assigned to them.

David Frankel's story is worth a mention here. In 1972 he was twelve years old, and wandered over to the CERL building from Uni High where he was in school. "It just seemed to be a regular university building at that time," he says, "it was open, unlocked, and one could wander the halls, and there was all this computer gear in there. And so I started to just make myself at home, and started to experiment with things that were there. It was a laboratory environment and although there were some specific classrooms that were used specifically for formal teaching classes, there were also other rooms, and again not knowing better I just made myself at home, started to experiment."

To Frankel's surprise, rather than being kicked out, people in CERL

PLATO KID. David Frankel, 14, a junior at University High School, sits at PLATO's computer display terminal with the typewriter keyboard in his lap.

He is the youngest "systems programer" for PLATO computer on the staff at University of Illinois.

Uni High Student Speaks Computerese

By STEVEN HERSHBERGER
News-Gazette Staff Writer

"There's not that much to talk about. It was no special deal," 14-year-old David Frankel said.

He was referring to programming the PLATO computer

fessional types like the board of trustees."

David said WCIA had talked with Bitzer about using PLATO for election returns. Bitzer then told his staff about the idea. David said he volunteered for

lunch time from Uni High at the computer lab. He admitted he often finds his contemporaries have "little to say."

At the computer lab his colleagues are graduate students.

David said he has traveled

the result of electrical engineering But I haven't done much with getting inside of the computer," he said.

His office includes, besides the bulk of the computer terminal, a portable

David Frankel in the news

were enthusiastic, interested, and helpful. According to Frankel, the initial interactions would go along these lines:

> CERL PERSON: What are you doing here?
> FRANKEL: Oh, you know, well, it's *neat.*
> CERL PERSON: It is, isn't it? Here, you want to try this, you want to try that?

"I remember Rick Blomme gave me some programming problems," Frankel says. Blomme kept encouraging Frankel, and Frankel kept coming back, having made CERL his new hangout, an hour or so each day after (or during) school. Frankel's insatiable curiosity made him absorb everything he laid eyes on: tips, techniques, know-how, even what others were doing. He became a gadfly, to many a brat, to some Bitzer's pet. Even so, people could recognize how smart he was. If he couldn't find out something on the computer, he'd rummage through the trash, looking at printouts and memos and whatever else was in there; that led to more clues and hints about how PLATO worked and what else he might do on the system. He was a mouse running loose through the four floors of CERL, a hungry gremlin determined to know everything he could about the system. Soon he discovered that some PLATO files were password protected. Hacker kid that he was, the fact that the passwords were kept secret drove him crazy. But then he discovered he didn't need to know them. "It became apparent," he says, "that there was a master password that overwrote all the other passwords. So for quite a while, many months, I think, it became my quest to find out what the master password was." Frankel and the systems staff fell into a cycle: he would discover the master password, they would change it; repeat. Then they changed the mechanism by which the password was stored, only to have Frankel figure *that* out. Repeat. "I learned a whole lot through that process," he says. So did CERL.

Frankel had found a way to be of use. He was a pest, but a useful one, breaking PLATO IV's weak security over and over, even when it had been strengthened. His rummaging through trash bins paid off, for that was where he found the master password, hard-coded right in the middle of the code to an editing program. The one dumb thing a programmer is never supposed to do, and they had gone and done it—put the master password, the One Password That Ruled Them All, right in the code, plain as day, for anyone to read. One of the master passwords was "infinity," and another was "mausoleum." Neither survived long with Frankel on the scene.

"For me the whole deal there was just learning how conducive the place was to learning. As I would discover these things, people never got mad at me, and I don't think I ever did anything particularly malicious. These were just sort of intellectual challenges and the staff there always sort of looked at it as, 'Okay, that's great, now we'll move on to the next challenge.' It was fantastic. I started in '72, spending time there, and much of my time was just sort of learning and amusing myself and then I started trying to do more useful things, and I had a couple of projects for various people that were brought to my attention somehow and of course I started to flirt with the idea that one could actually get a job."

One tiny problem with getting a job at CERL was that he was twelve years old. In a normal university department, or any business around town for that matter, except maybe delivering papers on a bicycle for the local newspaper, that would have been the end of the story. But this was CERL, a lab run by Bitzer. Bitzer loved the fact that a bright twelve-year-old kid from Uni High just would not go away but kept finding ways of being useful. Bitzer gave him a job. He just couldn't, by law, *pay* him, due to Illinois' child labor laws. Such administrivia meant nothing to Bitzer. Frankel was *in*. "I would make deals as I worked on projects for various parties," Frankel says. "I made various deals for different things. I remember I got a ski jacket for some work, I got an electric typewriter—there was no such thing as word processors at the time. I'd make deals for all sorts of things. But then when I turned fourteen I made $2.31 an hour." That was December 1973, and CERL held a party to celebrate Frankel's being old enough to be paid—as a systems programmer on the technical staff.

Urbana High School—the home of the local JETS chapter—continued to be another source of enthusiastic kids eager to learn more and start hacking on PLATO. For tech-curious kids, having so many PLATO IV terminals being installed in CERL and other campus buildings was nirvana. "Hell," says Doug Green, who attended Urbana High, "the administrators had no clue whether we were supposed to be there or weren't supposed to be there, or allowed or not allowed, professor's kid, whose professor's kid, or just some rug rat off the street, you know, there was just no way to discriminate between who was who. And we all knew each other."

Perhaps what Bitzer saw in these kids was a bit of himself. "In a way he was like a big kid," says Nancy Risser, who worked at CERL in the 1960s and 1970s. "He loved sharing the excitement, he loved sharing the ideas. . . . He had no need to be secret, he had a much bigger need to share ideas and work with other people and have fun with it. And bringing kids in was consistent with that."

Many came to learn. Many came to program. Often that led to programming projects and possible income. What they worked on, on the side, having secured authoring access to PLATO, was sometimes another matter altogether.

13

The Big Board

A new wave of PLATO kids continued to pour into CERL month after month. Somehow, if you were the least bit curious about cool new computers and graphics and technical things, you would find PLATO or PLATO would find you, and call to you until you made the pilgrimage.

Doug Green was a senior at Urbana High School in 1972. "Urbana had three tracks," he says of his school. "The college-bound track, the maybe-college-bound track, and the blue-collar track. And they put me in the middle one. So I got some college-bound courses, 'just in case maybe this stupid kid will grow up to have a brain.'" It was November 1972 when Green first heard the siren call from this exotic new PLATO computer. "I was seventeen but looked older," he says, "and was sneaking into off-campus parties." At one Saturday night party he heard someone say, "Hey, let's go play with PLATO."

"What, modeling *clay*? You're kidding."

"No, man, it's a computer over on campus."

PLATO until then had been completely unknown to Green, but, he says, "from the description I knew the building it had to be in. I parked on Goodwin and there was a sign I'd not noted before, 'Computer-based Education Research Lab.' Huh. Had to be the place." Green's memory of the experience continues:

Inside the east doors, an arrow pointed to PLATO upstairs. I followed my nose up and around and down a long dimly lit corridor toward the sound of something I'd never experienced. I stood in the doorway . . . getting my bearings. The room lights were off. Cigarette smoke thick in the air, the ceiling disappeared in the gloom. Odd metal boxlike structures lined the room, jammed into rows on hefty library tables running all around the perimeter of the room, then jutting out into the center in an L shape, so from the door it was like walking into a spiral maze. Dozens of people in the room, sitting in groups of twos and threes, hunched over each of the boxes, their faces weirdly lit with a strange orange glow coming from some sort of non-TV screen on the front of each box. Surreal as hell, never seen the like. Doing things with some sort of typewriter keyboard, pointing at the screens, laughing and yelling instructions at each other. Suddenly somebody nearby yelled, "Got 'im!" and simultaneously across the room somebody else yelled, "Damn it!" Games! They're playing *games*! I realized.

He continued to listen and watch. There was a new game out, just that week, called Moonwar. It was, he would learn, a "Big Board" game and all the rage at the time. When you entered a Big Board game, you first saw a list of players already paired off in the game as well as potential opponents, people not yet in the game with someone. Your task was to pick one of those potential opponents by "challenging" them, and if they accepted, off into the game both of you went to play against each other. The goal of Moonwar was to fire a laser and hope it hit your opponent who was moving, you didn't know where to, so you had to not only guess the correct angle to fire at, but also guess where your opponent was going to be a moment from now.

Hours passed, days passed, and Green continued to come to CERL and watch, discovering other Big Board games like Spacewar and eventually Mazewar. "I was at a terminal playing Moon Lander when suddenly everything froze, all the screens went blank, and an enormous collective groan and sigh filled the room," Green said. "PLATO had 'crashed,' that would probably be all for tonight. Most people trickled out, but some diehards stayed and talked. I had questions, lots of questions. Finally made it home about the time the sun was rising." Yet another person had fallen under the spell of the Orange Glow. "I was hooked, instantly and thoroughly," says Green. "I knew this was for me. I didn't know what it *was*, but I knew I wanted it. I couldn't yet dream how far it would take me."

Green spent "every free minute" at CERL during the winter of 1972–1973. "This was before there was any kind of security system in place. Anyone could walk in off the street and press NEXT to begin, and simply enter the name of the game you wanted to play. And more. There were chemistry lessons, and math, and Esperanto lessons, and typing, and tons of computer science, and well, just what *is* there on this thing?"

His first "quest" was to find every single lesson there was on PLATO. "I had no idea of the depth" of this "daunting task," he says. "I had a notebook in which I kept track of what I learned." But beyond his discovering the riches of PLATO, he was discovering he was not the only person caught up in it. "I made friends, others like me who had no official standing and no real reason for being on PLATO except that it was fascinating, and we enjoyed the games enormously. An underground community was forming. And Rick Blomme took note of us."

By 1972 Rick Blomme had spent years cultivating the talented kids he found drawn to PLATO, through encouragement and programming challenges to test their abilities. If a new arrival showed promise, he'd find something for them to do. If they did what he wanted, and the code was clean and bug-free, he might give them some file space to tinker on their own. "I knew Rick at first as the tall skinny fellow from upstairs," says Green, "who wore the Fu Manchu mustache and the long stringy hair, who stood in the corner . . . and watched the game players with a sort of paternalistic air. A 'systems programmer,' I heard he was called, not knowing what it meant."

In the early days of PLATO IV, the system crashed quite often, and system security was still weak. In fact, the notion of unique user identity had not yet been implemented, making it relatively easy for anyone to enter "user mode" and run programs, be they educational or recreational. And it was possible to enter "editor mode," says Green, to view PLATO's TUTOR code just like millions would be able to do decades later with HTML on the Web. With that knowledge, you could figure out how to write your *own* stuff, assuming you knew how to scrounge up the file space.

Blomme had been around PLATO now for more than ten years. On PLATO III he'd written his first game, which he called Spacewar, which he'd brought over to PLATO IV soon after PLATO IV was up and running. As befits the oftentimes parallel universes of computer history, the PLATO game was not the same as "Spacewar!," created at MIT in 1962. What lands MIT's Spacewar! in the history books is that it was a "video" game and very likely the world's first: it exploited all of the speed and computing power of a brand-new, expensive DEC

PDP-1, and used every single feature of the PDP-1 to faithfully simulate Newtonian physics—including gravity, orbital dynamics, and star fields—all in real time, with high-speed animation. Put another way, the hardware dedicated to Spacewar! offered Alan Kay's "guaranteed cycles," all of the system's power, for a single user. PLATO III, on the other hand, was a time-sharing system running on top of the CDC 1604. Any game written for PLATO meant that it had to abide by all the rules and limitations of a program written on top of a time-sharing system. It didn't have the luxury of being the only program running on the computer, the way the MIT game had. One severe constraint was graphics. MIT's Spacewar! ran on the PDP-1's expensive console display. PLATO had only its slow and antiquated storage tubes and TV displays.

Blomme viewed his Spacewar as a "chase game," where two players, one marked as X and the other as O, chased each other around the screen. The goal was to jump on top of the other player. "I added some stuff in that made it really, really interesting because you had a limited supply of fuel, and you could change your update."

Blomme's design reflects not only an acceptance of the limitations put upon it, but an embrace of those limitations, as if to say, "Okay, so what can we do creatively *within* these constraints?" This design thinking would become a hallmark of PLATO games: compensating for the limits of the hardware and phone lines (which meant things displayed slowly on-screen) by turning them into features of the game. Thanks to PLATO's Fast Round Trip, a player could always count on fast keyboard responsiveness. Expert players learned to ignore what might be going on on-screen while pounding out a flurry of key-presses hoping that once the digital dust settled, the opponent had been vanquished. Says Brian Shankman, an early player of the game who confessed he often lost at it, "The screen was blank much of the time unless you waited for refresh, at which point you were probably a dead ship."

Oddly, there was one feature of Spacewar that wasn't even a part of the game proper, yet in some ways had the most lasting impact on the emerging PLATO gaming community: Spacewar's Big Board. Given that Spacewar required two people to play at different terminals, it made sense to give users a way of seeing a list of people who were waiting to play a game, as well as who was already playing whom. The Big Board feature in hindsight seems inevitable: basic, simple, a no-brainer. But at the time it was revolutionary: it was the perfect solution to eager gamers wanting to find opponents to play these hyperactive, two-person games. Spacewar would not be the only game that would utilize the Big Board. Indeed, it would not take long for other games to pop up, using the same idea.

Those future Big Board game authors started out as Spacewar gamers. One of them was Louis Bloomfield. Born in Boston, his family first had moved to Cleveland, and then, in 1970, his father was invited to become the founding dean of the new medical school at the University of Illinois. Louis started ninth grade at Urbana Junior High. His father was interested in computer-based education, and wanted the medical school to utilize PLATO. It didn't take long for Louis to be exposed to CERL, getting a tour of the second floor and seeing the room of now aging PLATO III terminals. These were PLATO III's last days and the beginning of PLATO IV with its futuristic new terminals sporting the Orange Glow. In fact, Bloomfield would witness some of the new Control Data supercomputer hardware being carted onto the freight elevator at the back of the building, including the new CYBER, which was so heavy that it exceeded the elevator's weight limits. (Not a problem: this was a building full of engineers, after all. Elevator posing a problem? Figure out a fix. Problem solved.)

Bloomfield got to know Blomme. He and two high school friends spent more and more time at CERL, not only playing Spacewar, but soaking up the ins and outs of TUTOR and all of the technology in the building.

Blomme in these early days of PLATO IV strove to be the top player in every game on the system, starting with, of course, his own Spacewar. Bloomfield also was good at Spacewar. Perhaps too good. "You're moving around your little spaceship and you try to get the right or left of the player, and press the shoot button. And if you did that, you won, you got like three hyperspace jumps." Hyper-jumps were essential to survival. "If you were getting desperate," he says, "if the other player was going after you, you could just randomly jump on the screen." One day Bloomfield hacked Spacewar and gave his little spaceship the ability to do a million hyperspace jumps. He became invincible. Blomme, no dummy, caught on right away. "He knew immediately I cheated, and he shut the game down for a while because he just got pissed."

—

Bloomfield was soon writing his own Big Board game, Moonwar, over in the new PLATO classroom in the medical school annex. (It helped that his father was the dean of the medical school and a PLATO supporter.)

In a day or two he had the basics of the program working. "The first version," he says, "I just wanted it up and running, you know, just a toy, just a game, I didn't want to invest huge amounts of energy and time. We were always writing new little programs, new planetary motion programs or whatever, you'd write them in an afternoon, the aim was to just entertain yourself, and your buddies, they weren't big life projects."

The advantage of being able to work quietly in an off-site PLATO classroom that most of the diehards hadn't heard about yet was that you could use two terminals to test out the multiplayer aspects of the game, without people standing over your shoulder asking questions and spreading rumors about a new game in the works. "I got to the point where I was pretty sure it was working right, but I needed somebody to play with, I mean, I was by myself . . . so I went over to CERL, and I just wandered into the author room, I guess it was, one big room, they had people there, and I found a guy, nobody there I knew at the time. . . . I ran into a guy named Gordon Peterson, was his name. I buttonholed Gordon to play Moonwar with me. The first game we ever played was me versus Gordon, and it worked, and maybe if it had bugs, I fixed them."

In Moonwar, after challenging an opponent listed on the Big Board and having your challenge accepted, you and your opponent entered the game. On-screen you'd see some circles with the word "mountain" written inside them. Then, somewhere on the screen was an "O," just like in Spacewar, and elsewhere there was an "X." The mountains were handy obstacles that blocked the laser beams that you'd fire at your opponent. Initially the game didn't know how to handle poorly aimed lasers that were fired into the "wall" or edge of the screen. Those were simply "absorbed" as if they'd hit a mountain. But that didn't last long. Eventually the game seemed to be influenced a tiny bit by a pool table, in that the edges of the screen became "reflective"—the game described them as "mirror-smooth walls" from which your laser fire would bounce at an angle, just like in a game of pool. Whenever you fired your lasers, you first needed to type in a specific angle, like 135 degrees, to aim the laser. Precision made all the difference. As did speed. The other player was always trying to fire a laser back at you.

Moonwar was like a multiplayer pinball machine, and the PLATO community *flocked* to it. Like a viral video on YouTube today, word of Moonwar broke out quickly and it became an instant PLATO hit.

"Within a day or so," says Bloomfield, "people started playing Moonwar, and the Big Board started becoming busy." Then something strange happened. Bloomfield came in one day and found that his source code to Moonwar had been edited. Not by him, but by the systems programmers.

What they had put in his Big Board code was, in Bloomfield's words, the "most incredible list of curse words I'd ever seen, all -putd-'d to nothing." (-Putd- was a TUTOR command for replacing a text string with another text string, or, in this case, nothing.) Moonwar had become such a big deal in such a short time that Bitzer had been told about it. And naturally Bitzer tried it out, and naturally he started using it in his demos. There was just one small problem. It was commonplace that the gamers would choose every foul four letter word or phrase they could think of as their pseudonym on the Big Board. It did not require a lot of imagination to envision Bitzer, up on a stage in front of a crowd of dignitaries, arms waving, hands gesturing, demonstrating his masterpiece, the PLATO IV system, typing commands into the keyboard, wowing the crowd with his magician's sixth sense, and then eventually entering Moonwar, glancing at the Big Board, and seeing a bunch of player pseudonyms like "Shit Kicker" and "Motherfucker" all the way down the damn screen.

"Bitzer was mortified," says Bloomfield. "So the command went out, There Will Be No Curse Word Login Names On The Big Board On Moonwar. They -putd-'d every swear word—I didn't even *know* almost half these—I was the most naive Goody Two-shoes, and so I never swore back in that era, and here was this *list*, a hundred words long." In the long history of attempts to block out curse words on computers, it is generally accepted as a law that such efforts never work, and often backfire. Moonwar's Big Board Bitzer Emergency Curse Word Fix was one more example of it backfiring. Sure, it filtered out anything that even so much offered a *whiff* of something objectionable, but in so doing, it filtered out all sorts of completely *harmless* things as well. "It made it a little challenging to pick a name," says Bloomfield, "because it would knock out any curse word *even if it was in the middle of a word*. Like, if you want to be 'Bass Fisherman,' the word 'ass' would lead to nothing, so you would be 'Fisherman.'"

—

Rick Blomme was a hero to the new wave kids. "We really liked Rick," says Bloomfield. "Rick Blomme was one of the cool guys who would talk to you and joke around—all these people in what was called the 's' group, these are the system programmer types, so they had all the power, and we had to go up, they were on the fourth floor. And once in a while we would get let into the *machine room.*"

As cool as the kids viewed Blomme, Paul Tenczar, on the other hand, "was much more circumscribed and a tougher cookie in general," says Bloomfield. "He was more all business. I think he thought some of us were nuisances. Surely we were in a way, we really pushed that machine to the limit. Whatever new command would come out, we would be in there using it, exploiting it. . . . We were always pushing the limits, trying to get as much compute power as we could."

Sometimes they pushed too far. "We would crash the machine, in fact the entire PLATO system, systematically, by trying something that no one had ever anticipated. And I can remember one day in particular where I wrote a chunk of code, and I executed it, and next thing I know, it says, 'Welcome to PLATO.' Everybody—all the ter-minals—I was at CERL—*all* the terminals are saying 'Welcome to PLATO'—that's a funny coincidence. So, once PLATO sort of came back up, I went and ran the code again. Immediately: 'Welcome to PLATO.' I thought, Oh: it's *not* a coincidence. So I went upstairs to the fourth floor, and there are all the people in the machine room, I can remember a kid named David Frankel in particular, who was one of the 'p' programmers, the junior system programmers, he was at Uni High, and he was there pointing a finger at me going, '*You, you, you, you, you!*' and they had my program printed out, and it was me."

Any young hacker-wannabe poking around PLATO long enough was bound to figure out how to break through the weak security. This had already been proven multiple times with the Springfield High School gang. But now, in the earliest years of PLATO IV, 1972–1974, there were throngs, virtual *armies* of these kids poking around, not just at CERL, but at remote sites on the new, bigger system. Bloomfield eventually stumbled on his own way to break the security. "I figured out how to give myself system privileges," he says. "I basically could do anything. And I thought . . . you know, after exploring the world where I wasn't supposed to go, because I could go into any program, any lesson, all the system lessons, I would go in there, and I would read them carefully. . . . You know, eventually it gets *boring*. I did everything I could do, without causing trouble. And I wasn't interested in causing trouble, I just wanted to see what was *up*, so I figured, well, okay, I'll go negotiate with Tenczar to see whether he'll give me 'p' privileges, Frankel and Woolley and Kim Mast, other people who were the next people down from the system programmers, the kiddie system programmers. . . . 'Will you let me be one of those, in exchange for telling him how I broke security?' "

Tenczar was no Blomme. His reaction: furious. "He hated that whole thing," says Bloomfield, "and he basically just beat on me to just tell him how I cracked security and I'll fix it. Okay, so I told him, it was using a -jumpout- command, I jumped right into the middle of, I don't know, was it the 'edit' program or something like that, and I managed to set my password to anything I wanted to, including things you could not type in."

Bloomfield had discovered that system passwords contained untypeable characters. "You can't type in 0 characters, right? They didn't exist," he says. "So I could jump in there, and the way in which they handled system privileges was, they put passwords on the lessons that they did not want any normal person to be able to enter, they would put a password that you couldn't type, it would just like say, 'system,' but it would be zero zero zero, the zero character and then 'system' at the right, something like that, and I would just jump there with that, with my variable set to that, and it was as though I typed in an untypeable value, it would accept my password, and I was in business."

Tenczar remained furious, but Blomme chuckled at Bloomfield's exploit. "He thought this was wonderful," says Bloomfield. Blomme realized Bloomfield had discovered a fundamental flaw in the way CERL's systems staff had set up password security on PLATO. The passwords were not secure enough, and there was a way to circumvent them due to an architectural quirk in CDC computers, a quirk that the teenage Bloomfield had figured out on his own. For Blomme, this was just the kind of thing he admired in this new wave of bright kids hacking away at PLATO.

One reason these Big Board games were so popular, in addition to the social aspect of playing against someone else, was that they were so *playable*. "Playability" also meant they were simple and lightweight, with rudimentary Xs and Os or tiny little spaceships or other tokens. The real guarantee of playability goes all the way back to the diagram of PLATO I: Bitzer's Fast Round Trip architecture, which required that a PLATO user could type something at her keyboard and the letter she typed would show up on her screen within about *one-tenth of a second*. PLATO, the ever-patient automatic teacher, was also supposed to be ever responsive. All through these years, from 1960 right up until the rollout of PLATO IV in 1972 and onward, the Fast Round Trip rule defined the PLATO experience. With the new multimillion-dollar PLATO IV system, things were no different. When the gamer kids discovered the Fast Round Trip in the late 1960s, and then armies of them rediscovered it in the early 1970s, it provided an unbeatable platform for fast—indeed, frantic—games that required not only skill but sheer keyboard speed to survive and thrive. All of the Big Board games, as did practically every other PLATO-designed game, took advantage of the Fast Round Trip. It was a key part of what made a PLATO game a PLATO game.

Bloomfield briefly became PLATO's game author celebrity, a hit-maker extraordinaire that everyone knew as "louisb." First Moonwar, and then Mazewar, another Big Board favorite, this time where the trusty X and O started out on different sides of an on-screen maze, and you had to try to get to the other side before your opponent got to your side.

Bloomfield wasn't the only one interested in writing games. Once it became known that one *could* write a game, everybody wanted to write a game. But not everybody had connections yet, a sponsor or mentor the likes of Blomme or even Tenczar. Most of these kids were gamers, thrill seekers, staying up all night in the crazy halls of CERL during 1972 through 1974. "CERL was a zoo in those days," recalls Erik Witz, another gamer kid addicted to the PLATO scene. "Gaming was rampant. Pizza boxes and other trash was stacked to the ceilings. The most popular games where all Louis Bloomfield's, I believe. . . . Then CERL cracked down. Everyone was required to have user IDs. Before that you could log on just by pressing NEXT to begin. The next day I hacked into the system simply by trying common names and hitting one that brought up an initial password page. A week later I'd given

IDs to all my friends. The ops were astounded when they walked in and saw all the kids still playing games after they had implemented the security."

In a way, there were multiple levels of games going on. At the base level were the actual PLATO games that you could see happening and becoming ever more widespread. But at a more meta-level was PLATO itself, or at least the act of using it—after first figuring out how to get access to it—and in a way *that* was becoming a game. Initially, gaming at CERL was unrestricted. The numbers were manageable. The drain on the system was tolerable. But as resources became tighter, the number of complaints began to rise, the number of parents calling or coming by wondering where the hell their teenage son was kept growing, and in general things began to get more and more out of hand. The PLATO Services Organization online consultant staff on the second floor, and the system operators on the first floor, had both unwittingly become Os in their own right, fighting the gamers' Xs. The operators were determined to gain the upper hand, upping the ante in this growing competition with ever-more-ingenious gamer kids. This forced further development of restrictions to access in various PLATO classrooms, banning unauthorized use outright, or more likely limiting use to certain hours of the day. Over the next few years, the CERL staff would need to resort to even more strict rules, including creating lists of games, notesfiles, and other recreational files that were again banned outright or during certain hours. They also built tools so that any site on the network (other universities, government installations, etc.) could manage their own rules for controlling access to the terminals and phone lines for which they paid considerable monthly fees.

"That summer it kept getting more and more brutal to game unrestrictedly," says one former PLATO gamer we'll call Sam. "Mostly because of the efforts of people like myself, who were there all day and half the night. I had a clearance from the U of I cops right next door, in those days, to be out after curfew. I'd ride my little bike over to CERL, lock the bike up, and go in and game, you know, maybe break for dinner, then stay until two to three or overnight sometimes."

Doug Green remembers that when he got started, most of the login security had yet to be implemented. "You just walk into PLATO," he says, "press NEXT to begin, and they asked for name and course [later changed to 'group'], but it meant nothing. You could type anything in. It was just a pass-through. They were in the process of attempting to figure out how to implement security and the argument was, Well, do we really need this, you know, it's all open right now, it's all community access right now, we want to keep it that way, *but* . . . too many people are destroying the commons, and so we got to control this thing."

One night in the early spring of 1973, Green arrived at CERL, "looking forward to a night of domination on the Big Boards of Moonwar and Mazewar, for I was becoming quite proficient at each," he says. "But stunningly, the room was empty. Some people I knew were sitting forlornly at various terminals, unable to proceed. A new security system had been implemented just that day. From now on, to use PLATO you had to have a 'name' and that name had to be registered in a 'course' and you had to have a 'password' that matched the name and the course. A few people were actually 'logged on' and were—gasp—working. . . . But the game players were shut out. And so was I. Crushed. Angry. Frustrated. And determined to find a way around the predicament."

The few, the sad, the gamers. Those unlucky few who stuck around the library-quiet PLATO terminal room simply could not tear themselves away, even if they couldn't get online. But was it really impossible to get on? The "there-were-many-ways-to-get-around-this" wisdom soon spread among the most determined gamers, including Green. "Sitting there pounding away at the keyset, anxiously and futilely, one of us suddenly whispered, 'Hey, I'm in.' And there he was, at the User Mode screen, somehow suddenly able to access games and all, just as before. How had he done it? Even he didn't know. We experimented, and experimented some more. And then we learned." It turned out that if you pressed NEXT to begin, like the screen told you, then pressed BACK on the next screen that asked for your user name, returned to the "Press NEXT to begin" screen and pressed NEXT, then BACK, then NEXT, then BACK, in rapid succession, something broke. "You could fill the keyboard buffer and overload your timeslice allocation," Green says. "And the timeslice overload error would bomb you right past the security system and onto the User Mode page."

The hours ticked by, and people straggled away one by one. Only Green was left, mesmerized by things he was finding by simply poking around the wide-open system. "I had the entire room to myself. And I had found a brand-new course whose owner had not yet set up his passwords. . . . And so I performed my first hack on a computer system. . . . I knew I was doing wrong, but I was desperate. So I put my own password on that course, and I set up enough names and passwords to log on every terminal in the room. I had every terminal doing something, anything, drunk with power and showing off to myself how cool I was. And as I sat at one terminal, making entries in my notebook where I catalogued the lessons on PLATO, I looked into the screen and saw the reflection of Rick Blomme standing in the doorway behind me, scratching his head. He must have been working late and checked to see who was on the system, and seen the anomaly. I was scared, but played it cool like he wasn't there, just kept on making notes in my notebook. Rick scratched his head a moment, then left and returned a few minutes later with another systems programmer whose name I didn't know. They stood in the doorway together, scratching their heads and muttering to themselves. I watched them in the reflection on my screen, pretending to be making notes. The other programmer left and Rick stood there alone, fingering his beard. Finally I couldn't take the suspense anymore. I turned my chair around and looked up and said, 'Hi, Rick,' and smiled a Cheshire Cat grin at him. He slowly walked over and sat down across from me and just sort of stared at me for a while."

The conversation, as Green recalls it, proceeded thus:

BLOMME: Uh, what, er, um, how . . . uh . . . did you do this?
GREEN: But Rick, if I tell you, you'll fix it and I won't be able to
    get on anymore.

(At this point Green recalls Blomme giving him "one of those long penetrating stares" before speaking.)

BLOMME: Okay, okay . . . suppose I give you records in one of
    my courses?
GREEN: Well, okay, uh, what about some file space? I'd like to
    try some programming.

"And he stared at me and slowly began to nod his head," Green says. "And that's how I got onto PLATO, legitimately. . . . Within days I was learning to program. I'd earned my pilot's license the year before, and my first program was a lesson on the fundamental forces of flight."

Blomme was impressed, and when Green graduated high school in the spring of 1973, Blomme recommended him to the Aviation Research Lab (ARL). He wound up getting a job there that summer. But he kept coming back to CERL. And he started working on a game of his own, influenced by all the aviation stuff he was doing over at ARL: Dogfight.

In the new Big Board game Dogfight, instead of an X and an O, there were two tiny airplanes on the screen, going at it, trying to shoot each other down. PLATO's limitations meant it was impossible to render 3D graphics and fancy photorealistic animations like ones we see today, so Green kept it simple, the way Blomme and Bloomfield had done in their respective games. He made it a top-down two-dimensional game, and limited the graphics to two tiny representations of airplanes, just like Spacewar had two tiny spaceships.

Dogfight went viral; people loved it. "For a while it was the most popular game on the system, racking up more hours of play than even Moonwar. I wrote the original version using trigonometry to figure the angles and directions, then Larry White modified it to use a much faster table look-up scheme."

Larry White was another one of these brilliant Illinois students who wound up at CERL as a systems programmer. Green held him in awe. "Anytime I got stuck with a coding problem or especially debugging," Green says, "he was so good at debugging, you know, he could just glance, oh here, a comma out of place."

The PLATO system during this era was constantly crashing. Unlike today, where the software companies routinely write new code on a test machine, *not* the live production machine, during the mainframe era of PLATO, code was written on the live production machine—the only machine CERL had—and then tested at night while the expendable gamers were in force. Gamers were tolerated at night because the games they played often utilized more TUTOR functions than conventional educational lessons, and with the games being pounded upon all night long, if a systems programmer made a mistake while improving or changing a TUTOR command, it could very well show up as a crashed program or even a crashed system in very short order.

One time there was a critical system bug, crashing PLATO "deader than a doornail," Green recalls. "And they could not find it. Rick Blomme and Larry White wrote a specific program to look for this bug. How they did it, I don't know—but when this bug occurred, they dumped the entire core [the CYBER's main memory] out to the big line printer." Green walked into the machine room on the fourth floor, saw that the PLATO system had crashed, and saw Blomme and White with the pile of "core dump" printouts. "Larry and Rick are sitting side by side and they are scanning down, turn the page, and one set of eyeballs is scanning left half, and the other set of eyeballs is scanning right half. . . . Insane, trying to look something up. So imagine, just a string of zeros and ones and ABCDEFG's and that's all it is and as fast as you could turn the page, they'd turn the page, and scan down, turn the page, and scan down, and turn the page, and scan down, and turn the page, and scan down. It took them all night long to find out *one bit* out of position. And the next day it was fixed."

In Dogfight White used the pseudonym "Jackal" (after the film *The Day of the Jackal*). Blomme went by "Red Baron." Another PLATO gamer named David Frye called himself "Fright Pilot." Even Don Bitzer played the game, and always went by the name "pig," according to White. A feature of Dogfight's Big Board was that it checked the signon of each player as they entered the program. If the signon matched an entry in a special list, then that user's preset pseudonym was automatically entered, saving them a step. "This allowed faster entry for the authors of the game and some of their friends," says White. One day one of Dogfight's authors added Don Bitzer to the special list so that his pseudonym "pig" was always available to him. "A few weeks later," says White, "I learned from Don that this was a problem for him. Don would give demos to lots and lots of people, and one of the demos he often did would be to demonstrate the graphics and interactive and inter-terminal capabilities of PLATO using a game program. Part of his presentation was about how each player could select their own game name—but suddenly he could no longer do this in Dogfight and thus he could no longer demonstrate one of the social aspects of inter-personal interaction on PLATO." Bitzer's entry in the list was quickly removed.

The Dogfight authors thought up an additional trick that other game authors soon copied. If you used the TUTOR -slide- command, you could instruct your terminal to select one of the 256 microfiche slide images that would then be projected through the glass of the plasma display and out to the user, with any orange text and graphics magically superimposed. At least, that was the intent. But leave it to the game authors to find another use: special effects. No need to insert a microfiche sheet into the machine. The -slide- command still worked, sort of. It would attempt to select a particular image off the microfiche, causing the mechanism to shake and make sounds, even illuminate the slide projector's lamp, which caused things to light up behind the plasma display. Run the code in a loop, and you could get the machine to shake and rumble like a deranged washing machine on spin cycle, along with bright flashes of light as a spaceship exploded. Brendan McGinty, another gamer, vividly remembers arriving on the PLATO scene in August 1974, walking into a room full of PLATO terminals, every single person there playing games, many of them Moonwar, which was also modified to support the -slide- trick. "Every time you shot somebody in Moonwar," he says, "the slide projector on the old Magnavoxes would go off. . . . These terminals are rockin' and shakin' and everything, turnin' green, and I thought, *What the hell is all this stuff?* I got hooked on the games."

The CERL staff were not so amused. However, in the spirit of Bitzer's optimistic philosophy, the -slide- hack pointed to what was becoming a steady stream of reasons why the gamers had become beneficial parasites to CERL's CYBER host: the symbiosis exposed weaknesses, flaws, design oversights, and bugs in the TUTOR language and the system hardware. The -slide- command would in time be fixed. Game authors would have to find other ways to push the system beyond its design limits.

One PLATO gamer, Silas Warner, didn't find Moonwar very challenging. "There was never anyone around to play it when you wanted to play it," he says. He decided to write his own game that was more challenging. In Moonwar, a player's shot would bounce off the wall (depicted via primitive animation) but it would hit some obstacles. Warner decided to write a new game inspired by Moonwar, but he got rid of the obstacles and just had a square open field. In Warner's game, instead of players deciding how they were going to move and how they were going to shoot, he thought, why not have a robot player that you could *program* to move and shoot a certain way? Instead of player versus player, Warner wanted to be able to play against the computer as an option when real players weren't around, hanging out in the Big Board. His game was called Robotwar, a game that would go on to huge success when reprogrammed to run on the Apple II and sold commercially a few short years later.

—

Nineteen seventy-three and 1974 were the boom years for Big Board games. They were easy to write, instantly popular, and as the code to run a Big Board game became standardized and shared like a plug-in component, akin to open-source software routines today, it made it even easier to get a new game up and running. But already, ideas were percolating among the new wave of kids. It was starting to dawn on some of these kids that there was more that could be done with the TUTOR language than just writing simple Big Board games pitting one player against another.

A lot more.

14

The Killer App

By the time America Online expanded to millions of users in the 1990s, it had become abundantly clear to keen observers that something had changed with computing. Applications like word processing, spreadsheets, and databases, along with desktop publishing and graphics tools, had transformed computing from a primarily number-crunching business of the 1970s to a thriving industry in the 1980s built around the rise of the personal computer. But as those personal computers became networked in the 1980s in workplaces, and then widely over phone lines and the Internet in the 1990s, mainstream society would finally realize that computers had become *inter-personal*. The PLATO community had already understood the phenomenon years before, as had the makers of telegraphs and telephones decades earlier.

Silicon Valley had a term for it: "killer application." The main reason people now used computers and digital devices and could not imagine life without them was not their love for a spreadsheet, or a word processing program, or any sort of productivity tool. The killer app that drew people in and got them hooked was communication, connection, belonging. Computers let people *yak*. A lot and all the time. Email, chat rooms, instant messaging, SMS: all people seemed to do was yak. The vaunted, long-heralded Information Age might better be described as an Opinion Age: everyone had an opinion on everything, and the computer was the new tool to share those opinions widely, whether they were informed, fact-based, or not.

The killer app made its appearance on PLATO almost as soon as the PLATO IV system began its rollout in 1972 at Illinois, which suggests that it wasn't that PLATO itself, or its creators, dreamed up the killer app, but, rather, it's something that just happens if and when the right ingredients are in place. Perhaps it was inevitable: the two-user PLATO II diagram from 1961 depicted two students: a boy wired into a terminal on the left, and a girl to the right, with the rest of the system components in the middle. What PLATO's designers had probably overlooked when they created that diagram is the fact that they're facing each other. PLATO's destiny was to bring people together.

Only a few minutes' exposure to PLATO IV was enough for people to realize that it upended the generally received view of computers up to that point, if one were to accept how books, movies, and media had portrayed them. PLATO wasn't the evil self-aware intelligence, first seemingly benign but, after something going horribly wrong, now bent on destroying humanity. It most certainly wasn't *2001*'s HAL looking back at you through a foreboding round red glass "eye." Nor was PLATO some boring number-crunching, tape-machine, card-reader, data-processing behemoth found in big banks and government agencies. PLATO was your *friends* looking back at you, through friendly orange dots.

The Big Board games sparked an interest in a side of PLATO not contemplated by Bitzer, Alpert, or the funding agencies. This brave new side of PLATO was its *social* dimension, a dimension we take for granted today but that at the time was unforeseen, both at PLATO and in popular culture. Think of the way computers had been depicted up to that point: HAL in *2001* and the *Enterprise*'s computer in *Star Trek* were both sociable, capable of communicating by voice, but there was no hint of interpersonal communication between multiple people *through* the computer. The computer was a tool, you asked it questions, it gave you answers. The killer app came out of nowhere, yet in hindsight its arrival was obvious: the social side of PLATO exploded as the result of a perfect storm of ingredients, timing, and user community. Combine the permissive culture of CERL, the rise of the new wave of kids pouring in to check out this cool system with its Orange Glow, and the availability of the right mix of programming tools, and the rise of the killer app was inevitable.

Besides the fact that they were fun, one reason for the sudden rise of multiplayer PLATO games was that their creators took advantage of a system feature called "common." "Common" enabled a TUTOR lesson to set aside a chunk of computer memory that could be *shared* with other lessons. Each lesson that had attached this "common" to its own memory could take turns reading from and writing to it, enabling each participating lesson to share whatever data the lesson's author intended. Without this one feature of TUTOR, it's doubtful that multiplayer games on PLATO would have ever taken off the way they did. Game authors could instead have coded their games to write data directly to a disk file, but compared to memory, disks were painfully slow (a PLATO lesson's disk use was measured in DAPMs—disk accesses per *minute*—and the number wasn't allowed to get very big). Having programs communicate with each other via a shared disk file would have ground the programs (and possibly the entire system) to a halt. The fast reading and writing in CYBER's computer memory gave games the ability to instantly share data among players. That's how each player knew where other players were in the game, what direction and speed they were moving, what their scores were, their level of damage, whether they had fired weapons, and so on. TUTOR's common was pure gold to a game author.

PLATO was not the only system that had this kind of capability, known as "inter-process communications" and "shared memory" elsewhere. Many time-sharing systems of the era also offered, or would soon offer, such features. Indeed, the FORTRAN language had a feature called "COMMON blocks" going back to the mid-1960s, which did something similar. TUTOR's "common" had been named after it.

Given how rapidly things evolved in the early years of PLATO IV, what with the rising tide of interest from students, not only did new Big Board games begin to sprout up overnight, but new functionality would appear in them as well. The Big Boards began to offer ways for players to type and send *messages* to other players. They could be in the same room or thousands of miles away. Sending and receiving messages via a game's Big Board opened up a vast new set of possibilities for PLATO. It quickly dawned on users that talking with other users online was enjoyable and useful in its own right. Why even bother playing the game if what you really were after was the attention of that cute girl who was currently listed as "available" in a Big Board game?

Doug Brown had graduated Springfield High School in 1970 and was now an electrical engineering undergraduate at UI working as a junior systems programmer part-time at CERL during the school year and full time during summers. Other Springfield High students, younger friends of Brown's like Marshall Midden, Steve Freyder, and Mark

Rustad, would also work summers at CERL on particular programming projects, but then they would go back to Springfield in the fall. Brown wanted to stay in touch with them more often and found it difficult to do so.

Then he had an idea. Why not create a lesson whereby a user could "talk" to another *through the terminal*? It was as if he took the concept of the Big Board to its ultimate conclusion: forget the game behind the Big Board, just have a Big Board, and enable users to join what would decades later be called chat rooms, wherein they could chat. That's it. The "game" was the chatting. Brown called his program "Talkomatic." It sported a feature that took advantage of the Fast Round Trip principle going all the way back to PLATO I a dozen years earlier: if a user types a character, that input signal should travel as fast as possible from the user's keyboard, through the terminal, across the phone line, into the computer, and back out of the computer into the terminal's screen, all in a tenth of a second. If the user typed "hello," then the "h" and then the "e" and the "l" and the "l" and the "o" would appear with as little delay as possible, enabling the user to interact at the speed of thought.

Experimental chat programs existed in various time-sharing environments prior to PLATO IV, even including a primitive TALK program on PLATO III. Most followed the line-by-line style of messaging programs that arose in the decades to come, including Unix's Internet Relay Chat, AOL's Instant Messenger, Apple's Messages, Google Chat, and Facebook Messenger. This meant that when you typed in your message to someone, the recipient did not see your message until you were done typing all of it and then sent it. Brown hated this type of typed communication, and was determined to design Talkomatic to exploit PLATO's Fast Round Trip. The result: character-by-character chat using TUTOR's "common" function to share one user's typed message with another. As one user typed some text, the other user saw those text characters appear *live*, one by one. If the typing user made a mistake, and deleted some characters and then retyped them, the other participant saw, in real time, the mistake being made and the sudden correction and retyping. Today most messaging applications continue to operate in the other way: you don't see another's message to you until the whole message is typed and then sent. If an application developer is kind, they may provide a clue that your correspondent is Out There somewhere, by saying "so-and-so is typing" with a little busy icon. In Talkomatic, the individual characters of each participant's messages served as their own "progress indicators"—in user-interface design terms, you cannot offer a more transparent or elegant solution. Talkomatic was the typewritten equivalent of a phone call or conference call.

Brown's original motivation—a way to stay in touch with his Springfield friends—limited the scope of Talkomatic's original design to one-on-one chat. But soon Brown's office mate Dave Woolley was collaborating with him, and the combination proved fruitful. Talkomatic soon had chat rooms in the form of "channels," like those of CB radio, and in each channel up to five people could participate in a group chat. Over time features would be added that enabled users to make a chosen channel private, or lock a formerly open channel so no others could join it. Inside a channel, the participants were each given four lines' worth of screen space in which to type. With five users banging away at their keyboards, a channel heated up and became a thing to behold for the participants: the text of each user kept changing, character by character. Letters formed words, words formed phrases and sentences, flying across the screen. It became instantly clear who was a fast typist and who was slow. (For some users the temptation to judge a slow typist a slow thinker was strong, however misguided such a judgment might be.)

Talkomatic became the new hot game, even though it wasn't a game. But then, it was. In fact, perhaps it was the most exciting game of all. For some users, it was an online "place" where a bunch of friends could hang out and chat online, make jokes, gossip, yak, talk. For others, it was a way to meet mysterious other users who connected via terminals in far-flung places like Chicago, Delaware, Hawaii, or New York.

For yet other users, it was a place to play tricks on other users. (Forty years later a term would arise for this kind of trickery on the Internet: *catfishing*.) For instance, pretend you were a girl when you were actually a guy. Given the highly skewed male-to-female ratio among the PLATO user community, if someone named, say, "Lisa" presented herself in Talkomatic, guys were unlikely to exhibit any doubt. "There was this one guy," recalls John David Eisenberg, "I don't remember his name, his first name was Craig, and he would always go in Talkomatic as 'Sally.' It was some guy who worked for some military installation out on the East Coast. . . . I know he was in some government, military thing. And he would always go in as 'Sally' just to find out what kind of reaction he'd get. That's the first instance I know of people sort of masquerading as somebody completely different."

In time, enough Talkomatic users were fooled that the prank would wear out its welcome. (Years later, when *The New Yorker* published its famous "On the Internet, nobody knows you're a dog" cartoon, PLATO users of a certain age could crack a smile tempered with "been there, done that" recognition.)

The CERL system staff saw Talkomatic, saw that the hold it had on users, noted its obvious utility, and acknowledged that this form of communication was powerful and useful. Systems programmer Dave Andersen was tasked with creating a program that provided a way for any PLATO author to instantly "page" another author and open up a two-way typed conversation at the bottom of the screen. It was one thing for people to enter lesson "Talkomatic" and chat in one of the channels there. It was another thing entirely to embed a "talk" capability across the entire PLATO system, no matter what you—and the person you wished to talk to—were doing online at the moment. Andersen used privileged, system-level TUTOR commands unavailable to regular authors to make PLATO send output to *two* terminals instead of one. If user A got paged by user B, and user A answered the talk request and typed "hi," the "h" and the "i" showed up both on A's screen as well as B's. Likewise, when user B began typing, her characters showed up on her own screen, one by one, as well as user A's. On December 19, 1973, Andersen posted a note online letting the world know that TERM-talk was officially available. "An inter-station communication option has been provided," the note said. "TERM-talk will allow an author to talk to another author at another station."

TERM-talk instantly became one of the foundational components of the PLATO user experience. It was the live, character-by-character typed equivalent of a phone conversation—with Caller ID even—at a time before cell phones and texting or instant messaging. As long as the person you wished to talk to was online, you could reach them. That is, unless that other person did not want to talk to you, in which case when they saw the TERM-talk alert blinking at the bottom of their screen, they could hold down the SHIFT key, press TERM, and then type "reject" at the "What term?" arrow instead of typing "talk." This enabled the callee to reject the call.

"TERM-talk was strangely useful but sort of weird," says Silas Warner, who was a PLATO user at Indiana University. "To be interrupted in your work . . . people would page you just to chat, which was distracting, to say the least, if you were deep inside some programming code trying to get something working. It was like having a total stranger right out of the blue, calling you up on the phone at night saying, 'Hi there. Wanna talk?'" By 1978 enough people had experienced this type of unwanted interruption via TERM-talk that CERL released a new feature, "TERM-busy," which made you unavailable from TERM-talks. If someone tried to page you, but you had gone into "busy mode," you would receive a message at the bottom of your screen saying "user name/group wants to talk with you" and that user would see a message saying that the user they wished to talk with "is busy, but has been told you called."

TERM-talk took PLATO by storm, and before long, a variant would be developed specifically to help CERL's PSO group of online consultants provide what would today be called "live chat with a support person." On PLATO the feature was called "TERM-consult," and it epitomized the can-do, how-can-we-help culture of CERL by enabling any author or instructor to get help from one of the PSO staff online in the equivalent of a TERM-talk. Instead of a PSO person hopping on a plane or train or bus and traveling to some remote PLATO dial-up site to help out a user with a technical problem, they could just make themselves available as online consultants ready to respond to a user's "TERM-consult" request.

What made online consults particularly remarkable was the fact that they took advantage of another new feature the systems staff added. TERM-talk required modifying the system code so that the typed output from one user showed up on another user's screen, and vice versa. Well, what would happen if you sent *all* of the output of one user's screen to the other user's screen? Today it's called "screen sharing," but on PLATO, decades earlier, the feature was known as "monitor mode." With monitor mode and online PSO consultants, it was possible for a TUTOR programmer to get expert help within seconds. For example, a programmer might be stuck trying to figure out how to use a particular TUTOR command, or perhaps the command was not doing what was expected, and the programmer's productivity had ground to a halt. A quick TERM-consult call could get a PSO person on the line, and instead of the programmer having to type a long essay on exactly what the problem was and what the section of TUTOR code looked like, they could switch into monitor mode with the consultant, allowing the consultant to see the programmer's screen. Suddenly, the consultant, who might be down the hall or ten thousand miles away, could see the very code causing the problem, and continue chatting, TERM-talk style, at the bottom of the screen with the programmer. A programmer learned quickly to fully test their code and consider every possibility before invoking TERM-consult, as there was nothing more humiliating than having an online PSO person point out a missing comma or other typo in the code. Monitor mode was so useful that it was added to normal TERM-talks; during one, either party could simply press SHIFT-LAB and take both users into monitor mode to share screens.

Nineteen seventy-three was a busy year for the rapidly evolving world of computer conferencing systems, primitive and crude as they might be, not only at the University of Illinois but at a handful of other campuses and research labs around the country. In addition to Discuss (the simple message forum that had gotten Stuart Umpleby in so much trouble), there were other places on PLATO to post messages and share information with other users. One lesson, Pad, was, like Discuss, a simple note-and-response system where users could post short notes and then other users could reply. Pad was a hangout with no particular purpose other than posting silly messages, observations, jokes, complaints, and non sequiturs. Naturally, it was very popular.

Even plain TUTOR source code lesson files themselves were used for communication. A whole series of lesson files, named "notes1" through "notes19," were simply source code files where instead of source code, users posted messages. But as everyone would discover, using a source code file for group communications was untenable and, in the long term, unmanageable. For one thing, it meant that only one person could edit the file at a time. As usage scaled, it became more likely that someone else was editing the file when you wanted to post something—so you had to sit and wait. "Reading and writing notes in this manner became a problem as the system grew," says Kim Mast. "There were limited cases of people destroying other people's notes. While users were encouraged to write the note in their own file and copy it into the 'notes' file, every once in a while someone would forget, and be editing notes for an extensive period of time, locking others out."

Any growing community using a time-shared computer that requires the honor system to facilitate the act of communicating between members of the community will eventually fail, because as the community itself grows, so grows both the number of people who won't abide by the rules and the number of people who forget the rules. The result is that things get out of hand. Because the "notes" files were open to all users, anyone could edit anything, innocently or maliciously. They could change what someone else said, post something and sign it as having been written by someone else, or even delete not just a single posting but all of the postings. It was clear that a *real* program, an official system program supported by the CERL staff, was needed that offered all of the benefits of these simple "notes" files but none of the disadvantages. One day in the early summer of 1973, another mass-deletion incident occurred in the "notes" files, and for Paul Tenczar it was the last straw. He assigned David Woolley, a junior systems programmer, a new project: go do something about the "notes" problem. What emerged on August 7, 1973, was a new feature on PLATO called, simply, Notes. David Woolley was just seventeen at the time, a UI freshman. He was one of the new wave kids, and like several other junior programmers at CERL, he had come from Uni High.

Notes would become Woolley's claim to fame. What started as a quick solution to an urgent problem would become one of the foundations of the PLATO online community, and would keep him busy for the next several years.

Woolley settled on a simple organizing schema for messages. A "note," which might today be called a "thread" or a "topic," consisted of a "base note," and one or more "responses," stored and read linearly in chronological order. The base note was the message left by whoever started the new note. The responses were messages left in reply to the base note (or in reply to subsequent responses). "I came up with a design that allowed up to sixty-three responses per note, and displayed each response by itself on a separate screen," says Woolley. "Responses were chained together in sequence after a note, so that each note could become the starting point of an ongoing conversation." Base notes and responses were both limited to up to twenty lines of text. If a user needed to be more verbose than twenty lines, they needed to enter the first twenty, then "reply" to their own text, and enter more text. But they'd better hurry, because someone else might—and often did—reply first.

That was it. Nothing fancy. But its simplicity made it understand-able, and that enabled the community to embrace it enthusiastically. Woolley also provided a simple notes index display showing the list of most recent notes, how many responses there were, if any, and the date that the base note was written. In keeping with the severe constraints brought on by disk space and screen space, base note titles were kept brief, just fifteen characters. The index display pages listed fifteen base notes at a time, along with the base note number, date it was written, title, and number of responses. The NEXT key paged forward, BACK paged backward. The result went far beyond the primitive notes1 through notes19 source code file editing the community had been rel-egated to before.

Tenczar would have been happy with a solution that allowed for a single response to a note, so that a systems programmer could indicate that the problem had been fixed or not. But Tenczar was considering only one use case: using Notes as it *had* been used among the systems staff to receive reports of system problems from the user community and then reply back with quick one-line responses. Woolley realized that a fully functional Notes program ought not be restricted to this one narrow "customer support" use case; rather, it should support con-versations of any type.

On August 7, Paul Tenczar posted the first note in the new "System Announcements" notesfile:

newnotes Note 1

8/7/73 11:07 pm CST pjt / s

Since you got here, you will undoubtedly note that we now have a new system of user/system notes. We hope that they will greatly speed up your browsing . . . and provide us much greater protection from note-destroyers!

Please direct any comments about these new notes to Dave Woolley.

Old notes are obtainable by editing files -notes1- through -notes19-.

The first release of the Notes program supported three categories: System Announcements, Help Notes, and General Notes.

"System Announcements" or =announce= (the traditional way PLATO users mentioned notesfile filenames online was to surround the name with equals signs) was the official word on new features, problems with the system, changes to the way things worked, and other news. Given that TUTOR was constantly changing (either one of the attractions of the language, or one of its biggest drawbacks, depending on who you spoke to), keeping up with the =announce= notesfile was crucial. A tradition developed fairly quickly that the person on the systems staff who actually did the work to create a new feature would get to post the news about the new feature.

"Help Notes" or =helpnotes= started out mainly as a place to get help related to PLATO but would become a fantastic resource to the community seeking help on pretty much anything: Need help programming something on PLATO? Trying to find a reputable automobile mechanic in Champaign-Urbana? Stumped as to which flea collar might work best for your dog? "Help Notes" was the place to ask your question. Oftentimes you'd receive a reply within minutes if not seconds.

"General Notes" was for everything else, and became known as "Public Notes" or =pbnotes=, for messages to and from the entire community.

Over the next three years, Woolley would continue to add features to the Notes program. By far the most notable change occurred in the early winter of 1976, when Woolley announced to the world that he was expanding Notes so that there would no longer be just three "sections" of the program, one for system announcements, one for help notes, and a general notes repository, but, instead, the program was being redesigned so that there could be any number of notesfiles, on any subject imaginable. The Notes program would become the engine that managed and presented these notesfiles, but there could be, and soon would be, thousands of notesfiles, each dedicated to a specific subject. A tiny sampling of some of the hundreds of public notesfiles that subsequently popped up included:

abortion	aerobics	anim
animbest	animations	announce

antic	apartments	apple
astronotes	audiofile	avatrade
bald	believers	bereans
bestwishes	bicycle	books
bowlers	brandx	calcnotes
cars	cerlnotes	chessnote
comix	commodor	consumer
donkey	dreams	drugnotes
ednotes	eenotes	elephant
empirenote	events	filmnotes
gamesnotes	garden	goldwing
hebrewnote	helpnotes	ibmnotes
ipr	jobnotes	jokenotes
karate	kidnotes	lawnotes
lunch	magicnotes	medieval
micronotes	musicnotes	nature
pad	pbnotes	petnotes
philosophy	photonotes	platopast
poets	recipes	relativity
romance	sayings	science
scifi	scouts	scuba
sexnotes	shuttlen	skinotes
skydive	soapnotes	spacenotes
sportnotes	startrekn	strange
tennis	theology	trains
trivial	tvnotes	unixnotes
videog	wantad	whonotes
wilderness	wine	wishnotes

One pair of notesfiles grew in notoriety over the years: =derfnotes= and =derfbest= (the latter being "the best of =derfnotes="). =Derfnotes= was a repository of derf messages. A derf (backward for "Fred") was a note written using someone's signon without their permission, usually because that person had gotten up and walked away from the terminal. To derf someone was to write a derf note, which might range from a quick "I are a derf" (often that is all the time one had to type before the derfed individual returned to the terminal) to long embarrassing essays on the risks of leaving one's terminal unattended. The more well known the person, the more tempting it was for someone to derf them.

While Dave Woolley worked for all of his undergraduate years improving and expanding the functionality of the Notes program, his old Uni High classmate Kim Mast had been assigned a parallel project, Personal Notes. Personal Notes was to be PLATO's email system. It worked very similarly to the Notes program: users would use the same text editor to compose a message, and similar keyboard commands to navigate among the messages. Personal Notes, or pnotes (pronounced p-notes), as they came to be called, was launched August 10, 1974. Now PLATO users, at least users with "author" and "instructor" signons (as opposed to mere "student" signons, who were using PLATO strictly for its instruction), could "email" each other.

One thing that particularly made Notes and Personal Notes successful is their exploitation of a subtle but hugely important aspect of the PLATO system's unique computer architecture. Thanks to the Fast Round Trip, the Notes and Personal Notes programs provided additional responsiveness to users. Even though a note's twenty lines of text might take a few seconds to plot on the screen, given the 1,200-bits-per-second bandwidth constraints of the time (agonizingly slow by today's standards), all the while the system was monitoring furiously for the slightest keypress from the user. The resulting user experience was quite something to behold compared to typical non-PLATO computer programs of the era. On other computers with display terminals, if you requested a display, a block of text on the screen, it was often very hard to interrupt and stop the display with any degree of accuracy or control. Oftentimes you had to press a certain key sequence, like Control-C or Control-S, multiple times to get the text to stop. And then who knows where you were in the program or whether the program would respond to any other input. Whereas on PLATO, as information began pouring onto the screen and within the first few words you realized this was not what you wanted to read, or that you'd seen it before, there were a variety of single-key keypresses that would immediately stop the display midstream, and go immediately to wherever it was you wanted to go. For example, if you pressed the LAB key while the text of a base note was displaying, PLATO immediately stopped, the screen instantly erased, and the text of the first response began to appear. Likewise, you could press LAB again, and jump to the second response. Then press BACK and jump back to the base note. Press the "9" key and it would jump nine responses forward. Press the minus ("-") key from there, and the system showed you the previous response, number 8. You could press a long sequence of keys in succession, staying ahead of the display but knowing exactly where you were, for example, LAB, LAB, BACK, 9, 9, 9, -, and PLATO was right there with you, stopping text from displaying the moment you pressed a key, and landing you on the twenty-sixth response to the base note. The entire PLATO user experience was this way: working at the speed of thought.

When a long list of notesfiles continued getting longer, the user community complained. Not because of the sheer number of notes- files, but because of how hard it was to a) remember them all and b) keep up with the conversations going on in them. You might visit a popular notesfile and stare at the list of notes: perhaps fifty new ones in the past few days since you last visited, and each of those fifty have maybe twenty-five or more new responses. People were overwhelmed. Where to start? The success of PLATO Notes brought on a happily solvable problem.

Dave Woolley added a DATA key option that enabled users to go through notes and responses chronologically. Another systems pro- grammer released a special -jumpout- feature that enabled PLATO authors to write their own programs that took advantage of a "cycler" tool that would roll through a given list of notesfiles and only show you what you had not already read. (The World Wide Web would respond to a similar need as the number of websites, news sites, and blogs exploded in the late 1990s and early 2000s, by offering RSS—"Really Simple Syndication"—feeds that kept track of what you had read and what was new.) Unfortunately, due to a poor implementation of the "cycler" function and overuse by enthusiastic authors, PLATO could not handle the load put on the mainframe hardware. Rick Blomme then directed John Matheny, another CERL systems programmer, to create a centralized, more efficiently designed, system-supported util- ity, which got the name "Notesfile Sequencer." It was an enormous jump forward—another catalyst that not only accelerated a PLATO user's productivity, saving them enormous amounts of time, but in a way contributed to the general "acceleration" of PLATO users themselves. As the sheer amount of information and conversations kept growing, users could not keep up, and needed new tools to help them cope. With the Sequencer, users could create a personal list of favorite, must-read notesfiles, be it five or five hundred long, and the Sequencer would then automatically step through every single notesfile and only show the user those notes and responses the user had not yet seen.

Around the time that Dave Woolley originally built Notes, in August 1973, Karl Zinn, a professor at the University of Michigan with a long-time interest in computer-assisted instruction, had taken on a project with NSF funding to build an online forum. Zinn was active in the CAI community, knew everyone, went to the conferences, and published research. He knew Bitzer well, and visited CERL numerous times. He had seen PLATO Notes and related his impressions to a graduate student he'd hired named Bob Parnes. Zinn and Parnes called their system CONFER. "I showed Bob these various things, and wanted to borrow ideas from other places, or get somebody to agree that we could adapt something and create it for ourselves. So I would have been planting ideas with him and I influenced—every time he came back to me and said he was doing it this way or that way, I would say we really need more of this or that, because I was thinking of classroom use and he was still thinking of committee use at the time." CONFER wound up having mechanisms for voting on issues, but was also organized in a "comb" structure like Notes had been. A "comb" conferencing system is one with individual discussion files, each of which can contain individual "notes" or threads, each of which can contain a number of responses. This was how PLATO Notes was arranged. There was one other conferencing system that also inherited this idea from Notes, called PicoSpan, which was a commercialized version of CONFER written by a developer named Marcus Watts, who along with Larry Brilliant created a company called NETI to market and sell PicoSpan. Brilliant was a friend of Stewart Brand's, creator of the *Whole Earth Catalog*, and around 1984 Brand and Brilliant started working a conferencing bulletin board service running on PicoSpan called the Whole Earth 'Lectronic Link—The WELL—a service that launched in 1985. Computer histories, even recent best-selling ones, often cite The WELL as the first online community with a conferencing system. While PLATO Notes came out a dozen years earlier and had a larger set of users around the world, the two communities that emerged on these two platforms, PLATO and The WELL, were remarkably similar in how they conducted themselves, no doubt influenced by the "comb" architecture of forums, threads, and linear responses.

It is important to appreciate just how far ahead PLATO's suite of communications capabilities was in 1973–1974, and would stay that way for a solid ten years. Non-PLATO users had no point of reference to understand what PLATO users did or why they were so excited about working on the system. The sense of living an accelerated "digital life" was something only an actual PLATO user could understand. One way to illustrate the point is to consider a typical 1970s family dinner conversation. Family members might talk about, say, what they had done that day or what was in the news. If one family member was a PLATO user, they might casually mention that they talked with people in Hawaii, Delaware, New York, Florida, Illinois, and other places all that day. Other family members might react with horror at the cost of the long-distance phone bills for all those calls, until the PLATO-using family member pointed out that, no, these were TERM-talks. "TERM-what?" would be the typical response. The rest of the dinner might be spent attempting to explain to the puzzled, increasingly concerned family members what exactly TERM-talk, Talkomatic, pnotes, notesfiles, TERM-consult, and the Notesfile Sequencer meant. Most likely, family members would have fled the table long before the explanation was over.

Contrast this with a hypothetical family today: if they're gathered around a single table at all, they probably all have a fork in one hand and a smartphone or tablet in the other, each family member lost in some private interaction on their personal screens, oblivious to the cares, concerns, or interests of other family members, everyone doing largely the equivalent of TERM-talk, Talkomatic, notes, pnotes, the Sequencer, and the rest.

Deep down Don Bitzer was a hardware person, an electrical engineer through and through. All of these communications features added to PLATO, largely by undergraduates and sometimes high schoolers, were not part of Bitzer's original vision. But even he could see the impact they were having on the system. Bob Yeager, who worked on the elementary reading project, had one of the more comfortable offices in CERL, which made it fair game for demos to visiting guests. But the main reason his office was chosen for frequent demos was that it was one of the main rooms for the elementary reading project, which, he says, "had all the gadgets. We had audio, we had slides, we did touch, and everything." Often Yeager would be working away at his terminal when Don Bitzer would appear with some guests and have him move aside so Don could "drive" and do his magic. "The early demos were always the lessons," Yeager says, "and he would show those various things. . . . But someplace around '74, '75, the demos changed. He started doing demos of TERM-talk, of being able to see somebody else's screen and work on it, Notes . . . and then he would do a little bit of stuff on the lessons, but I don't think he did that on purpose, but I think gradually he himself began to see the communication aspects that we now call the Internet and so forth, that he had there in PLATO. The lesson side, computer-based education stuff, became less important. He was so dedicated to that, that's been his life, but I think he saw this other opportunity of communication and I was able to watch that and it was just fascinating to see how, looking back, how subtly more and more time went to the communications and less and less time went to the lessons."

It is tempting to assert that the PLATO system was the birthplace of social networking and social media. The historical facts suggest otherwise; at best, the answer is a nuanced "not really, but early signs had begun to show."

Consider the impact of another PLATO system feature, Access Lists, on the online community. Access Lists were customizable lists of users for whom access should or should not be granted or restricted to some file on the system. The notion of access control had been around forever—starting with passwords on files to protect who could view or edit a file. Every time-sharing computer system had to deal with security features like these; PLATO was no different. With the explosion of new notesfiles on PLATO in 1976, it was possible to designate one or more "directors" of a notesfile, as well as who had and who didn't have read/write, read-only, or even write-only access to it (=psonotes= would be write-only to all users except the PSO staff, and served as a place to privately ask a question or report a concern to the PSO consultants). That led to a general-purpose Access List facility that could even be applied to a TUTOR lesson. A file's owner could specify custom definitions of access, which might have special relevance for that file only. This new emphasis on access control added a new dimension to PLATO that had not been there before. Naturally, the user community figured out uses for it that its creators had not dreamed of.

The way some users utilized Access Lists in notesfiles provides a tiny hint at the phenomenon of social networking that was coming in several more decades. Dave Woolley takes a different view. He differentiates what he saw on PLATO versus what we see today in social networks this way: with PLATO notesfiles, he says, "the first order question was 'what'—what do I want to talk about, what project are we working on. With Facebook (and Twitter and LinkedIn, etc.) the first-order question is 'who'—who do I know, who do I want to be in relationship with. Of course, both PLATO and Facebook provide ways of addressing both organizing principles, but their basic natures, or affordances, are different. In a 'what'-based system the access list is secondary; in a 'who'-based system the access list is primary."

In Woolley's view, the geography of the PLATO system influenced how people organized themselves around "whats"—particularly when it came to notesfiles. If you had an interest in movies, you gravitated to the =filmnotes= notesfile, the place for people with an interest in that topic. If your interest was in religion, there were numerous religious notesfiles you could visit. If your interest was music, there was =musicnotes=. If your interest was sports, there were countless notesfiles on various sports. If your interest was interpersonal relationships, lesbian issues, gay issues, or other sexual-identity issues, again, numerous notesfiles, some with the "Anonymous Option," were available. (The Anonymous Option enabled users to post notes and responses anonymously, the theory being that users might be more open to participating without the burden or embarrassment of revealing their identity.) Woolley argues that the center of the universe in PLATO was the "what"—be it a game, a lesson, a notesfile on a certain subject, or whatever. Present-day social networks like LinkedIn and Facebook are completely different, having architectures entirely focused on the "who"—you as user are the center of the universe for these services. You can "friend" or "follow" other people, and the system will keep track of them and aggregate their status updates on your "feed." PLATO did not have social networking tools like friending, following, sharing, or likes. The World Wide Web would have to arise to give birth to hyperlinks and those sorts of tools first. Says Ray Ozzie, who in the 1970s was a UI undergraduate and also worked as a programmer at CERL, "In my experience with PLATO, social and relationships were strong overlays that we experienced hand in hand with the technology. The system and the activities brought us together, but we maintained the social network in our minds, hearts, souls. Not in a friends list."

But as more and more PLATO users created private notesfiles over the years, particularly beginning around 1980 and beyond, there was a shift. Participants in these private notesfiles returned to them each day not because they were a "what," but more because of the "who"—that is, they sought out news and updates from those happy few designated as having access (thanks to the Access List) to that private place. These users belonged to small private groups of people who chose to connect to each other. It defied all of the later computer-science definitions of a true social network, but the people involved at the time didn't care: they were *connected*, privately, and that's what mattered. Cliques of users had been around on PLATO IV since the beginning—recall the private channels possible in Talkomatic—but they abounded as PLATO expanded around the world in the 1980s. Oftentimes these private notesfiles were made up of a couple in a relationship, or friends, current or former classmates, and current or former co-workers, all using PLATO as a way to stay in touch, exchange stories, observations, complaints, gossip, as a way to *connect*. To belong to a private notesfile consisting only of friends was a way to use a computer system to communicate comfortably in a trusted forum that outsiders could not see, invade, or ruin. Much like Facebook today. Likewise, the Sequencer can be seen as a primitive news feed or timeline billions of people use every day on social networks. Not in the sense of how information was displayed, but in the sense of how information was compressed and filtered, enabling a PLATO user to scan through vast amounts of messages to, in effect, find out "what's new" each day.

PLATO was emphatically not a social network as we would define one today. The term did not even exist then. But the way users began to organize themselves around *each other*—as opposed to how PLATO's creators thought users wanted to be organized around "whats"—shows that the online community had begun to evolve. They had embarked on a path that can be confidently viewed as baby steps toward what is commonplace online today.

15

Empire

The first thing they noticed was the ears. They were just plain *wrong*. They seemed tiny—really, really tiny—when they were supposed to be big and pointy. That would be strike one against him: the absence of that universal trademark, his Vulcan ears. And he wasn't wearing his regulation blue shirt, black pants, and black boots. No Federation logo in sight. No communicator, phaser, or tricorder either. And what was he doing with that scruffy beard? What was *that* about? It was like he'd gone AWOL. It was against regulations—and his hair! The whole *image* was wrong. What were those things on his face—glasses? *Glasses?* Gone was that distinctive, trim, straight-as-a-ruler, Federation-issue haircut so familiar to millions. Instead, this mere *human's* hair was longer, scruffier, one might say, normal?

Strike two against him was the distinct whiff of alcohol that followed him around during the visit, at least that's how many who were there at the lab that day remember it—for some of them it was the first thing that came to mind when asked about the visit decades later. "He was drunk," says Bill Golden. "There's no question." Even Bitzer remembered the alcohol smell.

Spock without his Vulcan ears. A few days' start on a beard. Smelling like booze.

It was Tuesday, May 7, 1974. Actor Leonard Nimoy was in town, on a press junket, meeting with reporters, grabbing a bite in the back room of a local restaurant (where Nimoy, more interested in talking about his serious acting, grew aloof at reporters' incessant *Star Trek* questions—didn't they realize the *Trek* series had ended five years earlier?), stopping by the television stations for even more meet-and-greets and promotional interviews, shaking hands with local dignitaries, all these hi-how-are-ya's an effort to fill the seats for his "raunchy" (according to one newspaper review) stage performance as the rebellious lead character R. P. McMurphy in *One Flew Over the Cuckoo's Nest* at the Little Theatre on the Square, an hour's drive to the south in the tiny Illinois town of Sullivan. The show had just opened a few days before and was going to run for a couple of weeks. Nimoy wanted the world to know he was a serious stage actor—and what a run he was on during this time: *Oliver!*, *Camelot*, *Fiddler on the Roof*, *The King and I*, and, now, *Cuckoo*—but the throngs who still watched *Trek* reruns religiously on TV were saying: *Not so fast.*

And now here he was in CERL. The voice was the same. It was him. But it wasn't. Those *tiny ears*! Nimoy and his entourage made their way down the first-floor hallway, past classrooms full of terminals, and then he spent a lot of time admiring the fish in CERL staffer Susan Rankaitis's office aquarium. The entourage, led by university vice president Barry Munitz (the same Munitz largely responsible for the university administration reorganization that led to Dan Alpert's removal as dean of the Graduate College), eventually made its way up the stairs to the fifth-floor penthouse, home of Sherwin Gooch's music lab.

They entered an office shared by Doug Brown and David Woolley, who watched as Nimoy wryly commented on an ASCII printout (of a naked woman) taped to the wall. Connie Brown, a UI undergrad at the time, remembers Nimoy being cranky—everyone wanted the green-blooded, pointy-eared, exceptionally logical *Spock*, not this mere man, this actor named Leonard-something, this down-and-out, can't-get-a-movie-deal, struggling actor who would, only a year after this visit, publish his bridge-burning autobiography entitled, fittingly, *I Am Not Spock*. Nimoy wasn't giving them what they wanted.

They sat him down in a chair in front of a PLATO terminal to bask in its Orange Glow and witness the future. They showed him games, including an early version of Empire, PLATO's multiplayer space war shoot-'em-up game with its tiny little Federation starship. Sherwin Gooch showed him how PLATO played music through his Gooch Synthetic Woodwind. Revolutionary at the time, this was a music synthesizer connected to a PLATO terminal through which one could write and play music, even view musical notes in their full graphical glory on-screen, and touch the notes and hear them played over speakers or headphones. It was inconceivable, like so much of PLATO at the time. So far ahead of a future people weren't ready to comprehend, they often simply didn't comprehend it.

A local newspaper photographer snapped a picture of Nimoy watching the PLATO terminal's screen. Along the office wall rose a stack of countless crates of empty soda bottles, which the denizens of the PLATO penthouse collected until someone got around to taking them all the way downstairs to the soda machine on the first floor. Usually no one got around to taking them downstairs, and so the pile grew until it stacked up to the ceiling. The stack itself was a thing to behold now, so why mess with it?

While Gooch and the others were demoing PLATO's music capabilities to the star, Nimoy mentioned to the onlookers that he had recorded albums of his own. The penthouse gang politely stayed mum about Nimoy's oddball, campy album *Mr. Spock's Music from Outer Space*, wherein "his serious tone and refusal to break character makes this a first-rate experiment in comedy," according to one review. "I remember I was too embarrassed," says Gooch. "We had the record there in the office, and he was *so bad.*"

And then came strike three. Nimoy could not play chess.

That was the last straw. "We had heard he was a big, real-life chess fan," says David Frankel, at the time the youngest member of the CERL systems staff. While Nimoy was seated down in front of the terminal one of the things they showed him was the PLATO chess program. No ordinary chess program, this. A marvel at the time, considering that the PLATO chess program featured a fully graphical display, replete with pictures of kings and queens, rooks, pawns, bishops, and knights, but in fact all that sizzle was nothing compared to the steak behind the scenes: PLATO's chess was a mere front-end program, like a web browser or smartphone app of today, a *client* program, connected to a remote *server* running "Chess 4.2," a then renowned chess program created by Larry Aikin and Dave Slate at Northwestern University that ran as a separate program on CERL's mainframe, independent of PLATO, but connectable thanks to a programming interface they'd hacked together. This was not a toy, this was the real thing, and when it came to chess, this computer was *good*. And, if you were not careful, hard to beat.

But to the shock and dismay of the gathered onlookers, the ultralogical Spock in real life knew nothing about chess. "I didn't expect Nimoy to actually compete with the computer," says Frankel, "but I figured he'd move a few pawns around and be amused that the computer could interpret his actions and respond. Plus our graphics were pretty sweet—most chess programs at the time were purely alphanumeric." Nimoy's Vulcan counterpart was celebrated as not only an expert at playing chess, but an expert at 3D chess. To discover that in real life the actor didn't know chess at all was devastating to the gathered *Trek* fans. "People kept staring at his ears because they looked really, really small," Sherwin Gooch recalls. "We had him there at the terminal, we logged him into chess, and he said, 'Oh, I don't know how to play chess. Everybody thinks I know how to play chess but I don't know how to play. . . . But I know how most of the pieces move,' and then we all of a sudden went like, *uggghhhh*."

"We had a tough time getting any real action out of Lenny," recalls Frankel. "He kind of nodded at the screen and our demo was very brief. He wasn't very talkative and he looked like shit."

It is somehow fitting that Empire, perhaps the most famous PLATO game ever made, a game deeply influenced by *Star Trek*, should have originated in Iowa, birthplace of James Tiberius Kirk, captain of the *Enterprise*. The creators of Empire were *Star Trek* fans, all from the Midwest, all entering college in the late 1960s and early 1970s. The original idea came from John Daleske, a student at Iowa State University. An education professor there had heard about PLATO and arranged to get a terminal installed. A friend of Daleske's then heard about the strange new terminal with its Friendly Orange Glow and urged Daleske to check it out. In the spring of 1973, Daleske did.

Born in 1953 in Des Moines, Daleske was playing around with computers long before college. In the eighth grade he built a tic-tac-toe–playing computer out of old relays. In high school he entered the science fair with his own design for a general-purpose computer. He learned to program in COBOL in the ninth grade and worked on an IBM 1401. He spent the summer after high school graduation in Finland on a student exchange program, then returned to enter Iowa State University in the fall of 1971, starting out as an aerospace engineering major. That fall he took a computer math course, where he programmed in BASIC on punched paper tape. He found the classwork so easy he'd finished all of his assignments within the first week of classes, and wound up being asked by the instructor to help teach the class. "Programming came easy to me," he says. Before the aerospace engineering classes began in the spring of 1972, the departmental dean had discouraged him from pursuing it as a major. The students had been summoned to meet with the dean one day to get the lecture: "Unless you are willing to work with a close friend on a project for a year or two, then turn around and stab him in the back, this program is not for you," the dean announced. The requirements: "You will not miss a single class in the sequence. Some of the classes are only provided once every four years. Miss one, and you're out."

"A very strict schedule with no alternatives," Daleske says. He switched majors to distributive studies, "kind of like an honors program," he says, winding up with "the equivalent of three majors: computer science, mathematics, and philosophy, with multiple minors." The draft was still on in 1972 and people were still dying in Vietnam. Daleske was drafted and wound up spending the summer in boot camp with the Marines, but managed to return to college in the fall of 1972 to start his sophomore year.

For all intents and purposes, Daleske had his own personal PLATO terminal. Few others had caught the PLATO bug yet at Iowa State, and no one else understood that the poor geek who spent his days and nights in front of that strange orange screen wasn't hiding from society—he was right in the thick of it. It just happened to be online. He began arranging his class schedule around PLATO. He would have an early dinner at the dormitory, then dash over to the education building before the guards locked it. "I would stay in the room, with the lights off so that the guard would not know I was there," says Daleske. He might sneak out for a pizza later in the night, and upon arriving back at the building find that picking the lock was easy, and he'd go back in. He did eventually get caught. "A couple of times," he admits. But he kept coming back.

Far more worrisome was the day he was summoned to the office of the dean of the School of Education. After his experience with the dean of the School of Aerospace Engineering, this was not something he was looking forward to. But to Daleske's surprise, the education dean was not there to bust him, but rather to *encourage* him. "The dean gave me a letter indicating I could be in the building anytime I wanted and also gave me a key to the building," says Daleske. "He said he didn't want to have to replace the lock because I was wearing it out."

At one point Daleske took a computer science course programming an IBM 360, a big, old, dull mainframe computer. Daleske wouldn't be the last student, already contaminated—spoiled—by PLATO, to take a computer science course and not only find it tedious, but also get frustrated by the inept, old-school, dead-tree fan-fold printouts and noninteractive batch-job punch-card method of doing course assignments. *My God, haven't you seen the orange screen? Here, let me show you.* He cut a deal with the professors, who let him use PLATO to do his homework. "The teachers loved it because this gave them a buffer of time to distribute to needy students," he says. And in the process the Friendly Orange Glow cast its spell on a few of the CS professors as well, intrigued by the graphics capabilities of PLATO. *This isn't like our computers.*

Soon he was back to spending all night on PLATO, alone in the education building. When CERL's preventive maintenance period struck around 6 a.m. and kicked him off the system, he would stumble home and sleep a few hours until his classes began mid-morning. New games were beginning to spring up on PLATO all the time, as more and more people caught the game bug—if not *playing* games, then *writing* games. (One of the first commands you learned in TUTOR, especially if you were writing a game, was -randu-, which one used to generate random numbers. The -randu- command took on a bigger meaning over the years. It was the god of chance and luck. More than one PLATO person has been known to utter a wishful statement only to follow it with "Randu willing.") Daleske tried out all the Big Board games. But he had also learned that people were *writing* them. At the time he was taking an education class that required students to do a project. Originally he was going to do a dull project like every other student taking the class. But as he explored PLATO it was clear he needed to change his project to be something PLATO-related. He needed disk space. Disk space on PLATO was so expensive that it was very carefully guarded, and it was difficult to get. Unfortunately, Daleske didn't have access to any, or so he thought at the time (in fact, Iowa State would have provided him some, had he but asked), and he couldn't simply go create some. He thought he needed outside help. So he went online and contacted a gamer in Indiana who was spending much of his life in front of the Orange Glow as well. His name was Silas Warner.

—

Warner was born in 1949 in Chicago and moved to Bloomington, Indiana, in 1958. A fascination with computers beginning in the sixth grade continued through his undergraduate years in the late 1960s at Indiana University in Bloomington. "In theory I was a chemistry major," Warner says, but he ultimately graduated with a degree in physics in 1970. He stayed on at the university into the early 1970s, working as a computer programmer for the psychology department.

One day, a "big iron box," as Warner remembers it, arrived at the physics building. It was an early PLATO IV terminal without a touch screen panel. Another terminal arrived about two weeks later, for cannibalizing parts, in case the first one failed. "I got onto the PLATO system," says Warner, "and started making it do things, and of course discovered, as many other people about the same time were discovering, that they *thought* they had invented an educational computer, when in fact what they had *actually* invented was the greatest pinball machine ever created up to that time."

Four other terminals eventually wound up on campus, two in the education building and two in the journalism building. The journalism department was enthusiastic about PLATO, and had begun developing a number of instructional lessons, including some on cropping news photos, using PLATO's microfiche slide projector and superimposed graphics. The slide mechanism in the terminal wasn't very accurate and students first had to move a cursor onto some landmark in the photo to calibrate the terminal, and then do the cropping.

Warner had other issues with the microfiche projector in the terminals. "The real problem," says Warner, "was that the slide mechanism required compressed air. . . . Nobody had ever thought of putting a compressed air supply in a computer room. So the solution was to go to the chemistry department and get a couple of surplus gas bottles which originally held 3,000 psi of air, but were now down to 150 psi, screw a regulator on them, and attach each one to a plastic hose up to the slide projector. Well, of course, the air bottles wouldn't last very long, if they were kept in continuous use. We had to tell students to turn on the valve and watch the pressure go up to the green mark before they started their lesson, and turn it off at the end. And of course they always forgot to turn it off at the end, so the air bottles ran out." Warner got tired of having the air bottles run out on him. When he came in at night, he found the bottles empty. Or students would come in and find that the lessons wouldn't work because there was no air pressure in the bottles. Out of frustration, Warner then invented what he called the "Backup Non-Automatic Microfiche Propulsion Module." He took the hose off the slide projector and hooked it to a bicycle pump. Then he put a sign up on the wall:

IF SLIDE PROJECTOR STOPS,
TAKE FOUR STROKES ON BICYCLE PUMP,
AND GO BACK TO YOUR LESSON.

It worked!

And then there came the night that John Daleske contacted Silas Warner online via the Pad lesson. Daleske needed file space. Did Warner have any? "He was talking about this great game that he wanted to build," Warner says. "Big Board games were the only thing out there and he wanted to get away from the Big Board concept altogether . . . a sort of mini-universe that would run more or less continuously."

The idea had just come to Daleske one day, "just popped in," he says. Silas gave him some file space and Daleske then spent five solid days designing and writing the first version of what he called Empire. He had to finish it in five days, in time for grading for the education class. He got it done, announced it online on PLATO, and played it for a couple of weeks. "Silas liked it," says Daleske. "He provided some comments on the design, so we added a few features. This version was played during the summer while I redesigned parts to make it more interactive."

But there was a problem: Daleske had only one lesson file (a fixed amount of disk space for a PLATO program) to make his changes, and that was the version that was live on the system available for people to play. "This made it very difficult to design, write, edit, and test, because this implied kicking off players periodically," he says. Such was the nature of PLATO that when a lesson was updated or modified, it couldn't be tested without first terminating the version of the program currently in use. To avoid annoying users he worked late into the night, just like CERL's system staff, writing new code and fixing bugs. He would then test the system after the nightly PLATO system shutdown at 10 p.m. when it would stay down for about fifteen to thirty minutes, then come back up, in a new form—the "non-prime-time" form—with a different login screen, full of warnings about system maintenance and possibly unstable conditions meant to scare off students and instructors alike. This was development time. (And *game* time.)

The first version of Empire was primitive: one screen containing one "universe" consisting of eight planets. It only supported eight people at a time, but that was six more than any other game of the era. (Indeed, this first version of Empire may be the first graphical, interactive multiplayer computer game anywhere that supported more than two players simultaneously.) If you weren't one of the lucky eight players, you had to settle for "lurking," standing in line, watching the game from the outside.

That fall, Daleske and Warner decided to undertake a journey, what was widely known to PLATO people as *making the pilgrimage to Mecca.* "Mecca" in this case was CERL. If you were a PLATO-holic around this time, no matter where you lived, no matter how much sacrifice you had to make, no matter what the cost in time, relationships, jobs, academic standing . . . you simply had to make the pilgrimage to see what this damn thing was all about. Daleske arranged to take a bus in Iowa east to "Chambana," as some called Champaign-Urbana (also called "Shampoo-Banana" by some). Silas would hop on a bus from Indiana and head west. They would meet at the Illini Union, the student union building in the heart of the UI campus. Daleske, at nearly six foot three and broad-shouldered, was stunned to meet the imposing figure of Warner in person. Daleske recalls Warner having to "almost bend down and turn a bit sideways" when he went through the doorway at the Illini Union, "and these were Union doors. I felt significantly dwarfed." Warner was six foot nine and weighed over three hundred pounds.

They went and found CERL and discovered rooms full of PLATO terminals. "It was like going to computer-geek heaven," says Daleske. The keyboards on some of the PLATO IV terminals at Illinois were newer, with little bumps added on the F and J index-finger keys for touch-typists (many keyboards today still have those bumps). Silas would have none of it. He hated the bumps, and they had to go. For every terminal he sat down to work on, he'd chip the little plastic dots away, shaving them off until the keys were smooth.

Warner and Daleske discussed Empire game design ideas but mostly spent the time play-testing the game and observing how other PLATO users played it. Daleske noticed that whenever Warner hit a lull when he was working on PLATO, he would reach into a pocket, pull out a book, and read a few pages. Then he'd put it back, and do some more work. After a while he would reach into a different pocket, pull out a different book, and start reading from that. This CERL pilgrimage marked the only time that Daleske and Warner would meet face-to-face.

Back home, and as time wore on, they began having differing ideas of what direction to take Empire. Silas preferred sticking to the original design, evolving it over time, but not changing the basic gameplay. He wanted multiple universes, not just one, so that more people could play the game, and there'd be more game to play. They decided to make a copy of the game, and Silas's version would be called Conquest, letting it take on a life of its own while Daleske took Empire in a different direction.

"In the first version of Empire," Warner says, "eight players could play against each other. They were not allied in any way." No Klingons, Feds, Romulans, or Orions—all that would come later. "Just players, one to a planet. They were in fact controllers of each planet." Like later versions of Empire, the first version of the game had little icons representing the spaceships. But instead of piloting the spaceships, players would simply direct the spaceship to go from one planet to another, and when a ship arrived at another planet, a player could trade with that planet, or fight with that planet, or drop bombs, and so forth. Spaceship combat was automatic: if two spaceships got within a certain distance you would either have the choice of passing or fighting. If they fought, the battle was automatic. "That version of Empire was actually continued after the second version, under the name Conquest," Silas recalls. His version evolved to support six universes, each a sort of separate level of the game, where you could jump from one to the other. "Usually, universe 1 was always full," Warner recalls. "Universe 2 was sort of halfway there, and there might be a pickup, an arranged game, in universe 4."

Daleske kept at it, redesigning and improving the game, sometimes making changes that players didn't like. As time went on, the Iowa State campus got a few more PLATO terminals, the education building having two, and the computer science building having its own

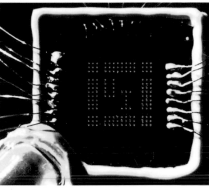

An early prototype of the PLATO plasma display, c. 1967 *(Courtesy Paul Tenczar)*

ERL building, late 1970s. Donald Bitzer's office occu-
ied the second-floor room in the legs of the tower.
'hoto by David Woolley)

Demonstration of PLATO IV terminal's touch screen, c. 1973 *(Courtesy Paul Tenczar)*

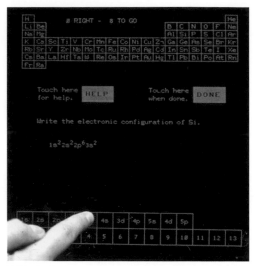

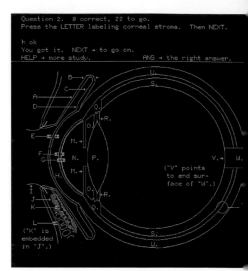

UIUC chemistry lesson involving the use of a touch screen, c. 1976 *(Photo by Ruth Chabay)*

EYEQ, a review and self-quiz of eye anatom by L. C. Helper and John Silver, UIUC Scho of Veterinary Medicine, 1974 *(University of Illino Board of Trustees)*

Fruit fly genetics lesson by David Eades, Gary Hyatt, and Paul Tenczar. It was one of the most frequently demonstrated lessons on PLATO. *(Courtesy Paul Tenczar)*

CERL's PLATO operators, 1982 *(Courtesy Jeff Johnson)*

Jo Bultman, Tim Halvorsen, and Bob Rader working in CERL's
fourth-floor machine room *(Photo by Paul Tenczar)*

Richard Powers sitting by the sole remaining PLATO IV terminal
in the foreign languages lab, 2003 *(Photo by Brian Dear)*

Title page for Moria by Kevet Duncombe and Jim Battin, c. 1976
(Courtesy Kevet Duncombe and Jim Battin)

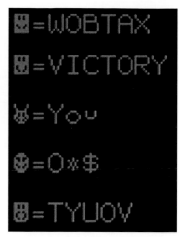

Welcome screens customized for holidays were a PLATO tradition. In addition to the one for Valentine's Day, there were screens for Halloween, Thanksgiving, Christmas, Easter, the Fourth of July, and other holidays. *(Courtesy Paul Tenczar)*

Examples of letter combinations that, when superimposed, created PLATO's unique form of emoticons *(Screenshot by Brian Dear)*

David Graper at the University of Delaware, c. 1977 *(Photo by Bill Lynch)*

David Woolley at CERL, c.1973 *(Photo by Paul Tenczar)*

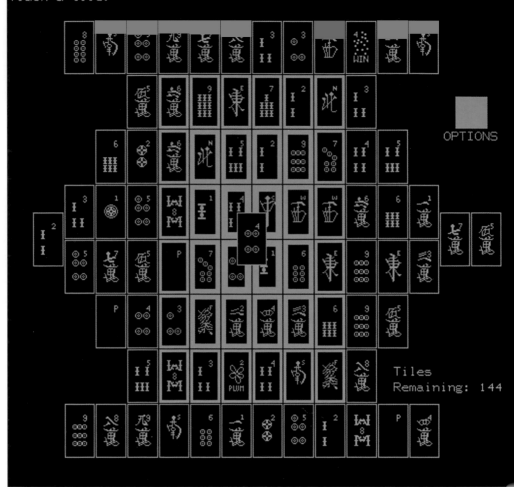

Brodie Lockard's Mah-Jongg game for PLATO, c. 1983 *(Courtesy Brodie Lockard)*

Jock Hill's prototype CDC portable plasm_
terminal built inside a Samsonite briefcas_
Each key on the keyboard was designed to be _
seperate plasma display to support keys in a_
language. *(Courtesy Bob Morris)*

Donald Bitzer troubleshooting international phone lines in
Moscow, USSR, November 1973 *(Photo by Paul Tenczar)*

Paul Tenczar, surrounded by Russian interpreters, at a PLATO
demonstration in Moscow, November 1973 *(Courtesy Paul Tenczar)*

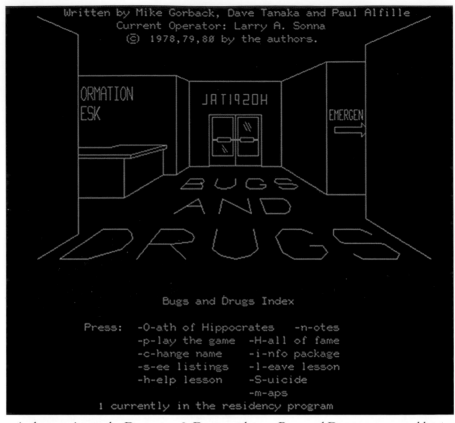

A clever twist on the Dungeons & Dragons theme, Bugs and Drugs was created by a group of pre-med students in 1978. *(Courtesy Mike Gorback, Dave Tanaka, and Paul Alfille)*

President Ronald Reagan touches a terminal screen at a Control Data Institute PLATO demo, 1981. *(Courtesy Charles Babbage Institute, University of Minnesota Libraries, Minneapolis, Minnesota)*

two. Other Iowa State students had caught the PLATO bug, including Mike Rodby, Chuck Miller, and Gary Fritz.

Miller had entered Iowa State in the fall of 1970. Fritz arrived in the fall of 1974, and started using PLATO by using someone else's signon, not realizing that was a serious no-no. "I was so naive," Fritz says. "At one point I found out who Mike Rodby was and I was talking to him and everything, and I said, 'Oh, here, I've got an author login, you want to use it?' And he said, 'Wait a minute! That's mine!' "

Daleske taught them TUTOR. "I sat by his shoulder," Miller says, "and said, *Why'd you do that?* and *Why'd you do that? Whatcha doing there?* and kind of learned how to program on PLATO from him." Fritz went through a similar apprenticeship. "It started out with me being the wide-eyed newbie at his knee," says Fritz. "He was explaining all the concepts of game writing and stuff like that, and I remember really clearly the time when I suggested a way to do something and he said, 'How do you do that?' and I said, 'You mean you don't know how to do that?' and he said 'No, I've never heard of this.' And I said, *Wowwww.*" For Fritz it was like graduating to be Daleske's equal, his peer. From then on, he says, "we taught each other a lot of stuff.

"We had a really good connection," Fritz says. "There was something that we had been trying to come up with a solution to something. . . . We had just been talking along for hours and just kind of ran out of steam and we were just kind of sitting there staring into space and frustrated and all of a sudden at the same second we both looked up and I said, 'Hey! What if!' and he said, 'You're right! That'll work!' and I said, 'Let's go do it!' We had the exact same idea and we never said a word, but we knew what we were talking about."

Daleske ran into another problem: getting printouts of Empire's TUTOR source code. They kept magically walking away. If you wanted to print out your code, you had to submit a request through the system, which would eventually cause a printer inside the CERL building to spew forth a stack of fanfold paper with your code printed on it. These printouts would be put on a desk and eventually picked up if you were on campus, or mailed out if you were elsewhere. But, given the nature of the thing being printed in this case, and the fact that some of the operators at CERL loved Empire, the Empire prints would often disappear, never to arrive in Iowa, or would arrive very late. For some, a printout of Empire was as good as legal tender. "I was rewriting at one point and did not receive a listing for about two months, until I complained and learned what was happening," Daleske says. Legend has it that many PLATO games got started after their authors first got hold of a printout of Empire's source code.

The printout delays were so bad that Daleske finally had enough, and started memorizing the code. Every single line. "I recall one bizarre episode where I was sleeping and I think Gary called in the middle of the night with something needing fixing," Daleske says. Right from memory he told Fritz which exact line of code to go to in which exact section of the program. "Gave him the fix and went back to sleep."

Daleske continued to expand and modify Empire into what some players felt was an increasingly complex game. One version was far more strategic, supporting a hundred simultaneous users, but some players, including Chuck Miller, missed the arcade-like aspects of an earlier version. Miller decided to create a more action-oriented Empire in a lesson file he had available called "michelin." By 1976, Daleske saw that Miller and now Gary Fritz were onto something with Michelin, and the decision was made to create a new version of Empire that merged the ideas of Michelin with one of Daleske's more action-oriented versions from a few years earlier. The result was Empire IV, and this turned out to be the biggest hit of all. At least, it had the potential to be: but the PLATO system was not quite designed to handle the demand the game put on the CYBER mainframe.

PLATO being a time-sharing system meant that everyone using it was allotted only a small slice of CPU time at any one moment. For students using PLATO for its intended purpose this was generally not a problem, as most PLATO lessons were fairly modest in terms of the computer resources they consumed. The system had what could be described as a "governor," to limit how big a CPU time slice was allocated to each user: that way no single user could monopolize the precious resources of the system. The limit was set at 20 TIPS (thousands of instructions per second) per user. There were two processing modes: foreground and background. You could switch between the two modes by doing a TERM-foregnd or TERM-backgnd. "Foreground" mode attempted to guarantee the user a certain level of CPU performance as long as your average TIPS didn't exceed the limit. "Background" mode users got what was left over from the foreground users on the system. Most of the time, that was the thing to do if you were a gamer, so most gamers did a TERM-backgnd as soon as they entered a CPU-intensive game like Empire. To help users switch into background or foreground as needed, the Empire authors added TERM-b and TERM-f to the game, so they wouldn't have to type as many letters.

Users needed to maintain an average of less than 20 TIPS in order to work without sudden interruptions. For PLATO game players, maintaining an average of 20 TIPS was always a problem—there was just not enough horsepower. The bane of PLATO game players was the dreaded autobreak, which is what the "governor" would do to the program you were running once it exceeded the TIPS limit: your terminal just froze. It might freeze for just a fraction of a second, or it might freeze for several seconds at a time—you never knew beforehand. When you're surrounded by hyperjump-happy oreo cookies (as the Orion ships and torpedoes were often called) as well as Fed torpedoes and phasers from Klingons and the action's furiously hot, the last thing you want is for your terminal to lock up, especially when it's just to share the central computer's scarce resources with students and people doing serious things on the system.

As things kept getting better for PLATO, in terms of more funding and more students and more educational institutions signing up for it, and more instructors using it in their curricula, things kept getting worse for the gamers. As PLATO grew in the mid-1970s, with more and more simultaneous users vying for time, the system engineers at CERL instituted a cyber-austerity regime, lowering the TIPS limit to 15. "This forced another rewrite," Daleske says, as Empire had become sluggish, nearly unplayable, with stutters and stops and exasperating pauses from auto-breaks occurring at the most inopportune moments of combat.

While Daleske kept working from the PLATO terminal in the education building, Chuck Miller and Gary Fritz worked from the computer science building, which now had two PLATO terminals of its own. "Pretty much Daleske, Fritz, and I monopolized the terminals as much as we could," says Miller. "Pretty much all night . . . and we'd get a gallon-sized thing of Spanish peanuts, and bring in either a six-pack or a gallon of Kool-Aid, and waste our lives away."

Miller loved the game but grew frustrated at Daleske's lack of progress in rewriting Empire to behave better under the tighter TIPS constraints. Daleske's excuse: a girlfriend, who was "taking a lot of my focus," he says. So Miller and Fritz got more involved, with design input from Mike Rodby. (Rodby's PLATO fling was brief. "He's the only person I know who got interested in PLATO, learned everything he could know, and then got bored with it, in six months." Rodby eventually went on to work on databases for the NASA space program.)

Then CERL lowered the limit again, to 10 TIPS. It was like an insult, a direct affront against every Empire player on PLATO. How could they? What did they think this thing was? A computer-based education system? Didn't they realize the universe needed conquering? Empire at 10 TIPS just didn't cut it, and the game was at risk of losing its growing audience and, by now, dominance, over another hugely popular game of the time, the former cream of the crop . . . Airfight.

Daleske and Warner and the others at Iowa State were not the only PLATO game authors in those days who'd become minor deities to the growing horde of gamers. Brand Fortner, located right at CERL, had already written Airfight, which seemed destined from the very start to be an insanely popular game. There had been nothing like it before. It was another PLATO first, in the long, long line of PLATO firsts: a first-person-perspective, multiplayer, shoot-'em-right-out-of-the-sky flight simulator. And until Empire came along, it had ruled the PLATO gaming world.

Fortner had stumbled upon a simple PLATO game called Air Ace, where you could type in some parameters, press NEXT, and "about ten seconds later," says Fortner, "it would redraw line graphics of the cockpit and you would see outside of the plane. And I thought, Well, that is an interesting idea, but gee, wouldn't it be nice if you could fly a lot faster and shoot down other people?"

By today's standards, Airfight's graphics and realism, like every other PLATO game, are hopelessly primitive. But in the 1970s Airfight was simply unbelievable. These rooms full of PLATO terminals weren't "PLATO classrooms," they were PLATO arcades, and they were free. If you were lucky enough to get in (there were always more people wanting to play than the game could handle), you joined the Circle or the Triangle teams, chose from a list of different airplane types to fly, and suddenly found yourself in a fighter plane, looking out of the cockpit window to the runway in front of you, with the control tower far down the runway. . . . You'd hit "9" to set the throttle at maximum, "a" for afterburners, "w" a few times to pull the stick back (using those PLATO arrow keys again), and then NEXT NEXT NEXT NEXT NEXT NEXT NEXT to update the screen as you rolled down the runway, lifted off, and shot up into the sky to join the fight. It might be seconds or minutes, depending on how far away the enemy airplanes were, before you saw dots in the sky, dots that as you flew closer and closer turned into little circles and triangles. (So they weren't photo-realistic airplanes—it didn't matter. You didn't notice. This was battle. This was Airfight.) As you got closer and closer to one of these planes, the circles and triangles got more defined—still small, still pathetically primitive by today's standards—but you knew you were getting closer and that's all that mattered. As you got closer and closer you hit "s" to put up your sights, to aim. Eventually, if you were good, or lucky, or both, you would be so close that you'd see a little empty space, an opening, inside the little circle or triangle icon. That's when you were close enough to see what players called "the whites of their eyes" and that's when you let 'em have it: SHIFT-S to shoot. SHIFT-S again. And again. Until you'd run out of ammo and KABOOM! It was glorious.

And it was addictive. People stayed up all night playing Airfight. If you went to a room full of PLATO terminals, you'd hear the clack-clack-clack-clack-clack-CLACKETY-CLACK-CLACK-BAM-BAM!-WHAM!-CLACK-CLACK! of everyone's keyboards, as the gamers pounded them, mostly NEXT-NEXT-NEXT'ing to update their view and their radar displays (another innovation of this game—in-cockpit radar displays, showing you where the enemy was).

Airfight was a TIPS hog, but Empire was even worse because it was juggling far more information in real time than Airfight: four teams instead of two, more than two dozen planets, not to mention all those armies, multiple weapon types, torps firing in every direction. At some point around this time, a very few, very lucky PLATO gamers discovered a neat little trick to cheat the growing autobreak problem. They called it "TIPS cooling." "Very few" and "very lucky" because to successfully cool your TIPS required a near-impossible set of conditions: a) you were lucky enough to find a PLATO terminal tucked away in some tiny little room in some obscure building somewhere on your campus; and b) for some reason the room was off-limits to the public without some sort of permission; and c) you had keys and could keep the room locked; and d) you were able to crawl out of bed early enough and get to the building and up the stairs to the room and get in and fire up the terminal and sign on; and e) you were disciplined enough to just sign yourself on, go into Empire, scoot over to some far corner of the galaxy where no one would want to come to shoot you down, press SHIFT-TERM, and leave the terminal just sitting there with its "What term?" prompt with its infinite Skinnerian patience just waiting for you to type a term, but instead you would get up, leaving yourself signed on in Empire with the TERM prompt sitting there, locking the door behind you, so no one else could enter the room and use the terminal. If you could pull off all of those things, were you in for a treat. By evening, your average TIPS were down to zero point zero. This was Pass GO and Collect $200, Get Out of Jail Free, and Free Parking all wrapped up in one. The hours and hours of inactivity would flat-out fool the "governor," or whatever the electronic auto-break cop was called somewhere deep inside the mainframe, and you could play Empire and raise hell without the fear of any auto breaks at all. It helped if you were good.

Games have always been one of the most resource-hogging, CPU-intensive types of applications written for computers. Today's PC and videogames use huge amounts of memory and superfast processors to accomplish that 3D realism and smooth animation speed that players have come to expect. To understand how limiting 10 or even 20 TIPS are in present-day terms, think of it this way: if your Mac or Windows personal computer ran at only 10 TIPS, it might take a week just to boot the machine, let alone do anything with it.

Empire was just the sort of thing that pushed the envelope of the PLATO system's capacity. The game had to keep track of who was in what ship; which ships belonged to which teams; where each ship was located in space; how fast each ship was going; how hot the engines were running on ship; how well each ship's shields were holding; what to display every time you replotted your screen; how many armies were on each planet; which team those armies belonged to; whether a planet was neutral, at war, or at peace; whether any given player had declared war or peace with another team; where all the photon torpedoes were at any instant and what direction they were each heading; the lifetime of each torpedo (they self-detonated after traveling a certain distance); whether or not a torpedo had come close enough to an enemy ship to explode; which photon torpedoes were detonated, and how long their detonation explosion icons should be displayed. The list went on. Empire had to juggle all of this and do it in such a way that nobody noticed. With 10 TIPS that was nigh impossible. A major code rewrite was sorely needed.

"The biggest piece of code was the section that tested for collisions between objects," says Daleske. "This required a comparison of x and y coordinates for a square in which the object was the center. . . . To determine if a point was within that square took up to four compares. These compares and the branching (like 'goto') code took a lot of the CPU time."

Fortunately, it just so happened that the CERL system staff had been working on a new series of TUTOR commands for managing arrays, or lists, of data, as well as tools for searching for particular values within those arrays. Two TUTOR commands had come out around this time, -find- and -findall-.

-Findall- seemed particularly interesting and had the potential to solve Empire's TIPS problems. "I was determined that if we could use the -findall- command, the TIPS count could be lowered," says Daleske. So Daleske, Miller, and Fritz began working on figuring out how to use -findall- such that if you fed the right data to the -findall- command in just the right way, it would spit back some data that told you exactly what you wanted to know. They camped out in a classroom in the education building, filling the three full-size blackboards with notes, drawings, and formulas.

The -findall- command is one of the trickier ones in TUTOR. With it you could search for occurrences of a particular value within a list of variables. If you wanted to know how many students got a perfect score on a test, you could feed -findall- the list of scores, tell it you were looking for all of the 100s, and it would tell you the total number that had perfect scores, and even which entries in the list those were. The command had an optional "mask" feature, which you could use to search only certain *bits* of a variable, rather than the whole thing.*

---

* Given the severely constrained memory resources on PLATO, TUTOR programmers had no choice but to treat every literal bit of memory as sacred. A byte of memory in CYBER was six bits, not eight like most computers have today. If you could store meaningful information in less than six bits, you could use the remaining bits for something else. Today, a program like Microsoft Word uses more memory than all PLATO systems in history combined. In the 1970s, programmers did not have the luxury of cheap, plentiful memory.

This masking feature was key to the Empire programmers. If only they could figure out how to lay out the array's bits, and what kind of mask to use.

"We filled those boards with various ideas of how to do masking," says Daleske. They worked for weeks on the problem, finally figuring out a solution for masking an arbitrary square. "We came up with a way," says Fritz, "of expressing an object's position in space in a 60-bit pattern such that you could do a -findall- on this array, and boom, there you had a list of all the objects that were in your proximity, and then go ahead on that reduced set, and do all the distance calculation." This bit pattern included your ship type, whether you were at peace or war with other teams, your x and y coordinates in space, and which display magnification mode you were using, all encoded in such a way that you were essentially asking -findall-, "Show me everybody within a certain distance of where I am." It was a brilliant hack, amazingly arcane, yet it worked beautifully. "It made an incredible difference in the performance of the system," Fritz says. And it was almost impossible to explain to anyone. "The mapping mechanism even after I documented it would still take a half hour to explain it to anybody," says Miller. "And it would still take me another half a day to sit down and figure out what the hell it was I did again."

"It seems magical to have a single mask be able to return a list of object candidates, but it worked," says Daleske. "In my opinion, this saved the playability of the game."

They didn't stop there. Always on the hunt for reducing the TIPS load, they rewrote parts of the game so that the processing was distributed among the different users playing the game—in effect, spreading the TIPS load out over the whole player community rather than putting the load on any one user. The code was designed such that each user in the program would update the movements of torps and ships and check for collisions. "This in effect made the design of the game like a time-sharing operating system." It would not be the first time that authors of multiplayer PLATO games would create system architectures that modeled themselves after the PLATO system itself. It just made too much sense to not do it that way.

In the summer of 1976, Daleske, Fritz, and Miller went to Des Moines to attend a *Star Trek* convention. Each brought a printout copy of Empire's source code, and each got, in a "We're not wor-

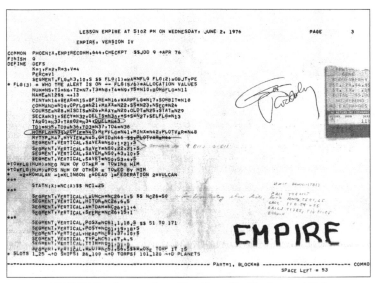

EMPIRE printout signed by Gene Roddenberry, 1976

thy" moment, their printout autographed by none other than Gene
Roddenberry. (Roddenberry had visited UI in November 1974 for a
standing-room-only auditorium lecture.)

By August 1976, Daleske had run out of money for school and had
wound up taking a job at Control Data Corporation in Minneapolis.
But before he left, he worked out an arrangement with Gary Fritz and
Chuck Miller to be the custodians of Empire, to continue maintaining
the code. Around this time, copyrighting PLATO lessons was all the
rage, now that CDC had begun its effort to market PLATO commercially, so they filed for a copyright. Ultimately, the agreement was that
Gary Fritz and Chuck Miller would own the copyright, but Daleske's
and (at Daleske's insistence) Warner's roles would also be recognized
and all four would share in any future royalties.

A key to the fun of Empire was its messaging feature, something
that so many other multiplayer PLATO games before and after also
offered. Empire users could send messages to all players, to a single
specific player, or to an entire team. "Of course, you had to stop fighting to type out a message and send it," says a former Empire gamer,
"so this didn't happen in battle, and you could turn messages OFF
in case (as was thought of) a teammate would flood you with mes-

sages* during a heated battle, slowing down your terminal processing speed. But the point about messages was, you dogged your opponent. Get a kill, send them a message. You could be as nice or as cruel as you thought was wise, knowing they'd be right there in two minutes to battle with you again. It was a lot, mentally, like basketball, and no coincidence a lot of the guys that were Empire heads were also hoops heads."

One day Bill Galcher and Tim Halvorsen, both part of the rat pack of PLATO student programmers at CERL, were sitting at their terminals down in the basement of the foreign language lab—a vast sea of some eighty PLATO terminals, the largest PLATO classroom in the world—conveniently located near their apartment. Galcher and Halvorsen were just doing their normal PLATO thing when a group of strangers, neither Bill nor Tim recognized, stopped by.

"Excuse me, are you Bill and Tim?"

"Yuh, hi. Who the hell are you?"

"Oh, I'm Gary Fritz."

"Th-th-the *Empire* Gary Fritz? *Wow!*" A true PLATO god, right here in the foreign language lab!

Fritz introduced the others, all of whom had come from Iowa as well. Tim and Bill couldn't believe these guys had come all the way from Iowa.

"We *had* to come."

"Why?"

"Well, it's *Mecca!*"

It was a night that started out just like any other Empire night on PLATO in the mid-1970s—the room full of terminals jam-packed, gamers flailing away at each other, all's well with the Empire universe—when all of a sudden—holy Randu, look at that—a giant orange carrot appeared on-screen. Desperate messages began pouring out from everyone in the game:

---

* If a player were a fast typist like Andrew Shapira, he might attempt to flood others in the game with messages, often very brief ones. Each message would be handled as a separate message by the game, forcing recipients to take time to read them all. An alternate strategy involved flooding other players with empty messages simply to "flush the buffer" clean of other meaningful messages the player might have—a sort of "denial of service attack" on an opponent's list of messages.

"What the @(#*^&%* is THAT!?!?"

"I don't know, but it just destroyed my ship!"

"Me too!"

"HEY!! It went into the Fed home system, and Earth is GONE!!!!"

Over the 1976 Christmas break, Chuck Miller and another friend who'd become hooked on PLATO, Jim Battin, modified Empire's code to add a surprise element to the game—something right out of one of the most popular *Star Trek* episodes, "The Doomsday Machine." Just as it had been depicted in the TV show, this Doomsday Machine was miles long and it ate whole planets for breakfast, spaceships for snacks, and like in *Moby-Dick*, upon which the original television screenplay had been loosely based, this was the story of a ship's captain obsessed with defeating a monster that could not be defeated. Except now, in Empire, there were thirty ship captains, all with the same obsession.

It didn't take long that first night for everyone in the game to realize that the orange carrot didn't care if you were a Fed, a Rom, an Orion, or a Klingon. It didn't care if there were 101 armies or one army to defend a planet, or if the planet had a pretty-sounding name, good location. Nothing mattered, not to this monster, this relentless planet killer. It knew only one thing—like the great white shark in *Jaws*, the blockbuster that had come out two summers before, it lived to *eat*. The thing would just decide to go after some planet, and when it reached it, it would first wipe out the armies, huge gulps at a time, and then in the blink of an eye, the planet was just *gone*. Then it would turn and head for some other planet—there didn't seem to be any pattern, it might be another nearby planet, or one across the galaxy. The unpredictability just added to the sheer terror of the thing.

Players abruptly suspended any sense of team loyalty, battle-just-for-the-sheer-fun-of-it, or mission to conquer the galaxy—they set aside everything that made Empire *Empire*—and they all joined forces to take on the killer carrot together.

The desperate messages continued flying by—Empire's program memory only had space for thirty one-liner messages for the whole game, and when it filled up, the way it was supposed to work was the oldest message was erased, and the newest message added. In a normal active game, it might take a minute for the buffer to recycle, but now the messages were flying so fast, the buffer was filling and recycling itself in a mere ten, twenty seconds.

"Where is it!?!?"

"How do you kill it?"

"Don't shoot at me, man, we've got this thing to get!"

"What have you tried . . . ?"

"Did it work!?!"

"It seemed to weaken!!!!"

Multiteam collaboration in Empire? Never. But sure enough, all the players were now working together, a first, all of them converged on *it*, that *thing* that might have looked like a ridiculous orange carrot on a PLATO screen to a passerby but brought sheer terror to Empire players, and *then* they discovered it appeared that it could not be destroyed. The combined phasers and torps from every single player in the game were flat-out useless against the orange monster. It just kept eating everything in sight.

Finally, someone remembered the *Star Trek* episode. *Yes, of course. Brilliant.* Little eureka light bulbs went off in everyone's heads—the orange killer carrot was the Doomsday Machine from *Star Trek*. Of course—followed by echoes of dialogue from the show:

> SPOCK: Captain, you're getting dangerously close to the planet killer.
>
> KIRK: I intend to get a lot closer: I'm going to ram her right down that thing's throat.
>
> SPOCK: Jim, you'll be killed, just like Decker.
>
> KIRK: No, no, I don't intend to die, Mr. Spock. We've rigged a delay detonation device, you'll have thirty seconds to beam me aboard the *Enterprise* before the *Constellation*'s impulse engines blow.

Everyone lined up their ships—thirty ships all in single file, a level of cooperation unprecedented in the history of Empire—and they proceeded, one by one, to fly right into the giant death maw of the Doomsday Machine. Not recklessly like Commodore Decker, but with the smooth confidence of Captain Kirk—each ship captain ramming their ship right down this thing's throat, initiating the game's SHIFT-q ten-second self-destruct sequence juuussstt so. This was *Star Trek* all over again—if the PLATO terminals could have piped in music right then, it would have been Sol Kaplan's unforgettable Doomsday Machine Death March from the original *Trek* episode's soundtrack, blaring away at full volume: *DUH nuh DUH nuh DUH nuh DUH nuh!* One by one, the Empire ships marched in, and one by one they were destroyed, each player gulped up by the indestructible Doomsday Machine, each captain's goal to *ram this ship right down that thing's throat*. Each player trying desperately to time their self-destruct sequence juuusst so. That was the theory, anyway: if you timed your ship to explode at the right exact spot . . . KABOOM—at least in theory. But, try as they might, the Doomsday Machine lived on, while each self-destructing ship was destroyed in turn. Players would die, and then quickly reenter the game—a game with fewer and fewer planets as the minutes went by— and fly their ships over to where the Doomsday Machine was, line up in front of it like lemmings at a cliff's edge, and if they didn't die right away, once again try to self-destruct juuuuust so, over and over again. Some people trying, dying, reentering, dying, reentering, dying, over and over dozens of times.

And then, after much carnage and insane self-destruction, a UI student named Brian Gilomen (his player name, amazingly, "Commodore"—named after Decker?) managed to time his destruct sequence exactly right—a tiny, exact window of time, and exactly at the right spot at the entrance to the maw of the beast—and it was destroyed.

There was much rejoicing.

Gilomen recalls that everyone was "scurrying around in terror that night." He had remembered the TV show and then it dawned on him: "If the programmers had derived their design from that *Star Trek* episode, there was a good chance that they would let the beast be killed the same way. I didn't want to do it, because I had a goodly number of kills that night, and I felt there was an excellent chance that I'd just go up in smoke and have to start my kill list from scratch. But I finally decided to go for broke. I was flying as a Fed, had my shields up as much as possible, and headed right towards its maw." Moments later the screen changed to some sort of "You Won!" message. The game restarted for everyone, and Gilomen was rewarded with a large number of kills.

Gary Fritz, to his eternal regret, missed out on the entire episode— he'd gone home for Christmas break. "We gave it parameters that were similar to the show," Chuck Miller says of the Doomsday Machine. "You had to be pretty much right in the mouth." In fact, a successful Doomsday Machine opponent had to be within fifty units of the Doomsday Machine's x, y coordinates at the exact moment of your ship's self-destruction—a near-impossible feat, considering that your ship was moving, the Doomsday Machine was moving, and while you were in the self-destruct sequence, you had absolutely no control over your ship whatsoever; you merely watched a countdown and crossed your fingers.

The players congratulated Gilomen and then teams were teams again, armies were armies, planets were friendly or they were hostile, and everyone began duking it out once more. Or so they thought. Suddenly the Doomsday Machine came back, and the mayhem started all over again.

It reappeared four or five more times—one of these times it was destroyed by a ten-year-old kid, much to the consternation and embarrassment of the Empire elite. The bright orange killer carrot known as the Doomsday Machine only stuck around for about a week and then it vanished. Players knew how to kill it, word traveled quickly, and that meant it wasn't as much fun anymore. It would never make another appearance, and the monster would become a legend, passed down from one player to the next, for decades to come.

It became a tradition that the Federation team players, after another all-nighter playing Empire at CERL, would head over to Sambo's, a twenty-four-hour diner, after PLATO's "non-prime-time" ended sometime around six in the morning. Biologically, this was supposed to be "breakfast," but "after being up all night," says Patrick Clifford, "the breakfasts ordered tended to look more like dinner fare than breakfasts, with steak sandwiches, fried shrimp, and other such heavy stuff predominating. . . . We used to drive the chefs nuts." The Feds would gather in a corner booth and begin "debriefing" what had occurred during the night's Empire run.

One early morning, Sambo's was surprisingly busy, but the Feds still managed to file in and take their corner booth and begin the debrief. They'd been assaulted by Romulans who outnumbered them all night. "We held them off all evening and when the system went down. . . . We still held Earth and the Fed system," says Clifford. "The discussion became particularly spirited with comments like 'I thought they had Earth when they cleaned us all off of Earth, but Jerry cloak-hypered in on them and dusted them off before they could beam down armies!' and 'There were at least six Roms headed straight for Earth and Funke suicide-hypered on them and got all of them and only brought down his shields to 30 percent!' This had been going on for about an hour when this little old lady—had to be at least eighty!—who had been eating breakfast with her husband, walked over and said in all sincerity, 'I don't understand what you guys were doing, but I want to thank you for saving Earth!' We all looked at each other and burst out laughing."

16

Into the Dungeon

You have just left the Wilderness. You are now in the City, where, unlike the Wilderness, it is safe to be alone and unarmed. In the Wilderness there be monsters, though they are annoyances, for the most part. After wandering around out there for a while and defeating the occasional goblin, you now have over $3,000 worth of gold. There is only one thing to do with gold in the City: spend it. If only you could find the Weapons Store.

To delve deep into the Wilderness is one thing; to go beyond, deeper, into the Forest or further terrains, it's much too dangerous without some weapons and supplies. Now, where is that Weapons Store? The corridors and rooms in the City all look similar. You don't have a map handy, which means you have to roam the maze and hope you find it by accident. Forward you propel yourself, forward, forward, right, forward, through a door, forward through another door, forward, left, forward, door, forward, onward, ever onward. You come across a Water Store. Not what you want. Then, right nearby, you find a Magic Shop, but it will only let you sell things, not buy (what kind of shop is that?), and you have no spells to sell. Then you find an entrance to the Wilderness, which means you've either retraced your steps in the maze or found a different Wilderness entrance—word has it there are a few of them around the City. You haven't been careful to map your way around, so you are not sure which Wilderness entrance this is. Oh well. Forward, you keep moving forward, forward, onward, and bump into the same Water Store and Magic Shop and entrance to the Wilderness. You are going in circles.

Then you remember. Someone told you once that you have to look for hidden doors. Walls that look solid sometimes aren't. So you begin to push your way against walls, and eventually, sure enough, one gives way, and you find yourself somewhere new, a new hall. Forward, forward, left, forward, right, through a door, into a room with numerous doors in the distance. You run over to one of them, you find the Weapons Shop. Finally.

"Welcome, Sire," says the shopkeeper as you enter the shop. Unlike most shops, this shop somehow already knows how much money you have in your pocket, and of course it is intent on taking all of it. You can buy, sell, price, or simply get information on the weapons. Unfortunately, unlike any normal store, you cannot see what is for sale in this store. You have to know what you want before you enter the store. Or you'd have to go to the HELP lesson and print out what you see there or write it down, and then come back to the store. A hassle. So you stay in the store and try for two common items you remember everybody starts out with.

"Which item, Sire?" the shopkeeper asks you, once you've indicated you're ready to buy.

"Sword," you say.

"Buying, eh, Sire? Well, how's about $134?"

"$40," you say.

"$40? Can't accept that! How's about $128?"

It was worth a try. At least the shopkeeper's willing to haggle.

"$90."

"Deal at $90, then?"

You feel like a fool. Clearly you could have gone lower. You try to.

"$70."

"Offer is too low, Sire!"

No matter what you offer below $90, the offer's always too low. Defeated, you say, "$90."

"Deal at $90, then?"

"Yeah, whatever."

"What a bargain, Sire!"

The empty "WEAPONS/ITEMS" column on the left side of your screen suddenly shows "Sword" in it. You own it. The amount of gold you have drops by $90.

You still have a lot of gold, so you decide to buy a mace.

"Make it $1,297?"

"$400."

"$400? I'll go broke! How about more like $1,135?"

"$700."

"$700!? I have children! Make it $760?"

"$750."

"Deal at $750 then?"

"Yep."

"What a bargain, Sire!"

You now own a sword and a mace and the shop owns $840 of your gold.

The conversation above comes from an actual experience in the PLATO dungeon game Moria, created by Kevet Duncombe and Jim Battin at Iowa State University in 1976. The conversation exemplifies the kinds of rich approximations of natural-language interactions possible on PLATO thanks to the TUTOR language. A clever programmer could create seemingly natural and effortless interactions between user and program, not just for games, but for educational lessons as well.

Upon close inspection, note that in the Moria shopkeeper's conversation with the player, each of the buyer's responses boils down to three typical types. There's some form of positive word uttered, perhaps a "yep" or "yeah," or negative such as "no" or "nah," or a monetary amount given. Experienced Moria hands would sometimes tell newcomers that as long as your response started with a "y," Moria interpreted that as affirmative, while starting with an "n" indicated negative. Battin and Duncombe spent a lot of time working on the TUTOR code that ran the Weapons Shop. Says Duncombe, "That was very sophisticated answer-judging stuff using hundreds of -concept- statements and every reasonable answer we could think of. You could type fairly complex expressions that did not contain 'yes' or 'no' but indicated assent or dissent and have it match, but eventually people would get bored with that feature, like real quick, and get down to just wanting to type 'y' and 'n', so eventually we had to add 'y' and 'n' as yes and no synonyms to the concepts."

Duncombe had started out as an electrical engineering major in 1974 at Iowa State, but along the way took a computer graphics class. "At the time," he says, "the PLATO terminals were the hot ticket for doing graphics at Iowa State. . . . I got my first taste of PLATO through this class and the classwork, and it came with an author signon because we had to write some graphics programs, so that got me my first taste there, and I was quickly hooked. It was a very addictive system, especially when you had the run of the system with an author signon."

Duncombe quickly discovered the games PLATO offered, including Airfight and early dungeons and dragons games like Pedit5 and Dnd. Each game had a distinct personality of its own, reflecting the idiosyncratic sense of design, humor, and mischief of its authors. In one dungeon game, if you encountered a monster but chose to flee rather than fight, your little graphical on-screen character was turned into a chicken.

Over the course of long days and nights in one of the few PLATO rooms on campus, Duncombe would bump into the whole crew of fellow enthusiasts, including Jim Battin, Chuck Miller, Gary Fritz, and John Daleske. A group called "ames" (named after the town Iowa State University was situated in) had been created on the system for volunteers willing to code TUTOR lessons for Iowa State professors. Naturally, this crew of self-described "addicts" all signed up to be in it. A more advanced authoring group, named "amesrad" for Ames R&D, was eventually created, and most of the same crew graduated to that more prestigious group. The advantage of being in "amesrad," says Duncombe, was that "you didn't really have a lot of deadlines or a lot of projects that had to be met, it was mostly just fiddling around with research & development."

The "research & development" for Daleske, Fritz, and Miller was Empire. For Battin and Duncombe, the R&D took a different form: the Tolkien-influenced game of Moria. Moria arose, as did so many games, as a reaction to what its authors saw as deficiencies in previous games, in this case one called Orthanc (another reference lifted from Tolkien). "A bunch of people were working on it," says Duncombe, "having a lot of trouble getting it to work properly, had a lot of bugs in it. I was just not understanding why it was so hard to write one of these games, it didn't strike me as a terribly difficult thing." Other PLATO dungeon games up to that time typically relied on TUTOR's -randu- command to randomly generate numbers that the game authors then used to generate the different levels of the dungeon. Then each would have to be stored in a data file, and data files on PLATO were expensive and rare. Duncombe wondered if it were possible to instead randomly generate the maze but not *store* it: it was generated on the fly. He showed a prototype to Jim Battin. "He thought it was kind of cool," says Duncombe, "and we started fleshing it out into a game."

They worked in the tiny room, no larger than a closet, that had two terminals in it: the same room that Empire's authors were using.

Things got pretty cozy. "We were all competing for terminals, but it wasn't cutthroat competition. Typically, Jim and I could be working on one terminal, and Gary and Chuck would be working on the other. So we'd be sitting side by side up there in the author room, working away on our games, occasionally bouncing ideas around. There's some of Fritz's code in Moria and there's some of Battin's code in Empire. Not a whole lot, but there was definitely cross-fertilization among the game authors there. Those two terminals turned out to be the most heavily used on the entire CERL system."

Duncombe claims that neither Tolkien nor any other published materials influenced the creation of the game. "It was all strictly based on the predecessor games of PLATO and fairy tales and our imaginations and whatnot," he says. "When we finally got something that was playable, that we were going to release, we started asking around, the other guys in the area there, and Dirk Pellett [another Iowa State student] actually came up with name. He was a big Tolkien fan and suggested that 'The Mines of Moria' was a good name for the dungeon. So we grabbed it."

Moria took off. "We started getting more and more hooked and adding more and more features," says Duncombe. "Hacking away at all hours." Moria, one of the first multiuser dungeon games on a computer, went on to become one of the greatest games on PLATO with fans around the world. Like nearly every PLATO game, it offered a detailed "help lesson," itself a massive labor of love, introducing newcomers to everything about the game, which it described as a "world of underground rooms and corridors where the characters in the game will spend their entire life trying to survive." You started out in the Wilderness, near the City of shops and shopkeepers. As you ventured out beyond the Wilderness, you could find sixty levels of Cave, sixty levels of Mountain, of Forest, and Desert. (Sixty levels of an extremely hard-to-find secret Ocean level was also added but rarely found.) The authors created a "Group" feature enabling up to ten players to travel together, one of the players designated the "guide." Key to Moria, like most other multiplayer PLATO games, were ways to communicate with other players, including "whispering" within the room you might be in. "Guilds" were another feature of the game. Each Guild had special powers. If you were in The Brotherhood you could raise the vitality of an entire group of players who might be facing ferocious monsters. The Circle of Wizards could teleport an entire group back to the

City. The Union of Knights members were more resistant to damage and could behead monsters. The Thieves Guild could find additional items in treasure chests that other players might miss. Over time players would accumulate gold, weapons, spells, and other magical items throughout their adventures. They could give them to other players, sell them back in the City, or trade them for things they didn't have.

The long hours turned into long months, and then years. Moria continued to evolve and consume Battin's and Duncombe's time. Duncombe would eventually flunk out of Iowa State. He was not the first, and was by no means the last, PLATO-addicted college student to do so. But in the long run, he would discover that flunking was not as bad a thing as it might have seemed at the time. "I learned more on my own doing Moria," he says, "than probably any of the classes. The games are always pushing the edges of what's possible. And by doing things that are difficult or are operating-system-like in a lot of respects, there's all kinds of multitasking things you have to think about, and interlocking things you have to think about, and it's really, by the time you've solved all those problems, you're working on a fairly advanced level." Duncombe and Battin both ended up with decades-long software careers.

In the history of online computing, one pattern continues to repeat: new programs, websites, phone apps, and games of all kinds will arise as a *reaction* to previous programs, websites, phone apps, and games—the new developers attempting to apply new ideas to what they've seen in existing applications. On PLATO across the 1970s, this pattern would not only repeat over and over, but at an ever more frequent rate. Witness first the Big Board games, then Airfight and the space war games like Conquest and Empire. Perhaps the greatest progression and evolution of programs exemplifying this pattern is that of the dungeon games: Moria was in the middle of this multiyear evolution. Simpler, cruder games had appeared prior to it, and more complex games would come after.

It is instructive to take a look at some of those simpler games, especially ones written by pioneering authors who lacked the luxury of any prior art, who lacked the luxury of being able to react to that which had come before. What one finds is that the original influences were offline: Tolkien's *Hobbit* and *Lord of the Rings*, certainly, but also, in

1974, the release of a tabletop fantasy role-playing board game called *Dungeons & Dragons* by Gary Gygax and Dave Arneson.

Perhaps the earliest dungeon-related game on PLATO was one by Reginald "Rusty" Rutherford III in a lesson file called "pedit5." (Legend has it that an earlier game, with the equally obscure name of "m119h," pre-dates Pedit5, but no one remembers who wrote it or when exactly it appeared—probably 1974. Like Pedit5, it suffered the fate of being disallowed by the higher-ups and was soon deleted from whatever PLATO account it belonged to. No physical evidence of it has been found.)

Rutherford did not fit the usual pattern of a new wave PLATO kid: he was already in his mid-thirties in 1975, seeking a PhD at Illinois, with a family and one child (named Reginald Rutherford IV). He had a job working in the PLATO lab in the basement of the physics building, writing TUTOR code for Professor Paul Handler, who, while nominally a physicist, had taken a great interest in the world population explosion and ran a group called "Population Dynamics" that was pumping out impressive PLATO lessons on the world population crisis. "I got paid effectively an hourly rate, not tremendous, certainly not by today's standards, but it was very, very interesting and there was a fair amount of free time between assignments." During that free time he got involved in the brand-new tabletop *D&D*. "About the spring and summer of '75, my friends in the gaming group at Illinois had gotten deeply involved in this. . . . It was a fierce game. It was awfully easy to die at level 1. Well, we kind of got really deep into it to the point where not much else was happening in our lives." They weren't quite pulling all-nighters, "but 2 a.m. was not that unusual," he says.

One day Rutherford thought to himself, "Gee, could I put this on a computer, using these wonderful visuals you have available on PLATO? And I sat down and I figured it out." As Handler's TUTOR programmer, Rutherford had access to the PLATO account for the Population Dynamics project, and found a number of lesson files created with the names "pedit1" through "pedit5." He grabbed them and got going. The result was a simple, single-user "dungeon crawler" game, much influenced by *D&D*. "It is the year 666—the year of the Beast," the introductory page of the game declared. "In the country of Caer Omn, near the town of Mersad, stands the ruined castle of Ramething. Beneath the castle lie the terrible dungeons of Ramething, an incredible maze of rooms and corridors, occupied by horrid monsters and piles of ancient treasure."

You were invited to create a character, name him, and enter the dungeon, initially in a corridor. The display was simple: on a PLATO IV terminal, the screen was nearly all black, with the word "CORRIDOR" on top and "EXIT" below. The corridor was depicted in a top-down fashion, with a crude little knight character standing in the hallway. One wall of the corridor had a doorway. You could not see what lay more than a step or two ahead, as if the corridor was dimly lit. Nothing else was seen on the screen. Like so many PLATO games, you "moved" by using the arrow keys. "W" took you forward, "d" to the right, "x" back, and "a" to the left. Upon pressing "w" you advanced and noticed more doorways along the walls of the corridor. Eventually you would reach the end, at which point the corridor split to the left and right, with more doors in the walls. Behind those doors were rooms, or more corridors. At any point you might encounter monsters.

The game was a hit, despite the fact that Rutherford never formally "released" it to the PLATO community. "News of the existence of Pedit5 simply spread by word of mouth," he says. He then discovered the age-old problem of popular lessons on PLATO: it was hard to make changes and "condense" the TUTOR code because people were always playing the game, and he would have to kick them all out to run the condense, annoying the players. Soon he resorted to adding code to the game that immediately kicked out users with certain signons who were routinely running the game during the day. Another problem: having only two lesson files, Pedit4 and Pedit5, severely restricted how much game and player information he could store away. "The available storage space only allowed for a single-level dungeon with forty to fifty rooms," he says. "The dungeon design was the same for every user, but the monsters and treasures were random—created at the same time as a new character, and stored with the character record. Only about twenty characters could be stored; when the game became popular, this turned out to be a real hassle."

In an interview nearly forty years later, Rutherford still laughed at the ruthlessness of his game's design. Pedit5 was downright lethal. There were no limits to the wrath of Randu, the TUTOR command used to generate random numbers, which determined where the monsters were when your game character was created. "The idea was, go in, make a reasonable score," he says. "If you get wounded, get the heck out because you can't take the next fight. Until you get a reasonable level, you don't want to try for the depths of the dungeon. And besides, I set it up so it got tougher as you went further in. On the other hand, you might run into a dragon in the hall on the first corridor. Tough luck. Unfortunately, that terminated your character."

Pedit5 would disappear soon after it was created, due to its shady unofficial status as an unsanctioned game created in the serious Population Dynamics account. Then it would reappear, only to disappear again. The very existence of the game became a random, unreliable thing. For the growing wave of PLATO game authors, this would not do.

Two such authors were undergrads Gary Whisenhunt, majoring in psychology and political science, and Ray Wood, majoring in electrical engineering, at Southern Illinois University (SIU) in Carbondale. In the 1970s some universities around the country were experimenting on a small scale with PLATO, often leasing two or four terminals for use on their entire campus. SIU was such a place: there was a single PLATO terminal in the basement of the library. "It was common for one person to run the terminal and another five to ten people to be in the room with you watching the terminal," says Wood. "Gary and I would meet during these 'joint viewing sessions.' He seemed to be more normal than the rest of the people—he actually had outside interests. . . . He liked messing around with PLATO, but he knew there were other things to do. Gary is probably the 'proto-geek' that all other geeks were designed after. He was into technology, but he also had a lot of interests outside of technology. Gary had a girlfriend who was relatively normal. He listened to rock and roll, drank beer, etc. The other guys didn't. So we kind of naturally gravitated to each other. It was great to be around someone who may not have had a life outside of computing, but at least knew that there was life outside of PLATO."

Still, they kept coming back to PLATO. They played games, among them Empire and Pedit5. Like everyone else, they grew tired of the repeated disappearances of Pedit5. Soon, they decided to do something about it: write their own game. The lesson file name for their game: "dnd." Like Internet domain names decades later, lesson names on PLATO were unique. When somebody grabs a domain today, that's it, it's taken. Back in 1975, Whisenhunt and Wood knew Dnd was a good lesson name. They grabbed it.

"I taught myself to program on the PLATO system," says Whisenhunt, "and liked it so much that I switched majors to computer science." Soon, they were pulling late-nighters and all-nighters, hacking away at the code. The goal, says Whisenhunt, was *fun*. They didn't share Rutherford's addiction to the *D&D* board game (though Wood was familiar with the board game and enjoyed playing it), nor was their new game anywhere near as lethal. Instead, it had in-jokes that fellow SIU students would recognize. The game started out with just two levels, the mazes of which they laid out by hand on graph paper. It took hours. Whisenhunt decided to write a "maze editor" online, which gave them a one-hundred-fold boost in productivity. They spent a great deal of time tuning the combat play, making each level as "fair" as possible, taking into account the player's overall score and skill level, and the level's difficulty: features the file-space-strapped Pedit5 had lacked.

Whisenhunt and Wood had another advantage over Rutherford. They controlled the account that the "dnd" lesson was in. Whisenhunt had become known to the library staffer who was supposed to be responsible for the presence of the CERL PLATO terminal in the building, but beyond that didn't know much about the system. Here was this student, Whisenhunt, who hung around the terminal a lot, seemed to know the system in and out, and was even programming on it. The library offered him a job, and soon he became the de facto administrator for the SIU PLATO account, including all of its files, of which Dnd was one. Unlike the travails that plagued Pedit5, "there was no chance," says Whisenhunt, "that Dnd was ever going to be deleted."

As was always the case on PLATO, word traveled fast when new games appeared on the scene. Dirk Pellett, who had been attending Caltech but then transferred to Iowa State, found the PLATO terminals there, got hooked like the other Iowa State crew, and eventually found out about Dnd and became not only a fan but also a contributor, offering a long-distance stream of suggestions and ideas for improvement. He never met Whisenhunt or Wood in person, but his suggestions were copious and good enough that in time, Whisenhunt and Wood, heeding the call of their college work (and grades), were happy to give access to Pellett to dive directly into the Dnd TUTOR code. Meanwhile, Dirk's brother Flint, enrolled in graduate school at the University of Illinois, and soon became an online collaborator as well.

In time the Pelletts took over the game and added extensive enhancements.

A sense of mischief and wit permeated the game's design. Every time you opened a treasure chest, there was a chance you might get injured or fail to open it, at which point the game program might scold you with a "Clumsy dolt!" message. There was a spell called "Kitchen Sink," which, of course, you could throw at a monster during combat. They named a monster "The Glass" after an obnoxious freshman student they knew. With a nod to *Star Trek*, the game included "transporters," which, to add some spice to the experience, were sometimes unreliable, sending you somewhere you didn't expect, or causing damage to your character's health. Says Dirk, "At first, I added a lot of magic items with their names and actions taken almost verbatim from *D&D*, merely adapted to the computer game Dnd, like 'cube of force' and 'horn of blasting.' Later, I deliberately avoided copying anything from other games, and made up new monsters and items and features that I had never seen anywhere else while keeping the humorous tone." Two such items were "orange glop" and "roving sludge."

Another compelling design consideration imbued in Dnd was the idea of consequences. Everything seemed to have a consequence, a price. You might find an incredible amount of gold in a treasure chest. You could be greedy and try to carry it all, but if you did, the game made your character grow tired far more quickly—thus making you more vulnerable to dangers in the dungeon. If you had a lot of money you could buy a "potion of revival" to use as a last resort if your character died. You could then use the potion, but doing so came at a price: you lost all the magic items you were carrying. "A great game gives the players the freedom to make a vast number of choices," says Dirk Pellett, "some of which are more beneficial than others, and some of which are disastrous, and lets them figure out which is which."

Dnd is historic not only for being one of the earliest dungeon games on any computer, but also for a particular game design feature that many other games would copy over the coming years and still employ today. In the computer gaming world it's a concept called "the boss"—it might be a monster, it might be one hard-to-obtain item, or it might be both. In Dnd the boss item was something called the Orb, and finding it was the ultimate quest for a player. Naturally, the Orb was hidden deep inside the dungeon at its bottom-most level. And, naturally, it was protected by the fiercest monster, a classic example of what would become known as a "boss monster": a Gold Dragon. Kill the Gold Dragon, and you could grab the Orb. (In a similar way, Empire's short-lived Doomsday Machine was a boss monster, the ultimate enemy requiring extraordinary skill and luck to defeat. A more recent game, *DOOM*, also has a famous, huge, powerful boss monster. For some games the boss monster appears as a sort of final test of a player's skill; other more recent games may sprinkle boss items and monsters throughout the experience of the game.)

Early on, Whisenhunt and Wood discovered that more players were successfully grabbing the Orb than they had anticipated, but they could not figure out how they were doing it. So they made a pilgrimage up to CERL to find out. They discovered that players there had discovered a way to cheat: step one toe into the corridor or a room, just enough to find some gold or other valuable item, then step back out to safety. Then step back in, just enough to see if Randu had deemed it time to deposit another item in their path; if so, they'd grab it and step back out. Players were doing this for *hours* at a time: sometimes many hours, all-nighters. It required a determination and obsession that took the game's authors by surprise. "We never thought that anyone," says Whisenhunt, "would have the patience and would spend the time to do such a simple thing over and over again to gain large amounts of wealth. I guess that was the precursor to videogame addiction."

Sometime in late 1974 or early 1975, John Daleske began work on another game at the same time the major effort was under way to develop Empire. His new game was called "Dungeon," and unlike Dnd and Pedit5 it supported multiple users. Daleske claims it was the first multiuser dungeon (MUD) game on PLATO, perhaps on any computer. Unfortunately, the game struggled to achieve popularity, possibly due to being too complex and not as playable as other games out at the time or under development. Responsiveness was crucial; the gamers would not tolerate anything that didn't run at the speed of thought. Another factor may have played a role in Daleske not devoting more time to perfecting Dungeon—and Empire, for that matter. In May 1975, a doctor diagnosed Daleske with terminal cancer, "which kind of stifled my creativity a bit," he says. "Thankfully," he says, "it was a misdiagnosis, but I find it did affect me for about a year."

Word continued to spread online around the PLATO system, and around the physical locations lucky enough to have terminals, that more and more games were becoming available. The same process that had attracted Whisenhunt and Wood, and the Pelletts, and Daleske and the rest of the Iowa State gang, was repeating ever more quickly and to ever more people. Not just the fact that there were games to play, but that it was *possible to write your own*—look how many people were making the leap—as long as you could scrounge up the always rare file space and didn't mind weeks of late-night coding.

Pedit5 evolved into a new game called Orthanc (another Tolkien reference, to Saruman's evil tower at Isengard), by Paul Resch, Larry Kemp, and Eric Hagstrom. In late 1977, a new game appeared, Oubliette, written by Jim Schwaiger, with help from John Gaby, Brancherd DeLong, and Jerry Bucksath. David Sides, a teenage gamer at the time, remembers Schwaiger as "probably one of the larger egos of anyone I had met at PLATO. He used to call everyone a 'pud,' that was one of his big words as I recall. He would walk around with a beret, he was a tall thin guy, real unusual character." Oubliette, another MUD, like Moria showed a small 3D maze on the screen (small because the maze walls could plot more quickly, livening up the game) and offered multiplayer capability. Like Dnd it offered a method of teleportation. However, if Randu was in a bad mood that day, it was possible to tele-port a player into solid rock, killing off the player instantly. (Players referred to the situation as "getting stoned.") Says Brian Blackmore, a UI student during the era, "If you teleported into rock, that was the end of your character. And you couldn't even get your items up. I mean, it was doomed. So teleporting into rock was like *the most feared thing that could happen to you.*"

Similar to Moria's groups, Oubliette provided mechanisms to travel in "parties" into the dungeons, and when encountering monsters the parties could fight them together, increasing everyone's odds of sur-vival. In fact, Oubliette, Moria, and Empire, where one played on one of four teams and needed the help of teammates to conquer the galaxy, were usually unwinnable without multiple players teaming up to fight and survive together. To help players find other players to join up with, Oubliette's clever authors had created various "taverns" in which you could meet other players and form parties. In a way, taverns were Big Boards in disguise, a similar mechanism for players to find others and go down into the dungeons together.

—

Renowned computer game designer Will Wright, creator of a series of bestselling videogames including *Sim City*, *The Sims*, and *Spore*, often speaks of two major themes in the field of game design: the balance of technology and psychology, and the "possibility space." Essential to making a good game, Wright argues, is striking a good balance between the technical aspects of the game's implementation (the game's technology) and the model of the game as rendered in a player's mind (the game's psychology). "The game is really happening in the player's mind, the player's imagination," says Wright. Early computer games were limited in display complexity, speed, and number of players. Compared to today's hyper-photorealistic, 3D shoot-'em-ups, it's amazing that anyone would find the early computer games compelling at all. PLATO IV's limited bandwidth between mainframe and terminal meant that while fancy graphics were possible, they took a lot of time to come across the wire and form on-screen. For game designers, this often forced a minimalist approach: display as little as possible to keep things moving. While you might think this would translate into uninteresting games, the Fast Round Trip made the games highly responsive and thus eminently playable.

By "possibility space," Wright refers to the sum total of possible "moves" or "actions" a player may take in a given game. With a board game like *Monopoly*, the possibility space is distinctly finite. But since the advent of the computer, games can have possibility spaces approaching the infinite. As multiuser games began to evolve on PLATO, many had no required goal or outcome. These games didn't care if you played the expected way, or did something on your own. They were "micro-worlds" that you could explore and do battle in however it made sense to do so. Thus, PLATO games like Empire and the many MUD games were renowned for large "possibility spaces" long before the term came into being.

The MUD games on PLATO engaged players just as deeply as today's popular photorealistic 3D video first-person shooter games like *Grand Theft Auto*, *Halo*, and *Call of Duty*. While lacking every bell and whistle that today's videogames offer, PLATO's games grabbed hold of a player's attention just as fiercely. Today's games require little to no use of a player's imagination, whereas PLATO's game authors recognized that the fewer a game's graphics and multimedia capabilities, the more the player's own imagination had to take over and fill in the blanks. PLATO's MUDs had many blanks to fill, but players effortlessly filled them in.

Between 1976 and 1978, two high school kids, Erik Witz and Nick Boland, created a PLATO game called Futurewar, which never quite took off as much as sòme of the other MUDs, but is historically significant by its being an eerily similar precursor to the massively successful science-fiction first-person shooter PC game *DOOM* that would come out from id Software in 1993. Witz, fourteen years old at the time he began work on Futurewar, designed the game not only with a three-dimensional maze that players would walk through, but also as a first-person shooter, a term that would not come into common use until the 1990s. "I wanted Futurewar to be different, so I gave it a futuristic theme," he says. The game offered a deeper 3D view into the maze than other games like Moria and Oubliette, and instead of preloading a large number of custom graphical characters used to display the monsters in the game, it used TUTOR -char- commands to dynamically load select graphics for monsters as they were encountered, enabling the game to have many more monsters than other games. Some of the monster graphics were even animated. "Like *DOOM*," Witz says, "a gun was displayed pointing in front of you. You could use different types of guns that would produce different shooting effects. With the gun you could shoot other players, or monsters. . . . Futurewar had terrains like water, fire, and radioactive waste, like *DOOM*. Futurewar had exit signs and elevators, just like *DOOM*. The bottom level of Futurewar was called Hell and featured the Devil, similar to *Heretic* [a 1994 *DOOM* offshoot] and *DOOM*. . . . Futurewar contained almost all the elements of *DOOM*." The creators of *DOOM*, contacted in 1997 regarding Futurewar as possible inspiration, denied any connection between the two games. "None of us ever used a PLATO system," said developer John Carmack. Futurewar also appears to have influenced the design of a 1987 Atari arcade game called *Xybots*—perhaps the first 3D first-person shooter arcade game—which takes place in a similar futuristic setting as Futurewar, with similar game design right down to one identical game monster: flies that darted around and were hard to shoot. *Xybots'* creator, Ed Logg (also the creator of *Asteroids* and

Centipede) claims no memory of Futurewar, but does say he was familiar with PLATO back in the 1980s.

Andrew Shapira, his reputation already firmly set as the top Empire player and fastest typist on PLATO, was also drawn to the new dungeon game craze. He and some friends, including Bruce Maggs and David Sides, began thinking about writing a new dungeon game. Unlike other developers, these three took their time.

Bruce Maggs had built a "maze runner" program, of which there were numerous in those days. As a programming exercise, it exposed you to interesting challenges and constraints. How big could you make the display without making the plotting speed unbearable? This had always been a problem with PLATO. Many very popular games, like Empire, Airfight, Moonwar, and Spacewar, wound up not displaying anything at all on the screen (often because users pressed STOP repeatedly, or because the game was too bogged down) when the combat was at its most furious. Other considerations: 2D or 3D? 2D had been done: Pedit5, Dnd. 3D was the future, as Moria, Oubliette, and Futurewar had already made clear.

Shapira saw what Maggs was up to and, he says, "thought it was pretty cool, so I asked him if I could build onto it a little bit." Soon, a new game was under way. There was a conscious goal to outdo every other dungeon game to date: bigger, better, more monsters, more players, more features, the works. They became fanatical about working on the game. But even so, due to the scale of the ambition, it took time. Shapira, Maggs, and Sides were all in junior high and high school during the development of the game. Years would go by.

An early version, named Darkmoor, was up and running for a while, but not open to the public. Unhappy with how it worked, they decided to rewrite it from scratch. Finally, the project approached being "done." They named it Avathar, a word right out of Tolkien's *Silmarillion*. Eventually, out of fears that Tolkien's estate or publisher might not be happy with them appropriating the Avathar name, they changed the name slightly, to Avatar.

Shapira and Maggs would camp out in two adjoined classrooms full of PLATO terminals on the first floor of CERL and hack away at the game for hours. Numerous authors were experimenting with all kinds of dungeon game designs, including versions where you could see infinitely far until there was something, a wall or other object, blocking the view. The problem with this approach was always display speed, so most experiments remained experiments and didn't survive into actual games. "That's really one of the hardest things about writing a dungeon game back then," says Shapira, "was getting it to run at a reasonable speed and also have it be fun."

Avatar would officially roll out in late 1979 or early 1980, and it was an instant hit. Nearly four years in the making, and it showed. This MUD was so deep, so ambitious, so complicated, it required a staff of volunteers, called "operators," to run and administer it. It had dozens of utility programs and editors that players never saw. Some friends of Shapira, Maggs, and Sides volunteered to take on the data aspect of the game: not only managing the names of magic items, weapons, and monsters, but also tinkering with their attributes, dependencies, and side effects, all of which added a level of detail that no other MUD on PLATO would match. Plus, the game supported sixty simultaneous users. It had a messaging capability that was so busy it was often nearly impossible to keep up with, like a busy Twitter feed today. Players could message individuals, groups, or all players.

The game offered teleportation capabilities like some previous dungeon games on PLATO, but there were new twists. Entire groups of players could be teleported if the leader of the group had enough powers and spells to pull it off. However, like in Oubliette, it was possible to get stoned. You might make herculean efforts to build up a great game character, and then suddenly you were randomly teleported into solid rock. Worst of all, the leader of a group might screw up, transporting the entire group into rock. Says Brian Blackmore, "Oubliette, if you get stoned, you're dead meat . . . but in Avatar you could get un-stoned." It might require what came to be known as "divine intervention," meaning, a player would seek out and beg one of the game authors or operators to reinstate your character safely back in the city. For a while, unscrupulous players would hang out at the bottom of the stairs from the city, and lure unsuspecting newbies to be teleported, often deliberately, into solid rock. Such shenanigans were not long welcome in the game, however.

When Bill Roper wasn't exercising his official duties being a grad student working for Stan Smith and his chemistry lessons, he was often in Avatar. "We had site manager privileges in the chemistry annex," he says, "and we would go over there and we would play Avatar. The way that Avatar was set up was, if you quit the game by pressing SHIFT-STOP on it, you would lose everything. But if the site manager backed you out, you wouldn't. And so you would sit there with the site manager and if we got into really big trouble, we would hit the button and get out of the game. Which wasn't exactly sporting, but did certainly improve our life."

Avatar had boss monsters like Dnd. One was "Reaper of Souls," probably the worst of the worst. "What a motherfucker of a monster," says Blackmore. "The thing blew me away so many times before I got to be really, really big. And it could blow away an entire party like *that*," he says, fingers snapping.

It had started at least as far back as Oubliette, and it continued in Avatar: real players paying real cash to other players for all kinds of virtual loot. "I know in Oubliette there was a fair amount of effort to cheat," says Josh Paley, a UI student at the time, "and I know that Jerry Bucksath was always amused that people would want to give him real money to get him to go change a bit here or there so they could have the best items in the game. The notion of changing one or two bits being worth one hundred or two hundred dollars to people seemed like a very amusing thing."

Buying and selling virtual items sometimes had a tangible benefit, making it understandable that some college students would participate in the market. If you were not planning on sticking around after receiving your diploma, then your time on PLATO often came to an abrupt end. If not your signon, at least your gaming persona. For dungeon gamers who had dedicated a lot of their college life to PLATO, an attractive potential payoff beckoned. Mike Wei, a UI student at the time, is but one example of what might happen next. "In '77–'78," he says, "I was already selling my top items to the poor zbrats and ybrats who needed a leg up in the various dungeon games. I do know that selling off all my chars and items as I left for the 'professional world' of programming at BYU in '79 helped finance my relocation."

Sometimes things got out of hand. Al Harkrader, who managed the operator team at CERL, remembers one Oubliette incident in which one of CERL's paid junior systems programmers was going into the Oubliette database and creating the items by using his system privileges, then selling them to all comers for hard cash. "I still remember the last day of chasing him around," says Harkrader, "because we were closing in, somebody'd told us what was going on." He and other CERL staffers were inspecting a modified data file they had restored, and while they were looking at it, it *disappeared*. And then the backups disappeared. The perpetrator was at another terminal, "trying to cover his tracks, ahead of us," says Harkrader. "It didn't work. We finally figured out, yes, he was doing it, and I think he also figured out that we knew he was doing it, because he was never seen again."

For dungeon dwellers who desperately needed "divine intervention," offering cash rarely worked, because the game authors and operators tended to be strict about keeping cash out of their games. Nevertheless, paying cash worked for many things. If you were stuck down in the dungeon and needed help finding a stairway back up to the city, or a rescue via teleportation, perhaps some kindly master player would help you out—for a fee. If you needed to buy or sell magic items, weapons, or even whole characters, there was a ready black market for such things. You simply needed to know who to ask. Sometimes experienced players would create a new dungeon character with the express intent of selling it. They would take it deep in the dungeon, build the new character up from nothing, fill it up with riches, weapons, and magic items, then sell it off to someone out in the real world. "If we found out there was someone trading money, we would delete [the character]," says Sides. "Because, for one thing, it was kind of dangerous. . . . I mean people trading money, and suddenly our decisions become much more important and I don't want people angry or threatening me because I'd do something that'd affect monetary gain on their part. Basically our policy, I believe, was that anytime we found that that there was money involved in a trade we would delete the items involved or at least it would become held up or held back or something."

There were several notesfiles dedicated to Avatar, one to discuss the game, bugs, ideas for improvement, posting questions for help, begging to be brought back to the city, or un-stoned, and so on, and another notesfile, =avatrade=, which turned into an eBay or Craigslist of sorts for trading anything of value in the Avatar virtual world. In theory the players were supposed to *trade* their items, but the game authors admit knowing that some people secretly used real cash when trading items and characters, and there wasn't much they could do about it.

It would be many years before the notion of "virtual economies" became an object of media attention and academic study with books, conferences, and journals dedicated to the subject. It exploded once the Internet rose up and multiplayer games had their second coming. By the 2010s, the marketplace for virtual goods had become a multibillion-dollar bonanza worldwide, including venture-backed start-ups providing online marketplaces for the exchange of items from games.

17

The Zoo

This room is never anything o'clock. Minutes slip through it like a thief in gloves. Hours fail even to raise the dust. Outside, deadlines expire. Buzzers erupt. Deals build to their frenzied conclusions. But in this chamber, now and forever combine. This room lingers on the perpetual pitch of here. Its low local twilight outlasts the day's politics. It hangs fixed, between discovery and invention. It floats in pure potential, a strongbox in the inviolate vault. Time does not keep to these parts, nor do these parts keep time. Time is too straight a line, too limiting. The comic tumbling act of causality never reaches this far. This room spreads under the stilled clock. Only when you step back into the corridor does now revive. Only escaped, beneath the falling sky.

The novelist Richard Powers wrote these words to open his 2000 novel *Plowing the Dark*. They describe "The Cavern," his imagined virtual reality room inside a computer lab in a fictional software company in Seattle. A "holodeck" in which anything, anywhere, anywhen is possible.

A real room in a real building had inspired Powers to invent The Cavern: a room that Powers remembered personally, for he had himself spent uncountable hours in it, crowded in with others seeking the same thing, usually in the dark—the fluorescent ceiling lights usually off, giving the impression of everyone glued to their terminals like attentive air traffic controllers—all of them there drawn for the same reasons, all of them there outside of time, outside of reality, deep in exploration of the limitless digital world of—and available through— PLATO. The place was known as "the Zoo," CERL's fabled, notorious, first-floor pair of adjoining rooms, 165A and B, where students often risked their chances of passing grades, to say nothing of risking wasting their lives away, by playing games, authoring games such as Avatar, reading and writing notes, yakking away in Talkomatic and TERM-talk, hacking code, and generally causing mischief all night long. Everything but *education*, the purpose for which the PLATO system had been designed at a cost of millions of dollars and years of hard work. It had been meant to revolutionize education and to deliver brilliant students more math, more science, more everything, sooner, faster, cheaper, and more effectively than teachers and conventional instruction. The government was counting on PLATO to pave the way for a bright high-tech future, in which, thanks to all that brainpower, a confident America maintained possession of the high ground of space and shiny little beeping orbs like Sputnik. The Zoo was the very same room in which the ILLIAC I computer had stood back in the 1950s and early 1960s, the very same hulking Iron Beast of a computer on which PLATO I was born. The room in which the ILLIAC's vacuum tubes had glowed before the Orange Glow even existed. The place from which the first bits of PLATO machine language punched into oily paper tapes and executed through wires and vacuum tubes of the ILLIAC I. All that was gone now, replaced by *what*, exactly? Officially CERL called it a "classroom"—that was enough to bring a wink and knowing chuckle from the gamer ghouls who crawled out of the woodwork to take over the room in the evening—no doubt they called it "classroom" to keep the powers-that-be happy, keep that educational story going. There was "Learning" going on here, with a capital "L." You never knew what dignitary or celebrity would be strolling down the hall with their entourage, getting the Tour and the Demo. But to the kids, the gamers, the hackers, the stoners, the punks, the nerds, the loners, the social misfits, the assembled addicts, *pilgrims all*, who had, down to a person, heard about this weird, cool, impossibly futuristic computer called PLATO, over in that crazy old building called CERL, from which emanated the Orange Glow, it was all about the games. To these "students," this was *the Zoo*.

Who could have predicted that this would be the fate of so hallowed a chamber as the very birthplace of PLATO, where brilliant, serious men and women dreamed of a better future for students (and teachers), a future in which education could be freed and transformed by a new apparatus that they themselves had designed and had built, and would teach anything to anyone because it was infinitely flexible, infinitely patient, infinitely knowledgeable? A machine that delivered

The Zoo at night

on the promise of Self-Pacing and Immediate Feedback. Only, for this crowd, those principles applied to *games*. And now, the ILLIAC had been carted away, dismantled, obsolete, gone, and forgotten. The machine room that was its home turned into a geek habitat surrounded by the same old cinder block walls, with fluorescent lights overhead and the ancient squares of linoleum tile on the floor below. The Zoo was now not so much a room as a throbbing darkened hive from which wafted an indistinct and not quite pleasant mix of vapors arising from unseen and long-forgotten food crumbs, sweaty body odor, pizza breath, onion breath, cheeseburger breath, Pepsi and Mountain Dew belches, and worse; on some tables and on the linoleum-tiled floor, here and there in darkened corners roach motels, some with vacancies, some not; the occasional crumb from a Pop-Tart, the essential nutrition of Zoo inhabitants when pizza was not available; and everywhere creaky wooden chairs situated in front of ancient wooden library tables underneath which decades of chewing gum and who knows what other grime had, like a boat's barnacles, accumulated, and on top of which rested bulky metal boxes propped up on "precision angle adjusters" (also known as four-by-four blocks of wood)—boxes out the back of which ran various thick cables leading to the fourth floor, where the CYBER hummed in its chilled-air sanctum. Connected to the front were equally bulky keyboards and heavily fingerprinted touch-sensitive display screens, from which emanated the Orange Glow. The Zoo, where instead of the incessant musical chimes of a casino's gambling machines one heard the nonstop clickety-clackety-*bang* of fingers on keyboards, frantically defending planets, firing phasers and photon torpedoes, shooting down enemy aircraft, attacking monsters in the dungeon, chatting and flirting in the Talko channels, and writing snarky messages in the notesfiles. The Zoo, which shared much in common with a casino: no windows, no sense of time, perpetual play. The only difference being that in the Zoo, the steady drain of currency was not from spent dollars and coins; the patrons of the Zoo came here and spent their *youth*, and the House (in this case, the Power House), *always* won.

Only when one stepped back into "the corridor," that narrow hallway that passed from one faraway end of CERL to the other, did one have any chance of escape back into the real world. *But why would you do that?*

B. F. Skinner had cherished his experience with Mary Graves, his legendary grade school teacher, whom he had described as "someone who listened to me, answered my questions, and almost always had something interesting to say or a suggestion of something interesting to do." In the 1970s, for many students who discovered PLATO and who became denizens of the Zoo, the PLATO system, its online community, its ever-growing catalog of lessons, notesfiles, games, and countless other activities, had become a different kind of Mary Graves. Instead of PLATO being the ultimate online teacher at the ultimate online Academy, rivaling that of the Greek figure for whom the system was named, a digital place to come and learn about everything from anthropology to zoology, PLATO had become, for so many young people, a place to come and learn about PLATO. A place to learn about each other. The system *itself* was the thing. If you had a voracious appetite for learning—if you had somehow, somewhere in your life learned how to learn, and derived joy from the experience of learning, then when it came to PLATO, the most fascinating learning of all was the system itself. The system, and its online community, had become a cyber-proxy for that long-lost someone who listened to you, answered your questions, and almost always had something interesting to say or a suggestion of something interesting to do. And it's been that way ever since. In 2017, PLATO may be gone, but the phenomenon is still here, only now gargantuan in scale, changing humanity, changing the earth, no turning back.

Who needs Mary Graves when you've got the Net?

—

Why did CERL tolerate the gamers? Why did Bitzer? They could have been banned in an instant. Many other PLATO classroom sites, both on campus and remote, had banned games outright. This was an expensive system, and sites that had PLATO paid dearly to utilize these precious resources. The mission was education, not entertainment. So why, then, did the Computer-based Education Research Laboratory, of all things, the very home of PLATO itself, not view games and gamers as a pestilence?

The answer lies with Bitzer and Alpert. They both shared an enlightened, permissive attitude that had benefited CERL and CSL before it. Alpert was not afraid of discovering bright young capable people and mixing them with senior engineers, physicists, and other researchers to collaborate on projects. Bitzer had the same philosophy, which had served him so well from the very earliest days of PLATO. Without Andrew Hanson's brilliant work as a teenager writing the multitasking operating system enhancements to PLATO III, Bitzer might have had a far more difficult time making progress in the 1960s. Likewise, the work of Mike Walker, Rick Blomme, Scott Krueger, David Frankel, David Woolley, Kim and Phil Mast, Mark Rustad, Doug Brown, Marshall Midden, Brand Fortner, John Matheny, Ray Ozzie, Tim Halvorsen, and dozens of other young people brought capability to the PLATO system for next to nothing in terms of cost to CERL's bottom line. Sure, sometimes gamers could get unruly, but sometimes the overnight madhouse gaming activity exposed flaws in the system hardware or software that could be conveniently discovered while the real business of PLATO—giving students an education—was at home asleep. By morning, problems could be fixed, to the benefit of those same students.

Not everyone on the senior CERL staff was happy with the gaming situation. Bruce Sherwood thought they wasted resources. William Golden, known as William "No" Golden for his often saying "no" to anyone asking for a signon or file space, was not a big fan of the gamers and game authors. But Bitzer had a more moderate view of the scene. He was impressed with the games, and from time to time played some of them himself. A good tech visionary knows that half his job is selling, and Bitzer was a master salesman: cool, confident, and knowing just what rabbit to pull out of his bag of tricks depending on the situation. Thus on occasion he had begun demoing games, notes-files, pnotes, TERM-talk, and Talkomatic. Like Steve Jobs and Apple would do years later with the iPhone, the sheer volume and variety of applications created on PLATO were dazzling in demo settings.

But still, not everyone was happy with the games or the gamers. "I would get a lot of reports," Bitzer says. "You know, 'We ought to do something about this.' 'We ought to do something about that.' But I didn't—if they weren't doing any damage, I didn't see any reason to stop creative people from expressing their creativity."

Ray Ozzie called Bitzer's philosophy "subversive innovation." It was an attitude that permeated much of CERL, starting with being open and welcoming to any visitor. If you were a student wandering into the building only to become fascinated with what you were seeing on a first visit, you would invariably come back, seeking out more knowledge. You'd learn it was possible to program in TUTOR. You might go the gamer route, you might not. You might be struck by how so many people were *building* useful things in the lab. "Not everyone was a gamer," says Ozzie. "Not everyone spent as much time in all of the different notesfiles that were out there." Indeed, Ozzie wanted to program more than goof around, and he invented a path for himself, discovering the next-generation PLATO V terminal that Jack Stifle's team was working on, learning that they had added a microprocessor to it, making it an early personal computer. Why not write programs that executed locally inside the terminal's microprocessor? "I went to Tenczar," says Ozzie, "and said, 'I can make this thing! Downloadable, you know? I'll do this, I'll do that, I'll do that, I'll do *that!*'" Before he knew it, Tenczar gave him the go-ahead, and Ozzie was being paid to work on a bona fide CERL project to create a new tool called "pptasm," the PLATO Programmable Terminal Assembler. "Tenczar was my entry point, and I convinced him that I could do this stuff, and I did. . . . I wrote an assembler and a loader and all this stuff." Ozzie found that Blomme was not keen on people writing assembly language to make things happen locally on a PLATO terminal, and pptasm did not make much of a direct impact in the long run. But it did spur Blomme and others to think about programming in a microprocessor version of TUTOR: what if one could write TUTOR lessons and have them run locally, instead of on the mainframe? In time the work done on pptasm would eventually lead to MicroTUTOR, which Control Data would market as Micro PLATO in the 1980s. It was a classic example of how CERL's subversive innovation led to important advances that eventually made a difference out in the world.

Likewise, the authors of many of the multiplayer games that kept the denizens of the Zoo glued to their terminals every night learned a great deal about programming and computers by creating those games. That knowledge gave them skills that they took into their subsequent careers. Those careers sometimes were at CERL or at Control Data, or elsewhere, including eventually Atari, Apple, Google, and other Silicon Valley companies. For Bitzer, for CERL, all the gamer fuss was worth it. Limit gaming to off-hours, but don't ban it outright. The new wave of high school and college kids that poured into CERL in the 1970s brought many brilliant minds to the lab, and that brilliance was put to use in ways no one could have predicted.

Richard Powers was a denizen of the Zoo during CERL's peak years in the late 1970s, drawn there for the same reason everyone else was drawn there. What else were you going to do? How could you *not* be drawn there? The same Richard Powers who in more recent years has become a major literary figure, a novelist who has garnered sweeping acclaim and won so many prestigious awards, including a MacArthur "genius" grant and, in 2006, the National Book Award for Fiction. Several of his most well-known novels, including *Galatea 2.2*, *The Gold Bug Variations*, and *Plowing the Dark*, directly confront issues related to computers and human-machine interaction. "Man," he would recall years later, thinking back on PLATO, "I spent more timeless hours in those rooms than I can say, and I remember coming back out sometimes after a long session, shocked to discover daylight or the lack thereof, just out the doors of CERL." The enticement of that Orange Glow could be downright scary. "I can remember coming back from a bunch of all-night sessions, and feeling, *This is a very dangerous, very powerful drug, and you better be careful.* I mean, the closest I've ever felt to addiction was probably sitting on that system."

But sit on that system he did, and for several years. "The psychology of computer use in all my digitally inflected books," says Powers, "derives from my initial experience of that strange public/private hybrid network mentality of PLATO. I so often felt that the years from 1978, when I left U of I to become a writer, until about 1993, when I came back [to Urbana] to write *Galatea*—which I think must be one of the very first works of fiction to describe a web browser—were spent waiting for the world to catch up to the possibilities inherent in the system that I had grown up on."

Powers had enrolled as an undergraduate at the University of Illinois in 1975, majoring in physics. It was there that not only did he find PLATO, but as was the case with countless other high school and college kids, PLATO found *him*. "I can without exaggeration say that the system changed my life," Powers says. "I can't say that I would not have become a novelist if it weren't for PLATO, but I certainly would not have become the novelist that I became. . . . PLATO was my first exposure to the man-machine interaction and it not only was the place where I was able to program, and was subsequently able to make a living doing that while becoming a writer, but it was the place where I first saw what might change in the human mind when given a kind of blank-slate playground in which to extend itself."

Like so many thousands of students who attended UI, he first came across the system through his regular classwork. A professor here or there would give his class an assignment that required visiting a special classroom somewhere on campus where students would interact with lessons, simulations, and quizzes online on the PLATO system. It was here that Powers succumbed to the spell of the Orange Glow. He discovered that not only was he learning his course material through a machine, through PLATO, but he also wanted to learn more about PLATO, the machine, the phenomenon, itself. This did not happen to every student who would enter those special PLATO classrooms. Most, in fact, came in, sat down, did their work, and were done. The Orange Glow may have tickled their fancy, but not enough to want to know where the glow came from and what lay at the other end of those thick cables coming out of the back of the machine.

"My memory is that my first discovery that there were more things under the hood," Powers recalls, "came about when sitting in these labs. There would be twenty or thirty terminals in a lab, and I would be working on a physics problem set, and the person next to me would be working on a physics problem set, and the person next to *him* might be playing this intergalactic game of planetary conquest. Of course, I wanted to know how he got on that. So there was a preliminary gradual discovery that a roomful of people could be involved in all sorts of mysterious applications, some graphical, and on occasion I would see someone there with a headset on, and there'd be clearly an audio component going on that I wasn't aware of."

The first forays, he says, were casual glances over the shoulders of others, causing him to wonder how they got to the activity they were in on *their* terminals and how could he conjure up that same activity on *his* terminal? What magical incantations were necessary? "I would lean over," says Powers, "and ask somebody, *How'd you get in there?* and they'd act very secretively, and say, well, you know, you just have to be *an author*, or you have to have X *privileges*, and I would try poking around, using my student account, and find out where it would take me."

Then Powers discovered *multiples*. Multiples were a type of PLATO user signon designed to be used anonymously—and simultaneously—by anyone. Pretty much every other PLATO signon was unique: only that one person could sign on using that name and group combination. But with a multiple, lots of people could use it. Typically created for demonstrations and for curious visitors, multiples were generic signons consisting often of "demo" as the signon's name and "demo" or something similar as the group, perhaps also with a password of "demo" if there was even a password at all. Entering the system using a multiple signon might yield access to catalogs of lessons that weren't available to the average student who simply was there to do his or her class assignment, be it in physics, chemistry, veterinary medicine, or French. Some catalogs would open doors to lessons in those subjects and more. Some catalogs might provide access to lessons about PLATO itself: how to use it, how it worked, what made it run, how it was programmed, and how you could program it too. These special multiple-type signons were the easiest to find out about. Powers would use them, take them as far as they would go, to explore "these little nested networks that would open up for a while, and you'd kind of tunnel in there and get hold of whatever information or secrets you could find and they'd close up again. The next time you'd come to the site it would say, 'Sorry, we don't allow that signon here.'

"One of the early fascinations that has stayed for me," Powers says, "is the discovery that the same interface that I looked at all day long to do my problem sets and my lessons could, with a certain set of key combinations, open up a little rabbit-hole, and produce surprise entry points into layers that were behind or alongside that. When it's your first introduction to the fluidity, the fungibility, and the multiple layers of computer interface, it's mind-changing."

Directly across the corridor from the Zoo was a small office where the PLATO system operators worked. Situating the system operators' office on the first floor of CERL while the machine room with the CYBER and communications equipment was up on the fourth floor made little sense at first glance. But the operators were responsible for more than just keeping the system running, routinely backing up data, and managing printout requests: they were also tasked with a job no one else wanted, keeping an eye on the Zoo. In a way, that was their most important job: they were the Zoo police, enforcing rules about who could be there and what those people were allowed to be doing. The rules were different during the day versus during the night.

As a group, a number of the operators had a reputation, deserved or not, for being stoners. The leader of the operators for a number of years was Al Harkrader, a perfect choice for managing that group: he had a quick wit, quicker smile, was liked by everyone, and took his job seriously. He also was no slouch at programming, and, seeing that some of the operators, particularly those on the graveyard shift, not only might be stoners but also might be high while on the job, decided to write a set of operator utility programs that made it nigh impossible for an operator to screw up. As Dave Fuller described it, Harkrader's software was "bulletproof and stoner-proof. The most important thing was stoner-proof."

The Zoo attracted an odd assortment of people during the night, people of different ages, backgrounds, personalities. What they shared was a compulsion to play games. One such gamer was Patrick Clifford. Who also happened to be a cop.

After receiving his undergraduate degree in aviation, Clifford found no relevant jobs in Champaign-Urbana. On a lark, he decided to take a job for deputy sheriff in Champaign County, and was hired. He was assigned a night shift that ended at 2 a.m. "PLATO was perfect," he says. "I didn't want to go home and go to bed—too keyed up—and there was very little else to do in C-U at that time of the night. I had already become a part of the PLATO community, [so] I spent most of my off-duty time over at CERL. Of course, I was required to carry a gun even off duty, so I had one with me all the time when I was at CERL. It got to the point that most everyone knew me and who/what I was and were pretty comfortable with it. It did cause some awkward moments, though, like the time I came over to CERL on the way home from work to pick up some prints or something and walked in, in full uniform, on a new operator . . . smoking what appeared to be a rather strange-looking and -smelling cigarette. He almost had a heart attack. . . .

"If something strange was going on, like the time the operator found a guy passed out in the elevator (she thought he was dead, initially) . . . or when a couple of Empire players would take getting killed a little *too* personally, someone usually dragged me in to help out. Actually, that's kind of how I got my PLATO pseudonym, 'The Enforcer.'"

A common critique of B. F. Skinner's behavioral experiments on rats is that if all organisms exhibited the same kinds of learning behaviors, then are humans no different from rats in mazes looking for cheese? Legions of critics have argued this for decades. With PLATO, whose creators had been, whether consciously or not, influenced greatly by the work of Skinner, it might be argued that the lessons and quizzes students were assigned to take were at least metaphorically not unlike Skinner's experimental mazes, and the feedback issued by PLATO during, and the grades received upon completion, were the cheese. What Skinner might not have anticipated, however, was that some-times an idea—an *itch* of curiosity, if you will—would enter the minds of these particular "rats," causing them to desire to occasionally climb up and peer out of their "maze," so that they could gaze upon whatever lay beyond. And what lay beyond? More mazes, more rats. Sometimes these other mazes were more complex, meandering in more interesting ways, leading the gazing rat to forget its own cheese and focus on the cheese in an adjacent maze. Pursuing that other rat's cheese might lead to even more intriguing mazes inside of which scurried even other rats.

Infinite mazes, infinite cheese, in every direction. This realization of the infinite potential and explorability of PLATO dawned on Richard Powers right away.

One day he found his physics professor, Dr. David C. Sutton, in the PLATO lab, sitting there hacking away at TUTOR code, authoring a new physics lesson. Sutton noticed Powers was interested. "There was a huge gulf between this lowly freshman in this class of several hundred and the lecturer for this class," says Powers, "but he kind of beckoned me over and I sat down and he said, 'You want to see how this is done?' and I said, 'Yes, indeed,' and he pulled up his authoring account and he began to take me through my first exposure to TUTOR."

Powers continues, "Burrowing into the little tendrils that would open and close in the system had a kind of physical analog." Each day that Powers spent in the PLATO lab drew him ever closer, like a euphoric moth, to the orange flame of the plasma pixels. Like so many students before and after, he would begin to notice, usually out of the corner of his eye, the Orange Glow elsewhere on campus: if you paid attention, you noticed right away that there were terminals all over the place. In time, like so many countless others had done before, he found CERL itself, the Ground Zero of the system, its birthplace, the Glow's point of origin.

But there were other rooms accessed from other mazes in other buildings too. Some rooms might be deserted, or tiny, or both, with just two or three or four terminals, or, if he was really lucky, just one. "What you would do," says Powers, "if you found a new terminal, is to sit down, and you would start going through the various signons that you had acquired—I hate to say it—either legitimately, or by stealth. And you would try them out, one by one, and see which ones would be accepted by the security on that system. And my God, I must have had tremendous memory at that time, because I could remember all these different signons and logins and some of them were multiples and some of them were private and there would be this moment of elation to discover that this machine would accept the signon that would get you into all of this material that the other sites were trying to clamp down on, the real CPU-intensive games, or what have you. And then you'd feel like, now you'd have *a little bit more power.* And now you have to kind of guard it from other people. You can't just let anybody know that machine will take this signon, because, first of all, you'll never get back on it, and, second of all, eventually the site monitors are going to find out about it and shut it down. And now I kind of understood after a year of scrounging around and following these physical trails why everyone was so cagey with me when I leaned over, in the lab, and said, *Hey, how did you get on that?* They were protecting their privileges and their hard-won entry into the system."

With his experience on PLATO, Powers saw firsthand how important *identity* was to this new digital world. Not just one identity, but many: the trick to the digital world was that you would have to learn how to juggle a whole *bag* full of identities, like a spy with a pocketful of passports, each enabling you to become yet another rat seeking yet another piece of cheese in yet another, different, maze.

"We kind of take that for granted now," Powers says, "but this first glimpse of the fact that I had an existence to other people without any physical connection, and they with me, was really novel for me at that moment too."

Powers's relentless pursuit of higher levels of access, beyond several student signons, the legitimate and stealthy demo signons, the word-of-mouth secrets yielding access to what he would later call "ill-gotten entry points," finally, in time, led him to take a summer job as a site monitor in the biggest PLATO lab of them all, in the basement of the foreign languages building. In his role as site monitor, he sat perched behind a bunker with a view out to a sea of orange: some eighty terminals tucked away in custom wooden carrels in front of which sat eighty students typing and touching and, one hopes, learning, and, sometimes, doing the same kinds of illicit online activities that he himself had done. Only now he was a site monitor, with the power to kick those ne'er-do-wells off the system and send them packing. Until now he'd been the enforced; Powers had now become the enforcer. *A little bit more power.*

Powers finally had an author signon. If burrowing into secret tendrils and stealing glances at unattainable vistas was enough to blow his mind before, then having an author signon now was truly arriving at the Promised Land. With an author signon, he would be able to write programs. "I began experimenting with TUTOR," Powers says. "I guess my first bit of code was a kind of concrete, interactive rendition of T. S. Eliot's poetry, using the text placement and rotation and interactive components of TUTOR to do a lesson introducing beginning students to poetry and reading and interpreting."

Joseph Dewey, a professor at the University of Pittsburgh, has written a book about Powers and his novels. "His use of computer science as a trove of metaphor," says Dewey, "is consistent throughout his work. His early training as a computer programmer served him when he turned to fiction. The metaphors of connection, simulated reality, and the inevitable loneliness at the heart of computer communication became essential metaphors for his take on late-twentieth-century computer culture. Rick uses science for its metaphoric impact in every novel: genetics, oncology, pediatric medicine, computer programming, virtual reality environments . . . the list is long." Dewey calls Powers's novel *Galatea 2.2* a "fictional autobiography," where "he recounts his own introduction to its premise."

Powers may have been a physics major, but he was drawn increasingly to the humanities, not because he lost interest in the sciences, but because the humanities helped him see the sciences in a new light. One day during Powers's junior year, Professor Robert Schneider, who was teaching an honors course, urged Powers to pursue the humanities with increased vigor. Dewey says that Schneider convinced Powers that "literature was the 'perfect place for someone who wanted the aerial view.'" Schneider was persuasive enough for Powers to drop out of physics and switch his major to English and rhetoric. "The struggle for me," says Powers, "was between some kind of legitimate generalist's pursuit and the specialization required to really excel in most disciplines. I felt that call for increased specialization certainly by my junior year in physics, and I think it was the kind of terror at closing doors that made me seek out literature as a kind of aerial view. And if I were to fit PLATO in there, I think for me the fascination with the system was that it did open out into so many different disciplines, that it could be a domain where the teaching, the exploration, the archiving of so many different disciplines could all come together in a kind of new medium. When I left school, after graduating with a master's degree in literature, it was computing where I made my first living."

By 1980, Powers was in Boston, working as a computer programmer at a data processing firm, working on an old NCR computer and writing in a COBOL-like language. It might not have been as elegant as the TUTOR language, and the NCR system lacked everything that made PLATO so inspiring, but it paid the bills. "For me," he says, "that struggle with my own identity, in selecting majors, and in preparing to have a profession, was fought over that issue of individual disciplinary management versus a kind of archival or broad-based curatorial relationship to lots of different disciplines. And PLATO was my Alexandria. It was my library, it was the place where I could attach myself to anything. And it was also the place—you know, I wrote code before I wrote fiction—where I realized that, where I first saw where careful construction of words, in the right order, could create actual things, real data structures that then had behaviors, and acted as objects in this open-ended world. And I think it was a kind of latching on to literature simultaneously with latching on to the possibilities of cybernetic projection that landed me in the world of professional data processing and then ultimately, after a couple of years, starting on my first book."

Powers's exposure to PLATO also led him to a fascination with microcomputers, first the TRS-80 and later the Apple II. It was an utterly different world from PLATO. If you'd asked the typical PLATO person what would top their list of things to take to a desert island, "a PLATO terminal and a phone line" would surely have been no. 1 and no. 2. Microcomputers, on the other hand, *were* desert islands. Who in their right mind would want to spend time futzing around with such a lonely, disconnected gadget? Sure, it could beep and bop and draw stick figures and display a small amount of text (usually in only barely legible capital letters), but it couldn't get you into Empire. No Moria. No Avatar. And you couldn't *talk* to people. No notesfiles to hang out in, no Talkomatic chat rooms, no TERM-talk instant messaging.

None of that fazed Powers. Microcomputers attracted because they lacked PLATO's *enforcer*, be it a human site monitor or the site control lessons that automatically kicked intrepid young explorers offline so "serious" users could do their work. "So long as your account was being monitored by somebody else," he says, "and you had to produce something towards not necessarily your end, or source base, or your ability to get access to a machine that would allow you to have that open-ended 'clay,' so long as that was the case, you couldn't go where you wanted to go. For me, the dream with the microcomputer, and I agree, it was a big step down, off the system . . . you cut the umbilical for two reasons. There's an upside and a downside. The downside is that you're no longer being fed and being connected to an online community. The upside is you can go it alone and you can do all those things that you weren't allowed in, under strict social control."

18

Red Sweater

A frantic voice cried out, "Don't turn on the lights!"

Brand Fortner, an undergrad and junior systems programmer at CERL, headed upstairs one evening for another late night of work in the little office he shared with another staffer, who used it during the day. As he entered, he was startled to find someone hiding inside in the dark.

"Excuse me?" Fortner asked the voice. "Red, is that you?"

"Yes! Don't turn on the lights! You have to find the red sweater for me!"

Bruce Parrello had been washing his clothes in the basement laundry of his dorm. It was one of the few times he ever took his sweater off. Someone approached him, got talking with him, and sufficiently distracted him to cause him to leave the laundry for a moment. When he came back, he discovered that his precious red sweater was missing. Now he needed help getting it back.

His name was Bruce Parrello, but everyone called him Red Sweater. For Parrello his whole life seemed wrapped around the sweater, which became a compulsion, an obsession. He *had* to wear it, indoors and out, every day—no matter what the outside temperature. When asked what the weather was like outside, he'd say it was "sweater weather." For Parrello, the sweater was his one reliable, constant companion, his shield against the world. "Bruce also complained of chronically dry hands that would break out," recalls another student, Al Groupe, "and so would always wear black leather gloves outside. Add to this ensemble a black briefcase, and you have Bruce." He could sometimes seem uncomfortable communicating face-to-face, recalls CERL veteran Marilyn Beckman Bereiter. "Couldn't look you in the eye. But you know, he was like an *artist* on the system."

Parrello had a thing for Orange Fanta soda. It was all he would drink. His friends were sure to have plenty of the orange stuff stocked in the fridge when he was visiting. Tim Halvorsen, Bill Galcher, and Donald Ware used to play a joke on Parrello by keeping an eye out for the soda machine man, so when he came by to refill CERL's machine, they'd have a talk with him first in an attempt to persuade him to switch the Orange Fanta with Strawberry Fanta. They only managed to pull off this prank a few times.

It was one thing to not have Orange Fanta on hand, but for Parrello to be without his red sweater was like waking up and discovering your arms and legs were missing. Especially awful, the night when Brand Fortner found him hiding in his dark office, was the knowledge that one or more troublemakers had *deliberately* stolen his sweater. "What really struck me," Fortner later observed, "was that he didn't want to be seen without it, he had to keep the lights off. I don't think any of us realized how important that red sweater was to him. We thought it was kind of a joke." Years later, when describing this incident, Parrello would admit, "I went completely insane."

Some forty years later, the perps who stole the sweater are still reluctant to discuss the caper. Marshall Midden, who then lived in Parrello's dorm building, was in on it. Or roped in, more like, according to him. Midden says another PLATO scoundrel, Dave Fuller, planned the stunt, grabbed the sweater out of the washing machine, and at the last second looked around and stuffed the soaking-wet sweater under Midden's shirt. "He told me to shove it down my pants," Midden says. Later, they dried the sweater, and Midden never saw it again. "Something about ransom. No Orange Fanta for a week, or something like that."

"Dave Fuller got other people to do things for him," says Midden. "It would not surprise me . . . him doing the planning, viewing, snatching, washing and drying, and then get someone else involved to throw the suspicion off of himself."

During the sweater's absence, Parrello reluctantly wore a yellowish gold windbreaker, leading some PLATO wags to temporarily apply a new nickname to him in the notesfiles online: "The Yellow Jacket." Parrello was not amused. He eventually got his sweater back, and he promptly resumed wearing it. The theft, and his reaction to it, became legend. Says Rick Simkin, another PLATO-using UI student of the era, "after that, he stood guard when he did the wash."

When Parrello entered UI in the fall of 1973, he did so as a sophomore, having managed to skip freshman year. This enabled him to snare a bachelor's degree in December 1975. "Five semesters was enough," he says, "because of a thing called the CLEP tests." The College-Level Examination Program offered students a way of leapfrogging lower-level course requirements by proving they already knew the material. Only six people showed up to take the CLEPs the day Parrello went, "which was ridiculous," he says. "If you passed all four you started as a sophomore." But even that wasn't enough for Parrello. He wanted to start as a junior. "I hunted down all the other proficiency tests the school offered, but I didn't do nearly as well and ended up with only twelve more credits. It was enough to save me three semesters of tuition, however."

In high school, Parrello had discovered the benefits of always buying and keeping on hand the same set of clothes, and whenever he needed new clothes, he'd buy more of the same once again. "I had a tendency to stick with one basic outfit—red sweater, maroon shirt, dark pants," says Parrello, "on the theory that it would take less time to shop for clothes if I knew what I wanted and didn't have to browse. This eventually grew into a sort of trademark, to the point where if I wore something else it caused a disturbance."

Parrello came across the PLATO community in October 1973, just as it was exploding. The initial version of Notes was up and running, as was Talkomatic. TERM-talk would be released in a few weeks. Pad was running. Stuart Umpleby's Watergate debate was at the time the hot topic in Discuss. Big Board PLATO games were rampant. Parrello discovered that people were using colorful pseudonyms, especially when they listed themselves in Big Board games like Moonwar and Dogfight. Rick Blomme was known to all as "Red Baron" in Dogfight, so Parrello decided as a joke to call himself "Red Sweater."

The name stuck.

Parrello says his monochrome wardrobe "bled over into my personal life." Some PLATO people weren't buying his explanation. One went so far as to write up an elaborate "Origin of the Red Sweater" legend, a tale that told of young Parrello, in high school, encountering an alien entity in a field. The entity took over Parrello's mind, so this story went, and forced him to wear this same red sweater every day of the year.

Red Sweater's reputation online grew as the weeks went by. He participated in Umpleby's impeachment debate in Discuss. "Red Sweater," as he identified himself in Discuss, defended Nixon and his cohorts, and refuted many of Umpleby's more liberal arguments. Unlike most Illinois undergrads in the early 1970s, Parrello was a self-described "right-wing nut," fiercely holding to conservative viewpoints that he was not afraid to dish out in generous portions when interacting with other PLATO users. "I was the only Republican," says Parrello. "I was very important to the process." He attacked Umpleby's assertion that Nixon and Watergate were somehow tied to the JFK assassination. But he was not without his own conspiracy theories. An example of Parrello's point of view from Sunday, October 21, 1973, as posted in Discuss:

> When Ellsberg or Berrigan, etc. break into someplace or a student burns a draft file, we are told that one's conscience supersedes the law. Ehrlichman and crew believe what they were doing was right. Is the criterion of Civil Disobedience and the end justifying the means only for the left? . . .
>
> [Richard Nixon] was engaged in a fierce budget battle with Congress. Notice how a fierce budget battle has suddenly been declared a loss by the Executive branch? You may draw your own conclusions.

Red Sweater also often hung out in Pad. He had an eye for odd newspaper stories and for stories that contradicted other newspaper stories, and these observations would prompt him to post something online. He once read about two separate studies on children and learning. One study showed that children in the sixth grade and younger could not learn grammar, while another study showed that students in the seventh grade and older were too old to learn grammar. "So basically," Parrello says, "if you put the studies together, which of course no one did, you could conclude that no one could ever learn grammar because they're too young to understand it, or too old to care."

Pad's limited message length was starting to become an annoyance. "I couldn't fit them all in," Parrello said. "So I decided to write my own lesson as a newspaper." He began looking for a way to get some disk space with which he could build his own Pad clone enabling him to write longer messages. The way to get some space in this era of little space was to steal it, or grovel to the CERL higher-ups like Bill "No" Golden. But if one looked hard enough, and asked around enough, one would sometimes get lucky and find an abandoned lesson or two. Parrello eventually found two such files, "fr2" and "temperatur." He'd typed "temperature" at the Author Mode, but the arrow on the Author Mode allowed for a maximum of ten characters, so it came out as "temperatur." Bingo. "They were just lessons that nobody was using," Parrello says. "They were empty disk space that hadn't been password-protected or anything, and that the owners forgot about . . . so I just moved in." He began coding. Sherwin Gooch eventually changed the "fr2" file name to the more reasonable "newsreport." Thus, the service—offered only through PLATO, but an online news service nonetheless, long before that concept would become familiar to the world—came to be known as "News Report."

What began as a way for Parrello to have more freedom to express his views online quickly became what may be the first instance of an online newspaper and first instance of a blog years before blogs existed. (If predecessors exist, years of research has not revealed them.) From a present-day vantage point, News Report was a hybrid: both online newspaper and blog, which is remarkable when one considers that years later these two phenomena would evolve differently, often at odds with each other; but in Red Sweater's case, right from the very start News Report had attributes of both. It was a curated, online source of news to PLATO users, and it was a digital soapbox from which blared the clear, unmistakable voice (crazed ranting, to some) of his Red Sweater persona. When one visited News Report, what one immediately saw was prominent branding for "The Red Sweater News Service." The resemblance to the core structure of many online publications today, particularly political ones like *The Huffington Post*, *Talking Points Memo*, and *Drudge Report*, is uncanny, although News Report's focus was more on the offbeat than the national headlines. Parrello used News Report as a way to poke fun at the daily absurdities of life. "We looked for the strange and the unusual," Parrello says. "If you could take a story that wasn't front-page, but was interesting and funny, that was sort of the ideal thing."

He started each day by going to the drugstore to buy some news-papers and then go over to McDonald's. "I would pore through them and find stories that I thought were different or interesting. I also subscribed to a couple of magazines." News Report started out with two daily columns, "The Morning Report" and "Today's Headlines." "The Morning Report was real news," Parrello says, "and Today's Headlines was generally three or four really bizarre stories that I had turned up. . . . I wanted it to be in place by a certain time each morning, so that people would know that, hey, if I log in at eight o'clock I'll be able to read the Morning Report and Today's Headlines."

Some days it was hard to face the inevitable chore, that many bloggers years later would face, of having to stare at a computer screen and force himself to post a new article as quickly as possible for a waiting audience. Some days he didn't want to get up. "I'd lay in bed," he says, "and hope that the whole system would just break down."

He'd have to get to CERL to use a terminal, and his preferred terminals were upstairs in 203A and B, the author rooms. Unfortunately, they were typically locked from 5 p.m. until 7 a.m., and he lacked a key. Worse yet, when the rooms were open, he found he was unwelcome: only authors doing "serious work" were allowed access. He was kicked out several times. "I considered it a responsibility, not a game," says Parrello, but others saw it differently. It did help that many of CERL's junior systems programmers were within a few years of his own age— plus and, in some cases, minus—and were fans and loyal News Report readers. "One systems programmer is really picky," Parrello would tell a *Daily Illini* reporter in 1975, "and whenever I get something wrong, he tells me. Makes you appreciate what a real newspaper goes through."

"We all eagerly waited every morning for his news," says Fred Banks, a UI student during this time. "He would spend about forty-five minutes with the local paper, and write his version of the news. It was very entertaining, I don't remember him demeaning anyone, or taking a right- or left-wing slant on the news. He was a great writer. One thing about the PLATO people, it could get very vicious if you have a grammar or spelling error."

Parrello loved offbeat stories. "There was this police station," he says, "that had a problem with a mouse that kept getting into the drug cabinet, and they tried putting rat poison and traps in it and he'd go straight for the marijuana and ignore the poison and the cheese." The station was in San Jose, California, and the cops named the mouse "Marty." In December 1974 the cops, having successfully "nabbed" the mouse, put him in a cage and kept him there in the narcotics department as a mascot. "We think he's earned the right to live," said a detective at the time. "He's being held without bail. No hearing has been set." The mouse lived for two more years. When it died, the police held a funeral, and two thousand people showed up. Newspapers across the country ran the story. It was the kind of story that Red Sweater loved to tell, and News Report became the platform through which to tell it.

News Report quickly developed a following, particularly because of Parrello's sense of humor. His literary theatrics did two things well: they attracted eyeballs and they raised eyebrows. His rants about some public figures, like Uganda's dictator Idi Amin, were "epic," says Dave Fuller. Looking back on his PLATO experience years later, Gary Fritz would say, "I remember RS's editorial copy projecting the image of a fire-breathing pit bull. Imagine Dan Rather on steroids and you start to get the picture."

"I was very conservative," says Parrello, "and I was surrounded by people who thought that people like me must be nuts. But in those days the political bickering was not so awful as it is today. I mean, I could disagree with people and we could still be friends. We weren't calling each other Nazis and stuff. . . . There were several people on the staff who were extremely liberal, I was extremely conservative, but we tried to keep the fighting out of it. And I insisted that all the articles had to be accurate."

Like many PLATO game authors, Parrello copied the practice that had begun on PLATO back in its earliest ILLIAC I days, of collecting extensive data about his program's usage: what would on today's Web would be called "analytics." He found that on a typical day he had about four hundred readers visiting, which, considering the size of the CERL PLATO system at the time, was a formidable accomplishment. (Today the rough equivalent would be to have a few hundred million people visiting one's website every day.) Keeping tabs on user behavior produced some interesting findings, as it always did on PLATO and does today on the Internet. "It was kind of a joke," Parrello says, "that when I didn't do my column, the circulation would go up, because people would come back to see if it was there, but late. So I used to joke that the circulation always doubled when I didn't do anything."

With News Report, Parrello had found an effective personal platform for expression. But the platform had its problems. One, the code was troublesome, "primarily because TUTOR was always changing," says Parrello. At first News Report was unreliable and prone to lockups and crashes. He struggled to find a way to organize the data such that he could be editing new articles *while* readers were busy reading old ones, without the program corrupting both the incomplete new article and the old articles. He had designed the program so that it knew articles were articles, and knew that letters to the editor were letters to the editor, and once you were done reading an article, you could proceed on to the next article in sequence, as opposed to inadvertently being jumped straight to a letter to the editor. It took two versions of the program before he figured out the right mechanism to make it all work.

File space on PLATO was always scarce, but Parrello was eager to snap any up that became available through legitimate means or was the result of further scrounging. Eventually the fact that he was squatting on two files that weren't his got out. "The owners came to me and said, We'll let you keep them," he says. "[News Report] had become a 'thing' at that point, and it really would have been noticeable to pull the plug."

"Periodically a space would fall through the cracks," says Parrello, "and I think News Report eventually ended up being three lessons. One was the main lesson itself, one was the database, and then there was another lesson that had some subroutines that wouldn't fit in the main one. And that last one was called 'god.'"

The Author Mode had an interesting behavior when it came to missing file names. "Choose a lesson," it declared, underneath which appeared an input arrow. So you typed in something at the arrow, pressed NEXT, and off you went. Except when you made a typo, or typed in the name of a lesson or file that did not exist. In which case, the Author Mode appended the phrase "no Does not exist" to the right of whatever you'd typed in. This led to much merriment by bored users. One day Parrello discovered that "god" did not exist: if he typed in "god" at the Author Mode, the system came back and said, "no," followed by spaces and then, "Does not exist."

"I went to the woman who ran the Hebrew department," he says, "and I said, 'Do you know if you type in the word 'god,' it says, 'no Does not exist'?'" Needless to say, he was successful convincing her that "god" should exist. Let there be "god," and suddenly, it existed, and Parrello saw that it was good. He even went so far as to write some TUTOR code so that if you ran lesson "god" normally, it displayed some legitimate Bible verse. But behind the scenes, "the rest of it was all News Report stuff," he says.

—

"As we grew," Parrello says, "I started having different types of articles: there were columns, and features, and each had specific capabilities that made them behave differently. Articles were supposed to be hard news, and features were supposed to be more cultural things. A column was like a single thing that would be updated periodically, and so it would not move in the database, but it would be constantly resequenced each time it was updated."

Now that Parrello had articles, letters to the editor, and columns, all coming in from different users and at different times of day, readers might discover content appearing at all hours. One never knew when something new would appear in News Report. Unlike print publications, but just like a blog or the online edition of, say, *The New York Times*, from the reader's perspective, new content just showed up. "That's what makes [News Report] the newspaper of the future," Parrello told a *Daily Illini* reporter in 1975. "Sure, radios can do this, but you have to turn them on at a certain time. With the News, you can have the new articles at any time you want." The public, let alone the newspaper industry, would not figure this out for years to come.

News Report would not remain a one-man operation for long. Parrello soon handpicked a small team of writers to help him out. Only work approved by Parrello would be published. "This allowed us to easily scoop *The Daily Illini* on stories we covered in common," says Al Groupe. "Of course, we all were rather busy, so the paper wasn't terribly 'thick,' but it was quite a trip when we reported on a fire on campus one weekend that the *D-I* didn't report until the following Tuesday."

Parrello, who served as editor, with his own set of special powers in the News Report program, gave a handful of users "reporter" access, the base level of access required to submit a story. Groupe and Parrello tried to get real press passes for News Report's reporters. "Since we were trying to be a legitimate news outlet," recalls Groupe, "I suggested that we try to get press credentials so that we wouldn't get hassled by the cops while covering a story. One day Bruce and I walked into downtown Champaign to visit the police and see about getting press passes. We ended up talking to the chief, but couldn't get him to understand the concept."

"You mean you have to go to a computer terminal to read your 'paper'?" the flummoxed chief asked.

Their reply: "You mean you have to go to a newsstand to get your paper?"

No luck. "We just couldn't get over that hump," Groupe recalls, "and we walked back empty-handed."

By March 1975, Parrello had rewritten the code for News Report four times, and he was so proud of the code, in a move that foretold the Open Source movement that would explode on the Internet two decades later, he made it inspectable by other users so anyone could see and copy it. Parrello gained a reputation as an excellent programmer. The service now had over one thousand regular readers and featured some twenty reporters, who operated with a lot of freedom, many of them using the program as a platform for their own creative ideas. John David Eisenberg, who was a full-time PLATO programmer and occasionally submitted material to News Report, considered the editorial policy "the most free of any newspaper around." Parrello considered News Report to be "a family newspaper" and policed for material he deemed "too dangerous." Parrello also considered himself sole editor, as he felt almost no one knew "how to spell or punctuate."

Tom Grohne was one "reporter" who submitted a popular series of humor stories called "Specs Nookno, P.I.," hard-boiled detective fiction in the manner of Mickey Spillane's Mike Hammer. Specs Nookno was named after the -specs nookno- TUTOR command, which "specified" to the system that it should not display the usual "ok" or "no" feedback messages to students after they typed an answer at an arrow prompt on the screen. Grohne's Specs Nookno was a tough guy, a private eye like Hammer, but one, says Parrello, "with absolutely no brains." Imagine Inspector Clouseau meets Firesign Theatre's Nick Danger. The Specs Nookno stories, none of which have apparently survived, were an instant hit, causing many laugh-out-loud situations in otherwise quiet PLATO classrooms. If there was one problem with the stories, Parrello says, it was that "there were never enough of them." The stories took a lot out of Grohne to write. He told Parrello once that it required "sending his mind into a completely strange place in order to do them." At one point, there was a long delay between Specs Nookno stories, and the reading public clamored for more. Some Cornell pranksters posted a letter to the editor saying that they had Specs Nookno and would not give him back until certain demands were met. "That went on for a while," says Parrello. "Finally Grohne was able to get his mind into the strange place again and Specs Nookno came back."

Other News Report reporters found their own successful niches. One contributor offered a column entitled "Saucy Wit and Droll Caprice," which covered, says Parrello, "idiotic behavior by politicians, which in those days generally involved hookers, alcohol, and car wrecks." The 1974 scandal involving Congressman Wilbur Mills and a stripper named Fannie Fox was one target of "Saucy Wit." Another was the 1976 affair involving Ohio congressman Wayne Hays and his mistress Elizabeth Ray. John David Eisenberg contributed a mock-pious column called "Sermonette" that offered funny essays on various issues and events like Thanksgiving. There was a "Poetry Corner," featuring works by various users, including Sherwin Gooch. ("Weird poetry," says Parrello, "otherwise we would not have allowed it.") There was a regular science fiction column. The legendary game creator Silas Warner, out at Indiana University, contributed an offline module to News Report to report the live standings for the Indianapolis 500 race. "I did a lot of Indianapolis 500 reporting for them," Warner said. Dave Fuller submitted various technical articles that attempted in a light-hearted way to explain the various components of the PLATO system. Fuller's articles emulated the manner of the popular "I Am Joe's Kidney" series of anatomy articles from *Reader's Digest*: "I Am PLATO's CPU," "I Am PLATO's Disk Drives," "I Am PLATO's Channel Controllers." Parrello himself contributed a regular column entitled ". . . and with a mighty crash." "Every time there was a major outage," he says, "I would go find out what caused it, and write an article about it the next day." Outages on PLATO during the News Report era were about as common as the appearance of the notorious "Fail Whale" on Twitter decades later. David Woolley contributed exhaustive movie listings. "Every week," says Woolley, "I put together a list of all the movies showing on campus and in theaters around town, along with show times and often a one- or two-line mini-review, if I had seen the movie. This wasn't too hard to do in a town the size of Champaign-Urbana—I think I spent maybe an hour or two a week on it." Parrello remembers Woolley's movie listings as being a great value, for they not only included listings that the local paper offered, but also one-dollar movie nights at on-campus locations. Readers loved Woolley's listings but wanted reviews as well. Woolley found that hard because he hadn't seen most of the films. Another user contributed reviews of *Star Trek* reruns and also detailed listings of when they were broadcast and on what TV channel. (This being an era before TiVos and DVR, some UI students made a determined effort to set up their class schedules based on precisely when *Star Trek* reruns aired during the weekdays, so as to be sure they would never miss a show.)

"Everything that a computer user needed," says Parrello about News Report, "movies and *Star Trek* episodes were right there! I also remember John Eisenberg did a review of the first episode of *Space: 1999* and we learned something important about the future: people talk without moving their faces." The core premise of *Space: 1999*'s premiere episode involved the moon having become an enormous storage depot for the earth's accumulated nuclear waste, which, unfortunately, would blow up with such force as to send the moon, and the hapless heroes at Moonbase Alpha, hurtling out of orbit into deep space . . . and onward to new and different adventures each week. Before the big explosion in the first episode, Martin Landau's character at one point uttered, "We are sitting on the biggest bomb man ever made." Eisenberg responded in his scathing review that Landau was "absolutely correct."

Not everything on News Report went over well with readers. "There was one disaster," says Parrello, "which was called 'Baseball Report,' which reported baseball scores. I thought that it would be something that people would find interesting. But I got so many complaints about it, people hated it, said it was boring." Things got so bad that readers began exchanging nasty pnotes with the reporter. "Eventually I had to pull the plug, which really annoyed the reporter who was doing it. But what could I do?" Years later, the audience reaction still baffled Parrello. "People *really* hated it. I'm not sure why. Maybe it was because we were all geeks and it was about sports."

Editorials were a favorite outlet for Parrello antics. He had a general rule that editorials needed to be on topics so ridiculous that no one could possibly believe they were real. "That was my rule," he says. One editorial advocated for getting rid of the earth's ozone layer, "because it was just too damn fragile." That spawned a series of replies, including one suggesting enclosing the earth in a dome and that they should hire R. Buckminster Fuller to design it. Another editorial, written when the gun control debate was raging in the national media, was based on a homicide study that had been done in Europe. "We came across a survey showing that 55 percent of murders committed by German women were done with frying pans. In the United States it was the telephone. But overseas it was the frying pan. So I did an article about how there was a deadly weapon available in stores and we needed frying pan control. And we did the whole thing, you know, registered communists got frying pans, and that shit generated a whole bunch of follow-on editorials." Yet another time, Parrello says, "a system programmer jokingly complained in =pad= that they ought to install a cattle prod, so that when someone asked a really stupid question, that they could send a jolt of electricity through them. And so I came out with an editorial against that: Keep PLATO Safe! And that got a bunch of replies too, and there was talk about bringing me upstairs and hooking me up to the machine so that the problem would be illuminated to the satisfaction of most people."

At least once, Parrello caused a panic among the powers-that-be at CERL. It was 1975, the year that Senator William Proxmire began issuing his notorious Golden Fleece Awards to identify government programs that Proxmire felt were fraudulent, self-serving, and wasteful. His very first award went to the National Science Foundation, the organization that had funded PLATO. In 1975, Richard Atkinson, who had once worked on CAI research with Patrick Suppes at Stanford University, had been appointed by President Ford to become NSF's new deputy director, and it was onto his lap that Proxmire's hot potato landed. "When he [Proxmire] delved into the social sciences," Atkinson says, "he found an NSF-supported grant dealing with an experimental analysis of love from a social/psychological perspective, and another grant concerned with a theory of love. At that time the *National Enquirer* was paying a $500 bounty to freelance reporters who came up with a story of this sort, and many writers would just scan the titles of research projects supported by NSF. The *Chicago Tribune* had a field day with the theory-of-love grant, and as if this weren't bad enough, they found a project titled 'A Theory of Necking Behavior.' We tried in vain to find this grant on NSF's list of social science projects. Days later we finally unearthed it among the engineering projects—the necking referred to was of a metal, not a human, variety. Several of the faculty grantees who were recipients of the Golden Fleece wore it proudly as a badge of merit and made the most of their notoriety on Johnny Carson's *Tonight Show*. This was serious business for NSF, however, because it played havoc with the foundation's public image and relations with Congress." Says Parrello, "Proxmire was coming down hard on the National Science Foundation. He was trying to force a complete audit of all the projects by the National Science Foundation, some of which he said were obviously ridiculous.

He wanted people to understand how important this was. I said [on News Report] that among the things supported by the National Science Foundation is a system called PLATO, which must be some sort of computer simulation of a Greek philosopher." Some of the CERL staffers who read News Report were aghast. "People thought that Proxmire was specifically targeting PLATO, and that caused a storm, and I had to put in a retraction immediately, and I'll always remember that as the importance of double-checking your pronouns, to make sure there's no doubt about which refers to what. I wanted to drive it home but I didn't want to cause a panic."

Groupe wrote a regular column for News Report called "Groupe's Gripe." "A general bitch column," he says. "Bruce always told me he hated the column, but couldn't get rid of it because it always got the highest readership of the paper." Groupe began investigating a student group called Students for Equal Access to Quality Education (SEAQE). His first SEAQE exposé appeared in a "Groupe's Gripe" in October 1975. Titled "SEAQE and Ye Shall Find," it wasted no time to accuse the group of campus violations:

> SEAQE . . . is currently engaged in canvassing dorms to solicit people to help with their fight against a proposed tuition hike. From what I am told, much of this canvassing is being done without a permit and is therefore in violation of University housing regulations.
>
> One of SEAQE's members stopped by my room to explain why he felt that we shouldn't have to pay for our education. When he discovered that I am not opposed to a tuition hike to keep the University going, he walked out and said, "It must be nice to have money." But that's what the tuition hike is all about. It would be nice if the University had money.

Groupe's article generated numerous online comments, some that objected to the planned tuition hike, and some that thought it reasonable. Those assailing the hike complained that those who didn't mind it probably had their tuition paid by "unnamed sources" such as parents. Several commenters, including Dave Woolley, countered that argument, saying they were paying their own way to attend the university and despite the planned hike, felt the education received was still worth every penny—and more.

In another "Gripe" column, Groupe went even further, declaring that he had discovered that SEAQE "was created by, and is a sister organization to, the Revolutionary Students Brigade, a Marxist organization whose goal is to overthrow 'rich and fat corporate heads' whose 'fortunes are based on the profit ripped out of the sweat and labor of the millions of working people.' " In December, another "Gripe" appeared, documenting how Groupe and Parrello had attended a supposedly open meeting of the Revolutionary Students Brigade in the Illini Union. "Here we learned that the RSB is linking up with the RCP (Revolutionary Communist Party) before the year's end. . . . They admitted that the formation of SEAQE was not only for the purpose they tell prospective members of SEAQE. They admitted that one of the main reasons for SEAQE's existence is to get students involved in the 'struggle of the masses' and in so doing, get students to go along with the RSB/RCP." Groupe dug deeper. "I found a copy of an official SEAQE membership list, which I was repeatedly told didn't exist, which lists the names and addresses of the active members. Just for fun I checked the Student/Staff directory to see what their majors were." What he found instead was that several of the members were not university students at all. Groupe's continued coverage of the student communists annoyed their organization enough that, Parrello says, "he was scared for his life for a while."

Unlike today, where blogs and online newspaper sites offer vast searchable archives, given the limits of disk space on PLATO, articles and letters had to expire after a short period of time, to make room for new material. (Unfortunate for historians: this meant the only way to read News Report archives years later was by viewing any surviving printouts; in Parrello's case, those printouts were lost long ago in a flooded basement. Nothing is known to have survived.) Parrello went a step further, in fact, by expiring "reporter" status on a user if he or she didn't submit new material within two weeks of the last submission (the policy instituted as a reaction to the initial onslaught of wannabe reporters who eagerly signed up when the "reporter" feature first was released, but never turned anything in, or were not reliable or perceived as being serious). One consequence of Parrello's design was that during UI's long Christmas break, the PLATO community discovered that News Report had purged *everything*: articles, columns, letters, as well as the reporters. The program did what it had been told to do, in order to make room for new content. It quietly nuked things, day after day, while everyone was home enjoying the holidays—away from PLATO terminals—until by the end of three weeks, News Report was an empty digital wasteland. The PLATO system kept running, delivering instruction to schools and institutions and other paying clients, but News Report ran dry and would have to be restarted when the next semester began.

One day, Mike Huben, a student using one of the rare PLATO terminals out at Cornell University, inspected the source code to News Report and found a vulnerability, which he then exploited. "Bruce was very proud of his coding," says Huben, "and left the inspect code off so that people could see the implementation." Huben used what was known to some as the "jumpout trick." He wrote a program that used the TUTOR -jumpout- command to jump from his own program over to a specific line of code inside News Report. He knew exactly where to jump to because Parrello had left the code inspectable to everyone. From there he was able to press SHIFT-DATA, causing News Report to return him to his own program. Once back in his own program, he changed the values of what in TUTOR were called "student variables," then jumped right back to News Report, but this time used TUTOR command -inhibit from-, which concealed to the target lesson the name of the originating lesson out from which the user had just "jumped." That caused News Report to think the user was returning from News Report itself, in effect causing it to drop its guard and retain the secretly changed values of the "student variables." At that point, Huben essentially "owned" the program memory that News Report relied on to perform its tasks. "I was able to pretend to be anybody I wanted," says Huben, giving him any access he desired.

Not surprisingly, once in, there was one important thing to do.

"The first thing I did was to lock out Bruce," says Huben. Parrello thought he'd been derfed, or that someone had stolen his password, so he asked a systems programmer to help him get back into News Report so he could set things right. A day or two later, Huben was at it again with another -jumpout- exploit. This time he published an article in

News Report, written by a fictional user named "patty hearst" of group "sla," that declared that the Symbionese Liberation Army had liberated the Red Sweater News Service, and, says Huben, were "threatening to destroy the Orange Fanta bottling plant" unless Bruce met their demands. Then he locked Parrello out again. Parrello was furious, and declared publicly that somebody had broken system security. "It didn't occur to him that it could be his own fault," Huben says, not even bothering to disguise his glee at successfully fooling none other than the famous Red Sweater, expert programmer, whose code was visible for all to see. What's worse, Huben had fooled him from a pesky Cornell University PLATO terminal eight hundred miles away. "After a few hours of laughing," Huben says, he TERM-talked Parrello and confessed. Laughs all around, and life went on.

Parrello was sometimes able to turn the tables and fool a would-be prankster. His prank was subtle and could easily be overlooked at first glance. Larry White, a student and CERL programmer who'd worked on Dogfight, recalls going into News Report every day to see what new and interesting articles were posted. One day, Red Sweater did something particularly noteworthy—something that would be hard to do on today's Web, but was a piece of cake on PLATO. He wrote an article about Rick Blomme, how he was good at the Big Board games, and how his game nickname was "Red Baron." "In the middle of this article," says White, "it mentioned that one of his main opponents that he really admired fighting against was 'white' of 'p.' " That was Larry White's signon. At the time, White had no reason not to believe that Blomme had actually been quoted as saying that, and that Red Sweater had duly noted that in his interview with Blomme and that he had then faithfully conveyed that fact to readers of the article. "I thought," says White, "that Red Sweater had written in 'white' of 'p' in the article. There was no indication to the reader of the article that anybody else saw anything different than 'white' of 'p.' And in fact I had played against Rick Blomme many, many times, Rick had beaten me many times, I had beaten him many times." White was proud of Blomme's public acknowledgment, particularly because Blomme was known to be a formidable opponent. The reality, as White would soon discover after having been duly duped by Red Sweater, was that *whoever* read the Blomme article saw their *own* signon mentioned as being the opponent most feared by Blomme.

—

News Report was enough of an achievement for Red Sweater to earn a seat in the PLATO hall of fame, but for a lot of the PLATO community, he's even more well known for something completely different: animations and emoticons. When we think of emoticons today, we think of primitive "smiley" pictures typed by lining up some punctuation points, like :-). Often they require tilting your head. These kinds of emoticons had not and would not come into use on the ARPANET for nearly another ten years, but on PLATO they had already been established as a new art form and were in widespread use. And, as was PLATO's wont, they looked completely different from the more primitive emoticons people use today. The only thing that visually comes close today are the Japanese emoji icons, now popular worldwide in chat applications and on Twitter.

To fully appreciate PLATO-style emoticons and the sorts of animation techniques that Red Sweater made famous, it is first necessary to step back and revisit the supremely odd architecture of the PLATO system. PLATO could be a Darwinian case study: here was a computer environment, a bizarre and impossibly improbable ecosystem, unlike any other in the world, developing entirely off on its own and left, literally, to its own devices, in the sense that nothing, including machines networked through the ARPANET, could connect to it. Within the confines of this curious environment, one that was created for educational purposes, all sorts of strange and wonderful and decidedly noneducational "creatures" quickly came to life and thrived: notesfiles, pnotes, TERM-talk, Talkomatic, multiuser games, News Report, etc. But there were also other strange "creatures" as well, emergent phenomena that could only exist because of PLATO's unusual system architecture. For example, PLATO's text emoticons.

On PLATO, text characters were not encoded in ASCII, or EBCDIC, or the more recent UTF-8, or any other scheme that was then or is now commonly deployed on commercial computers. Reflecting the underlying Control Data hardware design, the PLATO system transmitted text to terminals in 20-bit words containing three 6-bit characters and two extra control bits. ASCII was a 7-bit, later extended to 8-bit, encoding that in 8 bits has enough room in it for 255 different characters. PLATO's 6-bit encoding only had room for 64 characters, and not even enough room for capitalized versions of "a" through "z."

To capitalize a letter, PLATO needed two characters, the first encoded in such a way that it indicated that the next character should be displayed in capitalized form. (For years, when Control Data printers would print out TUTOR source code or the contents of a notesfile archive, any capitalized letters would appear as an up-arrow character followed by the letter. A full line of capitalized text was therefore a sight to be seen, riddled with alternating up-arrows and barely readable.) Likewise, to display a superscripted or subscripted character (for example, an algebra lesson that needed to display on-screen an equation like $a^2 \times b^2 = c^2$), a specially encoded character would precede the displayed alphanumeric, so the terminal knew to raise or lower the character. All kinds of "modes" were encoded in this fashion. But what made things particularly interesting was the fact that it was possible to type certain combinations of special keys and alphanumeric keys using the keyboard itself, with the result that a user could send the terminal into one of these special display modes and sometimes do very weird things. It took practice and often a cheat sheet, but once you mastered the codes, you could create what years later on the Internet would be called emoticons and smileys. Given the fancy display modes available for PLATO characters, one did not have to turn one's head sideways to see that the primitive pageant of a colon, followed by a hyphen, followed by a right parenthesis would become the Internet's attempt to draw a smiling face. With PLATO, one could hold down the SHIFT key, press the space bar, and while nothing looked any different on the screen (PLATO terminals lacking a blinking cursor), the next character of text you typed would appear *on top of* the previous character. While such a capability might not seem at first glance to offer any useful benefit, it turned out that superimposing one or more characters, through successive SHIFT-spaces, had the consequence of creating all sorts of funny faces, sad faces, scary faces, beer and wine glasses, martini glasses, and an infinite list of other unusual images. Thus were emoticons born on PLATO. Over the years they would be collected, cheat-sheet style, in a TUTOR file called "m4," whose inspect code was blank so anyone could go in and discover the meaning of character combinations like WOBTAX and VICTOR.

As with other features of PLATO, there was an educational benefit to having this strange capability of character manipulation. For instance, you could use it to form non-Roman alphabetic characters that were impractical using only a downloadable character set, as there wasn't enough room, or it required too much fancy typing on the part of the student or lesson author. Some foreign language lessons, including some in Hebrew and Hindi, used the ingenious solution of combining special character sets with what were called "microtables," similar to what in a word processor or spreadsheet today would be called a "macro," a recorded combination of keypresses that could be invoked or "played back" by pressing a much simpler sequence of keys (hence, the reason for the MICRO key on the PLATO IV keyboard).

Red Sweater helped popularize these digital hieroglyphs but, even more, helped popularize the spread of animated sequences of characters. "The animations were a result of a quirk in the PLATO architecture," says Parrello, "which was that several of the display control codes—mode rewrite, half-space, back-space, move up, move down—could be entered from the keyboard. You could therefore type in a string that when it was displayed did all sorts of neat graphic stuff. The basic bread-and-butter animation was an asterisk that moved from left to right."

Probably the most essential of the special character encoding modes that one could invoke by typing magical sequences on the keyboard was mode rewrite. Mode rewrite simply meant that the displayed character's pixels would rewrite over any pixels already activated "underneath" it. Now imagine, if you will, displaying a right-leaning slash (/), then typing SHIFT-space, then typing the mode rewrite sequence, then typing a vertical slash (|), then another SHIFT-space, another mode rewrite, then a backslash (\), and sitting back to watch the results. What would be the results? You'd have to watch quickly, because it would animate in the blink of an eye, but if you watched closely you would see an albeit tiny, and crude, but nevertheless twirling, baton. And if you added hyphens (a poor man's horizontal baton, if you will) with more mode rewrites all at the right spots in between the right-leaning, vertical, and left-leaning slashes, the animation was even smoother: now you had a more vivid animated baton. Repeat the whole process a few iterations and the baton just might twirl for a whole second or two.

But this was only the beginning. With hours, days, weeks of late-night trial-and-error, it was possible to discover all sorts of ways to draw not only twirling batons, but essentially any kind of pixel, line, or other image desired. The constraints, as usual on PLATO, were horrendous, like an army of gremlins determined to deny one's completion of a work of art. Perhaps the most obvious constraint was that a line of text only supported sixty-four characters, and since every special character in a mode-setting sequence of characters counted as a separate character, a determined animator would run out of space very quickly unless he or she figured out how to continue the sequence of animated codes using the next line, first repositioning the terminal's invisible cursor back up to the previous line and then getting on with the animation. This was painstaking work.

Red Sweater mastered the art of PLATO character animation, and showed off the fruits of his labor in notesfiles as well as through inline animations in his articles and columns in News Report. He figured how to use the microtables to store common emoticons and animations, and peppered his News Report articles with them. He had flying asterisks and bows shooting arrows, the inevitable twirling baton, and a little stick figure that would throw things, including batons, up into the air and then catch them on the way back down.

Not everyone appreciated Red Sweater's animations showing up in and polluting public notesfiles. A complicated animation might take the same amount of time—perhaps ten seconds—that a full page of text took to plot on the screen. Thus, a little stick figure twirling, throwing, and catching a baton might rob a busy notes-reader of ten seconds of his precious time. That rubbed a sufficient number of CERL staffers the wrong way that Red Sweater's continued use of PLATO was threatened. But right out of the blue came none other than Bruce Sherwood, the foe of every gamer and game author, the senior systems software staffer who was constantly fighting this or that battle over precious system resources. Sherwood was quite taken with Red Sweater's animation talent, and created a notesfile, =anim=, to serve as a proving ground of new animated creations and repository of collected works that had been posted to other notesfiles. All well and good, but viewing the animations notesfile was a whole different matter: such activity was considered "recreational" by many PLATO classroom monitors and therefore an outrageous waste of precious resources better spent on algebra lessons and whatnot, so many an animation admirer hanging out in the file would discover themselves quickly booted off the system and sometimes kicked out of the classroom.

These special "creatures" unique to the PLATO environment, namely, emoticons and character animations, could not exist elsewhere, and could only be viewed on a PLATO screen. Their continued existence, at least in the convoluted, arcane form they took on PLATO, was threatened if the environment in which they evolved and flourished was threatened. Emoticons on other computer systems using the more conventional, and vastly more widespread, ASCII character encoding, would evolve differently. The environments might have been more commonplace, but they were far less interesting in terms of quirkiness and capability. So users wound up doing "smileys" that required a sideways glance to understand. With networks of minicomputers and microcomputers to form the Internet and the World Wide Web, ASCII smileys proliferated, although without any sort of evolutionary transformation and increase in sophistication that was the hallmark of the PLATO world. The first recorded ASCII smiley, from 1981, looks exactly like the smileys that a billion Twitterers and Facebookers will type in the next twenty-four hours. Character animations, thanks to ASCII, never had a chance on personal computers and the Internet, but also thanks to ASCII, the primitive emoticons have survived and will be with us for years to come. PLATO-style emoticons and animations are long extinct, and have become, like the rest of PLATO, dusty artifacts for curious digital archaeologists to study. They represent another example of the inevitability of invention when a large, heterogeneous community of users get access to a system that not only permits, but encourages, creativity of expression.

As for News Report, it was ahead of its time by decades. "Bruce took News Report very seriously," says Al Groupe. "He was really trying to break new ground and compete with *The Daily Illini*."

The ideas and visions that News Report engendered were, in hindsight, brilliantly prescient and predicted the course of much that was to come. "At one time I had a dream of selling advertising," Parrello says. "I was told that was definitely not going to be allowed." When asked by a *Daily Illini* reporter in 1975 if there was anything wrong with News Report, he replied that it was the "lack of opportunity for profit." Said Parrello, "The problem with getting a newspaper profit on PLATO is that all the profits go to the CERL people." Parrello dreamed of more readers, extending News Report to everyone on PLATO, not just authors who'd heard about it and could access it from the Author Mode, but students, staff, faculty, and the general public (perhaps accessing it through public terminals in the Illini Union or the University Library). He told *The Daily Illini* he "would like to see the situation changed to allow anyone using the PLATO system" to use News Report. He even envisioned "classified ads and possibly a bulletin board system."

News Report was real, and it lasted as a vibrant online resource for some two years, but the idea of it being a sustainable, ongoing concern was just a dream. Like so many compelling PLATO-based projects built by college students, its future was doomed if for no other reason than the fact that Parrello was a college student, and college students eventually graduate (some, at any rate). He stayed on for a few more years at UI, working in the Foreign Languages Lab. "I would periodically want to make something more of it," Parrello would say years later, looking back on News Report, "to try and build it into something that had a future, getting press passes for the reporters, and things like that. But none of that ever really materialized. I mean, let's face it, my education was my first priority, and hobbies would stay hobbies." News Report limped on after he left the UI for a while at least, but without his daily input and editorial oversight the service withered away, particularly due to the built-in expiration of articles and reporters.

Parrello would continue wearing the red sweater when he moved to Northern Illinois University for his graduate work, and, when he got married in 1978, the thank-you notes he sent to well-wishers were stamped in red with a return address that began with "T.R. Sweater and Wife." The sweater issue would continue for a while longer. "It wasn't until I switched from mainframes to personal computers in the mid-1980s that the whole sweater thing started to die down," Parrello says. In a 1997 interview he explained his clothing compulsion this way: "Nowadays I have a whole closet full of identical sweaters, but twenty years ago I was extremely superstitious about the sweater thing and had only one. Fortunately, I'm much more mature now, and no longer believe a mere piece of cloth can bring you good luck." However, he was quick to add, "I still take three backups and sprinkle my PC with holy water before installing a Windows 95 application, but I think a lot of people do the same thing."

19

The Supreme Being and
the Master of Reality

In her 1984 book *The Second Self,* Sherry Turkle writes about the MIT computer culture of the 1970s and early 1980s. For Turkle the "second self" is the computer itself: she observed that as people stayed glued longer and longer in front of their computers, they saw something of themselves in the computer and something of the computer in themselves. What she saw at MIT were largely groups of young programmers spending way too much time on their computers. In her 1996 follow-up book, *Life on the Screen,* Turkle would explore a phenomenon that had either never arisen or had been overlooked during the research for *The Second Self:* the fact that perhaps the computer was *not* one's second self, or merely that, but rather, the computer enabled you to *create* a second self, a persona, *online.* Had Turkle ventured into any of the PLATO sites around the country during the 1970s, she would have discovered that many users had gone straight to this second stage: sure, many of the new wavers would identify as hackers and lived to program and make PLATO do things, but just as many, if not a lot more, found in PLATO a way to develop *lives on the screen,* sometimes adopting names different from their real-world names, and often adopting quite different personae as well. The Red Sweater was probably the most famous early example. But with the rise of Talkomatic chat rooms, multiplayer games, and notesfiles, and the wide range of tools available in the TUTOR language to build new programs, PLATO users found a multitude of ways to express themselves differently in the online context than in their flesh-and-blood context. All during the 1970s the notions of *identity* (who are you in the real world, and who are you online, and who might you become online) and *presence* (you not only exist in the real world, but you exist—there you are, you can see yourself on the Users List—in the online world) became an increasingly noticeable phenomenon. For some users with an inclination to write, PLATO was fast becoming a wide-open new platform for expression and for sharing ideas and stories whether brief or epic in length. Those ideas and stories might be told using one's own identity, or perhaps from the perspective of a persona other than the writer's own. News Report, Pad, Discuss, and many notesfiles provided such outlets for expression.

PLATO users, particularly teen and college-age users, began to utilize the system not just for fun, but as a new way to express themselves, either as themselves or via an invented persona. Initially the personae were but pseudonyms chosen for the Big Board game lists. But with the arrival of Talkomatic and notesfiles, something new, never seen before, was becoming available: PLATO was becoming a platform for expression the way the Web would become years later.

When David Woolley released in early 1976 a major upgrade to PLATO Notes, it ushered in a new era on PLATO where anyone with disk space could create a notesfile on any topic and allow or deny anyone access. Overnight, a slew of new notesfiles, eventually numbering in the hundreds and later thousands, were created.

In some notesfiles devoted to genre entertainment (science fiction, comic books, movies, and so forth), some users began creating what were called "round-robin stories," where one user would post a new note containing a few lines or paragraphs of some invented story, then someone else would come along and reply to the base note with more lines or paragraphs continuing the story like improv comedians. This would continue indefinitely, with a crazy patchwork narrative emerging to the great amusement of the notesfile participants. One such round-robin story began in =comics=, the notesfile devoted to comic books. The story concerned "The PLATO Patrol," a band of superheroes each of whom possessed certain special powers. The PLATO Patrol had started and had already made a name for themselves as a group of PLATO users, most of whom had never met face-to-face, in =pad= a year or two earlier, where they posted stories describing their fictional heroic exploits. They even printed up business cards, which included the names Quetzal, Red Sweater, Blueman, Fantasia, and Molecule Master. "No Job Too Small," the card declared, and underneath, "Universe-Saving Our Specialty," and listing CERL's actual main phone number beneath that.

"There were originally five people in the group," says Parrello, "and the stories took place in a weird mishmash of the Marvel and DC universes. . . . The original five heroes were the Red Sweater, who could project force fields around himself; Fantasia, who could absorb and generate almost any kind of energy; Quetzal, who could teleport things at will and project streams of quick-drying guano from his fingertips; Blueman, who could duplicate the powers of any other superhero; and Baron Vitellio Scarpia, whose powers were never fully explained but who had an interesting personality. Various others came in and dropped out along the way, including one guy called Fierdraken who could breathe fire."

Fantasia was Mary Ann Neuman. Quetzal was the creation of Bill Roper, a student at Southern Illinois University down in Carbondale. Blueman was one Kim Metzger out at Indiana University. The PLATO Patrol stories were such a big hit, says Parrello, that they eventually found a home in their own private notesfile, where the stories continued for several years. At least once, all of the members of the Patrol set aside a weekend to make a pilgrimage to CERL. Each of them showed up in costume and held what today might be regarded as a miniature Comic-Con convention.

With News Report, Red Sweater had already demonstrated how PLATO had become a platform for expression. Like online ventures that would appear decades later, success meant a need for more resources, particularly disk space. Where were all these content submissions to go? Disk space was beyond hard to come by, but soon, the user submissions were getting so big and unwieldy there was only one solution: jumpout.

Hyperlinks didn't exist on PLATO, and while eerily similar to online newspapers and blogs, News Report did not offer a way for a user to click on a word, or an image, and be transported to another page or even another site. But the next best thing was to employ TUTOR's -jumpout- command, which enabled one PLATO lesson to transport its user to another lesson. The -jumpout- command brought modularity to News Report. Parrello saw this as a way of extending News Report without having to keep building a bigger program, which would mean more disk space. Instead, he'd make the disk space problem someone else's problem. With -jumpout- he could announce in News Report the availability of something much larger than an article or column: he could announce the availability of an epic reading experience. One called *The Great Guano Gap*, if not the first, was one of the very earliest examples of interactive fiction or "digital storytelling" done on a computer.

Bill Roper had enrolled as a freshman as a chemistry major at Southern Illinois University in the fall of 1973. He took a FORTRAN course the next year, and continued visiting the computer center after that to tinker around with the machines. "I was writing Conway's Game of Life [an early computer game that mimics cells growing into little life-forms over many iterations] and somebody mentioned, 'Hey, have you seen the computer over in the library that can play games?' I said, 'No, that sounds interesting,' and so I headed out over there and found the PLATO computer. And before too long, got an author signon, because it turned out that writing programs on it was much more interesting than necessarily playing the games."

For the next three years, Roper would hang out on the PLATO terminal he'd found in the SIU library. "We had one whole terminal, and you could schedule an hour a day on it if you had an author signon, and of course everyone would try to scavenge the time that nobody was using." The one PLATO terminal on the entire SIU campus was available until midnight, at which time the library was shut down and students would have to leave. This was the same terminal that would soon be commandeered late into the night by the SIU students Ray Wood and Gary Whisenhunt, creating the Dnd game.

Compared to the situation at most PLATO sites at the University of Illinois campus, where disk space and signons were difficult to get and control of system resources was carefully watched, SIU was laid-back. "The guy who was originally the site administrator wasn't paying much attention, and we actually ended up with the inmates running the asylum," says Roper.

Roper became a fan of News Report and eventually submitted his *Guano Gap* story to Red Sweater. "When I was running the Red Sweater News Service," Parrello says, "people were always submitting stuff for publication. I hated turning people down flat, so I would generally tell them why I didn't want to use the material and what could be done to fix it. It was more polite than a rejection slip, but it had exactly the same effect, except once. A guy at Southern Illinois University and his friends had come up with a series of letters between two fictitious nations which started with a general state of tension and ended with the two nations being buried in bird droppings. The intent was to make an antiwar statement in a humorous way. The guy who submitted it (his handle was Quetzal and of course that's the only name I remember) had framed it with a *Star Trek* story in which Kirk and Spock briefly commented on the letters as they read them. It was intensely boring, so I did what I always do and gave the guy a bunch of suggestions: make the framing story bigger, relate it to the PLATO community instead of *Star Trek*, and so forth and so on. For the first (and last) time, somebody actually took my suggestions. The new framing story involved the antics of a godlike creature whose user ID was 'supreme being' of 'p.'"

Roper duly reworked the material, including the notes between the two countries, text originally written back in high school by him and some friends. He removed the *Star Trek* motif and made PLATO and the University of Illinois the new setting. Parrello provided editorial help as Roper went along. "Writing at a distance," Roper says, "I would get things occasionally wrong about the U of I. He had suggestions on that."

Parrello also provided Roper with an engine to simulate a TERM-talk. *Guano Gap* was an intensely interactive work of fiction, a science-fiction *Twilight Zone*–style production that could only exist on the computer. Much of the story was told by the reader watching a TERM-talk session between the narrator and various PLATO users, Red Sweater among them, but also "supreme being" of "p." Parrello's TERM-talk simulator code was uncannily realistic: characters would appear on-screen at random speeds, making it appear that they were really being typed by a person. (However, when "supreme being" of "p" was talking, the entire line of text would appear in an instant.)

Guano Gap started with this introduction: "Early one morning, the first page of text in the guanogap lesson began, I, Bill Roper, occasionally known as the Quetzal, walked up to my local PLATO terminal and somewhat fumblingly signed on." At that point a user pressed NEXT, and saw the screen erase and a simulated Author Mode appear. "Welcome Ypsilanti State University" was the satirical top-of-screen Author Mode message. The page indicated that there were unread personal notes. The narration continued: "Aha! Notes! I said as I hurried to read them." The fictional Bill Roper then went to his first unread pnote:

From: chancellor bosnia On: 07/23/72 14.32.35

TO: Minister of Diplomacy, Republic of Bosnia-Herzegovina
FROM: His Honor the Chancellor, Republic of Bosnia-
Herzegovina

Mr. Meshkernukelsniszerpblognufwazluzklawolski:

In recent years our nation's trade status has suffered due to the
fact that no nation in the whole goddam world knows we exist.
Our position is further hampered by the fact that even if they
knew they wouldn't care.

This strange message continued in the next two pnotes:

In an effort to improve this status, we have attempted to open
relations with many of the nations of the world: Armenia,
Northern Ireland, West Pakistan, Portuguese Guam, and the
United States of America. All of these contacts have proved
unsuccessful; usually resulting in our ministers being denied
access to the country due to illiteracy, or of our dispatches and
official communiques being used to make paper airplanes or
wrap garbage in.

In a more recent attempt, our nation gave diplomatic recognition to the Republic of Tibet. This means that there are now
two nations in the world who have done so: Us and Tibet. However, we have not been able to find the Tibetan government to
inform them. We have, however, made one unfortunate discovery: Tibet has nothing to trade that we do not have an enormous surplus of already. (Snow: 6 feet; Abominable Snowmen:
None, more than enough) I therefore ask you the question;
WHAT THE HELL ARE WE GOING TO TRY NOW?

—From the Office of the Chancellor

"This is definitely bizarre, I thought, as I noticed this note had supposedly been written in 1972," says Roper's character. "I decided to take
some positive action. It was time to call in Red Sweater."

At this point, readers continue to watch a representation of Roper's screen as if looking over his shoulder. The entire reading experience of *Guano Gap* is like a second level of monitor mode: we see what's on narrator Roper's screen as well as read his story along with him. We watch as Roper TERM-talks "parrello" of "uimatha," telling him he has something bizarre to show him, then takes him into monitor mode, so that he can see Roper's display as well. Roper jumps back to the first of the chancellor's pnotes, but it's not there. Instead, it's a mundane one-liner email from a friend at SIU.

"So what?" the Parrello character types, in the simulated TERM-talk.

"But that's not what it said," Roper's character types back. "It was a note from the chancellor of bosnia-herzegovina dated in 1972."

"This is a lousy joke," the Parrello character types. "Bye." PLATO displays "end of talk."

Frustrated, Roper's character returns to the Author Mode, only to find that he has more unread pnotes awaiting. He goes in and finds even stranger messages from the External Affairs Ministry of the Internal Di-Lama in Tibet, writing to the "Counslar" of Bosnia-Herzegovina. "Dear Counslar," the message began in broken and misspelled English. "The Di-lama grateful accepts the gift you sent of 2 horses." On it went, discussing potential diplomatic recognition between Tibet and Bosnia-Herzegovina, "but," the message said, "we must know one thing, where are you." Neither country's envoy knew where in the world the other country was. Tibet offers to open up trade between the two countries, offering giant pandas, snowmen, lots of rocks, stranded mountain climbers, and their equipment.

Roper attempts to TERM-talk Parrello again, and once again, his efforts to show Parrello these strange pnotes fail: they had vanished. "If this is your idea of something funny," Parrello types, "well this isn't funny roper bye" and ends the talk.

More strange pnotes appear. There were only two explanations, Roper narrates. "Either someone was playing a hideous, complicated, and thoroughly impossible prank, or else, I was going mad. To be continued."

Thus ended episode one of *The Great Guano Gap*. In episode two, Roper tells how friends hanging out in the PLATO room at SIU are amazed to see the chancellor's pnotes on Roper's screen. "What in the world is that?" Ray Wood asks. Roper is thrilled that someone else sees the message, convincing him he's not going mad. Another friend, Gary Whisenhunt, sees the strange message too. But a third, Mike Capek, thinks the whole thing is a prank and he's being set up by the others. Roper gets up and lets Capek sit down at the terminal. Capek signs on, attempts to condense his TUTOR lesson, they stare at the screen, and, instead of the normal system behavior, the screen says,

Call Your Instructor

Unable to condense lesson.

Author is a nonbeliever

at which point, so the story goes, the terminal's built-in microfiche projector blinds them all in a brilliant flash of light. When they finally can see again, they are shocked to discover that Capek has vanished from his chair in front of the terminal. They soon discover he's been somehow teleported into the PLATO system itself, and he soon appears as a tiny little figure on the screen, waving frantically and yelling, "LET ME OUT!" (Just a few years later, Bonnie MacBird, Alan Kay's wife, would write the story for the Disney movie *Tron*, which told the tale of a video arcade owner, played by Jeff Bridges, who gets zapped by a beam of energy and transported into the electronic circuitry of a vast tyrannical computer that he desperately wishes to escape.)

Things go from strange to even stranger as Roper suddenly sees a message at the bottom of his screen, indicating that he is being monitored by "supreme being" of "p," who, the system tells him, "also sees this display." Trembling and not sure who this user is, he presses the TERM key and begins to type. He asks, "Are you Frankel?"

For a CERL user in 1975 this would no doubt have elicited what on social media today would be a cavalcade of knowing LOLs. Frankel, the youngest systems staffer, at the time only fifteen years old, could sometimes be overbearing. *Guano Gap* was poking fun at him.

At first Supreme Being thinks this is some sort of joke, but, now angry, causes Roper's terminal to display all sorts of random noisy lines, which convinces Roper to show respect. The Supreme Being is not pleased that Roper plans to ignore any more Bosnia-Herzegovina messages, and indicates that Roper has been chosen to help save the world from a great danger.

Guano Gap appeared in serialized installments in late 1975. It represented perhaps the earliest example of a new genre of fiction: the interactive story told through software on a computer, navigated by the reader. (Around the same time, Priscilla Obertino was developing on PLATO the lesson *The Glumph*, an NSF-funded interactive multimedia story for elementary reading students, which also allowed children to touch underlined words to hear an audio clip of that word being spoken, and pick different outcomes of the story.) There was a meta-level to *Guano Gap* that made it particularly unusual: the story itself was one involving the PLATO computer, told through simulations of the very software and applications that the readers of the story were intimately familiar with. The result was a clever parody of the everyday experience PLATO users lived. It was something very new, taking storytelling in an entirely new, digital direction, where the only place to experience the story was online. The message required the medium.

Meanwhile, out at the University of Delaware (UD), which had by 1977 accumulated quite a few PLATO terminals, all of which were connected over expensive phone lines to CERL, a young undergrad named David Graper began writing notes about his experience with an interactive series of music lessons on PLATO called GUIDO. Anyone could sit down at one of the terminals in the PLATO classroom in the music building and use a special signon, "student" of group "udmusic," to try out the GUIDO material. One menu item that that signon offered to users was access to a notesfile, =guidonotes=. One day, Graper chose that menu item, entered the notesfile, and just began typing a story. Initially the stories were about GUIDO. Often referring to the program as "she," he concocted imaginary episodes of the strange effects GUIDO had on its users. Graper extended his subject areas to more general stories, which kept getting longer and longer, having nothing to do with GUIDO and certainly not appropriate for posting in =guidonotes=. His new stories centered around an umlaut-festooned alter ego with superpowers named Dr. Gräper, Master of Reality, who possessed a PhD in Humor.

Graper stories became a hit on the UD campus and to many users on the CERL PLATO system. One might walk into an otherwise quiet Delaware classroom full of terminals, all of them in use by students or staff reading Gräper stories, and hear snickers, chortles, and out-loud laughs amid occasional shushes by more serious users in the room. Invariably the laughs came from users reading the latest Gräper story. Word spread quickly at Delaware, Illinois, and even in Minneapolis, and a following grew. Fans would race to a terminal as soon as there was news of a new posting.

Graper would take snippets from his real-life college existence— boring classes in gigantic auditoriums, taking drugs and getting high, dealing with bad weather, snooty college girls, dumb frat jocks, dorks, losers, frustrating bureaucracies, corrupt teachers, and, worst of all, whiners—and craft them into humorous stories where his alter ego Dr. Gräper saved the world or sought revenge in the form of mischievous pranks on all his sources of real-world problems.

A taste of an early story: "Impossible Physics 201," posted online in 1977:

"Alright," the professor said to the massive clot of humanity gathered in the lecture hall, "let's get under way."

Thousands of notebooks opened to their first pages, and thousands of pens and pencils stood on the ready, preparing to copy down whatever the professor said.

"If you don't know already, this is Impossible Physics 201, and I'm Adolf Hitler."

"That's impossible!" a young man from the crowd shouted.

"Precisely!" the professor returned. "Physically impossible!"

The students murmured with appreciation at this first taste of impossible physics.

"Alright, students, let's try something else. How about this?"

The professor climbed on top of the central lab table and jumped off without falling. Standing in mid-air, he asked the class to explain what he was doing.

"You're . . . you're standing in midair!" an overly made-up female in the front row said.

"True," the professor said, "and what else?"

"You're denying the law of gravity!" shouted an exchange student from Nigeria who got a special grant from the people at Coca-Cola just to study under this professor.

"Right," the professor stated, beginning to pace back and forth in mid-air in front of the lab table. He walked back onto the lab table and then slowly stepped back onto the ground.

"What I just did can easily be explained by this equation," the professor said, pulling down a screen with an incredibly complex, all-in-tiny-print formula on it. "Study it hard, you're going to have a quiz on it in thirty seconds that figures for 97% of your grade!"

"That's unfair!" a girl in the front row objected.

"Hardly, Miss Mayfair," the professor grunted, pleased at Ms. Mayfair's look of dismay at his knowing her name. "It's impossible!"

Miss Mayfair's face turned red with embarrassment.

"Don't worry, Miss Mayfair. The only punishment for wrong answers in this class is disfiguration." To this Miss Mayfair laughed quietly and returned to her former composure.

"Alright, class, now for the quiz. Everyone take out a blank sheet of paper."

Thousands of sheets of paper ripped out simultaneously.

"Lay them on your desks, and prepare yourself for question one."

The classroom silenced.

"Alright, that'll be it," the professor said.

The students looked at each other in dismay.

"Your quizzes have been scored, graded, registered, modulated and transmitted to the unknown civilizations of the universe."

Suddenly, a rush of sound ran through the lecture hall. The students turned over their papers to find quizzes, written in their own handwriting, suddenly graded and scored.

"I didn't write this!"

"But it's your handwriting!" the professor retorted. "Class is over. I'll see you all yesterday."

The students again murmured among themselves about the sanity of their teacher, and filed out of the lecture hall. It was snowing. When they had come in, it had been summer, with 90 degrees in the shade.

David J. Graper was born in Grand Island, New York, an island in the Niagara River between the United States and Canada, a few minutes from Niagara Falls. He recalls that his science teacher in school once told his class that Grand Island would eventually float down the river and go over the Falls. "The teacher, perhaps a frustrated artist," says Graper, "put in the extra effort to actually paint his own 'artist's conception' of this future with our island poised halfway over the Falls and proudly showed it to our second-grade class, pointing out the street our school was on as a reference point. . . . His assurance that this future would happen several million years in the future didn't help much, since I was pretty sure I'd still be alive then and still going to that elementary school and like many of my chess club friends I spent the rest of my time eating in the part of the cafeteria that was furthest from the 'doomed' end of our school building."

Graper's father worked as a chemical engineer and computer expert for DuPont in Delaware, where the family moved in 1970. Graper attended Newark High School, where he became friends with classmate Dan Tripp in the school's television studio, which had received funding from the local public TV station. "I began as the weatherman for the intra-building morning television program, which was the standard lineup of high school crap—two people behind a counter reading math club meeting announcements, dreary recitations of the girls' field hockey scores, etc. The position of weatherman was the lowest of the three 'talent' positions available (the other two being 'anchor' and 'sportscaster') and was by far the most stupid position, since talking about the weather outside a building to people inside the building who could look out a window to see it for themselves seemed surreal. Hence the job was perfect for me."

Graper eventually enrolled at UD, also in Newark, and one day came across PLATO by accident. "I was using a piano in the music building and saw a small dark room with those glowing orange touch screens inside. I walked in and started playing with the system, since no one objected, and then I saw someone writing a note to a notesfile and that was it. I had discovered a new way to waste time, which is my true *raison d'être*."

He also found the GUIDO lessons, deeming them "excellent." "The software was truly masterful, I thought (and still believe), because it was a perfect application of technology to a specific educational task. . . . It took a truly dreadful, repetitive learning task (learning music intervals) and automated it. I never had to take a class requiring it but began playing with it because of the videogame aspect of it, later because I found I was able to use the skills it taught in actually hearing and writing down chords."

GUIDO led him to the =guidonotes= notesfile, set up for the purpose of enabling students to ask questions and post comments about the GUIDO lessons. Until Graper arrived, that is. Here was a notesfile, it seemed open to students, he was a student, so he sat down and he wrote notes. Initially, his notes were GUIDO related, but were not *quite* what the music faculty and staff had in mind. Fortunately, Bill Lynch, GUIDO's programmer and director of the notesfile, found Graper's stories amusing. They met and became friends.

Graper's stories gave him an outlet for his frustrations at what he viewed as the absurdity of university life, consumer life, television, marketing, urban congestion, dumb people, and a long litany of other complaints. What he had not expected was the reaction to his stories, the following of fans that grew, and the demand, unceasing, incessant, for more. In time, Graper was becoming such a weight on the otherwise quaint little =guidonotes= notesfile that Lynch decided to create a notesfile just for Graper. It was called =grapenotes=. Now Graper had a place all of his own to write as much as he wished. So he did.

For example, in 1979, Graper got a job at the UDPLATO project, whose headquarters were in an old two-story house everyone called "The PLATO Palace." He turned the job interview with Jim Wilson and Bonnie Seiler (she had come to Delaware after working in the elementary math project at CERL several years before) into a story:

Dear Friend:

I got interviewed for a job today at the PLATO palace, and like, I really think I screwed it up. Yeah. You've all got jobs, man, but it's because you always went to bed on time when you were kids and always handed in your homework neatly, with your names and the dates precisely printed on the top left hand corner.

As for me, my chances don't look too good. Let me replay the conversation for you:

[Bonnie Seiler and Jim Wilson come into the room where Dr. Gräper lies prostrate over a couch.]

Bonnie: Well, hello!
Dr. G: Don't you know about knocking?
Jim W: Oh, sorry!
Dr. G: Well? (Sits up)

Jim W: Well what?

Dr. G: When do I start getting paid?

Bonnie: Well, ah . . . first we'd like to find out exactly you'd like to do.

Jim W: Yeah. What are you taking in college now?

Dr. G: What business is it of yours!?! You're as bad as my old man!

Bonnie: Well, are you majoring in computer science?

Dr. G: Who do you take me for? Some calculator toting meatbrain? I'm taking introductory.

Jim W: Introductory what?

Dr. G: Introductory everything! Intro to Sociology, Intro to Basic Psych, Intro to introduction. . . . I've been taking intro courses since I was a freshman.

Jim W: So in actuality you are majoring in (giggles) introduction? An introduction major?

Dr. G: You could say that.

Bonnie: Well, ah, what would you like to program in?

Dr. G: I don't know. You tell me [heaving leg over nearby chair].

Jim W: Are there any subjects that you think could use further development? Any subjects not covered now by PLATO?

Dr. G: Well, I do have some ideas

Bonnie: [Getting out pen and paper] Shoot.

Dr. G: I think I should program in some lessons about rock music lyrics and maybe some dirty poetry.

Jim W: Well . . .

Dr. G: Just text blocks, you know?

Bonnie: [Putting away pen and paper] Well, ah, we'll call you . . .

Dr. G: When are my hours? I want them 6–7 on Wednesdays on months that begin with a "J."

Jim W: Well, we'll have to work that out. We'll call you.

Dr. G: Look, what's the bottom line? What position do I take?

Jim W: What position would you like?

Dr. G: That one. [Points to position name in PLATO brochure]

Bonnie: Well, Fred Hofstetter already has that position . . .

Dr. G: [Pounding table] Damn!

Jim W: Look, how about we call you later . . .

. . . and so it went.
Your Friend,
Dr. Gräper

No doubt the real-life interview went better, as Graper was hired. One of his eventual tasks was to empty out the "uddata" dataset file containing massive amounts of student data whenever it got full. Disk space was always scarce, and there was only so much of it. As a courtesy, he would post a note in =staffnotes= warning everyone that he would be reinitializing the file very soon, and if someone needed some data from it, get it out quick. The UDPLATO staff would discover that, what started out as routine, succinct, professional courtesy reminders, complete with the exact time and date that the data would be reinitialized, did not last long. Graper couldn't resist, and the routine reminders rapidly evolved (or devolved) into progressively longer, more elaborate stories, turning =staffnotes= into another outlet for his creative mind. A sample:

staffnotes / unidel 7/3/79 5:23 pm graper / udps / unidel

Dear Friends:
   "=uddata= is reinitialized," he said. The words echoed emptily in the abandoned music building room. He looked down at his lunchbox. In it were five hundred scraps of paper, all saying "Remember to reinitialize =uddata=!" He wrote himself a lot of reminders. He checked his watch. Written all over his arm and hand in ball-point ink was "Remember =uddata=!"
   It was 3:26. P.M.
   Now it was time for =uddata= to fill again.
   THE END
   Your Friend,
   David Gräper

*

staffnotes/unidel 7/17/79 10:03 am graper/udps/unidel

Dear Friends:
   Whoa Nellie, I just took a few winks here at the switch and

=uddata= is practically full! It's come rumblin' down the track full'a data, and if any of it's yours, get it out by 9:38 a.m. tomorrow cause I'm reinitializing it then and =uddata= pulls out again, ready for another load.

And at 9:39 tomorrow, when you're wiping the sleep from your eyes and you hear that lonesome whistle, don't come whining down here to the digital depot because, hell, the schedule's right there on the wall, can't you read it?

hell, gave you enough warning.

Your Friend,

Dave Gräper

Stationmaster

*

staffnotes/unidel 8/10/79 5:34 pm graper/udps/unidel

Dear Friends:

[Sound of gooey piano/orchestral music]

. . . Nature always provides the best way of doing things. Natural goodness comes only from the earth itself, and the most enjoyable life is the most natural way of being . . .

[Picture of people programming on terminals in the woods]

But sometimes, irregularity sets in. =uddata= gets filled up, and nature needs a little help.

[Programmers look at each other with concerned expressions]

That's why =uddata= is going to be reinitialized Monday, August 13 at 7 p.m. If there's anything in there you need, please get it out before then.

Reinitializing. It's only natural.

Your Friend, graper/udps

One time he reverted back to a simple, brief, professional note: "=uddata= will be reinitialized tomorrow at 12 noon," signing it, "Your Friend, Dave Gräper." Fellow staffers were nonplussed. "Just like that?" wrote Jessica Weissman. Other users begged for a real Gräper story. That night, Graper delivered.

staffnotes/unidel 9/12/79 10:00 pm graper/udps/unidel

[40,000 feet above Stuttgart, Germany]

[Flak bursting all around, plane rocks violently]

[Scene from inside cockpit]

Navigator: Captain Gräper! We've got to dump it now and get out of here!!

Captain: (Face illuminated by near miss) No way, buddy! We drop our load when I give the word!!

[Big explosion]

Captain: Damn! There goes engine three!

Navigator: We're only on two engines!! We've got to dump it now!!

Captain: I give the orders around here, mister!! We wait 'til the dataset is nearly filled, then drop it!

[Light on instrument panel lights up]

Captain: Alright! 54 out of 55 blocks filled! Reinitialize!

[Bomb bays open and tons of data fall to earth]

Captain: (Screaming triumphantly) Eat data, Adolf!!

[Plane lurches upward with weight gone, increases speed]

Navigator: (Sweating) Sorry sir, I guess I was . . . scared.

Captain: (Grinning) Only a fool wouldn't have been, son.

The Delaware use of CERL PLATO grew to the point that it was too expensive to continue having dozens of terminals dialed long-distance to Illinois, so the university bought its own CYBER system from Control Data. This had an interesting effect on Graper's legend. At Illinois, since most PLATO people were on campus, eventually a PLATO person would bump into the creator of this game, or that famous lesson, or some other mark of distinction. You were known. There was a real-world aspect to most CERL users. But the Delaware users, like users at any remote site connected to CERL's system, were less real to the CERL community because they were not people you were going to bump into. Likewise, CERL and Illinois were mythical, distant places to most Delaware users. This disconnect became all the more noticeable when all of Delaware left CERL and their terminals were one day switched to the university's own brand-new PLATO system, in 1978.

CERL users were fans of Graper, but now he was even more a mysterious entity who supposedly lived in Delaware. Some Illinois people had a gnawing doubt that Graper was a real person at all. Many believed he was actually a persona created by Bill Lynch, the name they saw so frequently as the person who actually posted the stories in the =grapenotes= notesfile (Graper would sometimes be busy, and hand Lynch a story he'd typed at home, asking Lynch to transcribe it into =grapenotes=, which had become a sort of "archive" for all things Graper). It's a telling hint at how early things still were, that an online persona in the late 1970s still needed to be physically present on campus to be taken for real, or at least, known to exist at some nearby location—somebody had to have testified to having seen this Graper person in the flesh. The Red Sweater, who had by the time Graper was an online celebrity left CERL to pursue graduate school elsewhere, was a legend around CERL now, but he was real, he was someone who for years could be seen sitting at a terminal in the foreign language building or at some other campus site, wearing his red sweater, and doing his thing online. He was long gone from PLATO by the time Graper arrived on the scene, but there were still plenty of people who remembered Bruce Parrello and so they had no trouble believing in the Red Sweater. Dr. Gräper, on the other hand, was out east somewhere—supposedly—in some landmass near the Atlantic Ocean, an incomprehensible distance away. CERL users outside Delaware never met him or saw him in person. Even decades later, they would still carry doubts that David Graper ever existed.

20

Climbing the Ziggurat

Andrew Shapira was determined to be the best. He approached Empire the way top athletes approach the Olympics. His desire to be number one even extended beyond Empire to less glamorous pursuits like typing. Typing lessons and contests (sometimes it was hard to tell the difference) were all the rage on PLATO. Like the Big Board game craze, for a while it seemed that every kid who'd snagged an author signon and some lesson space had to write a typing game. A typical design was as follows: Here is a paragraph of text. Type it without error, as quickly as possible. Here's another paragraph to type. Keep going as long as possible, and eventually you'll get a score. If your score is good enough, you'll be inserted into the game's hall of fame, the list containing the names and words-per-minute scores of the most accurate and fastest typists.

People did crazy things to get to the top of these lists. Cheating was not out of the question. Though even a cheater ran up against an absolute speed limit thanks to the relatively slow speed of the connection going from the keyboard, through the terminal, over the phone line, and back to the CYBER mainframe.

Shapira was in junior high when he started hanging out in the foreign language building's basement room with its vast sea of PLATO terminals. He'd found a typing test inside some utility program, and the typing test included a little racetrack game. "I guess I sort of got into it," he says. "The other guys typed pretty fast and I sort of wanted to get going, so I did." For the next six months he worked "really intensely on getting to be a good typist." Even when he wasn't at a terminal, when he was in class, he would be *thinking* about typing. "I would think about words that I would see and I would just type them mentally, think of what my fingers would do to type the things I saw. House, lights, light bulbs, things like that. And then, a few months later, I was typing faster than anyone." His reputation was earned and solid and unbeatable. He did 150 words per minute, some say 160, some say even more. All you had to do was chat online with Andrew in TERM-talk or Talkomatic to witness his freakish speed. "*Fast* would be putting it mildly," recalls Josh Paley, then another kid who hung around PLATO. "It was like looking at a human linefeed almost, I mean honest to God, you'd see a line full of characters, and then another line full of characters. . . . It was just *something*, I mean absolutely unbelievable. . . . The way he held the keyboard, his hands would come in from the side, so if you think about normal touch typing, where they tell you to put your thumbs on the space bar, and your fingers supposed to be resting on the *F* and the *J* or something like that, that wasn't how he did it." Ergonomics be damned, if he had to bend himself into a crab position, holding his elbows out, his hands coming in from the sides of the keyboard, so be it. "It was strange to watch but it was unbelievable," says Paley.

Being fast in typing lessons was one thing. Even climbing to the top of the hall of fame to gain recognition as the fastest typist on PLATO got you only so far. Pressing keys *fast*—pressing the *right* keys accurately—in games like Empire was a different matter. That might get you recognition in other ways. One Empire player, whom we'll call Mitch, says he, Andrew, and other friends "all had on speed typing drills that we wrote, just to get our Empire skills in order." A friend and game rival, and fierce Empire player in his own right, Brian Blackmore, even decades later held Andrew in awe for his typing brilliance. "That," says Mitch, "is because Blackmore is the second fastest human at a keyboard talking about the first." Blackmore, speaking in the late 1990s, said that if someone threw a PLATO terminal in front of him, he would dive back into Empire and climb right back up to number one in fifteen minutes, loving every minute of it. "And Andrew would hear about it, show up, and toast him," says Mitch.

Andrew was to Empire in the same way Donald Bitzer was to PLATO. Their reputations unshakable, set in stone. Andrew to the game, Bitzer to the system. Andrew's powers at Empire were superhuman, and extended beyond mere typing skills. He had a strategic and tactical sense, an innate understanding of timing, that gave him precious extra seconds to maneuver, flank, jump over, or otherwise surprise one or more opponents. Says Mitch,

Being a fast typist was no single guarantee! Newbies would arrive monthly . . . to try to take on the regulars. They would be carted off hours later in body bags. You had to be fast—but quickness alone was not enough, and there was an element of Zen magic to fighting. You had to know your opponent, study their moves, and play against their strength with your strength, or cover your weakness.

Andrew had no weakness. He was like Michael Jordan at this game. I have never seen such concentration in an individual, fingers flying across the screen yet conversely making hardly any sound. Sometimes he would take the keyboard and use it sideways, or upside down . . . just to prove he could still get a kill. Andrew could take people on one on five and win—good opponents!

Perhaps the second best Empire player was Brian Blackmore, the eternal Number Two in an eternal struggle to become Number One. He was fast. He knew how to play. He knew the strengths and weaknesses of all the regulars. "Brian put in the thousands of hours that a true devotee did," says Mitch. "Brian knew what it took to kill Andrew, and often did—but in a money situation, Brian would lose. Maybe he psyched himself. Once, Brian was on a roll, he had a terminal set up and he was taking on all comers. He was record shooting, killing people legitimately, going for the consecutive kills record of Andrew's. He'd been at this for over twelve hours. And still at it, running on Mountain Dew and chips. Andrew heard about it and entered the game, and within fifteen minutes had ended Brian's run at the record. Andrew was God."

Investing time to honing skills in Empire enabled the determined player to ascend a pyramid of glory that Empire's authors had built into the game: the hall of fame. Empire's authors went one step further: the hall of fame *itself* had a hall of fame. First there was the monthly hall of fame, and high achievers on that list were eligible for the other list, the all-time hall of fame. The monthly list was erased each month, after the entries on the list were examined by the program to identify who had earned a spot on the rarefied all-time list, if their scores were good enough. Over time, it was harder and harder to be good enough to get on the all-time list. But there were kids for whom this was their mission, their focus: to get all the way to the all-time top. Andrew Shapira was such a kid.

"Ineffable," Tom Wolfe called it, in his book *The Right Stuff*: that inde-
scribable mix of bravery, cool, and mastery of the forces of nature.
Only a few had this quality. They knew they could lose it at any time.
But if one managed to hold on to it, day after day, one might reach the
pinnacle of what Wolfe called "the Ziggurat," the pyramid of achievement, of glory, "to join that special few at the very top."

That was then, in the 1940s, 1950s, and 1960s. By the 1970s, times
were changing. The elite had changed. There was a new Ziggurat on
the horizon. The old one was *analog;* the new one something entirely
different, equally ineffable: it was *digital,* a pinnacle NASA's fabled test
pilots would probably have failed miserably climbing if they'd even
noticed it. The only way to see this new pyramid, and to climb it, was
to *go online.* To even know this new Ziggurat existed, you had to use
computers. You had to *know* computers. There were only a few places
around the country that had networked computers and were beginning
to offer something resembling this Ziggurat. But each system was different, developing a different vibe, a different culture. In the East you
had the hackers at MIT. In the West, more hackers at Stanford. And
then, out in the cornfields of Illinois, you had the emerging PLATO
culture, unlike anything else in the world at the time: a "digital Galá-
pagos" that had evolved into its own thriving ecosystem isolated from
the rest of the world. So far out in the middle of nowhere, nobody on
the outside gave it much mind if they even knew about it. The ones
who did had often rolled their eyes, especially at Bitzer, some insisting
that his flat plasma display was a hoax. *A fake. A trick. It uses mirrors.* Or
they would instantly dismiss PLATO as some uninteresting project
having something to do with that boring e-word: *education.*

But for those who had used the system, the education story was just
what the suits talked about when they submitted proposals to the government for funding. It's what the suits talked about when they shook
hands with the suits at Control Data Corporation. It's what the suits
saw when they gave demos to other suits. But for those in the *know,* the
new wave of kids who had lined up to PLATO like eager colonists to
Mars, and turned the system into a wonderland that extended far past
education, it was the teeming online culture filled with people hacking
and yakking and spending hours on games that made PLATO fascinating. This wonderland was quite real, it was exploding, and to them
it was very cool indeed. If only you bothered to spend a few minutes
gazing into one of those screens, from which emanated the Orange

Glow, poking around this vast new online world, you might begin to catch on, begin to realize you were gazing upon the future, the birth of cyberspace (at a time when it was powered by actual CYBER computers). Keep looking, and you would notice there was something here, some hint of a shape, an outline, a new challenge, a new pinnacle, that a lot of people were already climbing. If you wanted to begin, all you had to do was press NEXT. Beyond that single keypress beckoned a new digital Ziggurat.

The wide base level of this new Ziggurat was not hard to reach. If you could read and type, you were halfway there. The base level was simply *access*. We take the concept of access for granted today, with enough laptops and tablets and smartphones and other gadgetry, not to mention the ubiquitous Ethernet and LTE and 4G and Wi-Fi that it's all become a utility like electricity: you just assume the Net is there, out of sight, but always available. This was not the case with PLATO in the 1970s. Invariably, you had to go to *it*, it did not come to *you*. First you had to know it existed, then you had to know where to go to get it. You had to get to a terminal. Didn't matter where, they were all the same. It was what was inside that orange screen that mattered. As so many kids had done, you might randomly stumble upon it, wandering through the halls of CERL or some other building that had terminals. Or a friend might have said, "Hey, let's go play games on the computer" and dragged you to the Zoo, where your life might be changed forever. Or if you were not at UI, you might find terminals in whatever building had them at your PLATO site, be it in Hawaii or Connecticut or Delaware or Florida. Perhaps your professor had assigned your class one hour per week in a PLATO lab, to do lessons or simulations that counted toward your grade. If you were hooked by the system during your first exposure, you came back. Like Richard Powers, you learned to be on the lookout for what others were doing. You listened for tidbits of knowledge that might get you further, deeper online. Maybe you met a friendly Sherpa guide who showed you around, fired up a menu on the screen, gave you a tour of the system using one of the demo accounts. You got *access*. It was the first step, the foundation.

Sometimes you had to do a little breaking and entering to get access. Says one former PLATO user who, years out of high school, still prefers to be anonymous for the shenanigans he pulled on PLATO, "We needed physical access to PLATO terminals where we wouldn't be bothered after the 10 p.m. curfew for high school students. There were

PLATO terminals in the education building on campus, and somehow we got a key to that building and made a copy of it. After that we had unrestricted access to PLATO. We would bring in pizza and play Empire all night." Says Brian Redman, "I was the chemistry curriculum coordinator for a short period at the University of Arizona. I got the powers-that-be to let me take a terminal out of the library and put it in my lab for a summer so I could play Airfight twenty-four hours a day." This type of thing was going on all over PLATO, wherever PLATO terminals were installed. You did whatever you had to do to achieve access. Sometimes you then hoarded it.

"Keep in mind we were all under curfew age," says Lee Johnson, "but our parents and we didn't care. What nerds we were, sitting in a dark room in a musty old university building, lights out, squinting at the warm orange glow of screens. PLATO was a *big* part of my adolescence. I was too shy to make many friends or date girls. I lusted after girls, of course, but was not good enough for them, nor knew how to treat them, so PLATO filled up many lonely hours. My friends and I roamed all about the university campus in search of unrestricted PLATO terminals. We would find individual terminals in a study room on the fourth floor of some building, and immediately test it to see if it would allow our logins, and if we could play games. We would find ways into locked buildings, though I stopped short of actually jimmying locks or something. I'd just find an open door at an adjoining building, then walk over to the one with the terminals. I even got stopped by a prof once with a friend when we went to play some stupid chess or moon landing games at the Loomis Laboratory of physics. He called the university police on us."

David Soussan was a UI undergraduate who along with a few other addicts went to elaborate lengths to maintain undisturbed, after-hours access to PLATO in the music building classroom on weekends. "The lab closed around 8 p.m. Friday," he says, "but what they didn't know was the back wall of the cubicles along the back wall of the lab had been unscrewed and set in place. You could pull back this wall and crawl behind it, wait for the lab to clear out, and be locked up for the weekend. Then you wait another fifteen to twenty minutes in case someone forgot something before crawling out of your little hiding place and, *voilà*, you've got the whole lab to yourself." He would order pizza to sustain himself over the weekend. "Every now and then, twenty or so minutes prior to the lab closing you'd pull the cubicle wall back and find someone else already hiding inside. Both of you there for the same purpose, neither wanting to lose the option or be called out, you just made room and hung out silently together."

The ultimate in access was, of course, a home PLATO terminal. In the early 1970s, very few had them. Bitzer had one, as did a number of the senior CERL staffers. Louis Bloomfield, the developer of several popular Big Board games, had one thanks to his father running UI's medical school.

Two families on the East Coast had home terminals. The wealthy financier Arthur Lipper had heard about PLATO, was a big believer, and so got a terminal for his family, and even opened up a for-pay "learning store" in a mall in New Jersey with some PLATO terminals. Lipper's young son Chris became an Empire player, playing as an Orion all day long whenever he could. Henry Jarecki, a well-known psychiatrist and commodities financier, was friends with Lipper. He had heard Lipper had gotten this thing called a PLATO at home. Not to be outdone, he ordered one for *his* home. Obtaining a PLATO terminal at home during this era was as expensive as buying a new car, costing between $6,000 and $10,000, plus a few thousand more (possibly many thousands more) for the phone line per year.

Lipper lived in New Jersey, Jarecki in White Plains. "He and Henry Jarecki were buddies, good friends. Rivals, at least," says Garrie Burr, a CERL technician who would travel out to the East Coast to install and service their terminals. "It was always a case of one-upmanship. Henry lived in this great big huge mansion . . . this was a different world. . . . We were sitting in the Lipper family room one night after I'd fixed the terminal and we're sitting there watching TV and he turns to his wife and goes, 'How'd you like to go to Florida?' and she says, 'Sounds good to me' and he picked up the phone and ordered two first-class tickets to go down to Florida on vacation. This was just stuff beyond us."

At the Jarecki home, very young sons Andrew and Eugene wound up using the terminal for years, mainly for games. (Both brothers would in the 1990s become successful entrepreneurs, selling their company Moviefone to AOL for $400 million, and then after that both became successful *again* as award-winning filmmakers.) Says Eugene, who was five or six when the "big brown box" originally arrived at his house, "I guess my dad said he'd be the principal user. No one anticipated the child computer geek phenomenon at that time. So the fact that it was instantly easier for me to navigate my way around it, to type more quickly, to relate to the machine, I think had to do with the fact that I didn't have the baggage somebody like my father had had, of reading books for a lifetime and interacting with the world in a far less push-button-interactive way." Eugene would, like Chris Lipper, soon discover Empire, where he played on the Federation team. Even decades later, Jarecki described Empire as an addiction that never goes away. Eugene remembers chatting with inmates at the Menard Penitentiary in Illinois, a site participating in CERL's PLATO Corrections Project. "I had a lot of TERM-talks with a couple of guys at Menard," he says. It's doubtful the prison had any idea that its inmates were chatting via computer with kids in the outside world.

John Risken was putting in so many hours on the elementary reading project at CERL, he missed his family and they missed him. "My family was not seeing very much of me," he says, "so I went to Bitzer and said, 'Could I get a computer terminal at home, because then I could put in more time, and still be fair to my family?' So we got a PLATO IV terminal . . . and I did a lot of work from home in the evenings. . . . And even having gotten to be at home didn't make a whole lot of difference. There was a time when my older son was in first grade, he drew a picture of the house at one time, at school. He brought it home, and my wife showed it to me when I got home that day. There's our house, and there are three people standing in front of the house: my wife and two boys. And I said, 'Well, where am I?'" His older son pointed, and there, inside the house, he could see a person hunched over a computer terminal.

Access was so important, sometimes PLATO users would go to extreme lengths to get it. Bob Rader, one of CERL's senior systems staffers, had a PLATO terminal at home. "One time," says Rader, "I came home, and found not my son, but a friend of his," using the terminal. "And he was the only person in the house!"

Whatever it took.

There was more to access than just finding a physical terminal. Access was the prerequisite to everything that followed. All too soon your desire to see more, do more online, always *more*, meant that the freebie demos and temporary borrowing of someone else's signon was not enough. You could keep finding and stealing them. That was easily enough done. But soon enough people always found out. To really have access you had to have your own signon. And the signon didn't count unless it was an *author* signon.

"I don't know why it bit me so hard," Doug Green says. "I became what you could call a hacker, I guess. Just to get into it, I wanted *access*. I wanted to be part of this. I wanted to *experience* it. I wanted to *play* on it. And I wanted to *write* on it. Author games. It just *lit* my fire."

Being a PLATO author was the next step up the Ziggurat. There was far more to climb, and if you had any ambition, any desire to get near to, or become one of, those "special few at the very top," it was still a long, long way up. If you wanted a taste of TUTOR programming, you had to have an author signon. Having an author signon also got you a view high enough up to begin to see the lay of the land, this ever-changing digital landscape. With an author signon you began to see that "cyberspace" was a glistening, twinkling circuitlike sprawl, like that seen out of an airliner window at night when a plane has just taken off over a major city. It seemed every minute, every second was non-stop novelty, fun, controversies, arguments, games, and the goings-on of a vibrant community. There was more than any one human could handle, let alone comprehend. You had to learn to block a lot of it out. For PLATO users, this marked the beginning of what the majority of people in the world now cope with every day: the difficulty of filtering out all the noise. Once online with a PLATO author signon, the real world, real life, with all its nonsense of war, poverty, crime, scandals, failing economy, politics, and worries about school, all of it melted away, replaced by the frenetic online community of PLATO, where novelty was the operative term. Being online on PLATO was like being in a town that was growing quickly, becoming a small city on its way to becoming a metropolis, and in time, a nation without borders, built of not just zeroes and ones, but minds.

There was something else that you discovered once you had an author signon and began exploring the system. From that second step up on this new digital Ziggurat, you made a shocking discovery. *There were other Ziggurats.* You could see them. Plain as day. There they were. This new digital environment opened up all kinds of possibilities, all sorts of ways to be creative, to be of use (if that was your thing), to cause trouble (that too), or to accomplish something great and gain recognition from your peers, be they university staff or gamers or simply fans of whatever online persona you had created. Just like real life, cyber-life offered choices. What you chose to do, who you chose to

be, might determine your cyber-destiny. This was one of the key realizations that dawned on PLATO users. There was not just one path.
There were many. If you didn't like one Ziggurat, you could go find
another, or even start your own, then climb to the top, to join that
special few at the very top, or to be *the one.*

This was the great invention at the dawn of the information age:
pyramids of achievement, of respect, of accomplishment, of financial
success, of notoriety, of celebrity, can only hold so many people at the
top, so, invent more pyramids, and let people spread out and climb
the ones best for them. At UI and other universities connected to the
CERL PLATO system, as more and more students were assigned
coursework that included required PLATO terminal usage, as more
and more students with a technical bent found out it was possible to
get *paid* to use this thing—and not just university students, but high
school students from the surrounding area as well—the wave of young
people pouring into PLATO grew. They couldn't all be "s" programmers. They were going to have to prove their stuff to Blomme, Tenczar, Sherwood, and the others. Every day it was getting harder. There
needed to be more paths to glory. Now there were.

Some of these paths were more glorious than others. One easy path,
true low-hanging fruit for a new author eager to make a mark once he
or she finagled some lesson space and learned a little bit of TUTOR,
was to create an index, a directory, of other lessons, notesfiles, games,
and whatnot. These indexes had their equivalent when the World
Wide Web started twenty years later: out of nowhere, they popped
up like weeds. The more ambitious of the index makers strove to add
more features and conveniences as time went on, anything to attract
and then retain users who were free to go elsewhere (Yahoo would
wind up owning the spot at the top of the early Web directory Ziggurat). On PLATO, there were programs like "bigjump" and "llist" and
countless others all essentially offering the same thing: directories of
interesting things to do online.

One's identity on PLATO started with one's signon. A PLATO signon
consisted of two parts: a name and a group. The closest analog today
is one's email address, with the name to the left of the @ sign, and
some Internet domain to the right. As befits a computer world that had
evolved off on its own with similar but different solutions for every-

thing, the separator on PLATO was not "@" but a slash "/" to separate name from group.

To the new arrivals, the pilgrims flowing into PLATO every day, your group was your tribe. Your group told a story. Some groups announced that you were a part of the CERL staff, like "o," "pso," "p," and the top group, "s." Group "s" contained the systems staff, the programmers and software engineers responsible for PLATO's operating system, TUTOR language, programming editors, applications like TERM-talk, notesfiles, and TERM-consult, and other tools. "S" people like Paul Tenczar, Bruce Sherwood, Dave Andersen, and Bob Rader had unlimited privileges. "P" people, like David Frankel, Dave Woolley, and Brand Fortner, were the junior systems programmers (some of whom put up a sign in their office at CERL stating "Junior Systems Programmed While You Wait"). "During that whole period," says Fortner, "I figured that I would have died and gone to heaven to get a 'p' signon, and the epitome of everything would be to get an 's' signon. But, you know, getting that 'p' signon, I figured we ruled the universe."

Having system privileges gave one power. Fortner admits that he loved the power the "p" signon gave him. One time he tried to impress a girl with the PLATO power he had. "We were really big on that power," he says, "and I said, 'I can execute a systems command' and she's like, 'Oh, well, what can you do with that?' And I said, 'I can read files with my systems command.' Because at that time, you know, a normal program couldn't read and write files. And so, I added, I wrote a little five-line systems program that read a file, but I forgot to put in a couple of steps, and this is in the middle of prime time with thousands of students running, and I crashed the computer trying to show this girl how smart I was and I had to rush over to the computer room and try to explain things. I think they were unhappy."

Dave Woolley, who at seventeen had been hired by CERL right out of Uni High, was very much aware of the high perch on the Ziggurat that his "p" signon afforded him. "That hierarchy was very present," he says. "I mean, what can I say, I enjoyed it, because I was in this position of privilege."

Some users saw power turning into power trips. "In the old days," says Brendan McGinty, "it was a real power trip to have a group. And to have a signon. People would delete your signon, and it would be a big thing. If you cussed in a notesfile or something—I remember getting paged by David Frankel, 'frankel' of group 'p' . . . he was typing so fast, *you'll never use this system again!* blah blah blah, and it was like, *end of talk*. And I was booted off the system within about five seconds, I tried to sign back on, and it said that signon doesn't even exist. I'm sure it was such a power trip to these people, to have such control."

"He [Frankel] was a kid and he really, really rubbed people the wrong way," says Ray Ozzie, who at the time was another UI undergraduate spending most of his time on PLATO, but would later become a very successful serial tech entrepreneur. "He had a very big in with Bitzer, if you want to say that everybody had a mentor, Bitzer was his mentor, and Frankel walked around like he owned the place. I mean literally like he owned the place. He bossed around, he had that little shit attitude and he'd go round to Tenczar, to Rader. He'd go, hey Paul blah blah blah blah, hey Rick blah blah blah blah. He bossed everybody around, you know Paul and Rick and people would just like blow him off because he was just a kid, but he had an attitude like he owned the place, whereas everyone else had to in one way, shape, or form kind of always tread lightly because you knew you had privilege, in having your records, 'o' records or 's' records or 'p' records. I mean I lost mine at one point, various people had theirs pulled, it was highly, highly, highly political. Except for Frankel and he knew it, and he had this relationship with Bitzer."

Each of the junior systems programmers had a little play area, a system TUTOR lesson wherein they could test code out that they were experimenting with. Frankel would poke around these "p" staff lessons, "constantly looking for things you were doing," says Ozzie, and occasionally ratting to Tenczar or Blomme: "Notice this. . . . What do you think Ray's doing here in the third unit of tm8sys or ozsys? . . . He was always there to be a little pain in the ass."

Ozzie also remembers receiving pnotes from Frankel containing snippets of one's code with little comments about the code, "just to let you know that he was in there, just to dig you a little bit more."

Frankel's gadfly behavior got to the point that one day some of the other programmers grabbed him by the ankles and dangled him over the rail at the top of the stairwell, ready to drop him. Luckily for his sake, they didn't.

To the technical staff at CERL, your story preceded you. All the junior systems programmers "knew intuitively which path 'got you in,'" says Ray Ozzie, "basically, who was your sponsor. There were the

Blomme/Lee people like Larry White who came in via the bottom-up systems/research path. There were Tenczar or Sherwood people like me who came in more from the top-down TUTOR/lesson path. There were Golden or Rader people like [Al] Harkrader or [Ron] Klass who came in via the ops path. There were several Bitzer-sponsored people like Frankel. There were of course the brilliant 'music' people who all orbited around Sherwin."

If you were climbing the technical PLATO Ziggurat, the rest of the people above you wanted to know your story. "As a newcomer," Ozzie says, "you needed to earn respect and prove your credibility to the people who came in via a different path. The incumbents didn't know you, and always wondered how the hell the newcomers got their credentials, and whether or not they deserved them."

Bitzer held the perch at the topmost of the highest of the PLATO pyramids, but didn't sit and gloat there—nor did he even use an "s" signon very often. No, Bitzer either was making a statement acknowledging that he was not a software guy (certainly not a systems programmer), or he was simply too cool to use an "s" and instead preferred the relatively rare and decidedly less powerful "m" signon used by hardware and maintenance technicians.

Bitzer saw in Frankel the same brilliance and energy he saw in Andy Hanson and Mike Walker from years earlier. For Bitzer it did not matter how young or old a potential star contributor was. If you had an idea, then start working on it. His encouragement was so complete and authentic it was infectious. You never knew who was the next Einstein, so encourage everyone to be the next Einstein and increase the odds that the next Einstein would reveal himself.

Case in point: Doug Green's encounter with Bitzer. To Bitzer, Green was a nobody, just a PLATO enthusiast, a gamer, a game author. Bitzer was that elusive "father of PLATO" figure, usually out on the road doing demos and trying to find the next funding round. "I knew who he was," says Green. "Of course I knew who he was. The only time I talked to him, he was the only person around, and he seemed to have a free minute, and I asked him a question." Years later he's no longer sure what newbie question he posed to Bitzer, but it might have been about natural language processing. Perhaps if CERL considered this or that approach? "And he said," says Green, "'That's one of the questions

that we're working on around here.' And I said, 'Well, I thought . . .
you know, maybe if this . . .' and he said something to the effect of,
'Hmm, well, maybe you'll be the one to figure it out!' Blew me away.
Blew me away." If that didn't want to make you get cracking, coding
away on tough problems that would help the PLATO cause, and climb
the pyramid of glory in the process, nothing did.

The majority of PLATO people were located at the University of Illinois until the late 1970s, when the number not only off-campus but out
of state and affiliated with other institutions grew. Until that happened,
it was possible to bump into most members of the community face-
to-face in Champaign-Urbana. In the 1970s sitcoms and films often
made fun of the pickup line "What's your sign?" PLATO people at
a party would ask, "What's your signon?" It was common to know of
the names and history of nearly everyone at a party, but not recognize
any of them until you knew what their PLATO signon was—that was
the only frame of reference for most PLATO users. One's signon was
listed there every day on the Users List. One's signon was at the header
of every message posted in a notesfile. That was your PLATO identity.

Meeting people in person for the first time could be a shock. "I will
never forget this for as long as I live," says Ray Ozzie. He began TERM-
talking with another PLATO user who had always posted thoughtful
notes in the notesfiles and had interesting things to say in chats. "All
I knew about him," says Ozzie, "besides the fact that he was an interesting guy, was that he typed really slowly." They finally met face-to-
face and Ozzie was floored. The gentleman he'd been communicating
with, Gary Michael, who created a successful courseware-development
start-up company in town called Duosoft, was handicapped and in fact
had been typing with a mouth stick, one key at a time on the keyset.
"It was such an eye-opener for me, at the time, such an eye-opener. . . .
Nowadays it's easy to say nobody knows you're a dog or nobody knows
this or that, but talk about as complete a shocker to your prejudices or
your pre-judgments or whatever—it was probably the best thing that
ever happened to me. . . . Something I carry with me . . . the fact that
you're dealing with somebody's mind, not their body."

As the number of recreational lessons and games and notesfiles
proliferated throughout the 1970s, the lofty status of the systems staff
began to diminish. In the early years of PLATO IV, the most famous

PLATO users were the creators of the system itself. Having an "s" or "p" or even "pso" carried great clout at parties. But as time went on and the user community grew, and more and more people in the community began to distinguish themselves in different ways online, be it by creating an excellent lesson or instructional simulation, a popular game, or just running a popular notesfile, they began to gain fame and notoriety all their own. Oftentimes systems programmers at a party were no longer the coolest people around. Sure, you might have written the system code that displays the phrase "What term?" when you pressed the TERM key, but that no longer compared to identifying yourself as one of the authors of Empire or Avatar or any number of other games or recreational activities on the system. "People would think those guys had much more exciting jobs than us," says Al Harkrader, "and they were certainly more exciting since most of the stuff that we did was pretty esoteric, but Empire and Avatar and all those things were very tangible to most people so they could say, 'Well, you know, gee, I met Al Harkrader but I couldn't figure out what the hell the guy did, but I met Chuck Miller, the guy who wrote Empire, and that was totally cool.'"

21

Coming of Age

Ted Nelson is a genius, a visionary, a maverick, and somewhat of a Cassandra in computerdom. Born in 1937, he coined the term "hypertext" in 1963 and spent years designing and promoting his dream for the world: an elaborate, two-way hypertext online network, which he named Xanadu. (The world ignored it, and we have the Web instead.) In addition to his Xanadu work he's written numerous books, perhaps the most famous of which are a pair he self-published in 1974 in a single volume: *Computer Lib* and *Dream Machines*. To this day the twin books continue to be essential reading. Nelson understood the computer revolution before most people knew what computers were or what they meant for society, and even now nuggets of insight can be gleaned by leafing through their pages.

The books summarized both what Nelson felt the general public should know about computers (*Computer Lib*'s cover boldly declared, "You can and must understand computers NOW") as well as his travels across the country to discover what was going on with computers at the time. From 1973 to 1976 he was, in his words, "a lecturer and media maker, [in] various departments and auspices" at UICC (the old Chicago Circle campus of the University of Illinois). Early on during this time he visited several companies and university laboratories, assembling his findings and opinions in the books' makeshift, partly typed, partly handwritten, cut-and-paste pages with zest and urgency. It is fortuitous that Nelson was at a campus of the University of Illinois during this era, as it was inevitable that PLATO and he should find each other. Deep inside *Dream Machines* he devoted several remarkable pages to his visits to the CERL lab and to PLATO classrooms around the Chicago area. Despite his misgivings about education and computer-assisted instruction, Nelson marveled at the machines and culture that had grown up so quickly in the light of the Orange Glow of PLATO IV. "As a first taste of interaction on a graphical computer system, PLATO can be a thrilling mind-opener—especially to people who think computers can only behave loutishly or through printout."

Nelson was quick to notice one of PLATO's greatest and most essential architectural features, the Fast Round Trip, and did not hold back in his admiration for it: "The most basic underlying feature of the system, INSTANT RESPONSE, cannot be quarreled with. PLATO can respond . . . to a single key-pressing by a user, almost instantly; this feature is virtually impossible on IBM systems. This responsiveness is the system's greatest beauty."

But he also noticed that PLATO was a community of people, an online community unlike any he had seen elsewhere in his travels:

> Indeed, this extended Republic of PLATO—the systems people in Urbana, the authors and locals-in-charge throughout the network—constitute one of the maddest rookeries of computer freaks in the world. Where else would you find a fourteen-year-old systems programmer who's had his job for two years? Where else would you see people fall in love over the Talkomatic . . . only to clash when at last they meet in person? Where else can you play so many different games with faraway strangers? Where else can students anywhere in the network sign into hundreds of different lessons in different subjects (most of them incomplete)? Where else are people working on various different programs for elementary statistics, all to be offered on the same system?
>
> PLATO is one of the wonders of the world.

On any computer system there are two general populations of users: the designers and developers of the system and its software, and the end users of the system and its software. With PLATO it was no different. Within a few years of PLATO IV's arrival, the populations had grown to significant numbers. The population of end users—the students receiving instruction through PLATO—was massive, yet disparate. The other population was smaller, tighter, and deeply interconnected: authors and instructors, professors and their grad students, administrators, operators, managers, systems staff, and anyone else with an author signon. This smaller group comprised one of the most vibrant, diverse, and early examples of what writer Howard Rheingold would later call a "virtual community." It was certainly one of the earliest. Tragically, in one of what would be many cases of the saltwater view denying the freshwater view a chance to be known, the PLATO story went unmentioned in his essential 1993 book *The Virtual Community: Homesteading on the Electronic Frontier*.

The members of PLATO's virtual community comprised the lucky few having access to the notesfiles, pnotes, games, instant messaging, and chat room tools like TERM-talk and Talkomatic. This virtual community comprised perhaps a single-digit percentage of the overall user population. The vast majority were the students, the actual learners, assigned to take a course or series of courses online as an academic or job requirement. They might be young children, grade schoolers, high schoolers, college students, or trainees involved in government, industry, or professional organizations. They came in, they sat down, they did what they were told using these odd computer terminals with the Orange Glow, and then they went on with their lives. In and out. Done. Their numbers were in the hundreds of thousands, perhaps millions if one counts up every user who ever pressed NEXT to begin in the history of PLATO across all the systems that would pop up around the world in all the years PLATO systems were operating. But this vast majority were silent and invisible, largely nonparticipants in the virtual community.

Not to say that there wasn't leakage: some number of students, learners, and trainees would hear about the "other side" of PLATO, or even witness it, for example by looking over the shoulder of another user at another terminal (why is that person laughing? why are they typing so fast? are they *talking* to somebody? is that a *conversation?*). It was in the virtual community of PLATO authors and instructors where the most unusual and most notable things happened in terms of their place in the computer history. But it is important to understand that the primary, overriding purpose of PLATO was always education, and without all those student users, there would have never been a PLATO system.

CERL's PLATO community was not only growing, it was growing up. Many kids had discovered PLATO's secret dimension as high school students. They were the world's first "digital natives" (decades before that term would be coined). Now they were becoming college students, and in the seeming blink of an eye those college students were graduating or, ironically, flunking out. PLATO was maturing.

Metcalfe's Law states that the value of a telecommunications network is proportional to the square of the number of connected users of the system. The more people who used PLATO, the more useful it became, attracting even more people in a virtuous cycle, making it even more valuable.

In the span of just a few years, across the 1970s, CERL's PLATO IV system now had over a thousand terminals, and many tens of thousands of users, the majority of whom were in various clusters up and down the state of Illinois. But a significant number were out of state, from Hawaii to Delaware. By the end of the decade, there were multiple PLATO systems across North America with more on the way. These systems would be interconnected by telecommunications links, enabling users on one system to send and receive personal notes to people internally or on other systems. The total number of users—be they students, instructors, authors, gamers, or systems or administrative staffers—amounted to perhaps the largest online community in the world during that era, a number that would continue to exceed the ARPANET's combined user count until the early 1980s.

CERL's community now resembled a bustling, busy, large town, a microcosm of today's Internet. There one could behold the full gamut of human drama: sharing experiences and opinions, hopes, and fears, debating and arguing, laughing and crying, and everything that a human life is exposed to—birth, youth, relationships, dating, marriage, having kids, divorce, illness, and death. Death was a particularly new phenomenon for a virtual community to handle, particularly the death of a well-known, well-liked CERL staffer, Walter Brooks, whose signon like so many others had been a familiar sight online on the Users List. Brooks, one of CERL's system operators and a programmer at a local PLATO software company, was heading to work one day on his motorcycle and was fatally hit by a truck. The local community was shocked; grief poured out in notesfiles online. Some users, who knew Brooks only through his online self and had never met him face-to-face, flocked to PLATO to share their loss and exchange feelings with others. Today, when someone well known dies millions of people fill social media with messages of sympathy, grief, and hashtags. When someone died in PLATO's virtual community, it was an utterly new, unfamiliar experience that the community had to figure out how to handle. When Brian Larson, a well-known student PLATO programmer at CERL, took his own life in 1986, the community reeled in distress at the tragedy, creating almost immediately a notesfile, =brian-note=, to share grief, memories, and, in many cases, anger at Larson's suicide. Before doing the deed, Larson had not only sent private pnotes to numerous people, but he had posted a defiant suicide note in a public notesfile, stating that he refused to be a part of society. People from all over the campus, many of them fellow students, posted messages in =briannote=, many with questions asking "why." At the funeral a few days later, the church was filled with people, many of whom only knew Larson from his online presence.

Many of the extraordinary, unexpected outcomes produced by PLATO IV were thanks to the new wave of kids who had not been turned away thanks to the open, welcoming atmosphere of CERL and the PLATO system itself.

There was another unexpected outcome. At some point, a point that varied depending on the person, PLATO became more than a novelty in the lives of its more obsessed users. These users would cross an invisible line beyond which being on PLATO *became* one's life. There were countless examples of this. One was Mark Eastom, says Bruce Maggs, one of the authors of Avatar. Maggs roomed with Eastom during one of his undergraduate years, and Eastom became one of Avatar's operators, contributing by managing the monster data. "He was a real character," says Maggs. "PLATO was his life, he was one of these guys for whom this was *it*. This was all they had in their lives: their PLATO programming and PLATO game playing and PLATO friendships. There were a lot of people like that."

Living the PLATO life could turn into an addiction, a dangerous path to take. A PLATO-addicted college student risked grades suffering, possibly delaying graduation, or, worse, expulsion or dropping out. All of these outcomes were, sadly, commonplace.

It did not help that Don Bitzer himself had such a relaxed view on the gaming craze (gamers would argue otherwise). In 1975, he was interviewed for *The Daily Illini*, where he predicted that within ten years there would be PLATO-powered gaming parlors around campus, assuring the reporter that "such plans have already been discussed." He then gave an impromptu demo to the reporter and photographer.

"Oh, I see," he told them with a chuckle, "you want to get a picture that shows the director playing games during illegal hours." The article finished with Bitzer playing Dogfight:

> He chases his partner's PLATO plane back and forth around the screen, into corners and around in circles, constantly firing and missing with his PLATO bullets.
>
> "Ah, I missed him," he grumbles, continuing the chase.
>
> "Oh, you rat," he says.
>
> Suddenly, Bitzer traps his elusive partner. He squirms in his chair, eyes stretched open, head lowered closer to the keyboard, and jabs at the button to blow up his opponent. "Got him!" he cries out.
>
> Who says the games aren't addictive?

Thing is, they were. The whole system was. And it caused great turbulence to many a college student. When Dave Woolley graduated from Uni High and entered UI as a freshman in 1973, he wanted to study computer science. But that interest died quickly, he says, "because it was all engineering, and for any major in engineering you had to take like three classes in physics. What I was really interested in doing all that time was working on PLATO. That's really where I got my useful education. . . . I managed to get along. Some people couldn't handle the balancing act and flunked out. . . . I did all right, I majored in psychology largely because it was easy, didn't take a lot of concentration."

Alison McGee was an undergrad out at Indiana University–Purdue University Indianapolis (IUPUI), and was hooked on PLATO during her college years. "I got the worst grades of my life during those semesters," she says, "but no regrets. No way, no how."

Bruce Maggs recalls an experience his mother had while she was riding the bus in Urbana: "She was sitting by some students and one of them said, 'Are you still in engineering?' and the other student said, 'No, I played too much Avatar last semester and had to switch to liberal arts.'"

Many UI students from the 1970s and 1980s would in time confess to the havoc PLATO wreaked on their college careers. Michael Schwager was one. "I first saw Plato in 1977," he says. "I got accepted to the U of I in 1978 and became addicted to it, playing Empire till 6 a.m. In 1979 I flunked out of school, but I got good at PLATO."

David Sides, one of the coauthors of Avatar, stared into the abyss, grade-wise, a few times thanks to overdoing it on PLATO. "I know I got into a lot of trouble sophomore year, because I was ending up too long at the computer lab, I was there until three or four in the morning, and I had real grade problems that first semester of my sophomore year because of it. I didn't flunk out of anything, but I got a D in a midterm in chemistry, and that made a real major impression on me, and it was a real problem."

What made a stronger impression on him was what happened when he borrowed his mother's Datsun B210. He drove it to CERL, parked, went in, found a terminal, and in moments was down into dungeons gaming. "She had told me to not to come home as late as I had been, which was six or seven or eight in the morning." He emerged out of CERL into daylight at around eight o'clock, and the car was gone. He called home.

DAVID: Um, Mom, do you have the car?
MOM: Yes, you can walk home.

"I lived on the other side of Champaign," Sides says. "So that was a memorable experience for me."

In some PLATO circles, PLATO, partying, and entertainment were the priorities. Ray Ozzie describes his UI undergraduate days as "one big blur" of having fun, partying, gaming, and coding. A far lesser priority was classwork. "Nobody ever wanted to hear 'I've got this in the morning . . . I've got a test.' . . . You *never* heard that."

Steve Rose had enrolled as an electrical engineering freshman, and became a Sigma Phi Delta fraternity pledge in the fall of 1975. "One of the brothers," Rose says, "was the course director for the mechanical engineering department. He gave me an author so I could play games. . . . I had never seen a computer before and I was hooked." Rose tried out the usual list of games, but "it was Airfight that really got me," he says. "I remember many nights spent at CERL playing Airfight until they kicked us out at 6 a.m." *Games or grades: pick one.* That was the choice for many students during this time, and in Rose's case, games won, grades lost. He flunked out after missing many morning classes due to his CERL all-nighters.

Mike Carroll, the creator of the Pad lesson, the simple but popular bulletin board messaging forum on PLATO that was released before

David Woolley's Notes, was another victim. "I would come to school Monday morning and work on it pretty solid until Wednesday evening, when I'd have a date with my girlfriend. Then I'd be back Thursday morning for another session until Friday evening. PLATO went down every day at 6 a.m. for maintenance, so that's when I slept. Occasionally I would just fall asleep on the keyboard when the system was shut down, waking up with square keycap indentations in my cheek. I lived on Mountain Dew, KFC chicken livers, and Camel filter cigarettes. . . . Pad was only in existence for a year or so. It faded away in part because I flunked out of college."

Legendary gamer Brian Blackmore had wanted to go into engineering but lacked certain prerequisites, so entered the college of agriculture for the first year, thinking he could switch majors in the second year. "I basically flunked out of my first year of school because of playing games on PLATO," he says. "I ended up reapplying a year later, and got into the College of Liberal Arts and Sciences in the math/CS curriculum. Stuck it out from there, still playing games all the time."

John Matheny, one of CERL's junior systems programmers and developer of the Notesfile Sequencer utility, also ran into trouble during his undergraduate years. He took some computer science courses over in the Digital Computer Lab (DCL) across the street from Uni High, and like many students who had been exposed to the Orange Glow of the graphics screen and the snappy responsiveness of PLATO's Fast Round Trip, he was not impressed with the primitive punch card machines and teletype printers that the purist computer operators DCL forced upon its students. Author Steve Levy once described a similar situation at MIT, where operators of some off-limits big computers were known as "the Priesthood," those select few allowed to actually touch the mainframe and who made no promises to mere mortals when they submitted their punch card decks seeking computation. Being a PLATO programmer, Matheny decided to do what any sensible PLATO programmer would do. Rather than wasting time with punch card machines and Job Control Language "nonsense" at DCL, he wrote all of his code at a PLATO terminal in the TUTOR editor, then output his code to tape in CERL's machine room. He then took the tape over, got someone to read it into a tape machine at DCL, and sent the output straight to a punch card machine, which automatically printed out the deck of cards. His professors never knew.

But PLATO consumed so much of Matheny's time that his academics began to suffer. "I was expelled twice," he says, "because I was a really poor student, and it was because of PLATO." He was still on the staff at CERL and kept working over that summer, taking two courses he says he took seriously, and emerged with two A's. "They didn't let me back in for another year," he says, but things began to deteriorate again once he was readmitted. "I was spending all my time at PLATO doing work and none of my time doing coursework and so I got booted again." He managed to get a job at Control Data Corporation in Minneapolis, and eventually would finish his degree at the University of Minnesota attending night school.

Kevet Duncombe, one of the coauthors of Moria, had a similar situation, out at Iowa State. During an interview he gleefully described the intense time creating Moria. "We'd be up there hackin' away till all hours," he said, but when asked how PLATO affected his grades, he got quiet. "Yeah, well," he said, pausing, "I don't know," adding another long pause before admitting, "I eventually flunked out." He had been at Iowa State for two years, he says, "at which point I was invited to leave." Being asked to leave one's undergraduate enrollment was devastating, but losing access to PLATO—being sent back down to the base of the Ziggurat—was equally devastating. Like so many others, he hung around in town, as "they let me keep my signon for a while," he chuckled. "I was workin' fast food for a while there, just spending all my free time at the PLATO terminals." Chuck Miller had graduated, ended up at Control Data in Minneapolis, and one day recommended Kevet for a position. "That was wonderful, that got me my first professional gig. Without it I'd probably still be flipping hamburgers."

Says one former Uni High student we'll call Hayden, "When I went to Uni, PLATO gaming *was* high school, pretty much." His grades eventually were so poor that Uni asked him to leave. He ended up as a C student at a local public high school. PLATO gaming was the peak experience of his high school years. He made it through a year and a half of college before flunking out. But then something odd happened, similar to Matheny and Duncombe: he discovered that all those years he had invested in PLATO, gaming and hacking and the rest of it, had value after all. "Much to my utter consternation I found I had all these skills that were now being rewarded." He noticed that many of his contemporaries, Lindsay Reichmann, Mark Zvilius, and some from Urbana High School including Bruce Maggs, Bruce McGinty, and Andrew Shapira, were getting programming jobs, which gave them keys to the building so they could game at night. "I was like," says Hayden, "well, dang, I oughta get me one of those!" He found work on the side for a professor on campus, and like magic he got a key to the building. "So I could be a gamer again," he says.

On today's Internet, with billions of people connected, it is possible to find a busy, active online community devoted to even the tiniest interest niche one could imagine with forums, support groups, and digital hangouts. With billions of people connected, it is hardly surprising such would be the case. By the late 1970s, though, what CERL's PLATO system revealed was that with only mere thousands of users (several orders of magnitude less) they had already begun to reflect a similarly vast range of interests.

There was, for instance, a notesfile called =addict=, dedicated to PLATO addiction. In it, users could offer true confessions of their predicament: how PLATO felt to them, how being away from PLATO felt, and how getting back online felt. One user in 1981 described his PLATO experience this way: "When I do get on . . . blooie. . . . End of sanity. End to sense of proportion. End to perspective on what is important in life. When I first got on in 1975, I used to lay awake at night thinking, 'Gee, I can't wait until I get on tomorrow,' and getting an author signon was the greatest ambition I had." Another user expressed his PLATO predicament this way: "The orange dots are more personal to me than face-to-face encounters with people I don't know. This may be because when you leave a note with your signon attached, it is there for a long time, much longer than a spoken word is around, and therefore tends to be more thought out. Those who say computers are impersonal have never used a computer. They are far more personal than most people. P.S. Computer games are better than sex."

An electrical engineering student who was soon to graduate posted in =addict= asking for help on life after PLATO. "I'm unprepared to face the world," the anonymous poster said. "What's it like out there? Is there any way to avoid this?" Less than an hour later, another user replied, suggesting, "Get a fulltime job at the university. In that way, you can still manage to get a PLATO signon. That's how I'm still on this wonderful system."

In 1981 a freelance writer named P. Gregory Springer came to

CERL to explore PLATO for an article he was writing. Amid shady advertisements for Pseudo Caine ("an 'incense,'" the ad copy teased) and the Buzz Bomb ("mix cool nitrous with warm smoke"), the article appeared in 1982 in *High Times* magazine, which described itself as being "at the center of the cannabis counterculture since 1974." An article on the PLATO online community seemed out of place in such a publication until one realized the topic: computer addiction of the PLATO variety. Indeed, for so many users with author signons on the system, their experience was a kind of high, one that once you reached, you never wanted to come down from.

Ted Nelson may only have been being whimsical when he gave the title "The Cave of PLATO" to his drawing of a PLATO IV terminal in *Dream Machines*, but nevertheless it was a profound observation on a remarkably ironic situation. Bitzer had named this system PLATO, and his crowning achievement, the PLATO IV terminal with its gas plasma flat panel touch-sensitive graphical display, supporting internal mirrors and microfiche color slide projection, was indeed a cave of sorts, not unlike the cave in Plato's allegory. And like Plato's cave, this "cave" was one whose denizens preferred the depictions on its "wall," rather than the Technicolor reality going on outdoors. Not only that, but PLATO could be something you *belonged* to. Something you experienced with *others*, side by side or virtually through cyberspace. It was something you did, but it was also a *place* you could hang out in. Its more obsessed users viewed the actuality of the PLATO experience as dwarfing what one experiences in real life: more interactions, a sense of hyper-accelerated self, poring through content faster than possible in the real world. Like Plato's cave, the "drawings" on the wall were crude, shadows only, but still PLATO users were undeterred.

For instance, pornography. Another inevitable consequence of bringing a lot of people, particularly young people, together into a virtual space online and then providing a variety of open-ended tools at their disposal to be as creative as possible, is the inevitable arrival of pornography. The true nerds would digitize photos with a camera, then run special software to convert the image into crude orange lines and dots displaying on-screen. It might take agonizing *minutes* to plot a naked lady, but in 1975 taking minutes to plot a naked lady in the Friendly Orange Glow of PLATO was a thrill around which college

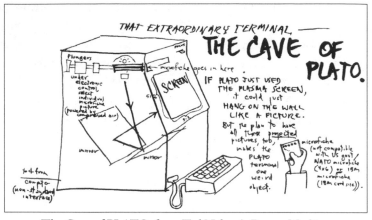

The Cave of PLATO, from Ted Nelson's *Dream Machines*

students would gather and cheer. There was even a PLATO Pornography Network (PPN), which, given its illicit nature, kept moving from one lesson space to another on the system, popping up like a weed that the authorities would shut down only to discover a copy had been saved and moved to some other file. Michael Gorback had a TUTOR lesson called "gorbackwk" that became infamous for its crude collection of porn images. But it was popular, and he added features to the lesson, including something called "porn-o-matic," an advice column, and touch screen games. There was even a private Talkomatic-style chat room. (Gorback's lesson was so well known that one day years later he happened to be walking into a Control Data office in Baltimore, and as soon as they heard his name, they knew who he was because of "gorbackwk.") At one point, much to the embarrassment of CERL and the University of Illinois, *The Daily Illini* published a multipage spread featuring screen shots of seedy porn pictures, mostly line drawings that even true cavemen ten thousand years earlier would have dismissed as hopelessly primitive. But it all had to start somewhere.

It wasn't all fun and games. Behind the Friendly Orange Glow of dots on the screen were real people, and some of them were hurting. The rise of the use of the "anonymous option" in some notesfiles like =ipr= (short for "interpersonal relationships") and =sexnotes= gave people a way to vent, ask questions without losing their privacy, or simply call out for help.

In November 1980 a student posted a note in =ipr= describing how depressed and suicidal he felt. David C. Barnett, a doctoral student in the counseling psychology program at Iowa State University, happened to be paging through =ipr= soon afterward, and came across the student's note. "I wrote a brief response," he says, "describing what help I thought counseling could provide. I acknowledged that going in for counseling could be a scary experience and suggested that the writer contact me with any questions about counseling."

Barnett continued, "As a counseling psychology graduate student, I found =ipr= very interesting—real-life discussions of the gender wars, romance gone awry, unrequited love, conflicts at work, roommate concerns, etc. At times, it felt like an Ann Landers column, but mostly it seemed like a community of people who wanted to offer support, ideas, reality checks, and encouragement to others who used PLATO. I found I tended to write a lot of responses encouraging people to get counseling or psychotherapy. After a time, it felt somewhat ludicrous to send this suggestion to the file so often, so I tended to send personal notes to people if the posting was not anonymous."

A few nights later, Barnett happened to be working on PLATO at around eleven when a TERM-talk request flashed on his screen. He had no idea who the user was or why the person was TERM-talking him. According to Barnett the typed conversation went like this:

CALLER: hi.
BARNETT: Hi.
CALLER: are you the one who wrote the note about getting
    counseling in ipr?
BARNETT: Yes.
CALLER: well, I want to die.

"At this point," Barnett says, "I gasped audibly and let out a mild oath. I certainly did not expect to be dealing with a suicidal person over a computer terminal. Bear in mind that all those cues that are so useful in counseling (the nonverbals and the nuances of the voice) were totally lacking. All I was confronted with was a black screen with print in orange letters."

Barnett, a trained counselor, gently probed the student with questions in an effort to get him to open up and share his pain. It turned out that the student's brother had died in a car cash, and this student felt responsible. "That night was the one-year anniversary of the accident," says Barnett. To add to this, the student had transferred from one college to another, where he now felt isolated and alone. He was flunking exams and had thoughts of ending his life. "PLATO had been one of the few sources of pleasure he had found on his new campus," says Barnett. "He told me of the many hours he had spent playing space war games on PLATO as well as many hours reading notesfiles. In the course of one of these frequent escapes to PLATO, he ran across the 'suicide' note in notesfile =ipr= that I had responded to with an invitation for questions. The original note echoed many of the feelings he was experiencing. He began thinking more and more seriously about wanting to die and took several steps." He went home to his parents, stole his father's gun, and brought it back to his dorm room in college. He wrote a suicide note, and planned to do the deed out in a vacant field. On the way there, he changed his mind and decided to sign on to PLATO one more time and reread the =ipr= note with Barnett's offer for help. It was at that moment he TERM-talked Barnett. The conversation went on for well over an hour. "I felt quite nervous the entire time," says Barnett, "and as he described more and more lethal-sounding clues, I became very fearful for him. I attempted to be as supportive as I could, despite the hundreds of miles distance between us, and tried to help him discover a local support system. After a nerve-wracking discussion, I finally uncovered his local crisis center, and he, in essence, put me on hold while he went to call them from a phone next to the terminal room. I waited for a long time—staring at an orange-bordered screen devoid of any communication from his terminal except 'please hold on.' By this time it was nearly 1:20 in the morning."

Barnett waited for twenty terrifying minutes, but the student finally reappeared and resumed the TERM-talk. He told Barnett he had spoken to the crisis center. Local volunteers were dispatched, and while the student was waiting for them to arrive, Barnett continued talking with him through PLATO. "After about fifteen minutes, he announced that the two volunteers had arrived, and he was going to hang up, with a promise to send a note to me soon. I said I would hold him to that, and he signed off the system."

A few agonizing days later, Barnett finally received a pnote from the student. The counseling center had saved his life.

There were other such occurrences on PLATO, nor were they limited to =ipr=. There were successful suicide attempts, though rare.

On more than one occasion, Donald Bitzer would receive a report of a potentially suicidal PLATO user posting notes. Police or crisis counselors might be called in, or, if the identity of the user remained unknown, systems programmers would then be directed to use system privileges—whatever it took—to figure out who the note writer was. Though an anonymous note in theory was anonymous, it never truly was. There were logs to study, system utilities that could monitor who was using the system, and other means for zeroing in on the at-risk student. Oftentimes it was possible to do so in real time while a troubled student was still online. One such incident involved a student who was posting violent threats in =ipr= against his girlfriend. A systems staffer was asked to find out who was making the threats. He contacted the instructor who managed the group the student was enrolled in, and the instructor then discovered that the user was sitting a few terminals away from his girlfriend, seated at another terminal, in the same PLATO classroom.

Bitzer tended to be the go-to person for these crises, though he did not like it. ("He avoided conflict at all costs," said one CERL staffer, "because he didn't want to have to deal with it.") In addition to death threats, suicide threats, pornography, hate speech, and other concerns, the number of incidents of hacking and security breaches increased over the years. Finally, Bitzer had had enough, and delegated the whole responsibility to a newly formed Security Committee comprised, reluctantly, of various CERL staffers.

PLATO fostered local and long-distance romances and relationships with equal frequency. Countless people met their mate through Talkomatic, or TERM-talk, or through pnotes or one of the notesfiles. Being female in the CERL community was a rarity, but it came with benefits for females looking for them. Alison McGee made the pilgrimage west from IUPUI to CERL one time to meet several online friends. "My friends in Indiana thought [that] was absolutely insane at the time. They thought me 'talking' to people on a computer was insane too, and I wonder what *they* do every day now, hmm?"

McGee went on to say, "The great thing about PLATO was that a lot of people who used it were very intelligent, and it was a great way to meet other intelligent people. And that would include, of course, men. We were heavily outnumbered by the men." She found herself "courted" by many of them, "and needless to say," she adds, she "loved the attention." Attention she described as follows: "One man sent a box of helium balloons to me and then drove from Illinois to take me to see *The Breakfast Club*. One served me dinner in his home, my first experience with rare steak (blech!), and tried to convince me that Dylan and Waits were better than Springsteen. One gave me really good memories of the top floor stairwell in the CERL building. . . . As I recall, I only made one trip there, which now seems almost impossible to believe. It's as if I can re-create every hour I was there, from my mouth dropping open at seeing the massive amount of PLATO terminals . . . compared to our three or four terminals in Indy maybe."

McGee eventually moved to Bloomington, where she met her future husband. "PLATO was a real sticking point between us for a while," she says. "My husband wanted me to get *away* from it. I still wonder when our society will recognize computer/email/cell phone addiction as a problem—needing 24/7 communication with another human *via a machine*. Back then, people just didn't get it, now they 'get it' too much."

Andy Otto was a student at UI Chicago Circle who witnessed an online romance develop between one of his best friends, Tim, and a girl in Hawaii whom Tim had met online through PLATO. "Lynn and Tim had long, deep conversations over 'talk,'" says Otto, "and tied up the only lab terminal for hours." Eventually Lynn came to Chicago and met Tim. They were married a year later.

Sometimes the romances began in a conventional way, two people meeting at some physical location, but if they were PLATO users, the system became an irresistible tool for keeping the relationship going when they could not be together. One couple, Gail and Jim, met in person first, but then continued communicating online as their homes were two hours apart, then wound up marrying. "We got to know each other through email," says Gail. "In fact, Jim proposed to me over pnotes, and timed it such that he was with me when I was reading my email."

The instances of people meeting and carrying on relationships through TERM-talk and Talkomatic are so numerous it can safely be said that it was a common occurrence. For instance, another couple, Jim and Susan Lawrence, became an item through Talkomatic. "She was in Baltimore," says Jim, "using a signon designed for recreation which allowed her to access lesson Talkomatic. I was in Urbana, using my author records, and purely by chance we happened to start talking one night in Talkomatic. Our discussions continued on and off for about two months via PLATO before I called her by phone, and we eventually met face-to-face. The rest is history, as the saying goes."

The CERL laboratory itself, intense as it was as a work environment, with people spending much of their lives there, became a site of numerous hookups, affairs, and romances. One former staffer recalls coming to work early on a Saturday morning and finding two other staffers going at it on the sofa in the lounge. More than a few CERL staffers were married to each other. Every now and then, such couples would divorce, then sometimes the two parties would remarry other staffers who had also divorced. These sorts of relationships occur in any workplace, particularly today with email, texting, Facebook, and the rest. But the online dimension was new in the 1970s, and CERL was pioneering in this aspect of human drama as much as it was in technology and computer-based education. Some even called the lab "PLATO Place" (after the TV soap opera *Peyton Place*). Says one former CERL staffer, "Let me tell you something. It was the 1970s. People were literally doing it in the road. It was part of the 1970s. There were a lot of young people. A lot of young, smart, active, healthy people. Stuff happened. And I can assure you that whatever stuff is alleged to have happened, it did not detract from, and probably contributed to, extra hours to work in the lab, so, you know, on that score I feel like, get over it."

There exists a wealth of stories about the romances and breakups, marriages and divorces that can be attributed to PLATO. Two PLATO marriage stories will have to suffice here. The first involved Greg and Judy Kirkpatrick, who had met at Florida State University, and thought it might be fun to have what they called a "computer wedding." "We didn't publicize it," says Greg, "we didn't make a big deal of it, in fact the folks there didn't even know what was going on. We had a few of her friends attend, there were about five or six people. We got married by a notary public. We had gone to the education library at Florida State, where there were a number of PLATO terminals, and we all sat down at various terminals, went into Talkomatic, and just chatted the ceremony there." The second involves John Daleske and his wife, Reina. Being an Empire author as well as a prominent PLATO personality, he had many friends and followers both in Illinois and in Minnesota, where he had wound up working for CDC. Daleske had met Reina Jones, who was based in Columbus, Ohio, online through

PLATO. Says Luke Kaven, who attended the 1977 wedding, "They met and struck up a romance and they fell in love right away, and I actually spent a few days with them with my girlfriend of the time that I met on PLATO. . . . So we agreed to go meet in Minneapolis and stay with John and Reina and hang out there, and we got invited to their wedding." PLATO people from all over came to the wedding. "There was *his* family," Kaven says, "and then there was *her* family, and then there were *the PLATO people*, and everybody was just absolutely *baffled* in their two families—they had no idea what was going on with this, and whether this [was] an illicit matchup because they had done it that way or was it a *cult*. [These PLATO people] all seemed to know each other, but they didn't understand. What did it mean to 'know somebody' through a *computer*? They had no concept of it. And they really looked at us like we were coming from outer space, I mean there was no precedent for them to draw on. I think they were worried—*Is everybody okay and is this marriage going to be good?*—and you know, we were pioneering that too, in the sense of mowing down attitudes and stereotypes of dating and relationships and so on."

A singular vision guided the PLATO project for its first twelve years, yet even during those years, a subtle trend began to emerge that also guided the project: give bright people a chance to be creative, get out of the way, and behold the innovation. The lab naturally attracted the kind of people who thrive when allowed to be their most creative selves, roadblocks pulled, nothing but a clear path to getting something new built. The Bitzer's Boy Scouts phenomenon, with the creative contributions of people like Andrew Hanson and Mike Walker, was but a mere hint of what was to come. In time, as PLATO III grew inside a larger lab with a larger staff, there came the point where people would simply take initiative on their own to go build something better based on the real needs—the real pain—PLATO users, particularly authors, were experiencing. Blomme and Krueger with MONSTER and CHARPLT. And then Tenczar with the real game-changer, the TUTOR language. Each step of the way accelerated the pace of innovation in the lab, the rate at which new things were now possible because of the innovations that had been realized. By the time the PLATO IV system became operational in 1972, the rate of acceleration was dizzying. While the adults went forth to create thousands of hours of instructional lessons in every subject imaginable for use in settings ranging from kindergarten to college, the floodgates opened and a new wave of high school and college kids poured in. They found TUTOR, they saw the amazing new flat panel display terminals with their Friendly Orange Glow, and perhaps most amazing of all, they discovered that they were welcome to give this technology a try and make it do something. Now there was an army of bright young people building games, building apps, competing with each other, outdoing each other, relentlessly exploring the boundaries of possibility with PLATO, only to find that someone had successfully knocked a hole right through a boundary, expanding what was possible even further. While the mission-oriented focus was to create and deliver computer-based education at ever more economically viable costs, the hacker kids were transforming that singular vision, now a dozen years long in the tooth, to one of collaboration, community, entertainment, and creativity.

PLATO IV was no longer just a CBE project, a big bet by the federal government to see what could be done with computers deployed as automatic teachers on a large scale in school settings. That vision no longer told the whole story. If anything, the new wave of kids had defined their own singular vision: the most interesting thing you could do with a computer network is fill it with *people*—give them ways to connect, collaborate, communicate, quarrel, yak, play. Over forty years ago the young people of CERL paved the way to the inevitable future that we take for granted today. Build a sufficiently broad and extensible set of interactive tools that allow for communication between users of a computer system, and the result is PLATO. CompuServe, the Source. The WELL. Prodigy. AOL. The World Wide Web. Google. Facebook.

Throughout the 1970s, the new wave of kids had transformed the PLATO system beyond its educational mission into something much more generalized: an online service, a concept that at the time was sure to trip up anyone outside the PLATO community. No one person within CERL had been responsible for the rise of this community— not even Bitzer. But collectively, they achieved something so ahead of its time it is still hard to comprehend given how early it all happened relative to what came later.

So what were they going to *do* with all of this technology? How could CERL transform all of this innovation, this virtual community, from a still relatively obscure, you-had-to-be-there experience centered in a small Illinois town surrounded by miles of cornfields, to something massive and mainstream?

Bitzer had known since day one back in 1960 that PLATO did not survive unless it scaled. *Go big or go home.*

It was time to go big.

Part III

GETTING TO SCALE

It is hard to fail, but it is worse to have never tried to succeed.

—Theodore Roosevelt, "The Strenuous Life"

Optimism is a bias. That is, you look at the world, and you get the odds wrong.

—Daniel Kahneman, "Bias, Blindness and How We Truly Think"

Everything added meant something lost, and about as often as not the thing lost was preferable to the thing gained, so that over time we'd be lucky if we just broke even.

—Charles Frazier, *Cold Mountain*

22

The Business Opportunity

All during the late 1960s and into the early 1970s, the number Bitzer had bandied about to describe the scale of the upcoming PLATO IV system was "four thousand ninety-six." That was how many terminals were going to be supported. The media reported this number for years. By 1971, before PLATO IV was even rolled out, a background paper prepared by Bitzer and Alpert that was widely distributed to members of Congress, academics, and think tanks, expected a four-thousand-terminal system to be operational "by the mid- to late-1970s." A Champaign-Urbana *News-Gazette* article that year announced that "the goal is to have 4,000 in use by the end of 1974." In reality, in all the years CERL's system would be operational, it would never reach even half that number, but that didn't stop the prognostications about what was coming next. In the span of only two years since PLATO IV had gone operational, a new number was starting to be mentioned—a number so huge as to defy belief. *Thousands* of terminals were one thing, but now Bitzer was starting to talk about a *million* terminals, an entirely different ball game. No wonder the "troika" of owners—Bitzer, Propst, and Alpert—of Education and Information Systems, Inc., the company building the multimedia peripheral hardware for the PLATO IV terminals, were so gung ho on the future: they stood to make a lot of money.

The notion of a million terminals was floating around in the public as early as 1974, when Ted Nelson mentioned it in *Dream Machines*, except he added another twist, stating that Control Data Corporation "is said to be projecting ONE MILLION PLATO TERMINALS BY 1980." It was never explained how this might be possible, where the demand would come from, or who would pay for it, but the rumors, exciting ones for sure, continued to swirl.

—

In mid-September 1975, Robert J. Seidel, an experimental psychology PhD who worked at the Human Resources Research Organization (HumRRO), a Washington, D.C., area government consulting firm, hosted along with the National Science Foundation a conference at the Airlie Conference Center—a posh, rural Virginia hideaway retreat, featuring its own airplane runway, an hour west of Washington. The event echoed back to the 1958 "Automatic Teaching" conference held in Philadelphia attended by the teaching machine and programmed instruction luminaries of the time, including Skinner, Pressey, Crowder, and Stolurow. Seidel's sixty invited participants included Bitzer; Suppes; Papert; Bunderson; Stolurow; J. C. R. Licklider, Marvin Minsky from MIT; Fred Brooks, author of *The Mythical Man-Month*; John Shoch from Xerox PARC; and Andrew Molnar from NSF. The purpose of the conference was to "identify key developments in computer and communications technology that are expected to have major impact by 1985" in the field of education.

Seidel had high hopes for the conference, including that the attendees would reach consensus on the need to include some fundamental, psychologically sound, practical, instructional theories into the designs of systems of the future. His aims were not achieved.

"My problem with all of these folks," says Seidel, "they're all *messiahs!*" The luminaries in the computer-based education field were all visionaries, each respective vision different from the others in important ways. The common thread seemed to be ego. By the end of the conference Seidel realized that these visionaries, including Bitzer, were all stubbornly sticking to their original visions. "He was a very pleasant guy, a very charming guy," says Seidel, "so long as you didn't get into questioning much about PLATO. That was like a closed book. And it stayed that way, as far as he was concerned. It was *the* answer. . . . When I talked to Minsky, he had his view of artificial intelligence and that was the answer for everything. And Papert had the answer in LOGO for everything. Don had the answer in PLATO for everything. It was his approach, and he wouldn't leave that plasma terminal either." Dave Liddle, who worked on the plasma panel project at Owens-Illinois and later joined Xerox PARC, shared similar opinions of Papert and Bitzer, the latter whom he measured at 700 "milliPaperts," explaining that "if the unit of arrogance is one Papert, okay, he's about a point-seven instead of Seymour."

His presentation was pure Bitzer: part genius display, part used-car sales pitch, part magic show, all fascinating and at the same time outrageously provocative. He and PLATO were just the kind of thing that drove computer scientists—and a lot of educators—crazy. He gave his talk an equally provocative title, "The Million Terminal System of 1985" (over the course of the year, the prediction had been pushed out five years), suggesting it was a done deal already. The vision he laid out shocked the audience, his statements brazen, confident, and not a little bit arrogant. After the presentation, one participant would say, "I think the size of Bitzer's plans probably does as much to inspire awe, jealousy, and fear as anyone who has ever done publicly funded projects in this area."

But Bitzer had more surprises for the gathered audience. The one-million number certainly inspired awe, but what he was envisioning was not only an educational system, but something that today we would recognize as the Internet. He had begun to believe that there was no way a large, centralized system was ever going to scale to a national level if it were dedicated *solely* to education. There had to be more to the system. Something had changed in Bitzer. This was not the same story he had presented when seeking funds from NSF not so long before. He had a sense that additional capabilities had to be available in any large-scale national system. Features, no doubt, that facilitated communications, collaboration, creativity, and recreation. There's no surprise that these were the very things that PLATO's own user community—particularly the new wave kids—had sprung onto the system, on their own, in the three years prior to Bitzer's presentation.

"I don't think that computers which are designed only for education will survive," Bitzer told the audience, ostensibly in reference to Stolurow's SOCRATES and Bunderson's TICCIT, but there was more to it than that. "They must have all kinds of applications to survive in the marketplace. We have computers, terminals, communications, facilities, and we have to provide a service. You can have the first three lined up, but if your service is limited, you can forget it. So just having technology isn't going to help you. You have to have desirable services available."

By 1975 Bitzer's vision for PLATO was still fixated on a large centralized mainframe computer to which were to be connected thousands of what he liked to call "graphics terminals." The only difference between what he envisioned for PLATO IV back in 1968 and what he was envisioning in 1975 was the scale. His revised, broader vision did not contemplate nor include microcomputers, which were still in their infancy. The same week Bitzer gave this presentation, microcomputer hobbyists, including Steve Wozniak, who had already begun working on what would soon be known as the Apple I, were meeting at the Homebrew Computer Club in Silicon Valley. Despite the fact that CERL and Xerox PARC had opened their respective labs to each other and offered demos and hands-on experiences and dialogue, none of what was brewing in Silicon Valley either in corporate offices or hackers' garages was affecting Bitzer's vision. He was a true believer in the PLATO way, and he wanted to see it scaled to gigantic proportions. *Go big or go home.*

"It is an entirely different game," he told the audience, "when you consider the million-terminal network. My forecast, based on our present plans, calls for, by 1980–1985, a million-terminal network, consisting of two hundred fifty central processing systems all tied together, communicating with each other and shifting loads if one starts to go down." As preposterous as it sounded at the time, one need only consider that major Web services today, be it Facebook or Google, have hundreds of thousands if not millions of rack-mounted servers in multiple football-field-sized data centers. Those servers are all tied together, communicating with each other and shifting loads if one starts to go down. The only major difference is what devices the end users would be sitting in front of.

When Seidel published the proceedings of the conference in book form two years later, one thing stood out: only Bitzer's chapter included not only the talk he gave, but also a transcript of a forty-five-minute heated question-and-answer session that immediately followed.

During that heated debate the audience attempted to absorb the staggering numbers and brazen vision Bitzer laid out for them. At one point Bitzer questioned whether society truly valued education, and Papert and Stolurow, aghast, were quick to respond:

BITZER: I don't think teachers, the parents, or the federal government are nearly as excited about improving education as they have led us to believe.

PAPERT: . . . Nevertheless, it might be true that a thing called education is possible, and some of the people believe in it, and it will be installed in the world.

BITZER: My feeling is this will progress because of interest for other than educational, direct educational purposes. It is by meeting these needs that you will be able to get back to education, which is the reason for all of this. But if we don't include these needs in some way we just won't make it, we will just become obsolete. We have to be very careful how we plan to get from where we are today to where we are going.

STOLUROW: This sounds like a much more radical change in position than I understood you to say before. What you are now saying, if I am reading it correctly, is that the direct instructional purposes of a computer-based educational system is [*sic*] the least important component of what you see the total mix to achieve this flexibility. And that the other things will carry the instructional component along.

BITZER: It depends on whose viewpoint you are looking at. If I were the parent buying the terminal, I would be looking at it on another basis. What is it going to do for me? What will it do for my wife, for my family, for my vacation, for my pay, all of these other things? Incidentally, if it happens to teach, that's great. From my standpoint I happen to think the most important impact it will have is making literacy commonplace in inner-city schools. But I just don't think that the rest of the world does.

Another bombshell Bitzer dropped on the audience was his belief that the days of the federal government funding large educational computing systems like PLATO were over. "My opinion is that we will have to count on industry, not the federal government, to support this research." While that notion sank into the audience there were questions about the inevitably huge costs of Bitzer's "million-terminal system."

Dr. Marty Rockway, the technical director at the time of the Human Resources Lab at Lowry Air Force Base, then asked Bitzer a question, resulting in this extraordinary exchange:

ROCKWAY: What is the probability that commercial industry will fund a $90-million-plus capitalization of the kind of system you are proposing?

BITZER: I think it's somewhere in the neighborhood of $500

million to a billion. I think there is an excellent probability, above 50 percent.

ROCKWAY: Why?

BITZER: Because I happen to know some things.

Bitzer's tease was not the best way to endear himself, or PLATO, to this esteemed audience of colleagues, but he could not help it: it was his way. Perhaps owing to his family's successful automobile dealership business, there was an entrepreneurial streak in him that made his burning ambition all the more visible at times like this. He was a true believer, a classic technology visionary, and one way or another PLATO was going to be his gift to the world. By 1975 PLATO IV was supposed to have grown to over four thousand terminals—or so he had promised for years going back to the mid-1960s. And yet there it was, the calendar said 1975, the deployment of PLATO IV was largely complete, but only about a thousand terminals were connected to the system. It was a magnificent feat, but it was nowhere near the scale he had boasted about in print and in person. Bitzer, who had not budged from Illinois since enrolling as an undergraduate in 1951, seemed to yearn for business success. Yet he struck people around him as hopelessly naive when it came to business. "He was so completely oblivious," says one colleague. Bitzer's optimism, car-salesman's confidence, and undeniable technical brilliance could charm any doubter, but the fact remained that PLATO IV never grew to anywhere near four thousand terminals. But even *that* did not faze him: in 1975 he was cooking up a deal with the maker of the CYBER supercomputer that powered PLATO IV, a deal so immense and boldly ambitious that it would take the old four thousand number and toss it right out the window. Four thousand terminals? Child's play. That was the past. That kind of scale might have *seemed* like a lot, but not anymore. Bitzer had hitched his reputation on a new goal, a *million* terminals—maybe more!—and the future of PLATO, and its Orange Glow, looked brighter than ever.

What Bitzer knew was that CERL and the University of Illinois were deep in talks with Control Data Corporation to license rights to PLATO—even the trademark—to the company so it could market the system commercially. In fact by the time Bitzer gave that "million-terminal" presentation in September 1975 at the HumRRO conference, CDC and UI were way beyond talks. CDC had by then recruited executives to drive the PLATO initiative, who in turn built teams, labs, offices, and even had secured a PLATO system of their own quietly running in a Minneapolis suburb. UI, CDC, and a reluctant National Science Foundation were attempting to pull off one of the largest federally funded technology transfers from academia to industry ever attempted up to that time. "There was no legal precedent," says Bob Morris, a CDC executive, "or any history of such a large joint research program between a university and a corporation." Nevertheless, CERL and CDC both believed PLATO represented an enormous opportunity. They just had to convince the university and NSF. CDC was going to make PLATO the most important mission of the company, and most of the company didn't even know it yet.

"You know, I just have a terrible feeling," says Bill Cole, who worked out of CDC's office in Sunnyvale, California, regarding Don Bitzer's frequent spouting of the million-terminal vision to anyone who would listen, "that Don might have picked up that number from Control Data."

One day back during December 1972, Bill Cole got a call from Bob Morris in Minneapolis.

"I've arranged to borrow you for three months," Morris told Cole. "Is that okay with you?"

"Sure, what are we going to do?"

"Well, it's hard to explain. . . . Come on up and I'll tell you right after New Year's."

Cole flew to Minneapolis. Morris told him about a new initiative Control Data was undertaking: the company planned to begin commercially marketing and selling PLATO worldwide. "That's the first time I ever heard of PLATO," says Cole. "He either had some documentation or loaned me his, so I spent a day or night or two trying to understand the system."

Then he, Bob Morris, and Morris's boss John Dammeyer went down to CERL for a visit. "We were pleased that they took an interest," says William Golden. "We thought we were doing things with their machine that nobody else was doing. We were taking every last advantage of what the machine could do and they agreed."

The usual dog-and-pony show ensued, the entourage of Bitzer and the Control Data VIPs trotting up and down the various corridors of CERL, stopping in here and there for visits and impromptu demos.

Says Cole, "I'm not sure to this day whether it was a Bitzer setup to take us to the attic and have David Frankel knock the socks off these three corporate suits." Bitzer, who in his spare time pursued a longtime interest in magic tricks, had in fact joined the International Brotherhood of Magicians. In a way his PLATO demos *were* his biggest magic tricks, and it was his job to knock people's socks off, or recruit an available CERL staffer to do so.

Bob Morris had no idea what to expect when he got to CERL. "I have to say I was somewhat overwhelmed by the creativeness in this team of people. And just delighted at the prospect of working with people like this. The lab itself physically was kind of a frumpy place. Well, it was worse than frumpy. Don Bitzer's office was in an old radar tower, as was Frank Propst's. The place was not elegant, but on the other hand the work that was done in there was just unbelievable."

It was the first of countless meetings and visits—CDC down to CERL, and CERL up to CDC. At one point David Frankel, the PLATO wunderkind, was asked to fly—the teenage Frankel had a pilot's license—Cole and another CDC engineering manager up to Chicago so they could attend the Consumer Electronics Show, with an emergency stop at an airfield in Rockford due to the engineering manager's airsickness in the single-engine plane.

Robert M. Price, Norbert Berg, William C. Norris, 1982

William C. Norris, CDC's maverick founder and chief executive, would receive Bitzer and other contingents from CERL into his office—he apparently hated traveling, particularly by airplane—and would never, in the long history of the Control Data/CERL relationship, visit CERL, tour the lab, or set foot on the campus of the University of Illinois.

But he was sold on PLATO.

His attention had been drawn to PLATO in its earliest CSL days, when Alpert's lab placed an order for the first 1604 computer, resulting in the fracas with IBM in the Illinois statehouse over competing bids, with CDC ultimately prevailing. Harold Brooks, the CDC sales rep for the Illinois territory, had sealed that deal, and then found a way to get CSL a second 1604, a used machine costing all of one dollar, that Bitzer could dedicate to PLATO, freeing up the other machine for CSL's other pressing projects.

In the mid-1960s Norris created Control Data Institutes, a network of job-skills training centers around the country, partly motivated by CDC's own need to hire more qualified workers. Entering the training industry gave him a new perspective on the importance of education in general, a perspective that would continue to grow as he watched Bitzer build up PLATO III and begin work on PLATO IV.

Control Data had emerged in the 1960s as the preeminent manufacturer of supercomputers—for a while the 1604 was the world's fastest. Then, with the help of the company's genius engineer Seymour Cray, it launched the CYBER 6400, 6500, and 6600 line of supercomputers through the rest of that decade. Bitzer decided one of these would be the perfect machine to power his 4,096-terminal PLATO IV. By 1969, CDC had reached $1 billion in revenue for the first time. Its stock soared. The annual revenue would continue to climb all during the 1970s. The company expanded, adding tens of thousands of employees. NSF funded PLATO IV, CERL bought a CYBER 6500 supercomputer, and the future looked bright. It was time to take PLATO to the world, something CERL wanted to do but was neither chartered nor equipped to do. Norris and CDC were ready and willing.

One day in late 1969 or 1970, a group of Ohio State University faculty members flew out to Illinois to visit CERL. In terms of computer-aided instruction research, Ohio State was a big IBM user: IBM 360 mainframe for the hardware and IBM's Coursewriter tools to write the courseware. Michael Allen headed up R&D for Ohio State's CAI lab and consulted on the side with IBM to help them develop their educational systems. In addition, he says, "I was teaching faculty members the principles of teaching. . . . The provost set up a program and said it was voluntary but it was like [a] real course, and you could only enroll in it if you were a member of the teaching faculty at the university, and I was their first professor that taught professors, so that was really fun."

When Allen and the faculty members arrived at CERL, they got the usual dog-and-pony-show demo, including the Fruit Fly and Titrate lessons, as well as having Jack Stifle demo a prototype plasma panel, wires attached all over and looking like a miniature Frankenstein experiment. The engineering was impressive, but perhaps too impressive. "We all looked at what they were doing," says Allen, "and we thought, it's too bad that their engineering community isn't coupled more tightly with an education community."

Many CAI researchers around the country shared this sentiment. It was acknowledged that PLATO was interesting, but why so centered on engineering and not on education? "There's some stuff that educators could use," says Allen, "but . . . they're really sidestepping all of what we considered were much more difficult, critical issues which had to deal with how people learn." Allen would admit that the IBM systems at Ohio were not as sophisticated or as capable as what CERL was building, and the PLATO IV system would make them even less so, but still, he thought that despite IBM's weaknesses, the Ohio researchers were able to support more effective and varied educational paradigms. "We saw them as a mirror image of us, they were just backwards, from where we were. We kind of rolled our eyes, thinking that gee, the people who were engineering these systems ought to be responding to educational needs, rather than trying to tell educators what they need."

Then there was the issue of money. Like politicians, academic researchers need funding to survive, and given the costs of computer hardware, software, and staff to run it all, CAI in the late 1960s and early 1970s needed a lot of money. Ohio State had a strong relationship with IBM, but it still had to look for outside funding, and like many labs around the country, they would run increasingly into a brick wall when they approached the federal government for support. "When we would write for grant money to the National Science Foundation, we were always turned down because they said, 'Well, we're spending all our money on PLATO and TICCIT.' And we said, 'Those things that you're spending money on are primarily engineering-based projects, and we think that you should be funding some more serious educational research projects.' To which they'd say, 'Well, those are serious educational projects.' And we could never win that argument."

NSF had simply bet on PLATO and TICCIT largely to the exclusion of all else. The degree to which they were PLATO and TICCIT believers boggled minds at competitive labs like Allen's. "I remember a conversation I had with someone at NSF," Allen says. "I was saying something about, well, they haven't even evolved their system to the point where we could test a particular educational issue. And his response back to me: 'They have because I have a terminal sitting right here at my desk.' *Oh*, no wonder we can't get you to think about anything else, you're sitting there using PLATO, and you guys have been wrapped around their little finger." He was happy NSF was pouring money into CAI, but he, and many other CAI researchers around the country, felt NSF was being nearsighted devoting so much of it to just PLATO and TICCIT. "They were kind of opposite ends and there was no funding of the middle ground," Allen says.

Then, as PLATO IV emerged, Allen's objections began to fade. He began to see what the fuss was about, including the impact that the new wave of creative young people had had on PLATO, transforming it in a short span of time from being a development-and-delivery platform for interactive educational lessons to being that *plus* a platform for rich and varied communication and collaboration. "Education is at its heart, communication," says Allen, "and the strength of communication that PLATO had was so strong that not only could you develop good courseware, but you could also have this interpersonal communication system and you could do beyond that even and have real-time games. Its ceiling was so high in the level of communication that it provided a good basis for the development of [a] big variety of applications. You would never develop those kinds of things on TICCIT, for example. TICCIT had again these blinders on: 'We've discovered the truth and the end of educational design, and now all we have to do is incarnate it in silicon and we've got it.' It was just incredible to me." Allen, like many others, including even some PLATO people, remained impressed with the video capability built into TICCIT, but the rest of what PLATO offered towered over the best of TICCIT.

At one of the many computer conferences in the early 1970s, Allen and Bitzer appeared on a panel together. Allen went first, and told the audience, "I think it's important for the research community out there to appreciate what Don Bitzer has done." Allen began to see Bitzer's view as not telling educators what they need, but, rather, building as open and flexible a system as possible, so that educators could do whatever they wanted. "Instead of waiting for the laborious research on education to define exactly what systems requirements there are, and then try to build those, they've gone ahead and tried to build almost everything imaginable that educators might want, and then said, 'Here it is, here's your laboratory! You can do many things unimaginable before, try them out and see which ones are really important.'"

After a while, Allen changed his mind about PLATO altogether. Perhaps having seen with his own eyes what the extreme alternative—

TICCIT—was actually turning into, and what limitations instructional designers were going to face trying to create courseware on that platform made him come over to the PLATO camp. "I started to be very, very appreciative of what Don was doing."

By 1973, Control Data was looking everywhere for an educator they could bring on board to help with a commercial rollout of PLATO—someone from academia with serious CAI credentials. They wanted Michael Allen. Michael Allen wasn't sure he wanted Control Data.

"They went after me," Allen says. "They had done a nationwide search for someone to come in and my name kept coming up."

Allen went up to Minneapolis for interviews. He told them he wasn't sure he was a good choice because he'd been outspoken against PLATO. To his surprise, that was exactly what CDC wanted to hear. "The more I stated that, that I'd been critical of it, the more people at Control Data said, 'You're exactly the person we need. We hear from all of these educators at the University of Illinois who have been involved in the system and you know this is just the ultimate in computer-based instructional systems. We want to make a business of this and we need someone who is willing to ask tough questions and yet has educational credentials so you're not just slinging mud or responding in an unknowledgeable way. You know what's going on. We really, really, need you.'"

CDC had hired the well-respected executive John Dammeyer away from IBM in the late 1960s and put him in charge of Control Data Europe. But then around 1972 Bill Norris handpicked him to lead the new effort to commercialize PLATO, and brought him to Minneapolis to form a team and get going. Dammeyer was keen on hiring Allen. "I was surprised at how receptive he was to me," says Allen. "I actually went to Control Data on a sabbatical. I never told anybody at Control Data I was here on a sabbatical. Because I intended to just learn what goes on in corporate life, and see how they were thinking about using computers in training, and then I intended to go back, because I was really happy [at Ohio State]. And Dammeyer basically said, 'What do you need?'"

Allen had been a mere graduate student a year and a half earlier—the ink was not yet dry on his PhD diploma. Now, Dammeyer offered him not just an office with a private bathroom, but an entire empty building. "Now, you got to staff up," Allen recalls Dammeyer saying.

"Who do you want? What do you need for computer equipment and stuff?"

"Basically, everything I had to scratch for and scream for at the university," Allen says, "Dammeyer just called a truck and delivered it, I mean, everything I wanted, he gave me. And so, by the end of my first year, I was thinking, 'Whew, this is a pretty cool place,' so I ended up staying on."

CDC knew that the PLATO system would need enhancements and new capabilities so it didn't simply reflect the needs and interests of people at the University of Illinois, but rather related to CDC's clients in the future. "That was kind of my job," Allen recalls.

But there was something else: an area in which he felt PLATO was weak was at the same time one he felt was crucial to CAI's success. "When I came in, I found out there was no activity going on with computer-managed instruction. All around the country we had seen that if you want to individualize instruction, or introduce computer-assisted instruction, you really needed the backbone of a management system in there. That was one of the things we started."

John Dammeyer had already recruited Bob Morris and Bill Cole to help with the overall planning to determine the feasibility and costs of launching a major PLATO initiative, as well as researching the market potential, marketing strategy, product strategy, and overall direction the PLATO project would take to maximize success for the company. Morris and Dammeyer flew out to San Diego for an off-site brainstorming session that lasted several weeks. More trips were made to CERL, and more CERL personnel traveled up to Minneapolis to confer with colleagues at CDC. In April 1974, Dammeyer and Morris presented their comprehensive business plan for PLATO to William Norris and his executive committee. "The program was large in scope," says Morris, and for the first time in Control Data's history required the concurrent and integrated development of a complete system (e.g., hardware, software, terminals, communication networks, applications, languages, and educational content), and the business strategies for its entry into the market. Norris gave the project the green light.

Over the following months, Morris and Dammeyer picked a formidable team of bright, experienced people to begin the work of trans-

John Dammeyer

forming a university-developed computer system into a commercially viable product. By 1974, they had gotten PLATO running on a part-time basis on a borrowed mainframe within the company, and a growing number of staffers had terminals. Bill Cole recalls a fad that emerged within CDC at its Arden Hills, Minnesota, facility, where much of the PLATO project was staffed. On one of his visits to Minnesota, he noticed that CDC executives and senior managers promptly turned into gamers at 5:15 p.m. each night. "Somebody said," he says, "Okay everybody, Airfight time!" There were "real addicts," says Cole. "I went and watched what they were doing, and what was happening was that over in the Tower [Control Data's headquarters building in Bloomington, where, for exercise, William Norris walked fifteen flights up the stairs each day to his office], a whole bunch of people, *including* quite senior vice presidents, were hopping on Airfight. And they were shooting each other down."

One of the people brought into the PLATO project was Jock Hill, a proud, bald-headed Scotsman with a magnificent Scottish brogue and the baritone of Garrison Keillor, the kind of voice you imagine leading men into battle. Born in Dundee, he grew up in Inverness, and after graduating college pursued a career in electronics and applied mathematics, alternating jobs at Rolls-Royce and IBM across the 1950s and 1960s. He wound up in Brussels, Belgium, in the early 1970s working for Systems International, a Rolls-Royce subsidiary, where he crossed paths with John Dammeyer, who had had a stellar career at IBM (including a major feature article on business management in *Life* magazine in 1967). When William Norris appointed Dammeyer to begin a small task force to evaluate the PLATO opportunity, Dammeyer brought in all the best people he knew: Jock Hill, John Cundiff, Bob Morris (originally a high school dropout who "attended the college of IBM" for a decade, including working for Dammeyer), and an engineering manager named Bob Moe who had joined CDC in 1960 and spent most of the 1960s working for Seymour Cray on the 6600 and 7600 supercomputers at Cray's fabled Chippewa Falls, Wisconsin, laboratory. A year after Hill joined Systems International, the company went bankrupt, and everyone, including Dammeyer, Cundiff, Moe, and Hill, was out of a job. CDC snapped up Dammeyer, who gave the rest of them an ultimatum, says Hill: "He'd cut us off dead, or we'd join him on his staff, so we joined Control Data together."

It is a testament to Norris that he chose someone as capable as Dammeyer for the PLATO job. Great people want to work with, and wind up attracting, other great people, and Dammeyer had his share of great people to pick from for the PLATO project: people he knew, people he trusted, people who would get the job done. "Dammeyer was a genius," says Hill. "Bob wasn't quite the genius, but Bob Morris was a scientist. And he intended to follow the guidelines and produce the most inspired leadership that I've ever been subjected to. Just incredible type of leadership. It was an exciting time to be around there." It was also fortuitous that CDC's fledgling PLATO team equally respected their counterparts at CERL: Don Bitzer, Jack Stifle (a hardware engineering genius whom Hill would describe as being able to walk on water), Dominic Skaperdas (another hardware wizard), and the rest of the hardware and systems software teams. Teams of good people all around. To an eternal optimist like Don Bitzer, his belief in a future of one million terminals and 250 PLATO mainframes seemed not only attainable, but already a done deal. What could possibly go wrong?

One piece of technology CDC had to evaluate in the business plan for PLATO was the terminal. They had discovered that the most expensive component in the overall PLATO equation—the thing that was keeping the Holy Grail of costs, cost-per-student-contact-hour, high—was the PLATO IV terminal, the Rube Goldberg box made from over three thousand parts. In particular, CDC blamed the flat panel gas plasma display. As friendly and familiar as its Orange Glow might be, it was simply too expensive. Morris wanted to know if a cheaper terminal could be built, so he tasked people to find out.

Around this time, Dammeyer and Morris learned that TICCIT was for sale. "MITRE Corporation came to us," says Morris, "suggesting that TICCIT was a superior technology to PLATO, and that we ought to consider putting our funding and our energies into TICCIT." Dozens of companies were interested to one degree or another. Of particular interest to William Norris was TICCIT's color television graphics and video support—features PLATO continued to sorely lack. However, in the end, Norris decided that to acquire the TICCIT project might come back to haunt them, in that the acquisition might be deemed anticompetitive—simply a means of eliminating a competitor. CDC passed. TICCIT wound up being sold to the government contractor Hazeltine Corporation.

Times had changed since Bitzer, Slottow, and Willson had invented the plasma panel a decade earlier. Moore's Law had indeed come true, and was on a tear all during the 1970s: the cost of integrated circuits was plummeting each year, while at the same time the capacity and power of the circuits were doubling. For instance, video RAM. Gone were the days of memory costing $2 per bit. PLATO's 512 x 512 display, which, if driven from video RAM with a CRT instead of running on a gas plasma panel, no longer would cost hundreds of thousands of dollars—per terminal—for the 262,144 bits of memory. Still, CDC now had access to this amazing plasma technology, and directed some engineers to study it and figure out what kind of business could be made of it. Patents resulted from their research, but in the end CDC decided not to pursue the plasma display market, nor use a plasma display in its own PLATO terminal. That meant they would need to do something else, and a CRT with video RAM seemed the best bet. Could it be done cheaply?

Bob Morris

"One of the disappointments to me," says Morris, "was inside of Control Data my PLATO team was not allowed to develop the new terminal." CDC had a division, the Terminal Systems Division (TSD), located in Roseville, Minnesota, devoted to building terminals for its customers. As is often the case when separate divisions of a company disagree with each other turf wars ensue. Morris's team was allowed to do the design, to deliver a set of requirements, but TSD insisted on building the actual product. "So we got the team together," says Morris, "we got the people from the Terminal Systems Division together. And I presented them with the requirements, and we used Jack Stifle to do functional design." Morris told TSD he wanted the terminal in nine months from initial sketches to finished prototype. TSD said that was impossible, that it normally took them two and a half years. Morris was adamant: nine months.

"We came to an impasse," says Morris. "By this time Jock had joined me, and Jock Hill had responsibility for the PLATO network and terminals, and we literally locked these people up for two weeks, every day, and we took, one at a time, every one of their objections about why it couldn't be done in nine months, and at the end of two weeks they agreed they could do it in nine months."

"The thing was," says Hill, "the Terminal Division had been bought from an outside company who made the big terminals that were used for flight control, and they were used to two- and three-year projects. They didn't have the mechanism for decision, and Bob Morris said, 'If you get decisions quickly within twenty-four hours, how fast could you develop it?' And they said, 'Development is not the problem, it's the decision making. We come up with problems, we need solutions, and we need fast decisions.' Bob Morris said, 'Okay, you got him, here he is.' And I was thrown in there."

"When the nine-month period came along," says Morris, "I was on one of my trips to Moscow, and I was due back three days before the end of the nine months, and when I got back to my office, the prototype was sitting on my desk, working."

"We had the biggest party at the end of that nine months that you would believe, right?" says Hill. "Up in Roseville we had shrimp until nobody could eat anymore."

If there is one constant in the eternal march of technological progress, it is change. The terminal Jock Hill produced epitomized the tumultuous change that the PLATO system would undergo in the coming years.

The most dramatic change: gone was the Friendly Orange Glow.

Instead of orange pixels, they were grayish white. The new terminal, called the IST (short for Information Systems Terminal), looked more like an early personal computer. A big, wide, heavy base, with a black grille in front, to which a detached keyboard was connected via a thick cable. On top of the base was a monitor, a special elongated CRT with a square display featuring exactly 512 x 512 black-and-white pixels and, mounted directly over the surface of the CRT's glass, a reflective, acrylic touch screen with barely visible gold wires crisscrossing across the display. During the nine months of development, the price of CMOS (complementary metal-oxide semiconductor) memory chips had plummeted even further than Hill had anticipated. "According to the really long-haired predictions," says Hill, "it was going to come down, by six or eight to one, and it came down about ten to one, right when we were doing our development. The result was that we could produce a memory-mapped video terminal, which as far as I know had never been done before, because it was cost-prohibitive. The technology wasn't complex; it was cost-prohibitive. And suddenly, it no longer was. And we were very happy to be *just there* at the right time."

The IST terminal

The terminal that Jock Hill designed looked like a terminal to people in 1975. The fact that today we would we look at it and see a PC is not surprising, because it was a personal computer. A *stealthy* personal computer. "We produced what in effect was a PC," says Hill, "in 1975." When one considers the year this machine was developed, and compare it to what else was available at that time, it is suddenly apparent that CDC had just leapfrogged over the entire microcomputer field. Here is Hill describing his machine: "[It had an] 8080 microprocessor, it had plugin cards, it had a separate monitor, with a cable going to the main box, it had a separate keyboard, it had plugin modems, plugin memory, plugin communications, and we even had a plugin disk driver, that wasn't part of the standard stuff, but we had it networked, so it was revolutionary. And our big problem was producing it at low cost. And we did that. That terminal came in with something like a $1,300 cost, in the first few terminals. And that was beyond everybody's belief."

By the time the IST was ready to be sold to consumers, the marketing people had marked up the price to over $8,000, says Hill. It was the beginning of a long line of very bad decisions at CDC. Hill believed the terminal should have been sold for $100 above cost. "If we'd done that, we would have flooded the market because people knew they could use it for other things. It would take loadable programs—we could load programs down from the mainframe into that terminal."

Jock Hill's technical skill was to Dammeyer and Morris's PLATO team what the "Q" character is to James Bond and MI6. Such was Morris's confidence in Hill that he gave him a little assignment on the side: develop a portable PLATO terminal using a plasma display. Hill prepared some designs, and delivered several fully usable prototypes that were toted around the world for the next several years for demos. They were bulky by today's standards, but compared to the early laptops coming out in the early 1980s—for example the suitcase-sized Osborne-1 computer—they were right out of a Bond film. Open up the case, and there is a full-sized 512 x 512 orange-glow plasma display, along with keyboard and even an acoustic coupler, phone headset, and modem, for dialing in to PLATO from anywhere in the world. Only a small number were built, and they never became commercial products.

While CDC was preparing the IST, Jack Stifle and his hardware team at CERL were building a new terminal as well. The difference between what CERL came up with and what CDC came up with is striking: CERL's was called the PLATO V. Gone was the compressed-air-driven microfiche slide projector and the internal mirrors to reflect the image back through the plasma display. With that out of the way, the terminal did not need to be as big, so it was half the depth of the PLATO IV. The sides were smooth, blond wood. The terminal had a Z-80 microprocessor in it, which took on some of the burden of displaying graphics on the screen. In a way it looked like a beautiful piece of furniture compared to the metallic, corporate, Sherman tank that was the IST.

One aspect of the Dammeyer-Morris PLATO business plan examined potential global market entry points. It was determined that CDC should go after the Soviet Union and several oil-rich countries around the world: places like Iran and Venezuela—places that pulled in considerable oil money, but lacked the educational infrastructure found in the U.S. CDC believed such governments would be willing to invest considerable amounts to improve education for their citizens.

Bitzer had already begun demonstrating PLATO internationally, including at a CAI conference in Bari, Italy, in 1972. In late 1973, a group including Bitzer, Tenczar, Jack Stifle, and Peter Maggs (Bruce Maggs's law-professor father, who spoke fluent Russian) took a handful of PLATO IV terminals—with the keyboards modified to use Cyrillic keycaps—to the Soviet Union, connected them up over a seemingly impossible patchwork of phone lines all the way back to Champaign-Urbana, and demoed PLATO at an educational conference at the exhibition center at Sokolniki Park in Moscow. It marked the beginning of a set of major trips around the world throughout the 1970s to demonstrate PLATO in other countries, and had the blessing and support of Control Data, which had already begun attempting to open trade with the Soviet Union in the middle of the Cold War.

Often the dial-up connection would fail. At one point, in what might be an apocryphal story, someone on the CERL team tested the phone lines and heard breathing on the lines. (The KGB was allegedly listening.) "We think they were recording everything that was going on," says Jack Stifle. "Once in a while there'd be clicks and there'd be noise, and things would go bonkers for a second or two, and we'd jokingly say 'Well, they must be changing tapes on the machine.'"

Bitzer has a reputation as a teetotaler, which made life in Moscow a bit difficult, as the CERL group discovered that Russians like to drink vodka. Seemingly all the time. Bitzer wasn't the only CERL staffer not interested in vodka, so elaborate steps were taken to surreptitiously replace the vodka in their glasses with water during the incessant toasting and clinking of glasses by the various Russian dignitaries welcoming the Americans.

Paul Tenczar recalls that the entire group from CERL had to be briefed by a U.S. government agent—one of the three-letter agencies, he couldn't remember which—before traveling to Moscow. "Somebody came through and talked to us all a little bit," he says. "We were to be fully open or whatever, but the one thing that we were not to mention *absolutely*—I mean under *dire* coming-back-and-thrown-in-the-slammer or something—to mention that you could monitor terminals, that a PLATO terminal could monitor another PLATO terminal. All of us were instructed by some agent." After extensive setup at the conference hall, the event finally opened, and attendees poured into the exhibition hall. "Within the first hour of the show being open," Tenczar says, "some guy came up to me, and he spoke pretty good English, starts talking to me, and the first thing he asks me was, 'Well, can one terminal monitor another terminal?'"

Later during the conference, Tenczar was approached again at the exhibit booth by someone who seemed to be recruiting him on the spot. "Some young fellow comes up to me and says . . . scientists are treated royally in Russia and, you know, if I would think of living there they could really arrange nice stuff, you know, women, and get me to academic centers and stuff like that. They thought I was somebody or other, and I said, 'Well, I kind of like Chicago' and of course he's never lived in Chicago and I said, 'Well I kind of like the skyscrapers there,' and he kind of smiled and walked off."

During the CERL group's time at the conference, which happened in late November 1973, the CERL staffers would occasionally check pnotes and notesfiles to see what was going on back home. One thing that was going on was the discovery of the eighteen-and-a-half-minute gap in the Watergate tapes, announced in a Senate committee on November 21. News from the West was blacked out in the Soviet Union, but PLATO managed to get the news past the Iron Curtain.

Other trips would follow. In 1974, a larger exhibition of PLATO was brought back to Moscow, this time led by a contingent of CDC people, including Bob Morris, Michael Allen, and Jock Hill. Larger

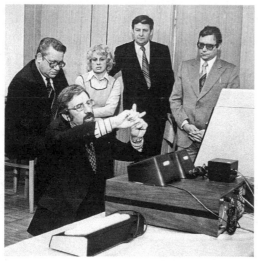

Mort Frishberg demonstrating PLATO
to Boris Yeltsin, Moscow, 1974

meetings were held with more Soviet counterparts, along with a stream of demonstrations of PLATO, including special lessons that had been developed in the Russian language (complete with a special Cyrillic character set). Among the dignitaries that stopped by for a demonstration was Boris Yeltsin, his hair not yet fully white.

One day during this time, Patrick Stubbs, a CDC employee back in Minneapolis, was working at his PLATO terminal when he suddenly got the familiar beeping call and message flashing at the very bottom of his screen saying someone wanted to TERM-talk him. It was Mort Frishberg. "I'm thinking, where's Mort, I didn't know where he was, I had no idea," says Stubbs. The TERM-talk went something like this:

> FRISHBERG: Hi Pat how are you doing?
> STUBBS: fine
> FRISHBERG: I'm doing a demo
> STUBBS: oh good
> FRISHBERG: What did you have for lunch today?

"I'm going, why the—you know—it took me by surprise, not the kind of question you usually get," says Stubbs.

> STUBBS: Uh, I had a ham sandwich and an apple
> FRISHBERG: OK good thanks goodbye

"I'm like, what?" asks Stubbs. "What was that all about? The next time I saw him I said, 'What was going on?' 'When I was in Moscow I was doing a PLATO demonstration, and they wanted to make sure that

Peter Maggs, Michael Allen, and Jock Hill, Moscow, 1974

you weren't a canned program on the other side. They said, 'Ask him what he had for lunch,' so I said, 'Ham sandwich and an apple,' they said, 'Oh okay.' "

CDC was determined to open up business in Russia. The U.S. government was not so keen, pointing out that to deliver a fully outfitted PLATO system to Russia meant delivering a CYBER supercomputer, disk drives, peripheral processing units, printers, terminals, telecommunications equipment, and of course a massive amount of proprietary software. CYBERs were used by the U.S. government in any number of agencies, many of which were secretive not only about their using the machines, but even mentioning the agencies' existence. Nevertheless, CDC persisted, continuing talks and PLATO demos for several more years.

Another of CDC's major foreign trade hopes was establishing PLATO in Iran. The company managed to open a Control Data Institute and some data processing business after receiving permission from the Shah's government. In May 1975, a group from CERL including Bruce Sherwood, Jack Stifle, and Peter Maggs, along with a CDC team that included John Dammeyer, brought PLATO terminals to Tehran for a series of demos. Numerous government officials including various top-level ministers were brought in and shown the system. The Iranians hosted the Americans as guests at a fancy dinner in a magnificently decorated ballroom, and in the center of each elaborate place-setting on the tables were Coca-Cola bottles placed in their honor. (At least, they looked like bottles of Coca-Cola. To their horror they discovered that the substance inside was largely sickly sweet, thick Coke syrup, sans the carbonation.) Efforts were made to demo PLATO to the Shah himself, but allegedly due to security concerns he never showed up. His son, Reza Pahlavi, a fourteen-year-old at the time, was also invited to receive a demo. The team was kept in the dark again at the request of security, until one day they were instructed to set up at a distant location. Reza and his entourage did finally arrive, surrounded by armed guards. He loved PLATO, especially the games. Back at the palace, he raved about the PLATO system, including to a family friend who happened to be visiting: the recently deposed King Constantine of Greece, aged thirty-three. Reza's praise of PLATO made such an impression on the king, he inquired if he too could get a demo of this wondrous American computer named after the famous Greek scholar. Constantine got his demo.

CERL and CDC created Persian-language support in PLATO as part of the demos, and eventually the Shah's government agreed to a deal. However, it required that the IST terminals had to be made in Iran (or at least have a decal with "Control Data of Iran" and Persian script on it affixed to the screen bezel). In the end, the Ayatollah Khomeni and the Iranian revolution ended CDC's hopes in that country. Several of the government ministers, including Prime Minister Amir-Abbas Hoveyda, who had attended the demos back in 1975, were executed. CDC personnel had to evacuate the country, and the company lost a lot of money. A number of the Iranian ISTs wound up at other PLATO sites, including the University of Delaware, where, amid campus rallies outside during the American hostage crisis, students inside pried off the "Control Data of Iran" metal labels in protest.

Iranian IST terminal

CDC marketed PLATO in Venezuela, another oil-rich country that Dammeyer and Morris had identified in their original plan. Happily for CDC, at one point a minister-level Venezuelan government official became very interested. "He heard about PLATO," says Bob Morris, "got all excited about it, wanted to come to the U.S. and visit us at Control Data and also visit CERL. Since this guy was at a ministerial level we felt we had to treat him right and all that stuff. He was a very hot prospect." The government minister was considering buying two PLATO systems and quite a large number of PLATO terminals, says Morris. "I went out and rented a Learjet, and he visited Control Data and he went through an executive symposium . . . and then we put him into the Learjet and he and I went down to the University of Illinois, to CERL, visited Don Bitzer and his team, and he saw the whole magilla there, and got back in the Learjet, back to Chicago, where he caught a plane and went home to Venezuela."

The next day Morris got a call from the top CDC person in Venezuela saying the minister wanted to buy those two PLATO systems and wanted Morris to come down and visit and bring a team of people. They attempted to make hotel reservations but found everything booked. When the minister heard of their difficulty finding lodging, he booked the presidential suite at the swanky Tamanaco Hotel. Morris describes it as an "obscene suite, with five bedrooms and a huge terrace and kitchens and living rooms and dining rooms and waiters all over the place and so on." Through CDC's contacts, they scheduled meetings with the minister. "We would show up," says Morris, "and he would not be there. This went on for a week. We were quite naive. Didn't know why this was happening. Finally, we found out through our contacts that what he was looking for was some sort of a payoff or a bribe to his relatives, and then based on that he would buy the systems off of government money. Control Data had a very strict policy on that, and we wouldn't do that. And so we wound up meeting with him for a couple brief meetings while we were down there, we spent two weeks down there, and came back with nothing. Well, in the meantime, this guy had gone public that he was going to buy these PLATO systems. He had exposed himself to the world, and an election was coming up, and as it turns out he had gotten the population of the country all juiced up about bringing in all this high-technology stuff and advancing the country and so on, and if I recall he didn't get reelected and someone else got reelected. There was a great deal of pressure on this someone else to make good on the promises that he had made, and eventually, I think as much as two years later, they did buy two PLATO systems."

Mike Smith was one of the regional CDC managers for foreign countries, including Iran, where he lived and worked for CDC for four years in Tehran until the revolution ended CDC's presence there, and then Venezuela, which he describes as "a can of worms."

"Venezuela was more corrupt than Iran, if that was possible. . . . In South America, the Venezuelans were known as the 'Iranians of South America' and not just for their oil reserves. You could get anything you wanted in Caracas—anything. Like many CDC international offices, CDCVEN [the acronym CDC used for its Venezuelan business] had its own guy specializing in local bribery and ours was good." This was CDC's fixer for Venezuela, "used for more local practical bribery associated with licenses, permits, getting employees and families out of scrapes, etc."

Smith describes two versions of the CDC/PLATO/Venezuela story: one version the company used for public consumption, and one was his own theory. The public one, he says, involved the Communist Party in Venezuela, which "became convinced it made them considerable political hay if they opposed the 'brainwashing-Wall-Street-running-dogs' PLATO system which would be certain to destroy the Venezuelan culture and rot the minds of the future hope of the country," he says. His own theory is that there was a certain Venezuelan individual who "could not tear off a piece of the action big enough to satisfy his greed," says Smith. "He crouched in the middle of a nasty web of influence, bribery, backstabbing, manipulation, greed, and certain malfeasance." Smith believes this individual colluded with a certain city manager to extract money from CDC to line their own pockets. "My short version," Smith once explained in an email, "is the PLATO buy became entangled in Venezuelan politics and did not survive the massive political infighting and jockeying for a bite out of it for all concerned (including two or more of our own guys). I do not believe we lost it because we did not bribe. True there was a corporate public effort to clean up our act (I have seen CDC bribe all over the world—even in places like Germany, supposed to be un-bribable) but HQ never backed off of doing business along those lines (anyway it was very difficult to stop the local CDC folks from making deals HQ did not know about). In a lot of countries it was the only way to do business. When the U.S. government started with pressure on U.S. companies to not bribe they started our downfall in the business world. . . . Anyway CDC had the right guys in CDCVEN to handle it. . . . For some reason PLATO never got its due anywhere."

He claims that he reported the situation to CDC's lawyers back at headquarters, and he says they simply told him never to repeat the story again. In the end, PLATO never took off in Venezuela, but one notable, wholly unexpected thing emerged from it: perhaps the most extraordinary, advanced, and unusual MUD game ever created on PLATO, developed by a CDC employee named Mark Johnson while he was stationed there: Drygulch.

Johnson, like Smith, had been stationed in Iran for several years, doing pre- and post-sales support for CDC's emerging data processing services—not PLATO related at all—including atomic energy, Iran's state television and radio, and other sites in the oil and gas industries. Everything was fine until the day arrived when he and his family had to evacuate, leaving all of their possessions behind and living for the next few months out of suitcases, first in London and then back in the U.S. Eventually CDC told him that Iran was essentially over, and they moved him and his family up to a hotel in Minneapolis. He got in touch with his boss Mike Smith, and learned that Smith was now in Venezuela. Smith hired him, and the Johnson family packed up and moved to Caracas. There, he found a PLATO terminal in a small makeshift data center CDC had started, its hopes high that it would expand into a major facility with multiple CYBERs supporting myriad PLATO terminals, not to mention offering other services to the oil and other industries.

Johnson's official job was to support the fledgling data center the company had set up for clients. There were no CYBERs or big hardware there, only terminals, printers, and a plotter. CDC had a single high-speed T-1 communications line back to Minneapolis. Occasionally the few CDC salesmen would land a deal with a company that wanted to run some program off of one of the CYBERs in the U.S., so they would come in and work from a terminal. But life was quiet, and Johnson settled in at the PLATO terminal with not much to do. He would bring his children in to play around with the educational games on PLATO, including many of the elementary math games, which they loved. "Drygulch" was one of the towns the little choo-choo train stops at in Bonnie Seiler's How the West Was One elementary math game, and Johnson got the idea for his MUD from there and from another game centered around horse races. He took the dungeons and dragons motif and turned it on its head. What if, instead of orcs, goblins, wizards, swords, and sorcery, he placed a MUD in a Wild West setting, where the "dungeon" became a gold mine, on the outskirts of a Wild West town? What if, instead of the various primitive stores in the "city" sections of games like Moria and Avatar, he let players visit various stores in the town? Johnson created a town hall, bank, general store, assay office (where you took the ore you'd dug up in the mine, to get weighed and converted into money, or, if you were up for it, you could shoot the proprietor), Kitty's Saloon ("known throughout the West"), a sheriff's office and jail, a stable, a hotel, and Boot Hill, the local cemetery. Visitors could enter and interact in any of those establishments; each had their own contextual menus of possible options, some of which helped your game character, and some of which might end you up in the jail—or Boot Hill. Johnson even went further, creating a local town government with a mayor, sheriff, and inspector, all roles that players could run for in elections ("voting takes place during odd-numbered years," the game explained). Beyond the usual Dnd player characteristics like strength and dexterity, he added injuries, hunger, thirst, carrying (how many pounds of items and minerals you could carry), and a separate set of vitality, hunger, thirst, and carrying values for the player's mule, which you could buy at the Smith & Sons Livery Stable for $1,000 of game money. Johnson even integrated a PLATO notesfile into the game, =gulchnts=, where players could announce that they were running for election, or read notices—automatically posted into the notesfile by the game itself—relating to election results, who's in jail, who died, and other news and gossip. The mine itself had its own set of "monsters"—instead of dragons and the usual Dnd fare, he had rats, bats, spiders, bears, and coyotes. Players could use the "f" key to attempt to fight the creatures, "s" to "swear" at them ("Swearing, if done right, is very effective, driving the animals away at once," said the Drygulch help lesson); "i" to ignore them ("When it works, it works well, but you must not move or show fear"), "e" to evade (if your dexterity was high enough), or "r" to run in panic ("a good last resort"). The great PLATO god of chance, Randu, was everywhere in this game like in every other game. Each time you acted, Randu considered your action, and the outcome might work to your advantage—or not.

All of this Johnson did on his own while stationed in Caracas, using a terminal connected to CDC's PLATO system 4,500 miles away in Minneapolis. After two years in Venezuela, he and his family packed up and moved back to the U.S.

The deals put in place in 1974 between CERL and CDC to commercialize PLATO were held up by lawyers and the government. "There was substantial resistance," says Bob Morris, "within the university and by members of the National Science Foundation to transferring technology to a profit-making corporation. Further, there were a number of fundamental principles which had substantial influence in the negotiations; e.g., academic freedom vs. proprietary protection, public domain requirements on government-sponsored research, profit making vs. nonprofit exploitation of the technology, etc. It took until spring of 1976 to hammer out all of the issues, and conclude with a set of legal agreements. There were hundreds of pages of legal documents in the form of five separate contracts covering different aspects of the research program and technology licensing. After more than two years of lawyers and bureaucrats haggling, the finished agreements were consistent with the frame agreement worked out by John Dammeyer, Don Bitzer, and myself."

One of the major sticking points at Illinois that held up the deal for months concerned questions on copyright and royalties. If Control Data were going to roll out PLATO all over the world, perhaps in a few years, deploying hundreds of mainframes and millions of users, making many lessons CERL staffers and UI professors had developed in hundreds of subjects available to customers, how would they get paid? *Would* they get paid? One thing UI decided to do was make every PLATO author sign an author agreement, to which they would

Donald Bitzer in the CERL machine room, 1970s

receive a PLATO author card. The royalty rate infuriated many at Illinois. Stan Smith, the chemistry professor whose lessons were so good they were a highlight of many demos around the world—including the NSF presentation that may have single-handedly won funding for PLATO—was apoplectic at what he felt were extremely low payouts for authors. He wrote a long letter to Don Bitzer pleading for him to do something, pointing out that for every student contact hour of lesson usage, the ratio of what CDC would make to what the PLATO author would make was more than a hundred to one.

Dan Alpert was by this time largely out of the picture and not involved in the decision-making process, much to his regret. He blames Bitzer for the stingy royalty amounts. "He negotiated very hard on the income for the plasma display panel, and for the PLATO license, but not for courseware," says Alpert. It would be a sore point among the authoring community and many professors at Illinois for years to come.

The effect the CDC-Illinois deal had on CERL was significant. "It became like independent little companies inside the laboratory," recalls former CERL staffer Lezlie Fillman, "who were protecting their product, because of this business that they might be able to get royalties from the stuff that they develop. So the math project didn't speak to the reading project, didn't talk to PCP, didn't talk to community colleges, and there was very, very little sharing of the ideas that these groups were coming up with and I'm sure it was because all of a sudden people said, 'Well, gee, if I develop this and put my name on it as an author, then I'm going to get royalties from it.'" People stopped cooperating, she says. "And there became a lot of infighting inside the lab about who was going to get at a particular concept first. Who was going to be able to put their product out there and begin to rack up student contact hours, which is where the royalties came from. Rather than being an all-lab push it became instantly factionalized."

Perhaps one reason why CDC was not generous in its royalty payments to Illinois is that it did not consider many of the lessons any good. CERL by this time had many thousands of student contact hours' worth of lessons in subjects from algebra to zoology, but the quality of the material varied. Even Bitzer would admit during this era that at least half the lessons were of uneven quality, often ineffective and poorly designed. Others were even less kind, sometimes citing Sturgeon's Law, which states, "90 percent of everything is crap."

CDC decided that it would create a new organization devoted to designing, developing, publishing, and selling new, original PLATO courseware in addition to the CERL offerings. Bill Ridley and Bob Linsenman were two VPs involved with this new undertaking.

Peter Rizza, a Penn State PhD who managed courseware design in the new organization, says, "You have to realize that in the early days of PLATO, there were no authoring systems, or learning management systems, or design platforms to use to create any of the PLATO lessons. There was only [TUTOR] programming, and since programmers were not educators, the initial results were technically impressive but educationally weak. It took a few of us some time to influence the management team that investing in good tools would help the technology become more useful and ultimately generate more business for CDC. It was five people who initiated the first move in this direction. Under the leadership of John Dammeyer and Bob Morris, people like Bob Linsenman, Mike Allen—and I will include myself as well—started looking at ways to make PLATO more successful."

Jim Glish, who had started out using PLATO at UI for animated graphics and multimedia and eventually moved to CDC, recalls that in the early days CDC's PLATO publishing group had difficulty with the notion of what Silicon Valley would call a "product-market fit." The group, he says, "was tasked by Norris to come up with potential course subjects to build courses around. And they had no clue what they were doing, they didn't have any training background, they were *publishers*. They waited for people to come to *them* with good ideas. How were they going to come up with a sample market? They didn't really have any market research experiences because they hired some PhDs who didn't know a market from a stick in the ground anyway. What they did—and this is a true story—they actually had a board, with probably two hundred subject areas on it. And they threw darts at it to pick subjects that they were going to develop courses for. That was the scientific marketing that was done."

Ted Martz, a longtime PLATO marketing person at CDC, recalls similar strangeness coming out of the publishing organization. Martz had been tasked to evaluate the courseware that was coming out of Control Data for quality control purposes. "It was classic. What was happening was I was evaluating the courseware, or my group was, and, yes, the lessons were working, they didn't bomb out, they didn't have any fatal errors, they were doing exactly what they were supposed to do, but what was frustrating was, no one was going out and ever really checking to say, *Is there really a market for this effort that we're doing?*"

Martz goes on to say, "There were things on dairy farming, growing tomatoes. . . . Primarily it was the agricultural group that Control Data tried to introduce . . . a whole thing towards helping the farming community, and PLATO was going to be a major portion of that. They were going to teach farmers how to better manage their livestock, how to grow specialty crops, so that they could get a better profit on their land. The only problem was, of course, that the farmers that they were thinking this was going to be a boon to were the people that were hard pressed to part with $1,000 a month for an access charge into this system. And so they completely misread their target market. I mean that was just an example . . . and then they wondered years later why they couldn't get anybody to sign up."

"The publishing group, I believe, was given license to steal," says Glish. "They basically were some higher-paid people in the company, they came out of the publishing background, and they were simply taking money from internal sources, to line the pockets of their division, and make a paper profit, regardless of whether the product they were producing were salable or not. . . . The whole emphasis on trying to make money with PLATO early on begged the issue of having to build the marketplace. You don't build a marketplace by throwing darts at a board and just seeing what's going to fly. Got to do a *little* bit more talking to customers and analyzing things. Now, eventually they ended up coming up with some pretty good markets like the robotics and the automotive skills training and so on, but that came out of direct customer feedback, rather than this early stuff, which was just hit-or-miss game playing."

At some point in early 1976, the Soviet Union, still interested in PLATO, sent a delegation to CDC's Arden Hills, Minnesota, PLATO facility for another round of meetings and demos. "They had been speaking with CDC," says Greg Warren, hired by Dammeyer in 1974 to join his PLATO team, "about using the system to help educate their folks at remote areas. Other than the communications challenges, they were really excited about using it and wanted to see it."

Warren describes what happened next:

The day approached where they were to be at CDC HQ for a full day of discussions, then to Arden Hills the next afternoon to see the computer system/room and for demos, etc. I recall the State Department was joining them for the visit and this was really a big deal for everyone. (Lots of special security and spooky-looking guys walking around Arden Hills—didn't want the Russians stealing technology was the word in the hallway.) Being a system controller, it was my responsibility to make sure the PLATO system was up by 7 a.m. every day, so I was in the office by 6:30. . . . About 7:30 the morning the Russians were due at CDC HQ, I got a call from John's son Eric [Dammeyer] saying that his father had collapsed getting ready for work and had been taken to the hospital—had no other information. I sat there stunned for a couple of minutes, then finally woke up and ran up the hall to Tom Moore's office and told him what had happened. The CDC delegation all stared in disbelief, and Tom asked them to consider this preliminary information and to keep preparing for the day as normal. Tom asked me to stay, and he called Bill Norris's office and asked if they had heard anything, but this was news to them. While Tom was on the phone with them, Eric and Maytie Lou [John Dammeyer's wife] called on another line and were providing information.

As for the meeting, Warren says, the "demo went on as scheduled and things ultimately worked out okay—the Russians were impressed." But as for John Dammeyer, he had suffered an aneurysm. "They got him to ER quickly enough," says Warren, "and he made it through surgery. His speech was a bit slurred and drawn-out for a few months, but ultimately he was doing quite well and pretty much a full recovery."

However, Dammeyer was out of the office for months, at a crucial time for CDC and particularly for PLATO. Tom Moore stepped in to run the PLATO project. Warren recalls that Maytie Lou believed stress "was literally killing him [Dammeyer] and that he should probably consider a career change. His doctors agreed." Dammeyer met with Bill Norris and Bob Price, another top executive, and everyone agreed that Dammeyer's reign over PLATO was over. He took on less stressful work, but eventually quit, packed up, and moved to Arizona.

"John, without question," says Greg Warren, "was the smartest man I have ever met. That is saying something. Not just 'smart' smart, but business and logic smart. The kind of guy who could join just about any conversation about a problem, listen intently for a few minutes, and then with about three questions zero in directly at the real problem. He was amazing. A good man, a good friend, and he did a lot for PLATO in the early days when it needed attention, commitment, nurturing, vision, and funding."

It is hard to calculate how much damage the loss of a decisive leader like Dammeyer did to PLATO's potential and its launch into the marketplace by CDC. The company did its best to line up a capable replacement, but they chose an outsider who lacked the history, the CERL relationships, and Dammeyer's passion and vision for PLATO. Incredibly, Bob Morris had been passed over ("We all thought Bob a shoo-in," says Warren), and nothing would be the same again.

The loss of John Dammeyer's leadership, and the passing over of Bob Morris to replace him, changed the destiny of PLATO within Control Data. "Bob Morris didn't have as much control over Bob Linsenman as he needed," says Jock Hill. "And I think that was one of the problems. John Dammeyer did, but Bob Linsenman and Bob Morris never quite melded in the same way. John Dammeyer was a very powerful person."

Dammeyer's aneurysm came not long before CDC went to New York to unveil PLATO to the media and announce its bet-the-company CBE strategy. Without Dammeyer, PLATO became more vulnerable. Says Hill, "The wolves came out, and they—*everybody*—they leapt all over it."

Eventually CDC did get a PLATO sale with Russia. As far as a U.S. business like CDC was concerned, Russian rubles were worthless outside the Soviet Union, so the plan was for CDC to trade CYBERs for shiploads of vodka and other valuables. Then the Soviet Union invaded Afghanistan, and all bets were off. "All of a sudden," says Bob Morris, "the State Department and the CIA just kind of clamped down on any kind of deals with the Soviet Union." CDC's hopes for a computer-based education joint venture with Russia, something Morris believes would have been valued at $80 million, were over.

—

In April 1976, William Norris presided over a CDC press conference at the Park Lane Hotel in New York City where he and Bob Morris took turns describing to the media the PLATO system and the company's computer-based education mission. "At this conference," Morris wrote in a memo afterward, "John Sheehan, Chairman and Chief Executive of Commercial Credit Corporation, announced his company's intention to build a nationwide network of learning centers offering PLATO educational services. Together, CERL, Control Data, Commercial Credit, and with the contribution of hundreds of content authors, created a revolution in computer-based education, information handling, and communications."

A handwritten 1980 internal memo in William Norris's archives provides a glimpse into the reaction employees across CDC had to this new and growing PLATO project. "There was very significant resistance inside CDC to PLATO," the document says, "starting with lower executive management's resentment to the short-term, negative effects which PLATO made on profits. Other well-intended attitudes developed, leading to severe jurisdictional pressures."

Morris viewed the unveiling to the public as having a salutary effect on damping down the internal resentment: "The organizations inside the company who were not involved with PLATO kind of viewed it as an R&D program that was very exciting and all that stuff, but they weren't sure that it was ever going to be an integral part of Control Data's business. It became very clear as a result of that announcement that it was a key strategic program for Control Data. . . . It was a very important event, because it was a point where Control Data made it known to the world that it was very serious about PLATO and computer-based education and totally committed to it, that it was an integral part of Control Data's business where, prior to that, it was always viewed as a research and development plan that Control Data was playing with. . . . PLATO is a funny thing. You can't talk about it and get people excited about it. You really have to see it, to feel it, to touch it. Some of those reporters and people like that who had heard a lot about it heard the name for years and years and years, actually got a chance to see it in action. To them it was mind-boggling. There was just a lot of acceptance."

—

Bitzer seemed to be on a plane constantly, PLATO terminal in tow, during this era, demoing PLATO to the four corners of the world. In 1976, hosted by Control Data Australia, he demoed from one end of that country to the other, including a live demo in Perth that was aired on national television. It was his usual demo, and included a few notable quips about the state of education, such as: "The elementary school system . . . turns out to be one of man's most inventive forms of inexpensive babysitting. In spite of how we complain about the price."

In October 1977, he and a group from CERL and CDC, as well as many of the usual suspects from the world of educational computing—Suppes, Papert, John Volk (for TICCIT), and Arthur Melmed from the National Science Foundation—were invited to testify before a House science and technology subcommittee on "Computers and the Learning Society." In addition, William C. Norris testified with a prepared statement titled "The Future of PLATO Computer Based Education." "It should be noted," he said at the outset, "that one of this nation's and the world's most urgent needs is better education. When one considers that three-quarters of a billion people in the world are illiterate and that illiteracy is rising, it might be the number one need." He continued,

> Control Data's top strategic priority is to apply computer technology to help achieve quality, equality, and productivity improvements in education. It has the largest single commitment of developmental resources in our company today. And that's how it should be. It is our conviction that if society's major problems are given priority, in the long run they will provide the best opportunities for business.

He ended saying that the PLATO computer-based education system was available already: "It is cost-effective now in many areas. Expansion into virtually all areas of education will happen, it will relieve the plight of inner-city and remote rural schools, it will move into the home, it will reverse the tide of illiteracy in the world. The principal uncertainty is the time required to achieve these objectives. It can be greatly shortened if the federal government will provide funding now for schools and universities to pilot CBE."

In case there was any lingering doubt, Norris had by 1977 made it crystal clear to everyone inside and outside CDC: he was betting his company on PLATO.

23

The Whim of Iron

When John Dammeyer and Bob Morris delivered their business plan to Norris and his executive committee in 1974, it already hinted at things that Bitzer would expand on in his 1975 "million-terminal" speech: very large-scale PLATO was not just about education. It couldn't be if it wanted to survive. Bitzer believed that PLATO had to have other uses in business and at home. It essentially needed to be an online service, more akin to what CompuServe, AOL, and Prodigy would become in another fifteen years. Education was a "feature" amid a conglomeration of services, some of which were useful for businesses, some of which might even be recreational. Dammeyer and Morris understood the educational vision of PLATO, but recognized too how the system had become what every Silicon Valley company today likes to say they've built: a *platform*, not just a product. PLATO in the mid-1970s was perhaps the most advanced online services platform in the world at the time, not just for the creation and delivery of education, but also for tools that enabled communication, collaboration, and also recreation. Consider the list of such tools had emerged on the system in just a few years: PLATO Notes, the Notesfile Sequencer, Personal Notes, inter-system notesfiles, inter-system Personal Notes, TERM-talk, TERM-consult, TERM-comment, Monitor Mode, and Talkomatic. Imagine what could be done if CDC spent some *real* money improving and expanding on those tools, many of which had been written by teenagers for next to nothing? Imagine if the TUTOR language were expanded, and perhaps other programming languages were added, and PLATO were given far more powerful database and search capabilities? What could the world do with PLATO then?

"Bill Norris was a great supporter of mine," Bob Morris told this author in a 1997 interview. "I'm a great supporter of his. To me, he was a very good personal friend and kind of like a father, in a way. In all the time we were doing this, I can think of only one area where Bill Norris and I disagreed, and it's still an area that we disagree. And that was, because of the cost of PLATO, and because we knew that with economy of scale, and getting more of it out in the market, we could continually reduce its cost, I had proposed that we start selling PLATO for business applications, for commercial applications. It was a system that would do things like airline reservations very well. It could do loan processing very well, any place where you had real-time transaction processing, PLATO was a system that had the underlying technology in order to be able to do these applications. And if we had sold PLATO as much in the commercial environment, as we were try-ing to in the educational environment, then we would increase these sales, we would increase the volume of production, we would increase the number of terminals on our systems and services, and that would help get the price down, and that would help make it more viable in education."

This helps explain why the IST terminal was not simply a cheaper-to-manufacture version of UI's PLATO IV. It was not even a terminal, but a fully fledged personal computer that could also *act as* a terminal. It had the hardware in it. It was expandable. It was built with the future in mind. Jock Hill knew exactly what he was doing. In essence, all it needed was a compelling set of software to make it incredibly easy to use online—or offline. In business, at home, *and* in education.

One may speculate that part of Norris's difficulty embracing the business and communication/collaboration side of PLATO was that he had been "imprinted" on the original vision of PLATO, going all the way back to PLATO II and III when CSL bought its first 1604. This was a good decade before the new wave of kids came along in the early 1970s, turning PLATO into something that was never part of the overall vision, but then suddenly was too good not to include in the vision going forward. Had Norris traveled down to UI and spent a few days at CERL he might have developed an appreciation for this more expansive view of what PLATO had become.

Morris tried to explain to Norris the benefits of pursuing business and education markets at the same time—charging *more* to business customers so they could charge *less* to education customers—but Nor-ris did not see it this way.

"Norris *logically* could see it that way," said Morris. "But his concern was, 'I'm doing this because I want to make a social impact on education. And if you guys go and turn your attention to selling in the business environment, you're going to start forgetting about education, and start forgetting about our end goal. I want you to concentrate on education. Okay?' And so based on that, we did concentrate on education, I still think today if we had sold into the business environment we would have been able to fund more of the stuff that was getting the price down and achieving the educational objectives that we were out to achieve."

When Norris set his mind to something and made a decision, however large or small, that was it. His word was law, his decisions etched into stone. Countless former CDC employees describe Norris's decisions and directions with the same word: *edicts.*

Norris was also the type of CEO with a ruthless memory that brought fear and awe to those who worked for him. "There was a man who *listened*," says Jock Hill. "He asked very, very meaningful questions, took a decision, and it *stuck.* And when he met you in the elevator, a week later, he was the sort of person who—or two weeks later—he'd say, 'Hi Jock. How are you getting along with that project now?' and he'd listen to the answer. Now, when you get a guy who's as high up as Bill Norris, taking that detailed a look at—I mean I was relatively junior, I was five, six, seven layers down below Bill—but, he knew what I was doing, and when he saw me in the street, he recognized me. That was nice. When the decision came that we had to either use plasma or CMOS memory with a CRT, I remember going to Bob Morris and saying, 'I'm being pressured by the outside organizations to use plasma, because that's all they ever used' and I said, 'It's not a good decision,' and Bob Morris asked me why, and what reasoning I had behind it, and how I could support it, and he went to Bill Norris with that decision, and Bill Norris supported me, 100 percent, and I was very, very pleased. I've spoken about this many times to other people, that you get such good support from a high-level guy. He kept his eye on what was going on."

One day, Norris had an idea. "He had expressed a desire to be able to solve problems of degrading urban environments and that sort of thing," says Morris, "and how does he get the professionals back into the urban areas, and I had told him about a concept a number of years earlier called 'research campuses,' where you would build a campuslike structure and have a bunch of central services like computer centers and so on. . . . We talked about that a little bit, and he said, 'Hey, why don't you just write me a short description of that,' and I was going off on vacation on the Bahamas, and I sat down there on the beach and wrote a ten-page white paper, describing these research campus concepts. And when I got back to the office, and I gave him a copy of it, he took it home over the weekend and read it, and then called me back up to his office and said, 'Bob, I want to do this,' and 'Give me an estimate of what it would take to put together ten to fifteen of these centers across the U.S.,' and the total came to about $150 million. And he took it to the executive committee and the board of directors, and then he asked me to come and make a presentation to the executive committee and the board of directors, to allocate $150 million to this idea. And he said, 'Bob, I want you to do this, I want you to run it.' And he was pretty serious about it. I told him I didn't want to, I wanted to stay with PLATO. He said, 'I'm investing $150 million in your idea, I really need you.' And what could I say?"

Morris and Dammeyer were the only two high-level managers assigned to PLATO who had been there since the beginning, spending a great amount of face time at CERL, getting to know the players there, getting to know the technology, the courseware, the vision, the dream. A dream they fully believed in. "When we started this," says Morris, "we were kind of autonomous, and there was little visibility. We, John Dammeyer and I, ran this thing the way we wanted to, very little interference from the outside. The more it became visible—that in Bill Norris's mind this was the most important thing happening in Control Data—all of a sudden there were two reactions: one was negative in the sense of 'Why are they getting all the attention and we're not,' and the other was 'This is Norris's pet project, I better get involved in it.'"

In 1977, Norris issued an edict to create the Control Data Education Company. He named John Lacey, who had been an executive vice president, to be president of this new subsidiary. "Bill Norris was adamant," says Lacey, "I really had to shepherd [PLATO] through Control Data. There was more opposition among employees in Control Data than there were positive attitudes about it."

Bob Linsenman described the culture as "snakes and mongooses. It was phenomenal. PLATO being such a critical part of Norris's thinking, and such a central part of his attention, it was a very political part of the company to be in. People were moved in and moved out I think sometimes on a whim. Some people considered being assigned to the education division as punishment. . . . Like being a point man in Vietnam. There was so much hope, so many dreams, from the boss, the pressure was sometimes unbearable."

"There were all kinds of power plays to get involved in it," says Morris, "to be a part of it, to trying to get the visibility and that sort of thing. . . . What they did was, in order to make this look like a fairly massive thing they took everybody in the company that had anything to do with education, and made them part of this company. Now, for some people, who were craving the high-visibility thing, and that sort of thing, that was great, that was the best thing that could ever happen to them. For other people, it appeared as you don't have anything to do so we'll put you in this organization. And to some extent some of them viewed that as punishment. And one of the problems was that when this company was formed, I think the total number of employees was around six hundred people, and within that six hundred people there were something like seventy executives. Of course, seventy executives managing six hundred people doesn't make an awful lot of sense. All of a sudden there are a lot of power plays and committees and committees on top of committees, and bureaucracy, and so on. It was a hard time to work through."

Dammeyer was gone, and now Morris was being pulled out of a project he loved. But Jock Hill believes that there were other forces at work to yank Morris out of PLATO. "Bob was forced out of a position of power where he was most useful," says Hill.

Another of Norris's edicts was that all management training in the company had to be taught through PLATO. Every manager at any level had to go through a minimum of forty hours per year of PLATO lessons. In Silicon Valley, this practice is affectionately called "eating your own dog food," and was generally considered a good thing. If the company that makes a product doesn't use it themselves, why should anyone else? Not only that, but how can managers tell their teams what to do if they themselves are not familiar with the product?

Steve Adkins, who had joined CDC in 1961 and stayed through to the bitter end, remembers the switchover from classroom training to PLATO. One consequence of the switch was, says Adkins, that "one lost the benefit of classroom training—meeting other new managers and sharing experiences. But the courses were outstanding, using the full animation capabilities of the PLATO terminal. Each course included a nice course workbook. I liked taking the PLATO courses. I would reserve a block of time in the PLATO Learning Center after hours to complete several modules in a short time at my convenience." One thing he learned using PLATO's courses had nothing to do with the subject matter of the courses. He realized some of the other employees had figured out how to cheat. "You immediately take the final test for each learning module," he says, "by racing through the test guessing at the answers. You keep taking the final test until you have discovered enough correct answers to pass. I never did this. . . . I like to learn."

Perhaps one problem with the way Control Data ate its own dog food was that they ate the wrong dog food, or they didn't eat different *varieties* of dog food. It's perfectly fine to assign managers company-wide PLATO for in-house training and personal development, but this was essentially exposing many CDC personnel only to the "student" side of PLATO, rather than the "author" side, where—ignoring the TUTOR coding for a moment—all the communication and collaboration tools (and games) existed. Such tools were widely used among the rank and file in the Education Company, but outside of that, usage dropped off considerably.

Problems arose when Norris stuck with impulsive decisions that were questionable, often based on bad information. Patrick Stubbs recalls how, while running a sales training course for the CYBERNET division of CDC, he would get senior executives from the company to be guest speakers. One such speaker was Norris's right-hand man, Norbert Berg. The attendees would pepper their guest with questions, such as "What's it like working with Norris" and "What's it like in a staff meeting with Norris," and he told them it was akin to visiting the scene of a traffic accident every day. "Somebody once described Norris as having a whim of iron," says Stubbs. "He would get an idea . . . he'd make some sort of casual remark, and I know this because I knew a few of the executives at the time, and they would tell me this: he would make some sort of casual remark, just as an observation, like, 'wouldn't it be nice if' or 'what happened to' or 'how about'—somebody would take that, and they would turn it into some kind of multimillion-dollar program."

Norris ruled Control Data with a very firm hand, says Stubbs.

"What he said, happened, it doesn't make any difference. If he set his mind on something, that was going to happen. Witness the Education Company. I'm sure a lot of people told him it wasn't a smart thing to do, and he did it anyway. He put a billion dollars into it, and what would Control Data have been if they'd put the billion dollars into integrated circuit research?"

William Norris was a maverick CEO, and as CDC expanded and grew richer in the 1970s selling their mainframes and peripheral equipment around the world, he put in place a variety of undertakings that other companies—and Wall Street—viewed as ill-advised, reckless, or worse. "Addressing society's major unmet needs" became Norris's rallying cry, a remarkably progressive mantra for a tech company in the 1970s and 1980s, and one that the rest of the industry and financial world regarded with befuddlement or derision.

Over many years of research for this book, including discussions on the Control Data culture with dozens of current and former CDC employees up and down the chain of command, one of the most common assessments of Norris was that he was a good-hearted, strong leader, a total believer in PLATO, but that he was surrounded by yes-men. Other researchers and journalists who covered CDC during its heyday would hear similar refrains. "His lieutenants are the problem," Bob Linsenman told an interviewer in the mid-1980s, while Norris was still in charge. "They try so hard to please Norris that they don't give a shit if the business falls apart."

In 1984, Randall Rothenberg wrote a profile of Bill Norris and Control Data for *Esquire* magazine. The article never ran. However, Rothenberg's recollections of the article's conclusions shed light on the predicament Norris and CDC were in, particularly with regard to PLATO. "Control Data," he says, "was an example of what we'd later call industrial policy; its expertise was in seeking government funding for technology projects relating to supercomputing. When the government market for supercomputing for military and economic applications began to dry up (because of, e.g., the advance of mini-computing), CDC, instead of adapting its business model, began to seek new uses within a government welfare structure for its existing supercomputing technology. Using the technology for training, small business development, etc., was a logical extension of this. What CDC could not do was diverge from a model predicated on powerful central control. The whole notion of distributed systems—in computing, in social welfare, in anything else, it seems—was totally foreign to it. So the inapplicability of its technology to the social-welfare aims it was seeking to address was something the company could not work around. Put another way, it had come up with the perfect Great Society solution—twenty years late."

Part of the legal agreement between UI and CDC called for an ongoing technology-sharing relationship between CERL and CDC, not only on hardware developments but also, perhaps more important, on system software. CDC installed its own PLATO system for development and testing in Minnesota, with its own team of "s" and "p" engineers just like CERL's team. Together they created mechanisms for a weekly sharing of the source code each respective lab was working on, merging all of the changes each site made to the code with the goal of keeping each site in sync with the other. Rather than shipping computer tapes back and forth, they developed a dedicated telecommunications line, known as the Link, which enabled file transfers between the CYBERs. The Link also enabled a user on the Minneapolis PLATO system to remotely sign on to CERL's PLATO system, and vice versa. Eventually the Link became something a number of PLATO systems had around the world, enabling the sending and receiving of pnotes on any connected system, along with support for synchronized, intersystem notesfiles. It was only a baby version of the Internet-like vision Bitzer had expounded on in 1975 when he spoke of two hundred mainframes powering a million terminals, but it was an innovative start.

Over time, several junior systems programmers at CERL who were also students at UI either graduated or dropped out, and wound up with jobs working at CDC. This cross-fertilization only furthered the two organizations working together, and helped CDC better understand the inner workings of CERL and who knew what. However, moving from CERL to CDC was an eye-opener.

"Working there was totally different," said David Woolley, who graduated from UI in the late 1970s and joined Control Data in Minneapolis. "At CERL it was basically anarchy, at least for the system programmers. There was very little control. Paul Tenczar was ostensibly head of software development, but I think that was because some years before there'd been some personality conflicts, arguments and stuff.... People didn't get assignments to do things, especially in the very early days, people just did whatever they felt like. . . . Sometimes there would be incredible arguments with people yelling and screaming and kicking garbage cans and stuff. I remember hearing one terrific shouting match between probably Dave Andersen and Paul Tenczar, where I think Tenczar ended up yelling, 'I'm going to delete your signon!' Andersen said, 'Not if I delete yours first!' It was a race to the console to see who could delete the other signon first. Things like that happened sometimes, it was pretty chaotic."

When Woolley arrived at CDC, he found a large corporate bureaucracy. "And at Control Data, everything went through channels, you got assigned a project, and you did it. Requests for changes came down from the PLATO Business Office, which ostensibly was communicating with customers and was deciding what customers needed and wanted . . . but in fact I think what they were doing was sitting around and thinking of things they thought were neat."

It took a period of several years, but CDC's commercial ideas for PLATO began to diverge from CERL's mission. Some of the systems staff at CERL grew frustrated at some of the changes CDC would make, or how time-consuming it was becoming to merge the different sources into a single code base. At one point Bob Rader, one of CERL's senior systems staffers, had had enough with the increased difficulties, and left CERL in early 1982.

One day CDC inadvertently introduced a bug into the code that did no harm on the Minneapolis machines but regularly crashed CERL's mainframe, causing outages during prime-time service to thousands of students. "It was just a fluke," recalls Al Harkrader, "it was one they didn't catch and it was a killer and it was takin' down CERL pretty regularly, until we tracked it down and then the guys at CERL got all miffed and they decided to tell the guys at CDC that they were gonna end code sharing, they weren't gonna share code anymore. And so they were gettin' ready to drop the bomb on CDC guys." Some of these dissatisfied CERL staffers had been teenage kids who'd gotten jobs as junior systems programmers and operators, and had stuck on for the next four or five years, so they were now in the twenties. They began to have second thoughts about "dropping the bomb" on CDC regarding the split—CDC being much bigger than they were, Harkrader recalls. But eventually they did drop the bomb, against Bitzer's advice, says Harkrader, who describes CDC's subsequent reaction as "Hey, cool. Nice knowin' ya!" CDC had called an astonished CERL's bluff.

Harkrader recalls CERL thinking that their CDC counterparts would apologize and do something when they "dropped the bomb," but instead, recalls Harkrader, the next time CDC sent a manager down for a routine monthly meeting, the manager, upon hearing the news about ending code sharing, told the CERL team, "Great, I was going to ask *you* guys about that."

At the end of the meeting, when the manager walked out of the room, Harkrader stopped him and asked, "Does that mean you're going to have more openings up there?"

Shortly after that, Harkrader packed up, moved to Minneapolis, and joined Control Data.

Steve Nordberg, an engineering manager at CDC, believes the time had come for CDC and CERL to end the code sharing, as it had become a roadblock to progress. "We were doing everything we could to understand what the customers wanted from the system and drive the system in that direction," he says. "We had new hardware to run that CERL didn't, I mean, for example, the introduction of that Extended Semiconductor Memory involved a lot of work on our part to make PLATO work—that CERL had no interest in. Over time the interests of the two groups drifted apart. And there got to a point where it just got too hard to continue to share code. And at that point we stopped it. Frankly it was the right thing for both sides because we needed to do what our customers were asking us for, and Illinois had both the research mentality and the internal customer base, and we just determined that the underlying systems couldn't be shared anymore and at the courseware level we were going to continue doing what we could to keep the TUTOR language in sync so that courseware could still be shared, and we were reasonably successful in that. But it was pretty much inevitable, anytime you have something coming out of a university, into a corporation, eventually you stop having identical systems."

Another difference between the CDC and CERL PLATO cultures was how they dealt with the online community. It had been born on CERL organically, thanks to the notesfiles, TERM-talk, and all the rest of the communication and collaboration tools, and was utilized heavily. CDC inherited all of those tools, but viewed them with trepidation. The corporate culture was not used to having a way for any employee—let alone a *customer*—to speak his or her mind and share opinions on any subject whatsoever to CDC and its other customers. The last thing CDC wanted to become was a common carrier, like the phone company, just a service that people used and what they used it for and what they said through it being completely outside the control of the company. So, while CDC PLATO systems had notesfiles, it took off more slowly. It did not help, from CDC's conservative corporate perspective, for all these young hotshots who had grown up as students on CERL's PLATO system and lived and breathed the online life, to get jobs at CDC and dive right back into the notesfiles and re-create the scene they were used to with CERL. However, that is largely what happened over the years (though eventually CDC employees were discouraged or directly disallowed from posting notes using their official company signons).

For example, around the time that =ipr= launched on the CERL PLATO system—causing such a stir because it used the recently introduced "Anonymous Option"—Luke Kaven, who was a student at Hampshire College in Amherst, Massachusetts, and had been exposed to CERL's notesfiles earlier at Cornell, decided to create an =ipr= notesfile on the Control Data PLATO system, using a professor's account over at the nearby University of Massachusetts. "This was kind of radical," Kaven says, "because of the content of it, sex and drugs and homosexuality, very personal things, and I thought, 'This is something interesting to study,' and I wanted to do an academic project studying the way that people disclosed things about themselves in this new context. And so I used the UMass account . . . to create a notesfile, the counterpart to =ipr=, on the Minnesota system. I remember I did that, I announced it in Public Notes, went home for the night, came back the next day, and I was in trouble. Somebody in CDC headquarters said—apparently it went all the way up to the top, to Bill Norris or somebody—and they said, 'Delete this notesfile immediately and whoever this person is, have them fired.' . . . Nobody in the company wanted to have that kind of discourse on the CDC system. So I became sort of notorious for that, and it took a few days before I convinced them to keep my job."

As CDC gained more and more customers, and built out multiple PLATO systems serving clients across the nation and into other countries, the company continued to bump up against an inevitable rising tide of a burgeoning online community of users, just as had been the case at CERL. The white-collar, three-piece-suit, corporate culture of CDC executives was not only antithetical to the concept of a free-wheeling online community, they considered it a threat to their business, something that could scare off potential customers who might be offended by things they read in those notesfiles.

In time, CDC brought in its lawyers, new to the concept of online communities, and the result was a modification to CDC PLATO systems such that every time a user signed on, before they got to the Author Mode, they had to read an infamous, stern disclaimer written in classic legalese:

> Certain message switching capabilities (computer assisted transmission of messages between two or more points) are provided as part of the PLATO system service offering. All uses made of these capabilities must be incidental to, that is, must facilitate or be directly related to, the development, proper use and application of instructional and related materials of the PLATO system. Use of these message switching capabilities for any other purpose is in violation of this policy and applicable law.

It seemed to make the lawyers and executives happy, while the user community largely ignored it and went on creating and participating in notesfiles on every hot topic imaginable, for years to come.

Over the years at CDC, William Norris had instituted various annual awards ceremonies for outstanding employees. The Shark Club was an event where Shark Awards were given out to the top-producing sales people, each of whom had very likely sold millions of dollars of CYBERs and other equipment in the past year. The events were typically held at resorts in warm locales. Another such ceremony, the Tarpon Club event, was held each year at a resort in Tarpon Springs, Florida. The Tarpon Award was an award Bill Norris handed out to distinguished employees working on PLATO.

Jim Glish, who over the course of his CDC career would be handed two Tarpon Awards from Bill Norris, recalls one year installing five PLATO terminals in a room for the benefit of board members of Commercial Credit, CDC's financial subsidiary and a major supporter of PLATO, so they could receive, as Glish puts it, "an experience of what this PLATO thing was like that they were spending a ton of money on. . . . Commercial Credit had deep pockets back in those days, and a lot of the Commercial Credit money was being used to underwrite some of the early R&D and marketing efforts for PLATO, especially in the education marketplace."

Glish was sitting at one of the PLATO terminals preparing for the board demos when a show coordinator tapped on his shoulder. "Have you got a minute?" the person asked him. "I've got somebody here who's interested to find out about this, I'm told that you're one of the main demonstrators of the stuff."

"Well, I'm not here to demonstrate," Glish replied. "I'm here to get the award, but sure, I'm always on call, and ready to show anybody anything."

Into the room walked none other than Mike Wallace from *60 Minutes*. Wallace was the guest speaker for this awards ceremony, and as he watched Glish's demo, he asked, true to form, probing questions, "giving me the third degree," says Glish, who says the conversation went like this:

> WALLACE: Well, do you think this is *really* going to be the solution to education? You *really* think this can help? Do you really think this technology can replace the *teacher*?
> GLISH: I'm not here to debate the merits of computer-based education, all I'm here to do is show you the technology, and I can tell you from experience that people like American Airlines and this, that, and the other company are using it.
> WALLACE [aghast]: You mean, I'm going to get in an airplane tonight and somebody who sat in front of a *computer terminal* instead of flying is going to be *piloting* that airplane?

"I couldn't say *one thing*," Glish recalls, "without him coming back with this very antagonistic, leading question, to which there really is no good answer. It's like, 'Do you beat your wife?' It was really quite an experience. . . . I think it was his first exposure to the PLATO system." It made enough of an impression on Wallace that he spread the word back at CBS, and eventually, Stubbs says, CBS News did a short, unmemorable feature on PLATO that did little to change the system's general invisibility to the mainstream public.

In actuality, both American Airlines and United Airlines successfully used PLATO for pilot training, a fact that CDC touted in its sales literature frequently. For years, Boeing 767 pilots in the 1980s sat in front of PLATO terminals for a part of their training. They still utilized full-scale, multimillion-dollar, full-motion cockpit flight simulators, and actual flying, of course, but PLATO played a key role in teaching the pilots how to use the various instruments and readouts on the then-new jet.

Another group that benefited from PLATO was the Federal Aviation Administration (FAA). When President Ronald Reagan fired the striking members of the air traffic controllers union in 1981, the FAA was suddenly forced to hire a large number of new personnel and train them as quickly and effectively as possible on air traffic control so the skies would continue to be safe for travel. CDC swept in and sold the FAA a PLATO system and a large bank of terminals that were installed at its main Oklahoma City facility to train the new controllers. The program was so successful—and the FAA's budget was so tight—that the aging CYBER computer (and the formerly white terminals, yellowing over time under years of fluorescent light) was still running TUTOR lessons twenty years later, well into the 2000s. Some of the original courseware developers would over the years retire only to be rehired by a desperate FAA needing them to fix or update the lessons they'd developed years earlier—since TUTOR programmers had become scarce.

One of the most successful courseware products CDC developed was a full Basic Skills Learning System curriculum, used in adult training centers, military bases, and other sites around the country. It was probably their most lucrative success in the education market. "One of the major contributions Basic Skills made to PLATO," says Peter Rizza, who was involved in developing the courseware, "was that it was the first 'Cyber Classroom' ever created. What this means is that entire third- through eighth-grade equivalent curriculum lessons were delivered to end users without the need of an instructor. The system ran itself." And like CERL had discovered, CDC found a market in prisons. PLATO's Corrections Project that Marty Siegel had led had been deployed in a similar fashion, and on a much larger scale. Siegel found CDC's Basic Skills instructional design somewhat lacking and its content superficial compared to Basic Skills lessons that his own group produced over the years: "You could sort of go from the beginning of the lessons to the end of the lesson and it would say, 'Congratulations, you've mastered this stuff,' and you really didn't. Because it was sort of like, uh, here, I'll give you a question, um, if you don't know the answer, I'll tell you the answer, then you type the right answer in, then you go to the next question. And you could sort of get to the end by not knowing anything, but just sort of copying the right answers in at each time. Our [CERL] lessons had a very different kind of instructional model underlying it, that they were really mastery based, and teaching underlying generalizations. The way it worked was that if you missed an item, that item would come back for review, but you'd get it in a different form. So you could never just memorize a particular item. It would always be a variation on a theme."

In time CDC would attempt to re-create the success story they'd achieved with Basic Skills by creating other large online courses for GED certification, and even an entire college-level engineering course called the Lower Division Engineering Curriculum (LDEC). The company also developed a successful series of courses in sales training.

After much campaigning, Michael Allen was able to convince CDC that it needed to create a line of tools to manage all of this instructional content, and the result was a product called PLATO Learning Management (PLM), a computer-managed instruction tool that today would be recognized as perhaps the world's first fully fledged learning management system (LMS), a type of software that has turned into a multibillion-dollar industry today.

"There were very few profitable markets out there because of the extremely high cost of the technology," says Jim Glish. "CERL tended to think that PLATO didn't cost anything and everybody could use it because it was cheap. But they had it for free. They didn't have to underwrite the costs or cost-justify everything they did the way a business did. Control Data was keenly aware of how expensive it was, and was constantly trying to find ways to make it less expensive for the end user, and CERL really wasn't very concerned about that, I don't think. I think they looked at it more from the academic purist point of view, of what it potentially could be, but the reality was it had to be cost-effective as well. Control Data was kind of caught between a rock and a hard place, because there were some of the factions within Control Data that kept it more expensive than it needed to be on the one hand, but then there were the market studies that were done and there were really only five key industries that could afford PLATO even up into the mid-1980s."

Those industries were aviation, including airlines like American and United; power utilities, including nuclear and electric; the financial industry; manufacturing; and telecommunications. "Basically it turned out that they were the industries that were regulated," says Glish.

One successful application of CDC PLATO in the power utility industry began by complete surprise. The evening of Wednesday, July 13, 1977, had started off normally for Luke Kaven and Michael Besosa, who stuck around the Control Data Learning Center on 42nd Street in New York where they worked in order to play some Airfight. Just after 9:27 p.m., the power went out, trapping them in the building. What they did not realize at the time was that the power had gone out all over Manhattan, in one of the largest blackouts in American history, caused by a chain of lightning strikes, transformer overloads, and finally a shutdown at the Ravenswood 3 steam generation plant, sending Manhattan into darkness and forcing La Guardia and JFK airports to close for eight hours. As a result of that disaster, "the state legislature in New York mandated that they had to have all their people trained and it had to be different and better training than they had before," says CDC marketing manager Ted Martz, "and so we developed some of the best simulations we ever did, some network simulations where a switcher out there would see something going down on the power line and have to make some judgments about a reroute of electrical traffic, and we simulated whether it would work or not work or blow up—very expensive, but it was the only thing on the market that could do what that was doing. And they valued it, to the point of when I think they went out and actually bought their own system." The simulation was based on a program called the Procedure Logic Simulator written by Luke Kaven, who was in college at the time while working for CDC and had his own IST terminal in his Hampshire College dorm room during the 1978–1979 academic year.

In the financial industry, training was required to get certification. For instance, one of the most successful uses of PLATO was with the National Association of Securities Dealers (NASD), the company behind NASDAQ. "We did all of the certification testing for them," says Jim Ghesquiere, who had been in the PSO group at CERL before joining CDC. "In order to be a stockbroker you need to be able to pass a couple of different examinations. And one of the things that NASD wanted to do, of course, was ensure that every time that they gave a test, it was of equivalent difficulty of other tests that they had given, so that I wouldn't get a test and just by the luck of the draw pull items out of a database and had a very easy test, and when you come, by the luck of the draw you have all hard questions. And you know, one person passes and the other person doesn't. So they had a large test bank, and when a person came to take the test, we would generate their test on the fly, and do it from a statistical analysis so that we would have pulled questions out of different topic areas, and balance the difficulty, so statistically everyone had exactly the same level of difficulty of the test. We could monitor all that online while they were taking tests, and we kept all the historical data and it was scrutinized a great deal by NASD. But it proved to be, I think, one of the early electronic standardized tests in the country. . . . Probably a couple million exams were delivered on that system." Merrill Lynch used PLATO for similar purposes. Certification testing turned out to be a lucrative industry for years for CDC.

In telecommunications, CDC found some market resonance, particularly with AT&T. Another large market, one CDC knew very well going back to the beginning of time, was the United States government. CDC had been selling scientific computers to various government agencies since the 1604—it was one of their primary markets. With PLATO, they had a new way to sell to these agencies as well as the military services, with their constant need to train personnel. "Within Control Data," says Jim Glish, "there were different marketing groups, so within the commercial marketing group, government and education weren't allowed to be factored in. That was probably another problem in Control Data . . . the segmentation of the marketing groups, and the lack of cooperation between them. It meant that some of the resources were duplicated, and some of the strategies were fragmented along the way. And the Government Systems Group had the main five Army PLATO systems, and the FAA, and so on."

Things were changing on the hardware front. After the success of the IST, CDC pushed for yet another revision to the terminal design. The result was a smaller, single-unit machine encased in molded white plastic. This terminal, called the IST-2 (which in effect caused the old IST to be forever referred to going forward as the IST-1) had no PC-style expandability like the IST-1, though it did have a microprocessor inside. The move from IST-1 to IST-2 seemed like a message from CDC, a rejection of the coming freight train known as the micro-computer revolution that was bearing down on the entire mainframe industry. "The IST-2 was brought out as a cost-reduced version," says Jock Hill. "We considered it a step backwards because they put every-thing on one board, but that was Roseville [the Terminal Services Division], basking in the success of the first terminal, they wrestled for control of the budget away from Bob [Morris], and when they got it, they went back to their old way of doing things, which produced a lot of terminals, at a slightly lower cost, but no technology advance, which was disappointing. Because Bob was looking for advances in technol-ogy all the way down the line."

CDC would later follow the IST-2 with the similar-looking but more capable IST-3, also marketed as the CD 110, a complete, stand-alone microcomputer workstation with the familiar corporate gray 512 x 512 pixels and touch screen, a gigantic eight-inch external floppy drive, and support for PLATO. Still no support for color graphics. The machine was very expensive, but gave CDC a platform for finally, reluctantly, moving the delivery of PLATO courseware from the mainframe to a micro-workstation, which many customers were clamoring for.

A common anecdote told among numerous Control Data veterans interviewed for this book involved said veteran walking into an execu-tive's office, and gently attempting to broach the subject of personal computers—that CDC should be doing something much more aggres-sively in this area, that it had a chance to own the market if it chose to. These anecdotes always end the same way: said CDC employee being booted out of the executive's office, amid shouts along the lines of *"We are a mainframe company!"*

One example of this attitude was memorialized in a handwritten memo from John Dammeyer about PLATO sent to Norris. After Dammeyer left CDC full-time, he consulted for CDC for a while doing research on the K–12 education market in the state of Minne-sota. The result, the memo stated, was "a major proposal to the Minne-sota legislature to place a minimum of two hundred terminals in small rural secondary schools for two dollars per terminal hour."

The proposal was politically mishandled, and along with the lack of curricular cohesiveness, the proposal fell through; but the study generated a broad awareness of computer-assisted learning. Unfortunately for CDC, the Minnesota schools, and especially Minnesota rural kids, APPLE move[d] in swiftly behind the rejected CDC proposal and, together with school administrators, perpetrated on the kids hundreds of the little APPLE toys.

The growing impact of such toys appears to have thrown CDC's PLATO people into a panic to produce their own "stand-alone" version of a functionally cut-down "little PLATO." This bodes the sacrifice of PLATO's relative excellence for near future revenues and shares of the educational toy market; and if this is permitted to happen, 20 years of WCN's [William C. Norris's] educational foresight and coverage shall have resulted in a tiny fraction of what he dreamed.

Unfortunately for CDC, these pesky little computer "toys" were not going to go away. If anything, their stampede onto the marketplace announced to all who cared to listen that what was going away were *mainframes*. In one of the great understatements of the computer era, in 1986 Bill Norris said, "We found the proliferations of Apples and IBMs a roadblock to PLATO."

Mark Ciskey was another bright software engineer whom, along with Kevet Duncombe, Chuck Miller had hired right out of Iowa State because, Miller says, "Kevet Duncombe had said, 'Smart guy, go get him,' so, sight unseen, I hired him too."

Ciskey had an Apple computer at home. "He told us one day, you know, 'What's so hot about PLATO, you can do anything you want on this Apple that we can do on PLATO.' We said, 'Oh yeah? You can't play *Empire*.' Well, that was on a Friday. Monday he came in and said, 'Watch this,' plugged his Apple in, and he had developed a PLATO terminal emulator. Granted, he had to really reduce the character sets and everything [Apple II's screen resolution was but a fraction of PLATO's], but it was PLATO. I said, 'This is *incredible*.' Because at the time CDC made its own terminals, at five thousand bucks a pop, and they're expecting schools to buy one for each student! My folks are both teachers, my sister's a teacher, and I know that if a teacher gets fifteen spare bucks per year, they're lucky. They can barely buy chalk for the room, much less buy five-thousand-dollar Control Data specialized terminals. Everybody had Apples, because Jobs was brilliant and gave them away."

Miller decided to call a meeting with some thirty CDC marketing managers, sat them down in front of this Apple computer that they were all painfully familiar with, and told them, *Watch this.* "Brought up the PLATO page," Miller says, "signed on, went into some courseware, and said, 'Well, what do you think of *that*?' And the head marketing guy—and this is the reason CDC died—sunk his face down in front of the Apple screen and said, 'That's *green*. Don't *like* green.' Quote, unquote. I was stunned, what the hell is wrong with you, and out of that room of thirty people, about three people came up to me and said, 'You mean I can walk into any school in the country with a disk in my pocket'—or briefcase, 'cause they were big then—'plug it in, and bring up PLATO *on their equipment*?' And I said, 'Yup.' The number two person heard that and said, 'Well, then we'll have to charge *more* than a terminal for that disk, 'cause it's *value added*. 'Cause that would have to be *six* thousand dollars for *that*.' *No, no, no*, you idiot, put them in Cheerios boxes, *give them away*. We don't want to make money selling hardware disks, we want to make money on *connect time*. That's what we are selling! And that idea never sank in."

Miller says, "The thing that hurt PLATO the most was the way it was rammed down everybody's throat as 'Thou shalt take PLATO and make it prosperous,' as opposed to 'Here's an opportunity, we have to change the way we do business.' CDC just never got over 'I want to sell a mainframe for $10 million.' They *never* got over that. Their motto was *If it plugs in the wall, it's way too small*. Got to be a mainframe, and going to a service was just beyond their comprehension. And they didn't have any compensation to make them think that way. It was just a mind-set that they were stuck on."

Even so, Miller and a band of other like-minded co-workers would not surrender to the corporate mind-set. Instead they took the idea of "PLATO-terminal-on-a-diskette" and their dream of turning PLATO into a home-based online service for plain folks—in essence exactly what Bitzer had been talking about in 1975—and they refined it, polished it, made a version for the IBM PC as well, wrote some documentation, put it in a shiny box, got a few well-placed allies within the company to bless or at least ignore it (explaining to any CDC manager who might raise eyebrows, *No worries, nothing to see here, we're just selling idle mainframe time at night, not going to bother anyone, move along, not the droid you're looking for*) and sold it as a new service called Microlink, eventually renaming the service Homelink. By the time it was squeezed through the corporate marketing apparatus, the pricing had been set to typical Control Data numbers: $45 to purchase the floppy disks and a few pamphlets containing documentation. Add to that an annual fee of $25. Add to that an *hourly* connect-time fee of $7.75. And if you wanted file space for your own notesfile, or to write some TUTOR code, the fee was 17 cents per lesson part per day. (For comparison, a game like Avatar used somewhere on the order of fifty or more lesson parts.) Bottom line: PLATO Homelink was prohibitively expensive. Miller estimates the subscriber count peaked at only around eight hundred customers nationwide. AOL, Prodigy, and CompuServe, the three big American online services that were about to explode in the late 1980s and early 1990s, never had to worry about Control Data as competition.

24

Diaspora

When Fred Hofstetter came to the University of Delaware in 1973, he was one of only a handful of music scholars in the country who had even dabbled with computers, let alone become fairly proficient at programming. Tall, thin, and bounding with energy, Hofstetter had first received a BA in music education from St. Joseph's College in Indiana, and went on to Ohio State University for a master's and a PhD in music theory. The research for both graduate degrees involved extensive use of computers. For his master's thesis, he wrote a program that enabled a person to type a musical score into a keypunch machine using a special, machine-readable code. The system then analyzed the music.

For his 1973 doctoral dissertation he constructed a computerized, musical equivalent of Professor Henry Higgins: given a piece of music, the system could determine what country it was written in by analyzing its stylistic traits. He took sixteen string quartets: four from Germany, four from France, four from Czechoslovakia, and four from Russia. "The dissertation raised a lot of eyebrows," he says proudly. By the time he finished the dissertation he had left Ohio State after accepting an assistant professor of music theory position at UD, where he rewrote the music-identification program to run on Delaware's Burroughs computer mainframe.

News of Hofstetter's computer projects spread during his first year at Delaware. He soon found himself on an educational computing subcommittee of the faculty senate. "No one ever dreamed what was going to happen, but during that year we started, we had some travel budget, and we traveled around the country, looking at systems. We went to all the main vendors." The committee's primary goal was to find an existing, already developed library of courseware materials so they could show people back at Delaware what could be done with computers teaching college students directly. "We looked at the Hewlett-Packard systems. We looked at the Digital systems. We looked at TICCIT."

TICCIT interested the committee at first. "The video aspect of it then was revolutionary within a computer," Hofstetter recalls. But they found TICCIT's built-in instructional design structure too rigid and restrictive, and the video capabilities that made TICCIT so attractive also meant they could not use it over long-distance phone lines, so they decided to continue searching.

They visited CERL and got the full PLATO demo. PLATO seemed to fit the bill: it already had lots of courseware, which meant not only might there be material Delaware could put to use with its own students right away, but also that many professors at Illinois had already bet on the PLATO horse and had invested great amounts of time and money in creating that courseware. It also seemed to fit the bill because the system was capable of supporting terminals at remote locations via regular telephone lines. Quite impressive was the fact that, in terms of keyboard and display speed, you couldn't distinguish any difference between a terminal connected ten feet from the mainframe and one dialed up from a thousand miles away. For Hofstetter there were other appealing aspects as well. PLATO was the only system with touch-sensitive screens. There were other systems with screen-marking capability, but those required light pens. With a touch-sensitive screen, the only hardware a student need know how to operate was his or her finger. Hofstetter was impressed. Among the many applications for music instruction that Hofstetter could foresee was a graphical representation of a keyboard on the screen, on which students could touch the "notes" to "play" a tune.

Hofstetter's enthusiasm warmed the interest of the UD administration, and the committee decided to green-light PLATO. In the fall of 1974, UD established the PLATO Project with Hofstetter appointed as director. By the following spring, the project had received its first PLATO terminal, connected by a long-distance telephone line to CERL.

Hofstetter had wanted to create a series of lessons to help music students learn fundamental concepts of music, including chords, harmonies, and intervals. He hired a student named Bill Lynch, who had entered UD in 1972 as a freshman, drifted from majoring in math to majoring in art, and casually taken an electronic music for nonmusicians course Hofstetter offered one semester. It was during that course that Hofstetter asked the class if anyone knew how to program computers, and Lynch volunteered. He began to code Hofstetter's lessons, first writing them in ALGOL on a Burroughs computer using a Tektronix 4010 graphics display. Hofstetter had managed to get one of Seymour Papert's music boxes from MIT, a crude machine that when connected to a computer could be told to make simple monophonic musical sounds.

Hofstetter grew frustrated at the Tektronix display because it was not selectively erasable. "If you wanted to erase anything on the screen you had to erase the full screen," he says. "PLATO had a selectively erasable screen, and PLATO had touch. I thought touch was the greatest thing for a musician because you could get away from typing codes and just have the students touching musical symbols." He went away to Minneapolis for two weeks in the winter of 1975. "I froze my tail off in January learning TUTOR!" he says. No one at Delaware knew TUTOR. Lynch picked it up quickly when in March 1975 UD's first PLATO terminal arrived from CERL, enabling him to move from a slow and unproductive ALGOL environment on the Burroughs to the far more productive, interactive environment on PLATO.

During the spring semester Hofstetter taught what was the first TUTOR programming course on the campus. A number of Delaware's first "PLATO veterans," as Hofstetter affectionately terms them, attended this course. "Of course they ended up learning far more than I ever knew about it. But I knew enough to help them get started."

Hofstetter's nascent Delaware PLATO project started soliciting proposals from various academic UD departments. Ten departments responded, indicating that they were interested in developing new courseware within their subject areas. In what was a near repeat of the widespread adoption of PLATO by the faculty at UI, the diversity of interests represented by those ten departments revealed much about the perceived potential for PLATO at Delaware: agriculture, art, computer science, continuing education, education, home economics, music, nursing, physical education, sociology. Not surprisingly, all ten proposals were approved in the summer of 1975. Eight part-time programmers were hired.

Soon another terminal arrived, and then two more. The number grew to eight, then sixteen. Professors began tinkering with the system, some hiring programmers to develop lessons in TUTOR for them. A number of terminals were set aside just for authoring, and were in heavy use to the point that it was hard to find an available one. A classroom full of PLATO terminals was opened in Willard Hall, home of the university's department of education. Not all the terminals there were hooked into Illinois. By 1977, the Willard site had a handful connected to Control Data Corporation's PLATO system running in Minneapolis. If a terminal wasn't in use, it just sat there, with "Press NEXT to begin" on its display. Normally a user couldn't tell just by walking up to a terminal what system it was connected to. So in order for people to know, little flags cut from construction paper were "flown" from the top of each terminal.

Soon the number of terminals expanded from sixteen to thirty-two, but that was not the end of the growth. "When it went to forty-eight," recalls Hofstetter, "we were paying more money to the phone company for leasing lines across the country than it cost to buy our own system." Armed with that fact, Hofstetter convinced UD that it was cheaper to buy a Control Data CYBER mainframe and start their own PLATO system. UD agreed and bought the system, including a CYBER 174 mainframe, which was delivered on January 31, 1978. By March all testing had been completed, and on St. Patrick's Day, 1978, CDC turned over the keys to Delaware. Another PLATO system was born. Over the next six years the number of terminals connected to the UD system would grow to over three hundred.

The lessons Lynch was programming comprised an ambitious new series that Hofstetter called GUIDO, an acronym for Graded Units for Interactive Dictation Operations. Named after Guido D'Arezzo, an eleventh-century Italian known as the first real music educator, the GUIDO package consisted of two main segments: ear-training skills and musical theory written skills. A unit in the ear-training skills series typically consisted of three parts: first, PLATO would display an "answer form" on the screen; second, it would send signals through a cable behind the terminal to a simple sound synthesizer that would then play a sound; and third, it would ask questions about how the student perceived the sound. Every week, the system would print out for the instructor a summary of each student's progress.

Lynch's art background and his strong personal sense of design gave the GUIDO lessons a distinct look and feel that, over many years, continued to make the lessons stand out. Many PLATO lessons developed at CERL had rather boring designs, full of text, with inconsistent navigation instructions and layout. GUIDO stood out with crisp displays, dramatic typography for the page headings (words like "QUIZ" and "PLAY"), and featured clean, easy-to-figure-out touch interfaces with a musical keyboard at the bottom of the screen, and a dashboard of buttons for students to interact with. The GUIDO lessons were the first TUTOR lessons Lynch had ever programmed, and he had no experience in interaction design or user interface design. But the results spoke otherwise.

GUIDO lesson

Back at CERL, Sherwin Gooch was meanwhile building a new music synthesizer box, the Gooch Synthetic Woodwind, which was exponentially more capable and better-sounding—with four-voice polyphony—compared to Papert's music box from MIT. Hofstetter would adopt this as the standard for GUIDO instruction. The UD music building had a small classroom equipped with a number of PLATO IV terminals with Gooch boxes connected to them, and a pair of headphones for each student. (Gooch would eventually build an even more powerful box, the Gooch Cybernetic Synthesizer, but Hofstetter decided that Delaware should build its own, the result being the University of Delaware Sound Synthesizer.) Walking by the often dark room in the music building, one might mistake the scene inside for an air traffic control center, with headphone-wearing students intently working at their GUIDO lessons, their faces dimly lit from the Orange Glow of the plasma panels, their fingers occasionally reaching out to touch the screen.

Music was not the only subject that found acceptance among faculty at Delaware. In just a few years, there would be hundreds of new lessons developed in accounting, student advisement, agriculture, anthropology, art, biology, business, chemical engineering and chemistry, civil engineering, counseling, economics, education, English, geography, geology, nutrition, family studies, Latin and foreign languages, library science, math, nursing, physical education, physics, political science, psychology, statistics, and urban affairs. The UD PLATO team developed more rigorous quality control standards for its lessons than were in place at the time for CERL-authored lessons.

Early University of Delaware PLATO users, including Lynch, would experience an interesting phenomenon that PLATO users at other sites would go through in subsequent years. By starting out on the CERL system, these users had seen firsthand the feverishly active, boundlessly creative, every-day-there's-something-new-going-on online community of CERL, with many notesfiles including technical ones where a question might be answered in minutes, if not seconds. For UD courseware developers new to TUTOR, having this support network enabled them to come up to speed far more quickly than if they were not able to tap into the community. But then, as UD grew the number of terminals on campus, it began to migrate terminals from CERL to CDC's own Minnesota system, a system that Lynch found empty, a desert island relatively devoid of the bustling activity he'd grown used to on CERL's system. When UD activated its own PLATO system and all of the terminals were switched over to it, the bustling community was gone, the online experience for those with author signons as quiet as a tomb. Delaware would have to start from scratch building its own community, but it had a distinct advantage: by 1978, enough UD people had been exposed to CERL before the switchover that they knew what the ingredients were for an online community. The tools were all there, the notesfiles simply needed to be fleshed out. Delaware also had the advantage of having over several years hired several people from CERL, including Bonnie Seiler, Jim Wilson, and Brand Fortner. In short order, many of the most popular notesfiles from CERL had been re-created on the UD PLATO system, and slowly but surely activity perked up.

The new UD PLATO mainframe also had the advantage of having a telecommunications link connecting it to CDC's machines in Minnesota. In fact, by the late 1970s, CDC had so many clients on its Minnesota PLATO services that it split it into multiple systems, one dedicated to internal R&D and other machines for customers, each machine supporting several hundred simultaneous users. All of these machines were also interconnected with telecommunications links. And, of course, there was a link between them and CERL. Add to this a series of powerful new capabilities enabling pnotes to be sent to any system so interlinked, and notesfiles being able to be interlinked as well, and the online community of PLATO suddenly started looking like a small Internet. Any system connected via a link to this growing network was able to take advantage of communicating with all the other linked systems. Any system that did not have a link was truly a desert island, its online community—if any—tiny in comparison to a linked system.

Delaware established a policy that other universities would copy in subsequent years: Control Data employees were not allowed on campus, period. Fred Hofstetter, in a 1980s interview, described the policy this way: "Our relationship with Control Data has been very, very professional. We want to be the user interface, we don't want Control Data to be interacting with our users. Control Data is not allowed to speak with our users. They are not allowed to solicit among the faculty here. They must contact me. And over the years a couple of times some overzealous salespeople have contacted directly some of our users and they were fairly soon after taken off our account. Because we just do not permit that. We don't want corporate salespeople dealing with our users. They come from a different environment. . . . The main market for PLATO is industry. And the Control Data sales force is much more industry oriented than they are education oriented. When these fellows come in with their three-piece suits and they take faculty members out for their three-martini lunch and so on, it doesn't have the academic feel to it that it needs to have to be successful on the campus."

Another site, the University of Maryland, Baltimore County (UMBC), installed a PLATO system of its own in 1982, and also banned CDC personnel from poking their heads in on campus unless they were there for a specific purpose such as repairing equipment. UMBC hosted PLATO service for a number of school districts, including nearby Baltimore but also Richmond, Virginia.

There was palpable antipathy to Control Data at some of these schools. In Baltimore, CDC and Commercial Credit, which was based in Baltimore, had originally run PLATO service in some schools and adult education centers. UMBC inherited that business as part of its purchase agreement, which in essence gave UMBC control of PLATO for any nonprofit, educational, or governmental body that wished to use PLATO anywhere in the state of Maryland. Schools that had interacted with CDC personnel prior to the UMBC switchover had not had uniformly positive experiences with the company or its people. At the Walbrook High School in Baltimore, the school receptionist would sternly interrogate any PLATO personnel from UMBC before allowing them any further into the building: "Are you now, or have you ever been, an employee of Control Data Corporation?" was the pointed question. A visitor asked the question had the clear understanding that to say yes meant leaving the premises immediately.

UMBC epitomized the desert-island type of PLATO system that popped up around the world in the 1980s. Their missions were clear-cut: deliver PLATO courseware as a service. The university administrations would buy these multimillion-dollar systems, hire technical and educational staff to deliver and support the services, and when (or even if) asked if they wanted a telecommunications link back to the rest of the PLATO network, typically would not even know what that was or why it could possibly be beneficial. And when shown the cost of such a dedicated line to Minnesota, the reaction was usually "no thanks." The mission for later-stage PLATO systems was typically to deliver *existing* courseware, rather than do what Illinois and Delaware did, enlisting the academic departments to find faculty members eager to create their own lessons for their respective courses. Control Data had by the early 1980s a large library of its own material, including the Basic Skills curriculum, which these remote PLATO sites were happy to provide to adult learning centers and other nearby institutions whose mission it was to help people improve their skills to get a better job. However, the astronomical cost of the CDC courseware would usually stop these universities in their tracks. UMBC closed down its PLATO system in just a couple of years, due to the sheer costs involved in offering CDC's courseware. The royalties for that courseware were so high, the university could not charge enough to the various educational and training institutions using PLATO to pay for it. The CYBER and all of its equipment would be disconnected and carted away. (Once asked about the UMBC debacle, Bitzer had this to say: "The reason they didn't have any more money was 'cause CDC screwed 'em." He says he bought the used UMBC PLATO hardware through a broker "and I put it all to good use.")

CDC sold PLATO systems to a number of educational institutions between the late 1970s and mid-1980s, including the University of Quebec, the University of Hawaii, the University of Western Australia, University of Alberta, University of Massachusetts, University of Connecticut, Florida State University, University of Nebraska, Royal Institute of Technology in Sweden, University of Brussels, and the University of the Western Cape in South Africa. Perhaps the biggest sale in CDC's history was to the California State University system in the mid-1980s for a reputed $50 million. In addition to the many CSU campuses having access, many school districts in California tried out PLATO. One, the Sweetwater School District in the San Diego area, found that troubled students and high school dropouts enjoyed using PLATO and often came back to finish their high school degrees because of it. The feedback the school administrators received from the students was essentially the same that prison inmates given a chance to use PLATO for instruction verbalized years earlier: they felt comfortable with the ever-patient machine, enjoyed the personalized feedback, and they were free from any potential embarrassment they might experience in a conventional classroom.

PLATO was also utilized in countless corporations for training, as well as a variety of military services in the U.S., Israel, and other countries. But despite installations all over the world, and pockets of profit here and there, by the mid-1980s, CDC's mainframe-and-terminal-based PLATO systems were rapidly becoming dinosaurs. Microcomputers were taking over.

25

The Crash Pit

They came from all over the world, to drink, party, and compete on a Waimanalo polo field on the windward side of the island of Oahu. It was February 2013, time for the three-day Kaimana Klassik: an annual Ultimate Frisbee tournament. Ultimate Frisbee is a mash-up of strategies and gameplay from basketball, rugby, and soccer, but using a Frisbee instead of a ball. The game, invented in the late 1960s, has hundreds of collegiate teams in the United States, and organized leagues and tournaments all over the world.

Thirty-two teams would compete, some having traveled thousands of miles to Oahu. Team names tended to be a bit deranged: Philthy, Beer4Breakfast, Stigmata Chelada, Freaks, Sarcastic Fringeheads, French Kisses, Skeletor, and Freshly Squeezd. One Oahu resident, a longtime Klassik spectator and Stanford alum, had come to the tournament to cheer on the Stanford Bloodthirsty team. His name was Brodie Lockard. He made his way to the field in a motorized wheelchair along with a 24/7 personal caregiver.

The Hawaiians named this island "Oahu," which any tour guide will tell you means "gathering place." In 2013, an auspicious gathering of people and coincidences would assemble in Waimanalo, at the foot of the towering, green, ancient volcanic mountain wall, known as the Ko'olau Range, looming over windward Oahu like a cresting, miles-long, thousand-foot-high wave.

While the Bloodthirsty played, one blood-red team jersey caught Lockard's eye. Above a large white number 32 was the name "MARCY." For Brodie Lockard, to see that name on the jersey of a Stanford team member, and to hear teammates call that name during the game, stirred painful memories. From the sidelines, Brodie did the math. *2013 minus 1979* . . . enough years had passed. And this Frisbee-flinging Marcy kid was a student at Stanford, as was Ted Marcy, a name Lockard knew well, back in the early 1970s.

Could this kid be Ted's son?

Ted Marcy was to gymnastics what Jimi Hendrix was to rock music. Marcy's specialty, back when he competed on the Stanford men's gymnastics team, was the pommel horse. He not only mastered it, he *owned* it. Some refer to him even now as a "freak," a "mutant"—terms that gymnasts use not to insult but to offer the highest form of praise. One physical fitness trainer who knew Marcy back then once blogged about him this way:

> I had read about Marcy for years during high school and NOBODY swung like him, no one had the hip extension he did. He made EVERYBODY look weak on the horse. . . . And what an inspiration! This guy TRAINED! Like an animal. Obsessive in the very best way. He would do ten routines in a row and EACH ONE LOOKED EXACTLY LIKE THE OTHER! He would do THOUSANDS of nothing but CIRCLES. Forever. . . . And they had to be perfect. Every, single, one. . . . The guy had control like hadn't been seen before, and his scissors were from another planet. So much better than everyone else, he looked like a man among children. He would also swim, run, and stretch like a maniac and he was the rarest of the rare: a pommel horse SPECIALIST. That would drive most normal men mad. He set the stage and the standard. . . . Once people had seen how it could be done, nothing less would do.

Marcy was a hero and Stanford legend to Brodie Lockard as well. Brodie had been a gymnast in high school in Tucson, and upon enrolling in Stanford in 1977 wound up on the same men's gymnastics team on which Marcy, who had already graduated, had excelled.

Gymnastics is an unusual sport. It's almost never a career. "Thing is, after college, you're done," says Jeffery Chung, a Stanford teammate of Brodie's. After college, you either go to the Olympics—a rare feat—or you get a day job and get on with your life (though some gymnasts have long careers with entertainment acts like Cirque du Soleil). Gymnasts do it because they love it: the required concentration, the physical stamina, the solitary laser-focus, the risks and dangers. Gymnastics is something you grab and never let go your whole life. It is a way of living. At Stanford, that level of dedication meant demanding, four-hour practices, six days a week.

December 6, 1979, fell during the week before finals, so some teammates were off studying. It started out like just another day of practice for Brodie and the team. But the day ended with Brodie in the hospital, where he would stay for the next nine months.

Most of what he knows of that day is based on what others have told him over the years. Like all gymnastics gyms, Stanford's had a "landing pit," or "crash pit," as gymnasts often call it. In Stanford's case it had been placed between the high bar on one side, and the trampoline on the other. The pit consisted of layers of foam rubber mats and foam gym pillows, which in the 1970s were still pretty primitive, says Lockard. In fact the whole pit was makeshift. The team coach, Sadao Hamada, "had put one together that was pretty homemade," Brodie recalls. "It had old volleyball nets on the sides, and it really didn't have enough foam in it." The nets were used to hold the pillows in place as gymnasts landed on them. Says Brodie, "The idea is that you could land in the foam in any position and not get hurt." He could practice dismounts from either the high bar or the trampoline, and land in the pit.

Brodie was dismounting from the trampoline into the foam, when—he's not sure but thinks he was practicing a twist while in the middle of a flip—something went wrong. "I just jumped too high and too far, and landed where there was like a foot of foam, which didn't do anything to break my fall."

Teammates were not the only ones who saw him fall hard. There was another witness in the gym that day: Ted Marcy. He should not have been there. He'd graduated several years earlier, and was by 1979 almost a full MD, in his fourth year as a medical student at Yale University. "I was visiting the gymnastics gym for old times' sake," Marcy says. "As I was waiting to get up on the pommel horse, I saw someone come off the trampoline into the crash pit and then not move. I had a bad feeling, and knew I needed to go over to check on him." Marcy remembers Brodie lying faceup, eyes open, and vividly remembers asking him three questions: "CAN YOU MOVE YOUR FEET? CAN YOU MOVE YOUR ARMS? CAN YOU BREATHE?" Says Marcy, "The answer to the last question was a minimal shake of his head." That prompted Marcy to shout for help.

What Brodie was told later, is that his hero, the one and only *Ted Marcy*, had been in the Stanford gym that day, why?—how?—and not only that, had even seen him *fall*, and not only that, had rushed over to offer help, and not only *that*, had given him mouth-to-mouth resuscitation before the paramedics arrived. ("Which is pretty cool," Brodie says, with a wide grin.) But beyond those sketchy details, Brodie remembers nothing of the accident itself or of the next four days. His spinal injury was serious and life-threatening. Often a victim of such an injury dies within minutes right on the spot, due to the paralysis affecting not only arms and legs, but of the diaphragm muscle. Lose that, and you can't breathe, so you suffocate and die. Had not someone knowledgeable like Marcy, not only an experienced gymnast but also a fourth-year medical student, been there, at the right time and at the right place, to recognize what was going on, waste no time making the right diagnosis and start providing the proper immediate care, Brodie would probably not have lived.

"The accident was devastating to everybody involved," says Andy Geiger, Stanford's athletics director at the time. "It was just unbelievable, and a miracle that Ted was there. Brodie, and Brodie's family, I will never ever forget as long as I live. It was one of the worst days of my life."

Jeffery Chung was one year older than Brodie, but they graduated from high schools the same year. While Brodie was a self-described "desert rat" growing up in Tucson, Arizona, Chung had grown up in Honolulu and attended twelve years at the Iolani School before entering Stanford. During his high school years, Jeff was dating a girl over at Punahou School. In her class was a kid named Barack Obama.

Lockard had entered Stanford as an English major, but then realized he might have more luck in the job market by taking math and computer science courses. Chung, on the other hand, pursued an intensive premed track heavy on chemistry and biology. Both Chung and Lockard joined the Stanford gym team as freshmen in the fall of 1977.

But before Chung moved to California, during the final months of his senior year at Iolani, he and some friends heard about a cool computer named PLATO, located over in a lab at the University of Hawaii's nearby Manoa campus. They had no affiliation with the university; they were simply drawn, like moths to flame, to the system like so many others had been.

Inside a campus building were four PLATO IV terminals sporting the Orange Glow. Chung remembers watching as others played games that he would learn were called Empire and Airfight. He was instantly hooked. He was soon sneaking out of his house each night, dashing over to the building, and playing online often until dawn. Then, he says, "I'd stagger home and fall asleep." To this day he believes his parents never knew about his nightly escapades—although his mother did wonder, *Why is Jeffery always so* tired *during the day?* "My mom thought there was something wrong with me," he says with glee.

Arriving at Stanford, he'd befriended Brodie and told him about PLATO and its incredible games back in Hawaii. "I was thinking, wouldn't that be a great company, a business idea, to have networked gaming?" he says. "There was no Internet or anything like that, so I didn't know how that was going to happen. I just imagined all these people calling in on phone lines, over modems, to play against each other. We used to play against guys in Champaign-Urbana or wherever PLATO was hooked up. It was the first time I'd played this kind of a networking game."

Chung dove into his studies, but did visit one of Stanford's computer labs to play "their version of Star Trek," he says. "It was nothing compared to PLATO, and then I just stopped."

The fun and addictive aspects of PLATO's multiplayer games had fascinated him—and yet now he was at Stanford, in the heart of Silicon Valley, largely oblivious to the very notion of such games. It would not catch on there for years, and when it finally did, it was as if PLATO never existed.

Months before his accident, Lockard had gone home to Tucson for the summer. Chung's frequent PLATO stories had gotten him curious about the system, which was unavailable on the Stanford campus, thanks in part, Lockard says, to Professor Pat Suppes, whose competing computer-assisted instruction work (and funding grants) might have been threatened by PLATO. Brodie's father was a professor of architecture at the University of Arizona, which, unlike Stanford, had embraced PLATO, leasing a number of terminals dialed up to CERL.

Professor Lockard planned to use PLATO in his architectural drawing class, but needed a TUTOR programmer to create the lesson. Brodie, who had just taken his first programming classes in assembly language and Pascal at Stanford, got the job, and over the summer created a tutorial lesson called "Edges," which introduced the concept of "spatial edges." One instructional page of Brodie's lesson included the following text:

> In addition to the surfaces in the environment, our evolutionary history has led us to pay attention to edges in the environment. It is from behind these edges that our enemies have always appeared, and over these edges that the more awkward members of our species have always fallen.

"Edges" shows a 3D drawing of a box with thick walls. The box appears to be empty inside, but as the student progresses through the lesson, a little man peeks out from the inside front edge of the box, his long nose drooping over the edge, Kilroy-style. Such was the ability of PLATO that the little man graphic could be superimposed in different positions in and around the edges of the box. Some positions made perceptual sense, and some didn't. The lesson was designed to help students learn that some edges are different—they're *spatial*—they indicate a spatial discontinuity. The brain has evolved to know that behind some edges things can exist and behind other edges it would not make sense. The student was supposed to touch those edges of the box that were spatial edges. When the student correctly touched a spatial edge, PLATO responded by thickening the line. If the lesson detected that the student wasn't getting the concepts right, the student was directed to a review section in which a simpler box was drawn. It was hard to get all the answers right, and easy to fall into that forced review section.

By the time Brodie got back to Stanford for his junior year, he was hooked on PLATO. He'd been on it all summer, with author privileges, climbing the Ziggurat, becoming part of the culture. He knew TUTOR, he had discovered the online community, the notesfiles, the TERM-talks, the games, and the sheer *presence* of people online. "I thought PLATO was one of the coolest things I had ever seen," Brodie says.

Back at Stanford, September 1979, and Apple Computer's Apple II personal computer sales were booming. The next big wave of microcomputers, from the Commodore 64, Radio Shack TRS-80, and IBM PC, was still months or years away. Steve Jobs had not yet made his visit to Xerox PARC—that would happen in December. Silicon Valley was ignorant of PLATO, focused instead on things PLATO users would have considered trifling. Silicon Valley start-up visionaries would not wake up to the online world for years to come. Most of the few who *had* seen PLATO brushed it off as an extravagant waste, a mainframe-based relic that had no relevance in the booming micro revolution that was under way.

Brodie had seen the future but now he was back in the past. He'd lost *access*. He wanted a terminal in his dorm room. Could he rent one? He contacted CDC and spoke to a sales rep named Cindy Poulos. The costs for leasing a terminal and phone line were astronomical. He might as well have asked to rent an exotic sports car. His hopes for PLATO were dashed, but his interest never left.

And then in December, the accident. He woke up in a hospital and began a long rehabilitation, including learning to breathe with an artificial breathing apparatus, the ability to breathe on his own now gone. He lay there with nothing to do. Jeff Chung would visit and they talked about PLATO and how Brodie dreamed of being productive again if only he had access. He asked his father if it was possible to get PLATO at the hospital. Maybe they would cut Brodie a break? He probably would never walk or move his arms or legs again, which meant no typing at a keyboard, but he still could *think*, he still had ambitions, he was still burning with ideas. On PLATO his disabilities would not matter. PLATO was a *meeting of minds*, pure and simple, and Brodie's mind was fine.

His father reached out to Cindy Poulos. She was so moved by Brodie's tragedy, she made a special trip to visit him at the hospital. Then, a surprise. "She brought her *personal* terminal from home for me to use in a hospital," says Brodie. "It was amazing." It was not a PLATO IV, but a big CDC IST-1. But it was PLATO, and with it he was once again connected. "People in my office would not have approved such a thing," Poulos says, "so I always just told them the terminal was out on loan."

"We had to find a closet to put it in, to lock it up," recalls Brodie. Then a phone line had to be connected. Luckily, the hospital went along with his requests, and let him have his way. (Brodie suspects the phone bill was probably tacked on to his gigantic hospital bill nine months later.) Knowing the terminal was right down the hall in a closet, and knowing that the PLATO community was reachable through it, gave Brodie a reason to hang on, at a time when most people might not have had the will or the strength to carry on. "It really kept me going," he says. "It gave me a reason to get up in the morning, and something to work on. . . . It was my main, my most enjoyable, activity in the hospital."

During Brodie's long hospital stay, well into the summer of 1980, part of his rehabilitation included learning how to use a typewriter with a mouth stick. Connected to one end of the long dowel is a rubbery grip like a mouth guard, which the user bites down and then, as he moves his neck and head up, down, and around, he positions the other end of the stick to tap any desired key on the keyboard. It was awkward, but by the time Poulos's PLATO terminal had arrived, he'd mastered the mouth stick.

He had also taken up an interest in a board game that one of the hospital staff introduced him to and that they would play every now and then: an old Chinese tile game called Mah-Jongg. There are many different versions of the game, with different rules and number of pieces. The version Brodie played consisted of 144 tiles, each tile having a certain symbol or number signifying which class or group it belonged to. The goal is to match one tile with another identical one, and remove them from the formation. To win, remove all the tile pairs. But it's not easy. Many tiles are placed on top of others, in layers, and there are tricky rules about which tiles you can remove and which ones you can't until others have been removed first.

Mah-Jongg is played in numerous ways, sometimes with four people, sometimes like solitaire. Brodie learned about a tile layout called "The Turtle," which involved placing tiles in such a way that a very primitive "turtle" was formed, the "back" of which, in the center, was the thickest, with several layers of tiles. Brodie liked the game and, once he got Poulos's terminal, he thought about programming a PLATO version of the game.

PLATO's 512 x 512 graphics and touch screen were ideal. The screen resolution was high enough to provide a crisp rendering of the 144-tile "turtle" layout, and the tiles could be big enough that they could be touched. It was as if PLATO were intentionally designed just for Mah-Jongg.

But how to represent *layers* of tiles? After all, the game requires some tiles to be laid on top of others, as in the board game. On the PLATO screen, with an essentially 2D top-down view, how would a player notice that some tiles were on top of others? Lockard's solution: thicken the sides—bevel the edges—of particular tiles. The higher up the tile layer, the thicker the border. The overall design not only worked, but produced perhaps the most striking, beautiful game display ever created on PLATO. To this day, it evokes wonder from people who view it for the first time, and sometimes, for the thousandth time. What made his design even more striking was the painstaking detail of the calligraphy and numbers on each tile. On an original PLATO IV terminal, the effect is a sublime orange vision. The thick borders of the higher layers of tiles made the center region of the screen glow like embers of a fire. He pushed the envelope on PLATO in ways nobody else had thought of, taking PLATO as far as he possibly could.

Incredibly, over months, he built the entire thing one tap at a time with his mouth stick.

To understand what a monumental achievement this was, consider that the 144 tiles each had different works of art on them, and each of those drawings, icons, Chinese calligraphic symbols, and numbers and letters was copied from the tile designs he'd seen in the hospital's board game, one by one, using PLATO's graphical character set editor. It took weeks just to create those graphics. He also wrote all of the TUTOR code for the game with the mouth stick. PLATO's TUTOR editor lacked modern scrollable windows: instead it required typing frequent commands just to move around in the code. A piece of cake for an able-handed person, a slow process for a mouth stick user. And yet, Brodie persevered.

Hours of addicting fun awaited players. Even more than solitaire, Mah-Jongg relied on memory and strategy. Speed got you nowhere, and usually nowhere fast. A wise player carefully examined the entire layout of tiles on every play. Mah-Jongg was not an action game, and though it required concentrated effort, the gameplay was deeply meditative.

The game monitored your every move, with enough Self-Pacing and Immediate Feedback that Skinner himself would have been proud.

"Touch a tile," the game would say. When you touched a tile, it told you to now find a matching tile and touch *it*. The program instantly evaluated whether the two tiles were indeed legal by the rules of the game, and if so, the two tiles disappeared, bringing you one step closer to winning. However, the game was quick to identify a tile that was not "free," meaning it was somehow blocked. What might feel like an easy, quick game could turn into a half hour, an hour, or more, the last thirty or forty tiles seemingly unmatchable. This is where the game's addictiveness revealed itself: just when all appeared lost, players drunk on the gambler's fallacy—*I'll figure it out this time*—found a surge of motivation and then, having discovered all was indeed lost, rather than quit, they'd load a new game and start all over again.

If you ever managed to remove all the tiles, the screen would erase and in large letters at the top would appear CONGRATULATIONS! and below, YOU WIN!, but neither of those were what caught your eye. In between the two headlines lay Brodie's pièce de résistance: a sigil-like drawing of a particularly scary Chinese dragon. Thanks to the slow phone lines of the day, it took about twelve seconds for the terminal to draw the dragon with its bug-eyed, berserk face, long tongue jutting out from an open toothy jaw, jagged-edge claws on four feet, and long arching back and tail, but the delay didn't matter. It was a rare moment, it was the fireworks earned through an achievement that might have taken a long time to reach, but it was worth it all just to watch the great, mad, bug-eyed beast come to life on the screen.

In one way—reminiscent of the architecture lesson he'd coded back in Arizona—the dragon was that "enemy" that appeared behind the "edges" of the Mah-Jongg tiles. But in Mah-Jongg, when you encountered the dragon it meant victory.

But in another way, this was the monster that lay beneath the crash pit that had so changed his life back in a Stanford gymnasium. He had found a way to conquer that monster, and through the game he gave the world a way to conquer it as well. But he was far from done with this game. Soon, he would figure out how to take Mah-Jongg to an entirely new level.

Meanwhile, there was the matter of liability. Was Stanford University liable for Brodie's accident? Should the university have made more of an effort at providing a safe environment for its athletes? Was the homemade crash pit insufficient and lacking? Did that translate into any form of culpability for the university, for the Stanford athletics program, for the coach? These were questions percolating among the Lockard family.

"The first thing Stanford's athletic director told my parents," Brodie recalls, "was that Stanford would not pay a dime unless we sued. Not to be mean, at all—he was just telling them how the process worked." It may have been how the process worked, but it wasn't a process that appealed to Brodie. "Suing was the last thing any of us wanted to do," he says.

Nevertheless, the family got a lawyer, who brought suit against Stanford, requesting, according to one source, $18 million. "It was a horrible time," recalls Jeff Chung. "Suddenly all the guys on the team were now either implicating Stanford gymnastics or not helping Brodie. And that was a terrible position to be in. I remember when they called me to the witness stand, and I just said, listen, the way I'm going to do this is just tell the truth, whatever they ask me, but people are going to have to ask me the right questions. . . . I don't think his legal team really asked enough probing questions."

Marcy was also called as a witness and, of course, Coach Hamada. A friend says that years later Hamada told him that Brodie's accident and subsequent trial were the worst time of his life.

Brodie was conflicted, embarrassed even, as the suit created an adversarial relationship with the very university he so loved, at which he was still, despite his physical hardship, in the middle of taking classes in pursuit of a degree. "This was my *coach*, my *school*, my *teammates*. The trial was beyond horrible."

For Brodie the ordeal felt like it went on for decades, although it was over in less than a year. It felt, he would say years later, as if "everyone within fifty miles of the gym" was asked to testify. In the end, Brodie, his family, and his legal team settled with Stanford for undisclosed terms, and the university agreed to pay his medical expenses.

After Brodie left the hospital, Cindy Poulos helped him enroll in a program Control Data offered called HomeWork, designed to give PLATO terminals to disabled people so they could work productively at home. If there was one thing Brodie had proven, it was that he was productive on PLATO. He developed something of a name for himself on the CDC PLATO systems. He hung out in some of the notesfiles like =sots=, the literary home of the "Save Our Tongue Society," and =forum=, an open forum for discussion about current affairs and issues.

While Brodie was working on Mah-Jongg on PLATO, he had resumed pursuing his bachelor's degree at Stanford, often having fellow classmates take notes for him for classes he wasn't able to physically attend. It didn't take him long to realize how handy a PLATO terminal could be for classwork—most students in the early 1980s still didn't have personal computers, or if they did, they weren't very useful. PLATO had text editors that, while not ideal, were better than what was usually available on micros. So he did a lot of homework on PLATO, printing out reports that often included very complex mathematical formulas, all relatively easy to produce on PLATO. By 1984 he not only finished a *double* bachelor's degree in mathematical science and English, but also finished a master's degree in interactive educational technology. True to form, the professors in the master's program— this being Stanford, i.e., Suppes's turf—steered clear of PLATO, but it did give him a broader understanding of computers in education. Brodie devoted his master's thesis to developing for his architect-professor father another PLATO lesson, this one called "Perspec," which taught students how to draw in perspective. Even after the accident, Brodie managed to attend Stanford gymnastics events as a spectator. "Brodie never lost his devotion to the gymnastics program," says Andy Geiger, "and I can remember him coming in his wheelchair and marveling at how well he could manage with a tongue-manipulated device to get around."

When in 1983 Brodie finally finished Mah-Jongg, he wanted to redo the game for another computer, thinking he might make money doing so. Was there an affordable microcomputer with a display close enough to PLATO's high resolution? He couldn't find one. Then in January 1984, Apple's Macintosh appeared. It raised the bar. As Alan Kay once famously quipped to Steve Jobs, the Mac was "the first personal computer worth criticizing." Any PLATO user would have probably felt the same. The screen resolution wasn't *quite* PLATO, but wasn't too shabby either, at 512 pixels wide and 342 pixels high. At least it matched *one* of PLATO's dimensions.

Sometime in 1985, Brodie got a call from a producer at the gaming company Activision named Brad Fregger. Fregger recalls getting Brodie's name from a recruiter, Caretha Coleman, the wife of his boss,

Ken Coleman. Brodie recalls sending his résumé out in 1985, looking for work, now that he had finished his two Stanford degrees. Caretha suggested that Fregger might want to hire Brodie. "I often worked with creative programmers who had 'day jobs,'" says Fregger, so it was routine to get a tip about talent he should try to recruit.

He and Brodie agreed to meet the next day at a wheelchair-accessible restaurant for breakfast. Brodie arrived, waited, waited some more, and finally gave up. Fregger never showed. For Brodie, something as simple as meeting someone at a restaurant was a nontrivial logistical operation involving helpers getting him prepared for the day, getting him into his wheelchair, into a car, then driving him to the restaurant, and helping him out of the car and into the restaurant. And then Fregger was a no-show. Says Fregger, "I forgot the appointment entirely." Brodie called Fregger later that day, asking what happened, conveying his annoyance in plain terms, and Fregger, mortified, apologized as best he could. They agreed to reschedule a new appointment on the Stanford campus later that very same day. It was Fregger's first face-to-face meeting with Brodie, complete with all the realizations of what Brodie's confinement to a wheelchair entailed. After a wide-ranging conversation, including Brodie's interest in computers in education—he was just wrapping up his master's degree—Fregger told him that while he didn't have an opportunity at the moment at Activision, he encouraged Brodie to call him if he ever developed a game, and that Activision would take a look and see if it was something they might want to publish.

On April 15, 1985—he remembers the exact date even now—Brodie got a full-time job at Stanford University as a programmer. The first thing he did was go to the university's bookstore and, utilizing Apple's educational discount, bought himself a Mac. His job made him productive, and got him interacting on a professional level with colleagues. "I worked seventy hours a week," he says. "And I loved it."

Almost as soon as he got his Mac, Brodie started writing a Mac version of Mah-Jongg. Between April and December 1985, on top of the seventy hours a week he spent working at Stanford, he spent nights and weekends building the new game.

That year he also got a Personics HeadMaster—a device that freed him from the mouth stick for many tasks with the computer. The

HeadMaster consisted of a lightweight headset that sent ultrasonic beams to a little box that fed signals to the Mac. The HeadMaster software worked with the Mac's operating system in such a way that virtually all programs became accessible, enabling Brodie to move his head *just so*, causing the cursor to move in a corresponding manner on the screen, and then, with the help of a little strawlike tube at the side of his mouth, puff into it to click or "type" a key. With this and, from time to time, continuing to use the mouth stick, he designed, wrote the code for, and drew all of the graphics for the new Mah-Jongg.

By 1985 there existed programs on PC and Macintosh that emulated a PLATO terminal. For Brodie this meant he could use his Mac to dial up to a PLATO system and get back in the community, with its notes-files and TERM-talk. While connected to PLATO he would copy the TUTOR source code for Mah-Jongg, then paste it into another text window on the Mac. It was a time-consuming process but still a huge time-saver, as it meant he didn't have to type the game in from scratch, and the Mac's editors had scrolling, something PLATO lacked, so it was far easier to move through the code. He also made screen grabs of the exquisite PLATO artwork he'd designed for the game tiles, and then, pixel by pixel, cleaned them up and converted them to the Mac version of the game. Once the code was all moved over, he converted every line of TUTOR into the C language environment of the Mac. By mid-December 1985 he felt the game was ready, and took up Fregger's offer from months earlier to get in touch if he ever developed a game.

"I have a game I want to show you," he told Fregger over the phone. On the morning of Christmas Eve, Fregger drove over to the house Brodie's mother rented in Redwood City, where Brodie had moved after being released from the hospital. She'd moved to be near him when he had his accident. She greeted Fregger and showed him in. Brodie was inside, in his wheelchair, with his Mac, but the Mac had been set up not facing Brodie but in a way that Fregger could sit down to use it and get a personal demo. But there was something else. Brodie had set up an arrangement of Mah-Jongg tiles on a table. They'd been arranged in the "turtle" formation, much like they were displayed in the PLATO version of his game. He referred to the version of the Mah-Jongg tiles on the table as *The Turtle*, and Fregger assumed that was the name of the game.

"*The Turtle* appealed to me immediately," Fregger would later say.

"The game was simple but wonderful, a compelling challenge people would probably want to play again and again. I said as much to Brodie."

Fregger wanted a copy of the game so he could play it more and check it out thoroughly. Brodie was hesitant to let his baby go away with an Activision producer. He asked Fregger for a personal promise that he would protect the game, and he also asked him to sign a nondisclosure agreement. Fregger agreed to both, and watched as Brodie's mother helped Brodie get positioned in front of the Mac and set up with his mouth stick. He deftly fired up a word processor program, made some tweaks to the document, and finalized the nondisclosure agreement while Fregger stood and watched in amazement.

"It was at this moment," Fregger would later write, "that the immensity of his accomplishment hit me. He had developed *The Turtle* in just this way—pressing one key at a time with a stick he held in his mouth. He had programmed the entire game, created all of the graphics, done it all . . . one keypress at a time. I was amazed. Later I would see a picture of Brodie taken before his accident. He was in position on the parallel bars, his form perfect, his body healthy. I felt a sadness he couldn't have always been that way, but I didn't feel pity. Brodie didn't allow pity. He had programmed *The Turtle* as therapy. He had *seized the moment*, he had accepted this unexpected, unwelcome turn of events, accepted the facts of what he could no longer do, and began discovering what he could do."

Fregger took the game home and let his wife try it out on their Mac. At five in the morning, he found her still glued to the Mac, playing Brodie's game. "I began to think we might have something here," he says.

The next day at work he shared the game with a colleague at Activision, under strict rules that he try it at home and that he not show it to anyone other than his wife. The colleague agreed. A few days later, he heard that the colleague and his wife had spent the entire weekend glued to their Mac playing Brodie's game.

Fregger suggested to Brodie and to colleagues at Activision that they call the game *Addiction*, which Brodie was not a fan of. Nor was Activision's marketing department, who instead suggested the name *Shanghai*. That's what they decided to go to market with.

The private reactions to Brodie's game, all positive and citing how addictive it was, smoothed the way for Activision to green-light the project. The major remaining hurdle was a contract. Fregger produced a contract for him to sign. "The contract was sixteen pages long, which floored me," says Brodie. "I was expecting two or three." Reading it through, Brodie noticed its provisions were stacked in Activision's favor. He describes it this way: "We get *this*. We get *this*. We get *this*. You get . . . *that*."

For Brodie it was not a good start. The two went back and forth haggling over the contract for a few rounds, but in the end Activision got most of what it wanted. It also got something it *really* wanted, much to Brodie's regret later. "I was naive enough to say, *Sure, have the copyright*," he says. *"What am I gonna do with it?"* What he hadn't considered was that color machines were coming; many more makes and models of computer beckoned as potential future platforms for the game. Plus, potentially lucrative *sequels* to *Shanghai*—a potential business franchise with a revenue flow for years. All of those things were tied to the copyright, and whoever owned the copyright owned a lot of interest in future versions, and derivations, of the game. Says Brodie years later, with a tinge of regret, "I just handed that over to them."

Then there were the *ports*, the work of translating ("porting") a computer program from one computing environment to another. Activision wanted to port Brodie's game to five other personal computer platforms, including the IBM PC, Amiga, and Apple IIGS. "The porting thing," he says, "was a little stingy as well." Activision would take about $5,000 out of his royalty payments, in advance, for *each* port they did, "to pay the guy doing the port."

The net result was that Brodie would not see a dime from Activision for nearly two years from the time *Shanghai* was initially published in 1986—peak years for the game's sales. Despite the fact that it became an overnight bestseller, winning every sort of award and media recognition in 1986 and 1987, Activision held on to every penny for a good long time. The reviews from that era praised the game for being not only deceptively simple but also addictive, not only to the reviewer, but also to his or her family and sometimes even to co-workers at the magazine that published the review. One reviewer bragged about how many times he'd been able to "see the dragon." The game's simplicity initially seemed to turn some people off, as if the game was a mere trifle, but then if they actually *played* it, it was another story altogether. Hours would pass quickly, as if the game put you under a spell. "We were both overwhelmed," wrote one reviewer of his and his wife's experience playing the Amiga port of the game. "Then the children were similarly affected. As of this writing, our *Shanghai* mania is of such proportions that I am beginning to fear for our health." In the June 1987 issue of *inCider*, *Shanghai* was written up as an Editor's Choice for the Apple IIGS port of the game. The editors explained that they ordinarily would not list a game as an Editor's Choice pick, especially since *Shanghai* "isn't even a new game, though it's new to the Apple II—it was a smash hit on the Macintosh last year . . . [b]ut our objections to *Shanghai* faded once we started playing—and playing, and playing. . . . The trouble is, *Shanghai* is unbelievably addictive." One *inCider* editor put it this way: "I don't know how a game with no spaceships, lasers, spies, villains, violence, bats, or balls could be so habit-forming, but *Shanghai* is. Once you get going, it's hard not to put the mouse down." In 1987, *Amiga World* gave the game an Editor's Choice Award, saying that one editor at the magazine "plans to name his first-born child after *Shanghai* author Brodie Lockard."

While Brodie may have felt his original contract with Activision was not ideal, it did after a while bring in considerable income. One day, sometime around 1988, Brodie got a call from one of the bosses of his Activision producer Brad Fregger telling him, "We struck a deal with Radio Shack a little while ago, and I have a check for you for $40,000."

"Could . . . you repeat that?" Brodie told the guy.

"So at that point," Brodie recalls, "things started coming in. And it was a lovely stream of income for many, many years."

Ever the entrepreneur, Brodie looked for new ways to channel his creative output to Activision, hoping they in turn would channel more money his way. Brad Fregger was producer for another Activision game, *Solitaire Royale*, and Brodie landed the job of doing the ports of *Solitaire* on the Mac and then on Apple's new Color Mac. Activision eventually came out with *Shanghai 2.0*, for which Brodie developed the Color Mac version. "The guy who did the Amiga port made it 3D and color and I thought wow, that looks great, so that's what I did on *Shanghai 2.0*." In addition, Brodie had developed an add-on to the game, hoping to make a little extra money. Activision decided not to publish it, but they did agree to include his flyer in their *Shanghai 2.0* packaging. "I made $4,000, which was enough to buy a scanner back then, so that's what I did with it," he says.

Activision then decided to launch a remake of *Shanghai* called *Shang-*

hai II, which arguably was *Shanghai 3.0* but with the roman numeral II. They hired someone else to develop it, but for one reason or another the developer and Activision parted ways. Brodie offered to do it, only this time he negotiated a contract more favorable to himself. Activision's CEO by this time was Bobby Kotik. Brodie negotiated a deal with him directly. "How about no advance of any kind, and my royalties go to 14 percent?" Brodie suggested. Kotik agreed. "And that turned out to be a very lucky move." It would make him a lot more money than the first *Shanghai*. "*Shanghai II* bought my house" in Hawaii, he says. From the living room window he has a commanding view of the nearby Ko'olau Range, looming over him.

According to Brad Fregger, *Shanghai* "became one of the most played computer games in the world, selling over ten million copies in all of its variations." For Brodie, who started the original *Shanghai* in 1985, the last Activision royalty check came somewhere around 2001. "So," he says, "I had a pretty good run."

At some point in the early 1990s, Brodie took a trip with his family to see one of his grandparents, who lived in Illinois. For fun, he asked that they stop in Champaign-Urbana to visit CERL—to make, he'd admit years later, a pilgrimage to the old Power House, to the source of the Orange Glow, where it all began. "We were just driving around and I just wanted to stop by and take a look. . . . We actually went in . . . I looked in the doorway."

With the Activision income dried up, and Brodie now based in a new home in Hawaii, he decided to get consulting jobs doing programming for three separate technology start-up companies. Just in case one of them went belly-up, he'd continue to have a steady stream of income. Plus, he wanted to do a new game. Consulting gave him a financial cushion while he explored working on a new game.

The new game eventually to emerge was called *Yucatan*. Instead of arranged tiles like Mah-Jongg and *Shanghai*, *Yucatan* was made of colored stone blocks, arranged to form Ziggurats. Once again, the same pattern reappeared, a pattern that had been recurring in every game Brodie had made since Mah-Jongg on PLATO. Perhaps all along the pile of tiles in his games had represented—and were his way of coming to terms with—the cushions of the crash pit back in the fateful Stanford gym.

Instead of the success he'd had with Activision, he found no interest from the newer generation of publishers. Times had changed. Even in a world now with billions of computer users, it was near impossible to make a hit game anymore. Players have too much choice. Smartphone apps are cheap or free. Web games are worse: developers might make a few pennies per sale. Brodie's *Yucatan* still exists on three casual gaming websites, and over several years he's only made a tiny amount of money, not enough to live on. He gets by on consulting jobs.

So there Brodie was, watching the annual Kaimana Klassik in Waimanalo in February 2013, spotting this kid on the Bloodthirsty team with "MARCY" emblazoned on his shirt. After the game, Brodie talked to him.

"Are you related to Ted Marcy?"

"He's my father," Jordan Marcy replied.

"Well, he kept me alive," Brodie said. "He helped keep me alive when I broke my neck until the paramedics got there."

"You're the trampoline guy!"

"Yeah."

Incredibly, Brodie then learned that Jordan's parents were, *at that very moment*, already en route to Hawaii, their first-ever trip. Their destination was Kauai, but they also planned a couple of days on Oahu—the "gathering place," indeed.

Jordan emailed, and, to be extra sure the message got through, texted his parents, who had already left their Vermont home. On a layover in Chicago they received the messages, and made arrangements to stop and visit Brodie.

"The three of us were able to meet there for a very magical lunch together," Ted Marcy recalls. "What are the chances that all of these things would align to make that time possible?"

Since then, they've struck up a correspondence. "The coincidences keep coming," says Ted. "My daughter, Ariel, also a Stanford grad, has designed computer games for educational purposes, and my son will be a software development engineer at Amazon Lab126 next year. What is Brodie? A computer programmer who likes to design games. He has corresponded with them to give encouragement. I am a pulmonary physician and spent some time during my fellowship learning about phrenic nerve pacemakers for people with high cervical cord injuries.

Brodie has depended on these to support his ventilation. I am also familiar with ventilatory techniques to assist with chronic neurological conditions. He and I have since corresponded on ways to improve his endurance, I think with some success, according to Brodie's very careful quantitative observations."

"After my conversation with Ted Marcy," says Brodie, "I adjusted my breathing equipment schedule. For the first time since 2005, I can now do something useful and engaging—talk, compute, drive—nearly all the time I'm up, without debilitating neck and shoulder pain. It's marvelous, and I'm enjoying every minute!"

Marcy says his lunch with Brodie was the highlight of 2013. As for taking credit for saving Brodie's life, he downplays his role, maintaining that he still doesn't recall providing mouth-to-mouth resuscitation after Brodie's fall, saying that everything remains "a blur." "The real credit," he says, "for how well Brodie has done in the intervening thirty-four years goes to the health care professionals at Stanford Med Center, to those at the rehab facility, to Brodie's family, and particularly and especially to Brodie himself. It is remarkable what he has accomplished."

To this day, Brodie is quick to acknowledge how Marcy and PLATO both played key roles not only in saving his life, and in giving him a new one. PLATO "got me out of bed," Brodie says, "when I couldn't do anything else. For day, after day, after day, for months and months. It was a wonderful social connection for me. For the first couple years after my injury, it was hard to do much. So I loved going into Dnd games and chatting with people in TERM-talk and all the other great things it could do. It was a marvel. . . . Besides being a great deal of fun, and socially enabling, it was a terrific tool for me. It helped me get a master's, it let me do my homework."

He no longer uses a mouth stick. With a tinge of regret he says, "My neck is kind of shot" from the all those years of moving mouse cursors this way or that, accumulating thousands, perhaps millions, of repetitive stresses on his neck and shoulders. He does still use the HeadMaster device, the same product that originally came out in 1985 and that he used when programming *Shanghai*. All these years later, these old devices are now very rare. "I bought the last two I could find," he says. "I just move my head like *that* and it moves the cursor. . . . It's worked wonderfully for me since 1985." Personics, the manufacturer, came out with a newer model in the late 1980s called the HeadMaster

Plus, which Brodie has as well. Unfortunately, it was the last version Personics produced. While numerous medical supply websites claim that they have HeadMasters in inventory, it turns out they don't. Brodie has checked them all.

Nothing newer, no shiny twenty-first-century technology, beats the original HeadMaster, he says, so he's holding on to the few HeadMasters he has left. "The people who make parts for it stopped making parts. So they don't manufacture the thing anymore. And when it goes away, when mine die, I don't know what I'll do . . . it will be horrible."

26

A Changing of the Guard

Bitzer's dream of a million terminals had become a dream deferred. Nineteen eighty, the time frame Bitzer first aimed at to reach the million-terminal mark, passed by in a blur, like a road sign glimpsed in a rearview mirror. Nineteen eighty-five, the next milestone, was in clear sight, but not the million terminals. There were sixty thousand employees working for CDC in 1984, the year that revenue reached $5 billion. In both cases, those numbers turned out to be peaks never to be attained again (by 1989 the company would only have seventeen thousand employees). Some of CDC's initiatives with PLATO were modestly profitable, but just as many were not. CERL and CDC had decoupled from their long-held code-sharing relationship, and were going their separate ways. UI authors were unhappy with not only the royalty rates they were earning from CDC publishing their lessons, but also the fact that the company decided to compete with those authors by developing its own courseware. "They never seemed to learn that lesson," says Bitzer. "They should have just taken our courseware directly. They had a license for all of the lessons. [It] would have worked out better than the way it was done. We had a much better remedial program for adult education than CDC had. In my opinion, they produced an inferior program at a very high cost because they had an organization that needed the work. They made an authoring deal with a professor from Penn State. His agreement produced more royalties than the sum of all the royalties for the UI authors. That didn't go over very well with the UI authors."

In addition, despite continued rosy statements from Norris (who in 1984 predicted that by 1994, PLATO would be bringing in $1 billion annually from the public schools *alone*), the world was growing increasingly skeptical that CDC was going to survive, let alone thrive, particularly due to PLATO.

"CDC's relationship with us by the late 1970s reached a point of benign neglect," says Bitzer. "CDC did their own thing, and we were doing ours. . . . We also helped create the state of benign neglect." The dream of a million terminals was still alive, but it was becoming questionable whether CERL would be running CDC equipment. Since the early 1980s CDC had been teaming up with other companies, including Texas Instruments and Atari, to develop software that would enable PLATO courseware to run on microcomputers from companies like those, but the market reception was weak. CDC partnered with competitor WICAT (World Institute for Computer-Assisted Teaching), forming a joint venture called "PLATO/WICAT" that attempted to chase the K–12 school market, but the market showed little interest. Meanwhile CERL was paying hundreds of thousands of dollars per year just to maintain its aging CYBER hardware. On top of this, CDC decided to invest heavily in developing a 64-bit processor mainframe— only to discover how completely incompatible with PLATO it was and then deciding not to convert PLATO to 64-bit. That was the signal to some observers that PLATO's future at CDC—at least at the mainframe level—was dead.

All of these factors were swirling at CERL and CDC. For CERL, it was becoming clear, thanks to Moore's Law, that with the cost of components dropping fast and the power of those components rising just as fast, it was soon going to be cheaper to find an alternative, non-CDC solution to running PLATO at Illinois. "I think they've realized," Bitzer would say in 1986, "that they've kind of dropped the ball. . . . And my guess is that they're going to go more into the cluster and standalones and look forward to us taking over the big central portion." Bitzer, on the other hand, was still firmly holding on to the ball. Problem was, the harder he held on, the more the ball lost air.

That "big central portion" Bitzer spoke of had a name, at least in the hallways and notesfiles of CERL: "Colossal PLATO." It was never clear whether the appellation was chosen to describe the project's ambitious scale or its folly in the face of a networked, microcomputer-dominated future. "That's a no-no name," said Tebby Lyman of the term in 1986. "There was an evolution within CERL itself," says Frank Propst. "PLATO began to be labeled a 'dinosaur' within our own staff." Beyond many of the key junior contributors, students like David Frankel, John Matheny, Phil and Kim Mast, Ray Ozzie, Tim Halvorsen, Len Kawell, Al Harkrader, and countless others, numerous senior people had left as well: Paul Tenczar, Bob Rader, Bruce and Judy Sherwood, Ruth Chabay, Dave Andersen.

Nevertheless, work began on the next generation of PLATO. Don Lee, one of CERL's longest-serving software geniuses, and an up-and-coming, stellar electrical engineer, Lippold Haken, were enlisted to design and build a custom mainframe, one that would free CERL from CDC and set it on its own way.

But there was another twist. Bitzer had decided that the future of PLATO at CERL was one that not only did not involve CDC hardware, but also in important ways would not involve CERL. He decided to start a company, called University Communications, Inc. (UCI). UCI would distribute service via satellite to send display information down to the terminals, and continue using phone lines to gather up a classroom's worth of simultaneous user input, and return it by a single phone line back to the central system. Bitzer's goal was to reduce the cost, which would have continued to be prohibitively expensive if he sent all the users' display information down to the terminals via phone lines. It was a kludgy approach but apropos of the PLATO tradition. By going the satellite route, Bitzer was forced to relax the strict tradition of the Fast Round Trip to accommodate many thousands of terminals connecting via cheap satellite connections (the laws of physics dictate that if you send a signal up to a geosynchronous-orbiting satellite and the satellite sends it back down immediately, that trip alone even at the speed of light still takes 240 milliseconds). UCI contracted with a company in Arizona called Compu-Sat to provide the satellite services.

Meanwhile Control Data grew frustrated at CERL's continued use of the term "PLATO," a trademark that the University of Illinois had sold to CDC as part of the 1976 deal. It was one thing for CERL to offer services under the PLATO moniker to educational and government-run organizations, but now it was morphing into something that was to be sold commercially by a for-profit company. CDC's lawyers sent repeated reminders to drop the name and use something else. Eventually it was decided that the new system, marketed by UCI, would be called "NovaNET."

UCI had three stated goals: 1) dramatically reduce the per-terminal cost of the service; 2) maintain or even improve the system's performance for each user; and 3) get to very large scale. Only now, the "very large scale" was no longer a million terminals. Bitzer's new number was 400,000. There was little mention of noneducational, commercial services like he had discussed back in 1975 at the conference in Virginia. NovaNET was a continuation of the pure, original PLATO vision going back to 1960.

Lippold Haken had been hanging around CERL since he was a young student at Uni High, taking a Latin class on PLATO III in 1971. During his teenage years he tinkered with a Gooch synthesizer box, and wound up writing one of the earliest computer-generated musical notation printing programs. On at least one of Bitzer's dog-and-pony PLATO trips in the 1970s, Haken tagged along, and during the demo his job was to sit on a basketball to provide sufficient air pressure for the PLATO IV terminal's pneumatic microfiche slide projector. "During the whole demonstration I'd sit on the basketball, and as they showed more and more slides the thing would deflate more and more," says Haken. By the mid-1980s, Haken had undergraduate and graduate degrees in electrical and computer engineering (he got his PhD under Don Bitzer), and had become a UI professor.

"Just because I had experience doing some hardware," he says, "I was drafted into this project to help build this computer which Don Lee had sketched out. When asked point-blank is this possible to build, I couldn't say no, on the other hand I knew it was a pretty crazy project to undertake to build your own mainframe computer."

Bitzer had searched for an existing product, but found no available computer with sufficient processor speed and memory capacity that would be cheaper than what he calculated it would cost to build on their own. Lee and Haken selected off-the-shelf components, used software to design the interconnections of chips and wiring, and got to work. The goal was to build a machine that was essentially a clone of a CYBER computer, designed to run existing TUTOR lessons unmodified but approximately ten times faster than any CYBER, and with a lot more memory. Haken named the machine the 'Zephyr' after once riding Amtrak's Chicago-to-San-Francisco rail line with the same name. "Basically it was cheap. It cost a million bucks to build a computer." People like Don Lee had real salaries, he says, "but a lot of it was slave labor." The machine was largely paid for from service charges to the many departments and institutions that had terminals connected to CERL's system.

—

The UCI business posed difficult questions for the University of Illinois. Ten years earlier, Bitzer had pioneered a major technology transfer from an academic institution to a commercial business when it licensed PLATO to CDC. Now he was presenting another pioneering scenario to the university: he would once again take technology created at CERL and transfer it to a for-profit business to commercialize it and scale it far bigger than was possible or appropriate in a university setting. But this time the issues were even more thorny: he proposed to run this new for-profit company *himself* as its CEO while continuing to be director of CERL, at least for the first few years. He also wished to raise venture capital for the UCI start-up. The potential for conflicts of interest was high. As early as 1984 he was having meetings and exchanging memoranda with members of the UI administration, presenting a variety of scenarios to the university to find one that would be acceptable to them so he could proceed.

Meanwhile, CERL's business office, which managed all of the accounting for the many local and remote sites that leased equipment and subscribed to CERL's PLATO services, had become a disorganized nightmare, and the person who ran it was dismissed. The business office reported to Frank Propst, the associate director of CERL, who then hired a new person to run it named Barbara Montgomery. Propst, married at the time, began a relationship with Montgomery, which led to him divorcing his wife and marrying Montgomery.

One Saturday morning Bitzer called Propst over to his house to discuss the idea of commercializing PLATO, now that it was generally agreed that CDC was failing to successfully accomplish that mission. Bitzer had also invited Dave Koffsky, an executive of University Patents, Inc. (UPI), a company that administered UI's patents. After being presented with the idea and the plans, Propst says he expressed great concerns about conflicts of interest, deeming them so complex and intertwined as to be irresolvable, due to UPI's complicated financial relationship with the university and with Bitzer and CERL. Propst claims that Bitzer clammed up from that point onward. "I was kind of out of the loop from then on," Propst says. "But during that time, Bitzer and UPI did in fact form this corporation, UCI." UPI managed to raise $2.65 million in venture funding for UCI.

Propst began quietly complaining and making negative comments behind Bitzer's back about his concerns around the growing conflicts and possible improprieties relating to Bitzer, UCI, and Compu-Sat. He also raised a red flag about how the arrangements were being set up with UCI vis-à-vis the CERL staff—particularly the disbursement of UCI stock, which he felt was being unfairly distributed, with some people receiving grants far in excess of their level of contribution to the PLATO effort over the years. "I kept telling them there is a *terrifically* difficult conflict of interest situation here," Propst says. Dillon Mapother, a longtime UI professor of physics who had become associate vice chancellor of research during the 1980s, became a sort of intermediary between Propst's concerns, Bitzer's efforts to get UCI up and running, and the university's own interests. "Bitzer concluded that the CDC effort was not going to be the grand launching of PLATO that many people thought it would be," says Mapother. "Bitzer's feeling was that the CDC people just hadn't done it right, and if it were really done right, [he] could achieve all the recognition, acceptance, and so forth, that they had hoped. So, I mean basically Bitzer said, what we want to do—and this became almost a fanatical thing with him—he wanted a chance to start a business that would demonstrate the merits of PLATO. Bitzer was really quite naive as a business person, and he spent the better part of, I think about two years, going around knocking on doors, giving presentations to American venture capital groups."

"I was caught in this situation," says Propst, "where I knew that the future of the laboratory was at stake because it was going to go broke, it was going to go deeply, deeply into debt, at this same time I was caught in this position of having to worry about the efficacy of the financial operation of the laboratory, and in addition to that, the issue of the relationship between the University of Illinois and this company that they were licensing, which was, in my mind—the agreements were insane. . . . After the damned agreements were executed, Dillon Mapother sent me a copy of them and said, 'Would you give me your comments on these?'"

Propst wrote a long memo identifying what he felt were all the problems in the agreements. "I took that memo down to Don," he says, "and I said, 'Look, Don, I've been asked to write this, and it's not real good, and I'll simply write him a note and say I'm too busy to respond, if you want me to. Or, if you want me to, I'll just go and send this memo through.' And Don looked at it and said, 'Well, I won't tell you not to send it, but I think you're crazy.' So I went ahead and sent the thing, and then it just kind of went downhill from there."

The decades-long relationship between Bitzer and Propst was by this point all but finished. At the same time, many in CERL were avoiding both Frank and Barbara Propst. In interviews years later, many former CERL staffers described how they saw the famed ebullient culture of the lab becoming a toxic workplace of suspicion, innuendo, and gossip. Learning of Propst's memo, Judith Liebman, a newly hired vice chancellor for research, got involved and wanted to know what was going on. She requested Bitzer to come meet with her in January 1987. A later court document described what happened: "Bitzer complained to Liebman about difficulties he was experiencing in working with the Propsts. Bitzer told Liebman that the mistrust between himself and Franklin seemed unremediable [*sic*], and that he doubted the three could continue working together effectively. In response, Liebman suggested that Bitzer consider removing the Propsts from their positions. In late January or early February 1987, however, Liebman, vice chancellor for academic affairs Robert Berdahl, and then-Chancellor Thomas Everhart decided that it would be inadvisable to transfer the Propsts without first investigating their allegations." According to Propst, Bitzer came into Propst's office a week later, informing him that he was being let go, but could not or would not explain why. Then the university engaged the services of a Chicago auditing firm to audit CERL's books. "We launched an audit," says Berdahl. "Propst alleged some irregularities in the operation of CERL by Bitzer. Claiming essentially that Bitzer was using the resources of CERL including financial resources to build his own private investment, his own private company, UCI, and this was misappropriating resources from CERL. That was a very serious accusation and as a consequence we ordered an audit. . . . [The auditor] was used to investigating criminality, basically, and he decided to treat this that way, and so he was not very disclosive [*sic*] of what he was investigating. So Bitzer knew there was an audit going on, but Bitzer had not really [been] informed of the allegations that had been made."

The audit finally was completed in October 1987, but by then Bitzer had begun drafting letters of resignation and had in fact threatened to resign at least three times. Berdahl was not fond of the auditor. "I found him to be very unpleasant and a person that I had absolutely no confidence in at all," he says. "And [I] actually asked that he be replaced as auditor at one point. He was treating a very distinguished faculty member as if he was a common thief. So it was rather interesting. We lost Bitzer largely as a result of some of that."

"After causing considerable disruption at CERL," one court document says, "the audit ultimately absolved Bitzer of any wrongdoing. Believing that the laboratory could not continue to function with the disharmony that had been produced, the administrators removed the Propsts from their posts at CERL in late November 1987. Barbara Propst was assigned to the position of assistant dean in the College of Applied Life Sciences, and Franklin Propst was returned to his previous position as a tenured professor of physics. Neither transfer resulted in any loss of compensation or rank, and both Barbara and Franklin received their scheduled salary increases for the 1987–88 academic year."

Frank Propst filed a federal lawsuit on November 25, 1987, claiming that he had been transferred in retaliation for what he believed were valid whistleblowing accusations regarding Bitzer, and that such retaliation amounted to a First Amendment violation. Barbara Propst subsequently filed her own similar suit. Thus began a period of great distraction and disruption at CERL and in the halls of the university administration, going on for the next several years. Numerous UI officials and CERL staffers were deposed. Rumors flew of alleged wiretapping of a CERL staffer's phone. At one point the Propst camp was given court orders to reveal the notes made by his psychotherapist. The lawyers on both sides dug in, and the fight dragged on.

Meanwhile, Control Data was collapsing. Revenue plummeted. Employees were let go by the tens of thousands. Parts of the company were sold off. And then in July 1989, it was announced that Control Data was selling PLATO. One piece—including the "PLATO" trademark—would go to an outfit called the Roach Organization, which would focus on K–12 school markets, ultimately delivering PLATO courseware on networks of microcomputers. Another piece eventually went to University Online (UOL), which would continue to run aging CYBER computers and deliver university-level PLATO (now called CYBIS) courseware. UOL would later rename itself VCampus. Another piece was sold to Drake Training and Testing, a firm that would continue the work of delivering certification testing of NASD brokers—what had been one of CDC's few lucrative PLATO businesses for years.

The luxury of hindsight makes it easy to look back on CDC's long saga to establish a lasting, sustainable business out of PLATO and wonder how they could have messed up so badly. Dave Woolley believes that CDC had a chance to transform PLATO into a huge online service for consumers. The very thing Bitzer spoke of back in 1975, a service that was not just about education, but would include commercial services, games, online community, and collaboration tools—whatever consumers might find worthwhile—all things that would help subsidize the educational dimension of the service. "I've often thought that if the upper management had a better understanding of what PLATO was, they had everything they needed to become Lotus and America Online and Macromedia all rolled into one. And they blew it, they didn't become any of those things. They went down the toilet. They had no concept of the value of the communication features of PLATO. Of course, the people working on the front lines, you know, the low-level grunts, understood really well how valuable it was, I mean, we used it, we used the notesfiles intensively to coordinate our own work. But, I think as far as the middle management was concerned, the notesfiles were just simply something that distracted their employees from doing meaningful work. And they were something to be squashed if at all possible, and certainly not anything of any value that you might consider developing into a product and make money off of it. I don't think that concept ever even entered any of their minds."

Soon, Bitzer carried out his threats, first taking a leave of absence, and then formally resigning. He had received an outstanding offer from North Carolina State University, in Raleigh, North Carolina, a state newly flush with money as businesses and banks were transforming the "Research Triangle" region into a Silicon Valley of the East, and he took it. The lawsuit, plus what he believes was unfair and nasty treatment by the University of Illinois, finally took its toll. A memo to the PLATO and NovaNET community was posted in =pbnotes= on August 11, 1989 (not even a month after Control Data announced it had sold off PLATO):

On November 17, 1989, I will retire from the University of Illinois. As many of you are aware, I have been at the University of Illinois since 1951 when I came here as an undergraduate student in electrical engineering. After retiring from the University of Illinois, I will assume a position as a Distinguished University Research Professor at North Carolina State University during the academic year. This unique position will allow me to continue to pursue research in a variety of areas including communications and computer graphics. Since this position is for the academic year, North Carolina State University has made it possible for me to spend the summer months at CERL. I look forward to my continued relationship with CERL, and working with its wonderful staff as they move forward into the 1990s. I am pleased that the new CERL NovaNET system is near completion and that it is already delivering education to many sites. Although much work is yet to be done, I am confident that the staff at CERL will continue to expand the NovaNET system and to be leaders in the educational research it makes possible. I personally hope to help contribute in improving the NovaNET system in the future.

Donald L. Bitzer

Suddenly there was a void left by Bitzer's departure, and, just as Jock Hill had described what happened when John Dammeyer had to step down from leading the PLATO effort at CDC in 1975, the wolves came out. While CERL was faced with finding a new director, forces within the university administration were circling with other plans in mind, including shutting down the laboratory altogether. The UCI/NovaNET controversy, the audit fiasco, and Bitzer's ultimate departure caused the university to rethink CERL's continued existence.

"To the extent that CERL was so much an extension of Bitzer's own ideas and energy and personality, probably his leaving led to some extent to the need to review whether or not CERL should remain a part of the university," says Berdahl. "The other thing that was going on of course was that we were going through some very serious budget reduction at the university. People in the university, particularly in the college of engineering, felt that CERL was getting very substantial subsidies from the university. . . . There were many who believed that the technology, the mainframe technology, with a specially designed computer [the Zephyr] at that point was an outdated technology and one that would be overtaken quickly . . . and that to continue to invest in that technology was shortsighted on the part of the university. And so there was a lot of pressure from the computer and engineering sector of the university to begin to withdraw that university subsidy and force CERL to sort of exist on either the revenues that it could generate through research contracts or the revenue it could generate through the UCI revenue stream."

Bitzer's decades-old vision for PLATO had had a long, productive, profitable run. But like so many technology visions—and stubborn technology visionaries—the world eventually moves on, and people conjure up new visions. The centralized vision of PLATO was overtaken by the new world of microcomputers and networks. Berdahl believes that CERL was caught up in a great technological divide between centralized computing and the new networked microcomputer movement, and that the lab was on the wrong side of the divide. "I think that what happened in that process," he says, "is that all of the CERL technology was considered to be obsolete."

CERL was assigned a series of acting directors, including Ned Goldwasser, a physicist famed for his work at Fermilab, and then Lorella Jones, another physicist.

"Don Bitzer disappeared very suddenly from the scene," says Goldwasser, "and somebody had to go in there and hold things together and it seemed reasonable and I was asked to do that. . . . I knew nothing technically about PLATO . . . nothing about its hardware, nothing about its software. And therefore I was not the kind of person who should be managing a project like that, a person who *knows* all of that should be. . . . I certainly felt like a fish out of water. I can remember when I first met with the staff, and tried to introduce myself and tell them what my interests were and my ignorance was and so on and so forth, it was a little clumsy. But I have to say I was welcomed, I thought very warmly."

However much Goldwasser and his successor Lorella Jones tried to resuscitate CERL, the forces arrayed against it were to prove too immense. Goldwasser and Jones found that much of their time was eaten up finding a justification for *not* shutting down CERL. The assumption, it seemed clear, at the highest levels of the university, was that CERL was to be closed. Any other more positive outcome would require a miracle. And the miracle worker, the man who had managed to keep CERL going for so many years, was no longer there.

Eventually, this controversy was handed over to another vice chancellor, Chet Gardner, who believed that funding for CERL should stop, that the lab had outlived its usefulness, was not performing any meaningful research within the university community, and was a growing budgetary drag on a university already forced to endure substantial belt tightening.

CERL staffers began a letter-writing campaign, pleading with anyone and everyone whose departments or organizations had used or were using PLATO and NovaNET to urge the powers-that-be in the university to spare CERL from shutdown. Dozens, perhaps hundreds, of letters were sent to the chancellor and various vice chancellors. They were unconvinced. Soon, a decision was made: a process would commence starting in 1993 to gradually shut down CERL. Staffers would be offered contract buyouts, or assistance with job placement elsewhere within the sprawling university campus. Not every staffer was offered a buyout, and the members of the lab remain resentful and bitter about this episode to this day.

During this time, another computer laboratory, the National Center for Supercomputing Applications (NCSA), a few minutes' walk from CERL and led by a new hotshot professor named Larry Smarr, had stolen PLATO's limelight. The lab, established with massive National Science Foundation funding in the early 1980s, had a mission to use supercomputers to aid in scientific research, including visualization of what years later would be called "big data." In 1993 two student NCSA programmers, Eric Bina and Marc Andreessen, released a new software program that would change the world more than PLATO ever did: a tool called Mosaic, described as a "browser" for something new and mysterious called the World Wide Web.

By the end of 1994, most of CERL's staff was gone. In order to finish out some projects, Lezlie Fillman was permitted to keep her position a little longer than most—all of the former staffers' phone lines were routed to her desk, in an office on an otherwise quiet floor of the building—and then her position was ended as well.

In January 1995, Charles Bridges, a longtime "s" systems programmer, went up to his office on the third floor, signed on to his PLATO terminal, and sent a pnote to Michael Walker, who recalls its opening sentence being along the lines of "By the time you read this . . ." Then, allegedly following instructions from a how-to book called *Final Exit* on his desk, took his own life.

Under orders from the university, the CERL technicians gathered up hundreds of terminals—including nearly all of the PLATO IV and PLATO V machines with their plasma displays—and had them destroyed. "I tore many of them up as part of my job," recalls Jim Payne. A tiny number of terminals survive to this day—some are now in museums—and the plasma panels, now over forty years old, still work perfectly, their orange glow as bright as ever.

The Power House, home to so many innovations over the decades, first from CSL and then from CERL, would soon be taken over by the neighboring mechanical engineering department. The fourth-floor machine room, home of the CYBER hardware for years, was emptied, the Zephyr and other equipment—including the now ancient and yellowing tiles that comprised the raised floor of the computer room—moved to a commercial space in downtown Champaign. UCI, perhaps in attempt to put an unpleasant past behind it, renamed itself NovaNET Learning, Inc., and retired the Zephyr after acquiring a remarkably small, affordable DEC Alpha computer and creating a complete emulation environment of a CYBER machine, with PLATO happily running as a job in the operating system.

CERL's old fourth-floor machine room became a lab for mechanical engineering students to study washing machines.

The Propsts' various lawsuits were consolidated and dragged on through the courts all the way to the Supreme Court in 1995, which denied hearing their cases, finally putting an end to a long, ugly chapter in the history of the university.

Bitzer was by then thriving at North Carolina State University, busy teaching a number of courses, earning rave reviews from students, as well as taking on a flock of bright, eager, graduate students. By 2003 he had developed an explanation for why he left CERL and Illinois: he had always loved working at CERL because only 5 percent of his time was spent with paperwork and administration, and the rest of the time he could have fun, do experiments, and actually build things. But then in the 1980s, it became 90 percent paperwork, lawsuits, and fighting the administration. And it was "no longer fun," he says. "North Carolina is fun." As of 2017, at eighty-three he is still teaching there, and has never stopped talking about the wonders of PLATO and NovaNET.

27

Leaving the Nest

One unfortunate but usually inevitable consequence of an academic lab like CERL was the transient nature of the community. Particularly the high schoolers and undergraduates, who would not remain in school forever. Most of them grew up and left, taking on newer, different career opportunities elsewhere. Some stayed in town, or at CERL, for long stints, but most were gone as soon as they got a degree. With ever-more-powerful microcomputers and the rise of online services like the Source, CompuServe, the WELL, and, later on, AOL and Prodigy, not to mention thousands of dial-up bulletin board systems, and of course the growing Internet, there were now copious alternatives to PLATO. There was also an increasing variety of ways to bring PLATO ideas and ways of thinking, the system's features, lessons, notesfiles, and games, to these alternative platforms.

Within CDC, hardware engineers like Jock Hill had tried to bring a microcomputing sensibility to the PLATO effort, but the corporate inertia stymied many of the chances CDC had for an early start on the competition. Likewise, many individual contributors, experienced with PLATO at CERL and CDC—people who fully appreciated the power of the communication and collaboration features of the system—began to see that there was nothing about these features unique to PLATO anymore. These kinds of tools were going to flourish in the future on other systems, whether there was a connection to PLATO or not.

Likewise, PLATO's head start on multiplayer games and game design in general was not something that could or would only exist on PLATO. These ideas were compelling, and it was inevitable that they would escape from the increasingly rarefied ecosystem of PLATO to find homes out in the heterogeneous world of networked computing.

The world was catching up to PLATO. What follows are stories of just a few PLATO people who moved on into microcomputers and the Internet, taking ideas that were commonplace on PLATO and bringing them to the world.

Brand Fortner, Bruce Artwick

Brand Fortner's popular Airfight game was remarkable in its time for its depiction of the physics of flight and three-dimensional view out the window of the cockpit. (It was certainly slow as molasses to replot the screen, but no one cared. The Fast Round Trip made the game incredibly thrilling anyway.) Fortner decided later to build a more formal flight simulator, called AirSim, which lacked the air combat arcade action, instead focusing on a more accurate, realistic simulation of flight in a number of different aircraft, chosen from a menu.

While Fortner was working on Airfight, UI's Aviation Research Lab was experimenting to create higher-quality real-time flight simulation graphics using parts cannibalized from a PLATO IV terminal, including its plasma display, which they hooked up to a more powerful PDP-11 computer. One of the students working on this other project was Bruce Artwick, who made it his master's thesis. After graduation he moved to Los Angeles to work at Hughes Aircraft but got the idea to revisit flight simulations using a microcomputer. He created such a game on an Apple II, formed a tiny company called subLOGIC to sell the game, and found insatiable market demand for it. In time he would develop more sophisticated versions for the IBM PC, which were licensed to Microsoft as the Microsoft Flight Simulator, one of the most popular PC games of the 1980s and 1990s.

Artwick once observed that his experience was the direct opposite of Fortner's: Artwick approached the flight world first by seeing simulators as training aviators, and only later did he get the idea to develop a game version. Fortner started with a game and then later built a flight simulator.

Silas Warner

Silas Warner, who while at Indiana University during the first half of the 1970s contributed to Empire early on and then created a number of other popular, innovative PLATO games, including Robotwar, eventually got a job at Commercial Credit in Baltimore, working on PLATO. He eventually joined Muse Software, a tiny game publisher based in Baltimore, as employee number one. There, he developed a variety of programs for the Apple II that did not sell very well, as well as a version of Robotwar, and his most famous and consequential game, *Castle Wolfenstein*. Like Mark Johnson's Drygulch, Warner took the dungeons and dragons game motif and placed it in a new, exciting context, in this case, a Nazi-controlled castle during World War II. *Wolfenstein* resembled the old PLATO game Dnd, in that it provided a top-down viewpoint of the dungeon mazes and rooms. Remarkable for its time was support for primitive audio clips associated with game action, including soldiers yelling to the player. Warner's game was hugely popular for a number of years, spawning in 1984 a sequel, *Beyond Castle Wolfenstein*, which Warner also developed. By the early 1990s, another company, id Software, independently brought the Nazi castle motif back with vastly improved technology. *Wolfenstein 3D* was a wildly popular first-person shooter game that in turn would lead to id Software's *DOOM* and increasingly photorealistic first-person shooter games right up until today.

Paul Alfille, Jim Horne

Jim Horne grew up in Edmonton, Alberta, and attended the University of Alberta, graduating in 1980 with a BS in physics. He stuck around the campus afterward, eventually getting a job at the university's PLATO project, developing CBE courseware, some with interactive videodisc players controlled by the PLATO terminal. Horne played the various PLATO games, finding the action-oriented ones like Avatar, Moria, Airfight, and Empire less compelling than pure logic games, including one, Freecell, written by Paul Alfille in 1978 on the CERL system.

Alfille and fellow medical students at the time, Mike Gorback and David Tanaka, had also unleashed a brilliant game onto PLATO called BND (Bugs 'N Drugs), which like Drygulch, took the dungeons and dragons motif and applied it to another setting, in this case, a hospital. As you walked the corridors of the hospital, you would encounter "monsters," but in BND they were bacteria or germs, and your "weapons" to fight them were various antibiotics. Choose the right antibiotic and you might defeat the monster; the wrong one and, well, things went downhill fast.

Alfille's Freecell was a single-player solitaire-like card game designed as a mental puzzler, where a player moves cards around the screen based on elaborate rules. Like most games on CERL, it featured a hall of fame, which only encouraged users to keep at it for far too long. The game was designed to be eminently winnable, though tricky: yet another combination that lured players in and kept them going for hours. Unlike CPU hogs like Empire and Airfight, in Freecell the vast majority of the time the player was simply sitting staring at the screen, figuring out what move to make next—a setting that was quite kind to the central processor.

By the early 1980s the game had been published by Control Data and was available on all PLATO systems, which is how Jim Horne discovered the game on the Alberta system. "I thought it was a fantastic game," he says. "It's the kind of game, man I wish I could invent something like this, because there's no luck involved after the original draw, after the original shuffle, and most games seem to be winnable. It intrigued me very much about whether all games could be winnable. And I wasted way too much time playing it."

He liked the game so much he decided to copy it to the IBM PC, writing a version that ran in DOS but took advantage of the color and primitive but built-in special characters for hearts, clubs, diamonds, and spades. "So I had everything I needed to be able to write a Freecell game." He created a little company, Frog and Peach Software, Inc., and in 1988 released the game as shareware on bulletin board systems and CompuServe, where in time a large audience of fans enjoyed it. He asked players for a donation of $10. That same year he applied for a job at Microsoft and joined the company, working on the team building Microsoft LAN Manager. "For years after that," he says, "people would send checks for $10 to my home address in Canada, where I used to live even when I had moved to Redmond to work for Microsoft."

During the time that Microsoft released Windows 3.0 and 3.1 between 1990 and 1992, most gamers still preferred the faster DOS environment for game playing and refused to use Windows because it was too slow. Bill Gates saw this and decided that Microsoft should make some games of its own to increase acceptance of Windows across this important user constituency. "He didn't want to start a whole project to do it," says Horne, "so he sent out mail to the developers of Microsoft saying, We're going to build these things called Windows Entertainment Packs. We're not going to pay you for this, but if you've got a game you've written in your spare time, we'll collect, like, the best eight of them or something like that, and throw them into this Windows Entertainment Pack which would ship on a single 5¼" floppy, basically designed to be copied and given to your friends'—there was no copy protection at the time. . . . But it was just to try to generate enthusiasm for games." A new, graphical version of Freecell was one of the games chosen to be included. Beginning with the release of Windows NT, Freecell was included with every subsequent version of Windows, including the gigantically successful Windows 95. Millions of people discovered and became addicted to Freecell.

The deal Gates struck with the lucky few employees whose games were selected for the original package was that he would grant each of them ten shares of Microsoft stock. "Which if I'd kept," says Horne, "it'd be worth a bunch of money now." Instead, at some point he sold it, buying a pair of high-end stereo speakers worth over $10,000. Windows had "telemetry" in the operating system that informed Microsoft of some rough usage statistics for what people were using on their PC. Horne discovered that Solitaire was one of the top three programs, and Freecell was around number seven, ahead of Word and Excel.

Before the game was published by Microsoft, the company required it to be declared free of any copyright, patent, or other intellectual property rights entanglements. "I had to make sure there was prior art—descriptions of very similar games, in books that predated by decades the PLATO version." That was enough for Microsoft, and they gave him the green light. In all the years of Freecell's success on the PC, Horne never called, wrote, emailed, or otherwise contacted Paul Alfille, nor did he ever include any sort of acknowledgment to Alfille's work in the game itself. Later, when Alfille, who finished medical school and became an anesthesiologist, learned of the existence of Microsoft Freecell, he was dismayed and shocked. "I assigned rights to U of I Board of Trustees or whatever we were required to do. When I saw the Microsoft version a few years later, I contacted the U of I, but they had no information or interest in pursuing the matter."

To this day, the two have never met, spoken to each other, or even corresponded. "I should probably send him a note," says Horne, "and thank him helping make me famous."

Robert Woodhead, Andrew Greenberg

Cornell University leased a couple of PLATO IV terminals from CERL in the 1970s, but never expanded beyond them in the way sites like Delaware did. Very few students were exposed to the system, though, not surprisingly, some of the ones who found PLATO so compelling and addictive would make pilgrimages out to CERL to see the home of the Orange Glow for themselves. Two such students were Robert Woodhead and Andrew Greenberg. Woodhead was an active, outspoken, and controversial participant in the CERL notesfile community. "Apparently I annoyed enough people to get a notesfile dedicated to my latest absurdities," says Woodhead. The notesfile was called =balsanotes=, in honor of the nickname everyone chose for Woodhead: "Balsa Brain." "One of the things I learned from the social interaction on PLATO—apart from learning how to be a merely obnoxious jerk as opposed to a totally obnoxious one—was not to take myself too seriously."

Greenberg and Woodhead were gaming addicts, playing Avatar, Empire, and Oubliette. "I'd been thrown out of college for a year—too much PLATO, not enough studying—and decided to try selling software," says Woodhead. "The PLATO games showed that cool games could be done on computers, so for me the question was, could I do as cool a game on the tiny PCs, in particular when the game would have to be a one-person one? I started work on a game called *Paladin* for the Apple II, and shortly thereafter by chance discovered that Andy had written a similar game for the Apple in BASIC called *Wizardry*. We put our heads together and decided to team up to rewrite *Wizardry* in Pascal and expand it considerably. Andy did most of the design, and I did most of the coding." While working on *Wizardry*, Woodhead says he was spending around $1,000 monthly to dial up directly to Minnesota to play Empire. "This was my relaxation while developing the *Wizardry* games," he says.

Wizardry: Proving Grounds of the Mad Overlord shipped in 1981 from a tiny software publisher called Sir-Tech Software, Inc. The game features a Mad Overlord named Trebor (Woodhead's first name backwards) and an evil wizard named Werdna (Greenberg's first name backwards) and eerily resembled aspects of PLATO's MUDs. In 1982 a sequel, *Wizardry II: The Knight of Diamonds*, followed with a half-dozen additional sequels coming out at regular intervals through 2001. The series was a huge hit, selling millions of copies. The games' designs influenced many other games for years afterward.

Many in the PLATO community were not happy, arguing—some going to great lengths to gather what they considered smoking-gun evidence—that *Wizardry* was a rip-off of PLATO MUDs, in particular Oubliette. Says Woodhead, "I've run into a few people who feel that way, but I think our true crime was that, often despite our best efforts, we ended up making money on it. We just wrote the game because we wanted to do something cool, and we hoped we might make enough money to pay back the time investment. We were stunned by how successful it was." He also argues that the Apple II being a standalone personal computer meant that *Wizardry* had to be a single-user game, which required rethinking many aspects of multiuser dungeon game designs that had become second nature on PLATO.

What *Wizardry* represented, Woodhead believes, was the continuation of a natural progression of ideas that had already been going on for years on PLATO: Pedit5 leading to Dnd leading to Moria and Oubliette, leading to Avatar. "I'm proud of what I was able to do," he says, "but at the same time I am happy to acknowledge the influence of PLATO in the conception of *Wizardry*, and do so often."

Paul Tenczar, Ron Klass

During the 1970s Paul Tenczar, the creator of TUTOR and one of the senior systems software managers at CERL, became increasingly convinced that the mainframe vision for PLATO was no longer viable. Perhaps some of the ideas he'd been exposed to on his trip to Xerox PARC had sunk in. But it took a while. "At first he was extremely negative," recalls Dave Andersen, a CERL systems staffer, regarding Tenczar's reaction to the rise of microcomputers, "and then he didn't talk about it for a while, then he did his own company." Tenczar's change of tune came about as he saw the inevitable, inescapable rise of microcomputers as the new direction to pursue. Once he became a convert, he still met resistance from people like Bitzer and Propst.

"When micros came out," says Tenczar, "all the empowered courses, all the computer science departments, all the major companies completely missed the microprocessor revolution. Some of us, Stan Smith was one, and then myself, began looking at the Apple II. I actually did homework, and I looked at that and realized that that has about ten times the processing power of a PLATO terminal—this was in '77, when the Apple II came out." He tried to drum up interest in the Apple II but, he says, "people thought I was nuts." After having a falling out with Bitzer and members of the senior staff, he founded, with some partners, a company called Regency Systems, designed to bring PLATO's CBE capabilities to a PC platform. "We had the first stand-alone authoring system in the world," Tenczar says. "It took people four or five years at PLATO to realize that the micros were the way to go."

In addition to Regency, Tenczar, Ron Klass, and Ray Ozzie co-founded an independent software company called USE—short for Urbana Software Enterprises. USE's mission had been to create a microcomputer version of the TUTOR language. The company operated out of offices of Regency, which would eventually subsume the USE project's TUTOR-like language, and call the new language USE as well. Regency would compete in the computer-based training industry for the next fifteen years.

In 1981 Tenczar left Regency and began Computer Teaching Corporation, shipping a new TUTOR-like language called EnCORE in January 1984. Trademark issues forced the renaming of the product to TenCORE later that year. TenCORE evolved into a powerful authoring language for PCs that attracted many of CDC's key corporate PLATO clients.

"Through until 1981 or so, I'd say PLATO was state of the art," Tenczar once explained in a 1987 interview. "Since then it's worse than that. . . . It's always been a history of fighting, of the current staff fighting the new changes. It's a real fight to have people change. And the use of micros—there's still the fight going on over at CERL. Bitzer truly believes that it's not the way. My company's success shows that it's a very viable way."

Michael Allen

Once Michael Allen's team at CDC had released the PLATO Learning Management (PLM) feature to customers, which he believed would address PLATO's lack of tools for computer-managed instruction, he embarked on a new project that would ultimately be called PCD3, for PLATO Courseware Design, Development, and Delivery. "PCD3 started as an effort to find a more cost-effective way to develop good courseware," says Allen, "because TUTOR programming was just too expensive. Looking at all the courseware and how much Norris [and CDC] was investing in courseware, it was just astronomical." And he wanted that cost brought down. Allen's expectation was that the result of this new project would be a better authoring language with more structure. "I started this because I wanted the quality to go up, I thought we needed better material, and I decided it was too hard to build complex courseware." He hoped that whatever emerged out of this new project would mean that "you wouldn't have to have years of TUTOR experience before you could build really good stuff. And what happened is that that message kind of got up to Norris that here's this guy that thinks that he can reduce the cost of developing courseware. And so these things kind of came together. I ended up getting my money. They're thinking that I'm trying to reduce the cost, and my thinking—that my primary objective—was to increase the quality."

Norris was generous with funding, and Allen formed a team to do serious research on authoring PLATO lessons. They wanted to understand what was holding back the development of a lot more CBE courseware. Why did it take so long to create courseware, and what could they do to make it easier to create? "We realized that TUTOR was the culprit," he says. "We also found out that interpreting educational or instructional designs into TUTOR was hard. You couldn't have an educator do it in most cases because they just couldn't master TUTOR enough in a short enough amount of time. So you ended up having teams all the time, you had the programmers, and you had the subject-matter expert, and you had the designers, and so on." After interviewing courseware developers and subject-matter experts, he concluded that both groups were talking past each other and not communicating their intents clearly enough. So he wanted to build a tool that would increase involvement of each side for a project without having the nontechnical people having to learn TUTOR. The result was PCD3.

Separately, another group led by Jim Glish developed a simpler, more form-based PCD1. Oddly, this came out after PCD3. To make matters more confusing for customers, CDC also had yet another team, out in California, create yet another product, PCD2, which so baffled Mike Allen he still doesn't know what to make of it.

"PCD3 I always considered to be a prototype of a product," Allen says. "As it turned out, it really wasn't another language like another TUTOR, it was more of an iconic-based approach. I thought it had some really good concepts in it and I wanted to go forward with it."

Unfortunately, Control Data's mid-1980s financial crisis prevented Allen from taking the PCD3 product any further. "It was the wrong time because Control Data was having so much trouble with its disk drive business and you know lots of its divisions ran into trouble and we were losing money and here I wanted to do something new." Right at this time, Bill Norris gave Allen a "best of PLATO award," which Allen described as "a great, big, huge, heavy, bust of Plato." "I was invited to speak to the board of directors . . . and I said we've got something here that I'm really excited about, it's a new way to develop software, and it has big implications for PLATO, and it can go more broadly than that, and I appreciate your giving me the millions of dollars that I spent by CDC's calculations, it was something over $4 million that I spent researching how to develop what a better tool might look like, a better authoring tool, and I expressed my appreciation for that and said we've got something here that's really going to take the market by storm, and lots of applause and so forth, and almost immediately after that, I talked with somebody in marketing and he said, 'You aren't seriously thinking about putting this out as a product are you?'" It turned out marketing had no money for another product—they had just spent another large sum on new four-color brochures and training materials for PLATO.

"Here's a company in desperate need," says Allen, "and we've got an opportunity here to do something that will really get a lot of attention and I think make a lot of money and so on, and he said, 'I can't. We can't do that. We've already printed our brochures!' And so I took that to Bill Norris, when I heard that, and I said, 'Listen, can you believe this? What kind of business strategy and thinking is this?' And he said, 'I don't want to admit defeat, but I got to tell you, things are not good in the company now, and support for PLATO is at an all-time low, and I really appreciate what you've done, and I think what you're doing is really exciting, but I think that as a product champion, you're going to have to take it outside Control Data and run with it, because I think it's going to die here and by the time we can get support for it and do something with it inside Control Data your technology will be dated, and so you got to go with it.'"

It took Allen over two years to get it out of CDC. Every time he tried, some VP would offer to help but then attempted to kill the product behind Allen's back. Finally he appealed to a high-level executive who ran the Government Systems Division, who was appalled at the treatment Allen was getting from various VPs. He forced CDC's executives to come up with a legitimate business plan for the next generation of PCD3, or stop blocking Allen from taking it to market on his own. "Of course, nobody had a business plan," Allen says. Once he got the go-ahead to proceed independently, lawyers raised new intellectual property roadblocks. "And so the agreement came out that I could have my know-how, but I couldn't take any documents. So that's what we did. When we started the company, I owed Control Data $2 million [for] my know-how, but if I paid them back earlier, why, there was a reverse-decreasing schedule, so if within the first year I could buy out Control Data's interest in my own company, for $400,000, then I would be free and clear forever, and owe CDC nothing more. . . . I had raised enough money so that in the eleventh month I paid $400,000 to Control Data and had the rights to do it."

The new company, formed in 1987, would be called Authorware. The product was called Course of Action, and was developed on a Macintosh. Eventually it was renamed Authorware. Before the product shipped, Allen had the team re-create Darts, How the West Was One, Titrate, and the Fruit Fly lessons from PLATO to prove that the new product could do all of those famous lessons—on a Macintosh. It was a dramatic step forward beyond what PLATO could do: here was a full mouse-controlled graphical-user-interface environment, rich with menus and icons, for developing compelling interactive learning content. The product, not cheap at $8,000 apiece for the professional version, nevertheless grabbed 80 percent of the authoring system/computer-based training market within three years, becoming the de facto standard in the industry.

Sometime around 1988 Steve Jobs made a move to hire Allen out of Authorware to join NeXT, Inc. "We went out and demonstrated the early version of Authorware to him," Allen says. Jobs was impressed. "He said, 'You like this, you're in this education stuff,' and I said, 'Yeah.' He said, 'Huge market for me, huge market for me. I need somebody who can bridge computer programming with user interface with education. You're my man, I want you to come and head it up for me.' And I said, 'Whoa, I just wanted to get your funding for Authorware.' And he said, 'Well, you think about it, but you can't turn this down.' He left, and it was about lunchtime, and I was wanting to go to have lunch with him and they said, 'Well, nobody goes to lunch with Steve in his office.' His team, the top guys that were there, took me to lunch, and they said, 'You got to think about this carefully, because, let me just tell you a few things. Every good idea you have is Steve's idea. You don't *intend* to win an argument with Steve. Even though you may actually *feel* like you've won it, you know, you only can *sell* your idea and it becomes Steve's and actually you were probably arguing *against* it. So you don't get credit for anything, except among us, you know, we all know that, and he's very temperamental, and he may stand behind you a long time and he may not. So, there's a big *risk*, but we'd love to have you.'" Allen turned down the offer. Steve Jobs was furious. *Nobody turns me down*, he told Allen.

In 1991 another hot multimedia company called MacroMind merged with Paracomp, makers of popular 3D graphics software. A year later, that combination in turn agreed to merge with Allen's Authorware company to form a multimedia authoring powerhouse called Macromedia, which subsequently had its initial public offering and was soon worth over $1.5 billion. In 1996, a struggling NeXT was acquired for $400 million by an even more struggling Apple. In 2005, Adobe Systems, Inc., acquired Macromedia for $3.4 billion.

Ray Ozzie, Tim Halvorsen, and Len Kawell

Ozzie, Halvorsen, and Kawell were three friends who worked as junior systems programmers at CERL while pursuing their undergraduate degrees at UI in the 1970s. In 1977, Halvorsen and Kawell had taken jobs at Digital Equipment Corporation (Ozzie, widely recognized by Microsoft years later as "one of the world's best programmers," had been turned down). Ozzie stuck around and cofounded Urbana Software Enterprises, working there for a while, but he was eager to move out and join the burgeoning computer industry like his friends had done. In 1979 he got a job offer at Data General in Massachusetts to join a small team led by a brilliant programmer named Jonathan Sachs.

"DEC had just started to develop this processor called the VAX," says Halvorsen. "I really wanted to be involved in the VAX. It was just starting. It hadn't really gotten anywhere yet, and I really wanted to jump on it because I was just excited about the whole idea." But Halvorsen immediately experienced the culture shock of leaving the teeming, thriving PLATO virtual community, with all of its tools for communication and collaboration. "Leaving PLATO left a gap," he says, "the gap being the entire experience of community. You lose that right off the bat. . . . I'm talking about the degree to which you're inter- acting with the rest of the community that's electronically available to you, and the constant availability of an always networked system, was, you know, addictive. Obviously it was. And so Len and I both went to DEC. And the first things we did—we went into the same group—the first things we did, I mean, I wrote TERM-talk for VMS [the new VAX operating system], and he wrote pnotes, essentially. Both of us basically wrote it as a way in which to provide some of that experience in our new computing environment. . . . That got us all started, and obviously everybody loved it because prior to that point there was no such thing as mail."

As a result of the success they had PLATO-izing the VAX environ- ment, Kawell decided to replicate PLATO's notesfiles as well. "It was all written in Pascal," says Halvorsen, "it was a simple little version, but it worked just fine."

Halvorsen and Kawell successfully imbued the VAX VMS operat- ing system with TERM-talk, Personal Notes, and Group Notes (the notesfiles), and VMS included them as features. "Essentially we were trying to see if some of that [PLATO] experience mapped and trans- lated into the commercial world."

"We were having a great old time trying to be real-life commercial programmers and whatnot," says Halvorsen, but the intense experi- ence of PLATO the three had gone through at UI kept giving them ideas for wanting to do something, a project, maybe a new company of their own. "Ray and I got the bug," says Halvorsen, "because we hung around all the time together, got the bug of wanting to start our own company. We made a couple attempts to try and start a company with venture capital and whatnot, and it didn't work out so well." Per- sonality conflicts, conflicting personal agendas, and a mediocre busi- ness plan for something they called Microcosm did not excite the VCs. Halvorsen is glad it failed. "I think it is really valuable experience when you're first starting out," he says, "is to fail right off the bat."

Meanwhile, Kawell and Halvorsen thrived at DEC.

Ozzie tired of minicomputer maker Data General, and was eager to enter the booming business of microcomputer software. He joined Software Arts, makers of the hottest software application at the time, a revolutionary spreadsheet called VisiCalc, where he was involved in developing a version for the IBM PC. In the meantime, Jonathan Sachs had left Data General to team up with a local software entrepreneur, Mitch Kapor, to create a new spreadsheet they were calling *1-2-3*. The company they formed was called Lotus Development Corporation. *Lotus 1-2-3* enjoyed explosive success in the IBM PC marketplace, and the company was riding high. Sachs kept contacting Ozzie to join the team ("Come to Lotus, come to Lotus . . . we need more good people," Ozzie recalls him saying). Ozzie was by now keen on doing his own new program relating to online collaboration among teams, which he initially called "MX," the ideas of which had been floating around for a while but were now beginning to dominate his focus. MX eventually got a new code name, "Echo," only to eventually get another, "Notes," named intentionally after PLATO Notes. Ozzie wanted to take the ideas he had seen work so well on PLATO—tools for team collaboration and productivity—and bring them to the workplace, where it was abundantly clear by the mid-1980s that workplaces everywhere were going to be filled with networked PCs. There was an opportunity to address a software product void that few even knew was coming.

Sachs continued to pester Ozzie, urging him to come meet Mitch Kapor and at least hear him out. So he did, first meeting him at a computer convention where *Lotus 1-2-3* was the hit of the show. He told Kapor, "I wanna do this thing, I know you wanna do spreadsheets, but I can't come to work here because I really wanna do this other thing." He told Kapor he and Halvorsen had tried doing Microcosm but had gotten nowhere with venture capitalists, which was frustrating, and now he wasn't sure how he was going to get VC money for his new idea. Kapor offered him a scratch-my-back, I'll-scratch-yours deal. Ozzie would help Lotus develop its follow-on product, called *Symphony*, an integrated suite of applications, one of the first on an IBM PC, and when that was delivered, Kapor would find a way to fund Ozzie's project. "He didn't know what the hell it was," says Ozzie, "he didn't understand what I wanted to do, but he knew what he needed. So we did a handshake deal, I went to work at Lotus." He set aside his "Notes" plans and worked for nine months on *Symphony*, delivering it in July

1984—ahead of schedule. "He was so happy because it only took nine months, he came downstairs, I remember it vividly the day it shipped, and he said, 'Okay, you did your part, you no longer have to do anything on this, start working on your specs for what you want to work on' and we basically discussed that Lotus was so flush with cash at the time—the cash was rolling in so fast we didn't know what to do with it—so Mitch figured that we would be able to do something Lotus funded. The moment that he raised hopes, like, hey, this could be a reality, I spent like a month putting up the specs and doing up some screen snapshots on an early Macintosh of what the thing would look like." Filled with excitement, he reached out to Halvorsen and Kawell, inviting them to drop what they were doing and join him in building this new product. Halvorsen and Kawell enthusiastically agreed to join if Ozzie could actually land the funding.

Ozzie did not want his new project to be situated in the Lotus office. "He wanted the thing to have its own culture," says Kapor. "He put it in a geographic remove, that all made sense to me." Rather than a conventional venture capital kind of funding, Kapor struck a creative deal with Ozzie that was more akin to a publishing deal: Lotus would advance Ozzie's company about $1 million from future royalty earnings, which were effectively locked up with Lotus. "It was a unique arrangement, as far as I know, and it was structured that there were incentives to work with each other and to work out issues rather than not to. And there's a lot of mutual trust. . . . There was kind of an understanding between Ray and me at a personal level that made the thing go."

When Ozzie finally presented his ideas for "Notes" to Kapor and Lotus's internal product committee, "they basically thought it was wacky, they didn't understand how I would do what I was saying, but they all nodded their heads, *yes, yeah, sure, what the heck, fund the guy.* Except for one guy, the VP of marketing at the time, Jim Manzi." Manzi told Ozzie he wasn't sure how he was going to sell this thing. "And so I went back," says Ozzie, "and Mitch said, 'Look'—he wasn't in the habit of overriding his people—he just said, 'Go change the stuff to make it a single-user product.' So I said, 'How the hell do I go and take what I've done and make it a single-user product? I mean . . . what it is, is a *group* product.' And so what I did was turn it into a 'personal' document management system as the spec, and I got everybody to nod their head yes, we got funded, and the first month that we were out,

Mitch approved a change in direction back to the group stuff and that's when it went back to 'Notes.'"

The new company would be called Iris Associates. "We were completely independent," says Ozzie, "we owned all rights to what we did, except they had, in exchange for providing all this funding, an option that they could exercise that would essentially, if they exercised the option, they committed to paying us a certain royalty rate. They got exclusive marketing rights—actually, they basically got all rights to the product and we got compensated by royalties—but they would have to use best efforts to market the product. They couldn't buy it and sit on it. Year after year went by and they just kept pumping more and more money but they knew that I knew Gates very well. So they knew that if they stopped funding it, I would immediately take the product, 'cause I owned it, to Microsoft. So it was sort of like they're damned if they do, damned if they don't, so for many years while they couldn't figure out how to market the thing, they just kept funding it, figuring it's better to keep it going than to give it up. Eventually in January of '88 they exercised the option and it became theirs, essentially the product became theirs at that point, and they were committed to marketing it and they were committing to an affirmative obligation to pay royalties, and so we still operated independently collecting royalties through '94."

Jim Manzi characterizes the Lotus-Iris deal as "very unusual, and it largely worked not because of the brilliance in the contracting language, but I think it largely worked because of the relationship between the principals over time."

Lotus Notes, the official name of the product when it finally shipped, offered email, calendaring and scheduling, an address book, access lists, document commenting, online forums, anonymous notes, the equivalent of a Notesfile Sequencer, a database, and programming tools to build custom applications within the *Notes* environment. The Iris team took a pile of PLATO ideas they'd lived and breathed at CERL and transferred them into a Microsoft Windows environment for the PC. But however impressive the final product, it was the kind of tool that required an entire organization to be trained on and commit to—it didn't work if only small clusters of employees used it. That meant an entire organization had to change their behavior and reengineer itself in order to fully exploit *Notes*'s features. Lotus decided that even though the product was for workgroups, it was not going to work well for *small* workgroups—who would install it? Who would administer it? No, it was better suited for an enterprise. To make that abundantly clear to the marketplace, the company set the starting price for the product at $64,000. Their first customer was Price Waterhouse, who were so impressed with the product they ordered a historic ten-thousand-user license, the largest single order for a software program in the computer industry up to that time. Other corporations soon followed with their own orders.

Notes was a hit.

Ozzie attributes at least part of *Notes*'s success in the enterprise space to serendipity: "In the late 1980s this country's economy was going through an odd transition, people were being laid off and there was this intense amount of restructuring, under the auspices of reengineering, that was starting to happen. And we were available right at that time as a tool that people felt could help reengineering."

Lotus finally acquired Iris Associates in 1994 for approximately $84 million, ten years after Ozzie, Kawell, and Halvorsen had launched the tiny firm with a dream of bringing a major chunk of PLATO to the wider world. Across that decade Lotus had undergone a stunning transformation from the company known for *1-2-3* to becoming the company known for *Notes*. The timing was fortuitous, because Microsoft had aggressively been out to kill *1-2-3* with its own product, *Excel*, which had by then become and has remained to today the leading spreadsheet in the market. *Lotus Notes* meanwhile became so attractive to large enterprises, and, once installed enterprise-wide, so difficult to migrate away from (a desirable situation venture capitalists term a "high switching cost"), that IBM decided to buy Lotus in its entirety in July 1995 for $3.52 billion.

The collaborative, group-productivity ideas embodied in PLATO—tools largely created by PLATO's own users, many of whom were teenagers at the time—finally had an impact on the world. It would be the largest financial transaction in history for a project deeply inspired by PLATO and developed by PLATO veterans, and, ironically, it had nothing to do with computer-based education. In time IBM's *Notes* would surpass 120 million users. Few of them would ever know of the product's deep connection to PLATO.

Ozzie formed a new company in 1997, Groove Networks, which took the ideas of *Lotus Notes* and reapplied them using a more decentralized, peer-to-peer fashion. Groove was acquired by Microsoft in 2005 for

$120 million. Ozzie became the chief software architect of Microsoft, laying the groundwork for the Azure project, which became Microsoft Cloud. In 2010 he left and formed Talko (named after Talkomatic), a mobile conferencing platform largely focused on voice messages and annotations to those messages. In 2015 Microsoft acquired Talko, which the company absorbed into Skype, the global Internet communications service Microsoft had acquired in 2011.

Epilogue

"Service interruption in 60 seconds . . ."

The familiar message appeared at the bottom of the computer screens of the twenty-two die-hard NovaNET users scattered around the country. It was 2:16 a.m. Central Time on the first of September, 2015. For these night owls, these service interruptions were a part of life, as predictable as the sun rising and setting. If you had been a user of the system for decades, possibly since you were a kid, a service interruption provided a welcome break, and sometimes the prospect of an exciting change, perhaps a new system feature or TUTOR command, when the system was brought back up.

What was different, this time, was that the service was not being interrupted. It would not be brought back up. It was being terminated. Here they were, holding out in the original cyberspace, the cyberspace that had originally been powered by computers actually called CYBERs, gathered to witness the termination of the last, longest-lasting, virtual place they'd called home for decades. This time the system was going to kick them off for good and then die.

Some of these users had stayed up late at night ostensibly to make one final set of backups of their old work dating back possibly to the 1970s. Others were there to hang out, see who showed up for the end of the world as they'd known it, and post a few messages to the others, knowing, with the end nigh, that no one would see these postings ever again. One user described the scene as "a band of fogeys who collected to roast marshmallows over the flames."

Others described how long ago they had gotten their start on PLATO, and continued using NovaNET over the years. "I just calculated," said one user, "the number of days from my first encounter with PLATO III to today: 16,285 days."

"Never found a better online community," said another. "One or two, almost as good. All things end, and unfortunately it is time for NovaNET to end."

This wasn't going to be like a TV station going off the air, the familiar script that begins with scratchy old footage of an American flag waving in the wind against a blue sky, perhaps a fighter jet flying by, while an orchestra played "The Star-Spangled Banner" a little faster than usual, followed by the issuance of a final station identification message, then a video color bar test pattern and a shrill audio tone thrown in for good measure, and then, click, off the air, followed by the video snow and audio static signifying nothingness, as the viewer gently dozed off. No, this was NovaNET, the computer-based education service and online community, and the twenty-two diehards staying up this late were invariably men, mostly white-haired, graybeard ghosts who quietly haunted one of the oldest and least known continuously operating computer networks in the world, going on some thirty-five years or forty-five years, depending on who was debating the NovaNET vs. PLATO IV origins. Some of these old diehards seemed to hang around online for no discernible reason, drawn perhaps out of nostalgia or loneliness, like lone regulars at a local neighborhood saloon with its flickering, fading neon beer sign out front and a handful of customers inside quietly drinking away the hours, days, and years.

NovaNET's official mission had always been educational, delivering self-paced, individualized instruction to students primarily situated in troubled schools or remedial education programs around the country: students who had not been successful in normal schools, but found a way to finish their high school diploma, or achieve a GED certificate, in an institution that offered the private one-on-one ever-patient digital tutoring in front of a NovaNET terminal or a Mac or a PC running the NovaNET software. An intimate, quiet setting—not dissimilar to CERL's and CDC's PLATO learning centers in prisons that had been successful for so many years—a setting where the student could focus on his studies and not worry about the pressure of peers or the disappointed gaze of a harried instructor. For years NovaNET had successfully delivered on its mission, educating hundreds of thousands—more likely millions all totaled—of students all over the country, helping them get on with their lives, helping them get better jobs, better incomes, enough to support a family.

But NovaNET had another community of users, and it was this tiny group that was there to bear witness to the end, the closing of a system whose origins go back fifty-five years. A system whose vision, both from educational and technical implementation standpoints, had not changed since day one. This other small group of users were the authors, the systems programmers, and former gamers and participants in the online social community that boomed in the 1970s and stayed active until the early 1990s, after which only the diehards would stop by each evening to check in, say hello, offer a quip or two, and sign off. Tonight was the ultimate sign-off: they were the final few, perched on the topmost point of the last Ziggurat, experiencing the final seconds of NovaNET's existence.

So much had happened since CDC had sold off PLATO in 1989, and CERL had shut down a few years later. The Roach Organization, which had bought the rights to PLATO from CDC, eventually renamed itself TRO Learning, went public (its stock symbol, TUTR, a knowing wink to PLATO's programming language), then renamed itself again to PLATO Learning. In 2012, in what seemed an admission that the PLATO name had become an embarrassing reference to a past the marketplace now knew nothing about, the company renamed itself once again, this time as Edmentum.

VCampus, one of the spinoffs of CDC's PLATO business, was long gone by 2015. Drake, the testing company, was acquired in 1999 by National Computer Systems, which then soon acquired NovaNET Learning. Pearson, the giant British education conglomerate, then bought NCS in 2000 for $2.5 billion. NovaNET, reputed to be the only profitable division within Pearson in the early 2000s, quietly blossomed, reaching a peak number of simultaneous users, 13,184, on June 18, 2008. That would turn out to be NovaNET's high-water mark. By 2010 the whispers were growing louder that NovaNET's days were numbered, and in October 2014, the news became official: Pearson told customers that NovaNET would shut down in 2015, and everyone would need to wrap things up over the next academic year.

Since the close of CERL, the world had been transformed by the Internet and the World Wide Web. During the PLATO era most people around the world had no idea what computer-based education was, or what an online community was, let alone the benefits of email, chat rooms, instant messaging, notesfiles, and multiplayer games. In an astonishingly short period of time in the mid-1990s, hundreds of millions of people would suddenly not only discover these technologies, but would soon not be able to imagine how they could live their daily lives without them.

Computer-based education—a dinosaur-era term not used anymore, now known by various buzzwords, including e-learning—is booming more than ever. Millions of people study courses online around the world. Massive open online courses (MOOCs) exploded on the scene after the Kahn Academy pioneered the use of educational videos on YouTube. In 2013, the CEOs of the two leading MOOC services, Udacity and Coursera, were asked if they had ever heard of the PLATO system. "No," both said, blank expressions on their faces.

Schools continue to spend billions on computers, software, and networks, but the question remains who is benefiting more, students or the vendors. Educational technology has become a hot area for venture capitalists, who now routinely pour billions of dollars per year into new start-ups. Much of the technology developed by CERL in the 1970s has never been replicated.

Part of the CERL dream was to build a system so flexible, so infinitely capable, with such a vast array of tools, that any educator, regardless of skill level, could sit down and author his or her own material for students. This model worked at the University of Illinois and the University of Delaware, and found some traction at other universities and large institutions during CDC's peak years with PLATO. Michael Allen took the idea to the Macintosh with Authorware, launching the next generation of powerful authoring tools designed to give any educator or trainer the ability to craft whatever they had in mind to teach their students. But even Authorware is no more; Adobe discontinued it years ago. There are still dozens of authoring systems today, but the notion of spending time and money to develop custom courseware using programming tools has lost almost all of its early energy, and only big corporations and governments have budgets that can justify custom development. The costs of developing one hour of courseware—actual interactive lessons developed on the computer and then used by the student via the computer—never came down. If anything, they have continually climbed since the 1970s. MOOCs are, in a way, a reaction to this authoring quandary: educators seem to have punted on the whole question of devoting resources to creating interactive materials the way professors like Stan Smith so eagerly—and brilliantly—did on PLATO. Instead, they plop a video camera down in front of the lecture hall, and record a lecture to students, then toss it on YouTube or one of the MOOC websites. This became the new definition of "courseware" after 2010. MOOC companies and educators are working to provide more sophisticated forms of interaction beyond students passively watching videos and then going through pages of multiple-choice questions, but they're starting from scratch: today most educators have never heard of PLATO or TUTOR, or that anything that powerful could have existed before many of them were born. PLATO has even faded from the institutional memory of the University of Illinois; professors there, like everywhere else, are trying all sorts of new tools to deliver education online, running into problems that were once solved decades ago. The field of educational technology, largely ignorant of its own history, seems eternally condemned to repeat itself.

If CDC had tried harder, could they have gotten millions of terminals in the schools? Bob Morris is doubtful. "While we were *very* ambitious about getting the cost of delivering an hour's worth of PLATO into the schools, it was still expensive. My belief is that we had gotten it down to the point where it was reasonable. However, what's reasonable to us is not necessarily to the customer. One of the things that we always fought with PLATO was we would bring people in from academia, whether it would be universities, primary schools, and that sort of thing, and give them a full briefing on PLATO, and by the time they left, they would be so excited they couldn't see straight. And you're convinced that within a week you're going to get an order. Well, the problem kept being that they had alternatives for using the money. And the school administrators are rewarded more for the size of their staff than they are for the technology they bring in. And if PLATO services or a PLATO system would cost them so much, they would say, 'How much bigger staff can I get by spending that same money?' And invariably we lost to their increasing their staff or their not reducing their staff."

Walter Bruning, a CDC executive, once explained to reporters in the mid-1980s what William Norris's deep, underlying motivation with PLATO was really about. "His fundamental philosophy is to operate the schools," he said. "He cares a lot about PLATO, the name. He cares a lot about courseware and what we have learned (about education technology). But when Bill talks about educational delivery, what he is really talking about is running the schools. . . . We are on a novel and maybe even revolutionary path. We are after the privatization of one of the largest public services in this country—the privatization of the public schools."

Since that time, the notion no longer seems so outlandish. As Mark Andreessen has famously said, "Software is eating the world." It seems that big business, including Silicon Valley, is more determined than ever to devour public education and turn it into a monetized business. Whether the passionate, caring, hands-on Mary Graves of the world have a place in a world of private, for-profit schools, no doubt loaded to the hilt with technology, remains to be seen.

"Tick tick tick," one of the last-stand NovaNET users wrote at 1:52 a.m. NovaNET normally shut down at 2:00 a.m. each night for maintenance, and there was no reason to doubt tonight was not like any other night. But at 2:00 a.m., the system was still running. August passed into September, and NovaNET was still up, twenty-two stubborn users still online and waiting for the end. "As far as the operator is concerned," John Hegarty, one of the last remaining "s" programmers on the Pearson payroll, posted at 2:01 a.m., "it is the same as ever. They don't know what's going on." Somewhere in Iowa, at Pearson's data center, an unwitting system operator seemed to have forgotten to press a few buttons to shut the system down.

A small contingent of the new wave, now decidedly old wave, waited for the end of an era, scheduled for midnight on August 31, 2015, and the end was now several hours late.

"I hope I don't have to use the console," John Hegarty said at 2:04 a.m. "I always hated that thing."

"All these old guys, staying up late so they can be on NovaNET a little longer, just like when they were kids," Don Appleman, another longtime PLATO/NovaNET systems programmer and regular participant in =pad=, posted at 2:11 a.m. "Me included," he added as a postscript.

And then, a few minutes later, it came. The backout message. First the sixty-second warning. Then, "NovaNET service interruption in 30 seconds . . ." flashed at the bottom of everyone's screens. There was nowhere to run, nowhere to hide, no point in signing off—the system was going to sign you off before anybody could—so why not sit back, enjoy the ride.

"The backout message—it burns!" Don Appleman posted at 2:16 a.m., the system only alive for a few seconds more.

"NovaNET service interruption in 10 seconds . . . ," the message at the bottom of the screen now said. For the twenty-two people online, it was time to take a deep breath, and kiss goodbye the decades of digital fun, conversation, and, for many of them, a career spent on a system they loved.

The screen went dark. Then, a pause, followed by some white text, centered in the middle of the black screen, announced the dreaded moment:

—NovaNET not available—

NovaNET was history.

In 1967, Mac E. Van Valkenburg, then chairman of the department of electrical engineering at the University of Illinois, a man who had been one of Don Bitzer's PhD advisors, had this to say about Bitzer's technological baby: "Any predictions made for PLATO are speculation. It may be forgotten, or it may be the development of the century."

Choose your answer, and press NEXT to continue.

Acknowledgments

The first person I must thank, a person without whom this book might have never existed, is Katherine Hourigan, managing editor of Alfred A. Knopf Books. In 1987, I dashed off a series of query letters to several editors who had edited a number of my favorite books. Shortly thereafter, she sent me a brief letter indicating she was interested, wanted to learn more, and to "please send the manuscript at your convenience." Twenty-eight years later, in late 2014, it was finally convenient to reply and send in a large portion of it to her. Thankfully she was still managing editor and took my twenty-eight-year delay in stride. In 2015, she shared my work with a number of editors within the overall Knopf/Doubleday Publishing Group. The result was an offer from Pantheon. To Katherine, thank you for hanging in there and giving this book a chance to get out to the world.

Equal thanks to my literary agent, Regina Ryan, and to my Pantheon editor, Keith Goldsmith, both of whom always believed in the project and have offered unwavering support, wise counsel, and encouragement all along the way. Thanks also to the rest of the Knopf/Doubleday/Pantheon team: Dan Frank, Altie Karper, Fred Chase, Kevin Bourke, Mike Collica, Oliver Munday, Michiko Clark, Anna Dobben, and Danielle Plafsky.

I began this book in 1985 not knowing what form it would take; at the time I was considering a magazine article comparing and contrasting the dramatic differences I found in the PLATO and TICCIT cultures: the personalities of the people were so different, the designs of their respective systems were 180 degrees opposite; PLATO had a vast online community of people all over the world; in contrast, TICCIT was strictly a local-area network of terminals running training courseware.

I quickly became hooked on PLATO as a freshman at the University of Delaware in 1979, only to discover that Delaware was a mere remote outpost to the main event. Thus in 1980 I made my first pilgrimage to CERL, visiting again in 1982 to play in an Empire tournament. In 1986 I went again, this time spending several days interviewing the CERL staff. Every single person I met with was open, friendly, enthusiastic, willing to answer any and every question. Thanks to Tebby Lyman for providing me what turned out to be an invaluable stack of newspaper clippings, press releases, brochures, and miscellaneous documents and reports during that trip. In fact, over the course of the thirty-two years of this project, it has been a genuine pleasure, a blessing, to spend time with and hear the stories from such a vast number of people from around the world whose lives were touched by PLATO in some way. In addition, I found practitioners in the fields of psychology, instructional design, computer science, cognitive science, engineering, physics, biology, business, and education who knew about PLATO and were uniformly generous in their time to answer questions and guide me in the right direction.

Special thanks go to Dan Alpert, without whose guidance and leadership PLATO never would have happened. He provided me with multiple interviews, priceless documents, and endless support. Dan lived to be ninety-eight, and never missed a ski season in over fifty years. Thanks also to the generosity, support, and ready availability of the extended Bitzer clan, who put up with my incessant questions over a period of thirty years: Donald and Maryann Bitzer and their son, David; relatives Earle, Ruth, Sandy, Carol, and Jack Leskera. To Don and Maryann especially I offer immense gratitude and thanks. I'm also grateful for the kindness and hospitality of Paul Tenczar (who made himself available for numerous interviews over the years) and his wife, Darlene; they let me rummage through a suitcase full of Paul's photographic slides.

Thanks to Andy Hertzfeld, Paul Resch, Bill Galcher, Donald Norman, Dan O'Neill, and Ray Ozzie for reviewing all or portions of the manuscript over the years. Thanks also to writers Steven Levy and John Markoff, whose inspiring writing, support, and encouragement have helped give me the strength to finish this book.

Thanks to Ralph Nader and William C. Taylor, authors of *The Big Boys: Power and Position in American Business* (Pantheon, 1986), who without hesitation gave me two boxes full of all of their research on Control Data Corporation, including raw interview transcripts and handwritten reporter's notes, which I put to extensive use in Part Three of this book.

When one works on a book across three decades, the list of people to thank grows immense. Thanks to everyone who agreed to meet in person, invariably extending their hospitality, kindness, and support, sometimes for many long hours of conversation; as well as to the many people who spoke via phone, sometimes for many hours; as well as to the many people who wrote and sent me long personal essays, or sent photographs, news clippings, and documents; as well as to the thousands of people who corresponded by email—sometimes over many years. Thanks to Mike Achenbach; Sam Adams; Alix (Gaby) Albert; Kathryn (Lutz) Alesandrini; Paul Alfille; Chris Alix; Mary Allan; Jim Allard; Brock Allen; David Allen; Michael Allen; Daniel Alpert; Mitch Altman; Ernest Anastasio; David Andersen; John Anderson; Richard C. Anderson; Thomas H. Anderson; Marc Andreessen; Tanner Andrews; Amotz Anner; Andy Anthony; Andrew Appel; Don Appleman; Jeffrey Armstrong; Brij Arora; Eric Artman; Bruce Artwick; Neil Ashby; Isaac Asimov; Tom Ask; Chip Aspnes; Richard Atkinson; M. S. Atkisson; Wagner James Au; Allen Avner; Scott Baeder; Ron Baerg; Bill Bahnfleth;

Charles Baker; Cheryl (Eklund) Baker; Frederick Z. Banks; Quentin Barnes; David C. Barnett; Richard A. Bartle; Michael J. Bartosh; Jeff Bauer; Mark "Adam" Baum; Kurt Baumann; Rhona Baverman; Marilyn Beckman; Carl Behmer; Brian Bell; Gregory Benford; Mike Benveniste; Michael B. Bentley; Robert Berdahl; Rudy Berg; Mike Berger; John Bittner; David Black; Brian Blackmore; Richard Blahut; David Blair; Douglas Laine Blair; Rick Blomme; Louis Bloomfield; Mark Boeser; Nick Boland; William Bonetti; Grady Booch; Eugene P. Bordelon; Alfred Bork; Frank Boros; Ben Borowski; Lyle Borton; Lisa Boucher; Martha Boudreaux; Marisa Bowe; Jim Bowery; Pete Boysen; Jim Brain; Craig Brannon; Mark Brennan; Andrew K. Bressen; David Brightbill; Tony Brinati; Geoffrey Brock; Doug Brown; Gary Brown; John Seely Brown; Mary Brown; Richard M. Brown; Joe Brownlee; Bertram Chip Bruce; Richard Brummer; Walter Bruning; Leif Brush; John Bryan; Art Brymer; John L. Buckingham; Rick Buckmaster; Jerry Bucksath; C. Victor Bunderson; Garrie Burr; Craig Burson; Gary Burton; Paul Buta; Michael Butler; Deborah (Skinner) Buzan; L. Leon Campbell; Lynn Campbell; Marc Canter; David Capron; James Capshew; Jim Card; John Carmack; Christine Batakis Carroll; Bruce Carter; George Carter; Soji Carter; Darren Casella; Ruth Chabay; Cully Chapple; Sylvia Charp; Darlene Chirolas; Frank Christ; Jeff Chung; David Churches; Pat Clifford; Mike Cochran; Al Cohen; "Eddie" Cohen; Jeff Cohen; Stephen P. Cohen; Bill Cole; Corey Cole; Jerry Colonna; W. Dale Compton; Don Connelly; Ted Conway; Jeff Corey; Priscilla (Obertino) Corielle; Kaywin Cottle; Chris Crawford; David Crockett; Rich Crooks; Larry Cuban; John Cundiff; Carol (Bennett) Dalenko; John Daleske; Andy Dallas; Bernard Damberger; Jeff Dammeyer; David E. Daniel; Ronald Danielson; Dave Darling; Steve Deiss; Tom Deliganis; Thomas DeLoughry; Stewart Denenberg; Carl Dennis; Sam Denton; Tom DePlonty; Diane Desatnick; Jack Desmond; William Deutsch; Dean Dierschow; Andrew di Sessa; Orlando J. Dona; Eric Downey; Scott Dudley; Sharon Dugdale; Jim Dumoulin; Kevet Duncombe; Wesley Dunn; Natalie DuPont; Jim Dutcher; Kim Duvall; David Eades; Mark Eastom; Fred Ebeling; Chaz Ebert; Roger Ebert; Ernst Eichmann; John David Eisenberg; Richard A. Ekblaw; Mary Eliot; Robert Elmore; Douglas Engelbart; Rupert Evans; Scott E. Fahlman; Peter Fairweather; Margot Fass; Martin Fass; Gerry Faust; J. Michael Felty; Wallace Feurzieg; Lee Fillman; Brad Fincher; Steve Fine; Brice Finkelstein; Chester Fitch; Declan Fleming; Estelle Fletcher; J. Dexter Fletcher; Jim Flinsch; Michael Folk; Bill Forsyth; Brand Fortner; Monica Fortner; George Frampton; Larry Francis; David Frankel; Marvin Frankel; Bob Fratini; Stephen Freyder; H. George Friedman; Duane D. Friedrich; Mort Frishberg; Sue Frishberg; Gary Fritz; G. David Frye; Jim Fuerstenberg; Chris Fugitt; David Fumento; David Futrelle; Owen Gaede; Eugene Galanter; Bill Galcher; Samwise Galenorn; Paul Garceau; Chester Gardner; Sean Gardner; Stephen Gee; Andy Geiger; Louis V. Genco; Melanie Geyer; Jim Ghesquiere; Andrew Gibbons; Don Gillies; Joel S. Gillman; Brian Gilomen; John Gilpin; John Giourdas; Gary Gladding; Robert Glaser; Jim Glish; Adele Goldberg; William Golden; Richard Goldhor; Jack Goldman; Edwin (Ned) Goldwasser; Sherwin Gooch; Mike Gorback; Kevin Gorey; Barbara Grabowski; Barb Grajewski; David Graper; Dave Grattan; Steve Gray; Douglas Green; Jay Green; William Green; Andrew Greenberg; Tom Grillot; George Grimes; Tom Grohne; Alan Groupe; Justin Grunau; C. K. Gunsalus; Yechezkal Gutfreund; Lenny Gutierrez; Eric Hagstrom; Lippold Haken; Ed Hall; Steve Hallberg; John Haltiwanger; Tim Halvorsen; Jonathan Handel; Cecil Hannan; Andrew Hanson; Al Harkrader; Ron Harstad; Michael S. Hart; Shawn Hart; Miles Harvey; Skip Haughay; David Haycraft; Mark Haynes; Rick Hazlewood; Harold Hendricks; Carole Herriman; Karl Hess; Jock Hill; Fred Hofstetter; Laura Hofstetter; Lenny Hoover; Lara Hopkins; Jim Horne; Mike Horner; Valli Hoski; Ernest House; Charles Hresil; Bob Hubel; Beverly Hunter; Tom Hunter; Jeff Huston; James Hutchinson; Daniel C. Hyde; Renee Inman; Aron K. Insinga; Mike Jacobs; Bryna Lee Jacobson; Greg Janusz; Eugene Jarecki; Gloria Jarecki; Henry Jarecki; Bob Johansen; Chris Johnson; James Johnson; Jeff Johnson; Keith Johnson; Lee Johnson; Mark A. Johnson; Roger Johnson; Douglas W. Jones; Jeremy Jones; Steve Jones; Richard Jorgensen; Ted Kaehler; Dennis Kane; Samuel Kaplan; Mitch Kapor; Goran Karlsson; Kent Karraker; Aaron Karsh; Milton Katz; Luke Kaven; Len Kawell; Alan Kay; Greg Kearsley; B Keefe; Mark Kehrli; Andrew Keller; Elaine Keller; David Kibbey; Dick Kibitt; Kaywin Kidd; Tom Kiester; Rayburn Killion; Greg Kirkpatrick; Ron Klass; Paul Kohlmiller; Nathaniel Kohn; Christopher Kolar; Rob Kolstad; Paul Koning; D. A. Kopf; Colin Koteles; Diane Kovacs; Glenn Kowack; Jim Knoke; Daphne Koller; Celia Kraatz; Jim Kraatz; Michael Krasny; Neil Krasovec; Peter Krause; Jody Kravitz; Scott Krueger; Richard Kubat; Mike Kulas; Fred Kutell; Mitch Kutzko; Helen Kuznetsov; Valarie Lamont; Carol Lampe; John Lampe; Murray Lappe; David Lassner; David Lawrance; Andy Lawrence; Sheridan Layman; Tom Layman; Lauren Leach; Ted Leach; Robert Lebowitz; Carolyn (Leech) Leeb; Gene Leichner; Jeannine Leichner; David L. Lendt; George Leonard; David LePage; Rick Levine; Alan Levy; Wayne W. Lichtenberger; Judith Liebman; John Limber; Sean Lincoln; Jon Lineweaver; Steve Lionel; Arthur Lipper; Greg Lipper; Howard Lipsitz; Larry Lipsitz; Terry Lisansky; Todd Little; Brodie Lockard; Eric Loeb; Ed Logg; Gary Loitz; Harvey Long; Chris Lopez; Ken Lopez; Scott Lucado; Phyllis (Hall) Ludwig; Arthur Luehrmann; Mark Lundberg; Tebby Lyman; Bill Lynch; Roger Macdonald; Peter Macinnis; Eddie Maes; Bruce Maggs; Peter Maggs; Kurt Mahan; Jonathan Manton; Jim Manzi; Dillon Mapother; Scott Marcy; Ted Marcy; Jed Margolin; Charles F. Marino; Rosalie (Dietrich) Marley; David Martin; Paul Martin; Ted Martz; Stephan Masica; Phil Mast; John Matheny; Melissa (Dammeyer) McCormick; Nancy McDonald; Brendan McGinty; Jude McLaughlin; Rob McQuown; Erik McWilliams; Terry Meadors; Russell Medford; Mike Medved; David Meller; Arthur Melmed; Bill Melohn; M. David Merrill; Marshall Midden; Beth Ann Miller; Chuck Miller; David Miller; Joe Miller; J. Michael Milner; Sam Milosevich; Robert A. Moe; Andrew Molnar; William Montague; Carl Moore; Catherine Morgan; Bob Morris; Ron

Mortensen; A. Kent Morton; David Moursund; David Mudrick; Barry Munitz; Mike A. Murphy; George B. Myers; Don (Black) Nelson; Mara (Stolurow) Nelson; Ted Nelson; Mary Ann Neuman; Bob Niederman; Jon Niedfeldt; Jurg Nievergelt; Douglas Noble; Steve Nordberg; Mats Nordin; William C. Norris; Bob North; Kevin Nortrup; Sal Nuzzo; Mike O'Brien; Terry O'Brien; Brian M. O'Connell; John O'Hara; Mike O'Hara; Dan Oberlin; Anthony Oettinger; Steve Okonski; Michael Olivier; Kevin Osterhoff; Ron Ozer; Ray Ozzie; George Pake; Josh Paley; Nina Paley; Mark Papamarcos; Seymour Papert; Bruce Parrello; Jim Payne; Dirk Pellett; Flint Pellett; Jack Peltason; Steve Peltz; Glenn Pence; Dean Perra; George D. Peters; Stephen Petrina; Frederick Peyser; Scott Pfeffer; Bob Philhower; Don Piven; Robert C. Platt; Jason Pociask; Cindy Polus; Dorothy Pondy; Richard Powers; Robert M. Price; Frank Propst; Peter D. Pruyne; Andy R.; Bob Rader; Manfred Raether; Max Rahder; Susan Rankaitis; Rick Rashid; Diane Ravitch; Brian Redman; Dennis Reed; Dan D. Reeves; Rick Reichenbach; Bill Reichenborn; Lindsay Reichmann; Alex Reid; Mark Reifkind; Paul M. Resch; Steve Rhodes; Brent Rickenbach; Rick Rickenbach; Pat Ridgely; John Risken; Nancy Risser; Grant Ritchey; Peter Rizza; Chuck Roberts; Paul E. Robertson; Eric Robeson; Alaine (Warfield) Robinson; Mike Rodby; Dan Rohner; Tony Rollo; Bill Roper; Steve Rose; CJ Ross; Robb Ross; Eric Rotberg; Andrew Rothenberg; Randall Rothenberg; Mary M. Rowan; Peter Rowell; Owen Rubin; Dave Runte; George A. Russell; Mark Rustad; Donna Saar; Jonathan Sachs; Jim Salmons; Stephen Satchell; Aaron Sawdey; Nate Scarpelli; Bill Schaeffer; Tom Schaefges; Grant Schampel; Jerry Schermerhorn; Dave Schoeller; Peter Schow; Jeff Schramm; Robert Schroeder; Foster Schucker; Donald Schultz; Jan Schultz; Steve Schutt; Michael Schwager; Bill Scott; Vince Scuro; Paul Sebby; Joseph Seeley; Alberto Maria Segre; Robert J Seidel; Bonnie (Anderson) Seiler; Chris Seitz; Gary Shaffer; Leslie Shaffer; Brian Shankman; Andrew Shapira; Brian Shears; Bill Sheppard; Kevin Sheridan; Larry Sherman; Chalmers Sherwin; Anton Sherwood; Bruce Sherwood; Judy Sherwood; Kaitlin Sherwood; Don Shirer; David Sides; Laurene (Ragona) Sides; Marty Siegel; Paul Silver; Rick Simkin; Steve Singer; Lisa Singh; Henrk Sjodin; Dominic Skaperdas; B. F. Skinner; Danny Sleator; Irene Slottow; Jeffrey Slovin; David Small; Larry Smarr; Anthony Smith; Kenneth Smith; Mike Smith; Shannon K. E. Smith; Stan Smith; Vernon Smith; Walter Smith; Jim Snellen; Joel A. Snow; Patricia Sokolove; Cynthia Solomon; Elliott Soloway; Larry Sonna; Robert L. Sproull; Rae Stabosz; Bill Stainton; Robert Stake; John Starkweather; Tom Starr; Rich Stawicki; Alfred Steele; Esther Steinberg; Lou Steinberg; David Steinberger; Rob Steinberger; Neal Stephenson; Ken Stetten; Robin Stevens; Jack Stifle; Lawrence Stolurow; Ben Stoltz; Maureen Stone; Lawrence Stover; Jeffrey Strang; Scott Strickland; Patrick Stubbs; Jack Suess; Maurice E. Suhre; Patrick Suppes; Rachael (Preiss) Susman; Alistair Sutherland; Bob Swanson; Scott Swanson; Nathan Syfrig; Mike Szabo; Joshua Tabin; Dave Tall; David Tanaka; Jack Taub; Bob Taylor; Dick Taylor; Frank Taylor; Dan Teitelbaum; Chuck Thacker; Jonathan Thaler; Jim Thomasson; Charlene

Thompson; Tim Thompson; Sebastian Thrun; Jim Tobias; Anthony Toma-sic; Timothy Trick; Dan Tripp; David Trowbridge; Jim Trueblood; Earl Truss; Paul Tucker; Sherry Turkle; Murray Turoff; Phil Twiss; Richard Twiss; Yarko Tymciurak; Judy Tyrer; Stuart Umpleby; James M. Unger; Al and Diane Urbanckas; Ken Vanderhoff; Wendy van Dijk; Julie (Skinner) Vargas; Steve Vetter; Emil Volcheck; Louis D. Volpp; Daniel Walczak; Bill Walker; Mike Walker; Silas Warner; Chuck Warren; Greg Warren; Marcus Watts; Gordon Weast; Harriett Weatherford; Charles Weaver; Larry Weber; Donald Wedding; Mike Wei; Scott Weikart; Morton Weir; Jessica Weissman; Steve Weyer; Ken Whaley; Gary Whisenhunt; Larry White; Charlie Wickham; Tom Wicklund; Ron Widitz; Claude Wilkin; Dan Wil-liams; Mike Williams; Richard Williams; Steve Williams; Tom Williams; Robert H. Willson; George Wayne Wilson; Susan Winter; Erik K. Witz; R. J. Wolfe; Marilyn Wood; Ray Wood; Robert Woodhead; Jim Woods; David Woolley; Alan Wostenberg; Darren Wright; Guy Wright; Kim Wright; Pam Wright; Allen Yang; Bob Yeager; Laura M. Young; Andy Zaffron; Paul Zarnowski; George Zebrowski; Matt Zelchenko; Peter Zelchenko; Dave Ziegler; Roy Zingg; and Karl Zinn. If I forgot anyone, the omission was inadvertent.

The following businesses and organizations helped in numerous ways: NovaNET Learning, Inc., which from the early 1990s all the way through to the Pearson Education years waived the expensive hourly connection fees so I could access the system for research. Control Data Corporation, for loaning me a complete PLATO workstation, printer, and online account to capture screen shots and printout notesfiles in 1987–1988. Hazeltine Cor-poration's Training Systems Center (later a division of Ford Aerospace) for allowing me copious use of its high-speed Kodak photocopier. I also owe a debt of gratitude to twenty years of search engines: AltaVista, InfoSeek, Yahoo, and Google.

Since grade school I have held both a fascination with and indebtedness to libraries and librarians, some of the most helpful people on earth. If there is a Twitter hashtag that epitomizes exactly how I feel about research librar-ians, it is #LibrariansRock. This book would not have been possible without the eager, enthusiastic support of the staffs at the Library of Congress; the University of Illinois Archives; the University of Illinois Library; Brigham Young University Library and Special Collections; the University of Wash-ington Library; the University of Maryland Baltimore County Library; the University of Maryland College Park Library; the University of California San Diego Library; San Diego State University Library; the University of New Mexico Library; the Ohio State University Library; and the Charles Babbage Institute at the University of Minnesota.

Thanks to the staff of the Computer History Museum in Mountain View, California, who in 2010 co-produced with me a two-day conference cele-brating the fiftieth anniversary of the founding of the PLATO system. Five hundred people attended what turned out to be the largest event ever held at the museum up until that time. Thanks to Ray Ozzie and Microsoft Corporation for funding the recording of the full conference in HD video, which remains available online at http://youtube.com/platohistory thanks to the generous support of Google, Inc.

Thanks to Mike Cochran, Bill Galcher, Paul Koning, Steve Peltz, Paul Resch, Joe Stanton, Steve Williams, and Steve Zoppi for their incredible work creating and maintaining Cyber1, a publicly available, fully functioning PLATO system on the Internet, complete with much of CERL's and CDC's published catalog of lessons and games, available at http://cyber1.org. It has been an invaluable research tool.

Finally, I would like to express eternal love and gratitude to my wife, Patricia, whom I met via a TERM-talk on PLATO. She happily read and reread countless drafts and willingly shared—and endured—every moment of an eventful thirty-two-year journey that was the creation of this book.

Thanks to you, dear reader, as well.

Brian Dear
Santa Fe, New Mexico
January 30, 2017

Interview and Oral History Sources

The story of PLATO as a technological and cultural history is unusual. Unlike most such histories, there are no existing major books, magazine articles, documentaries, or other common sources to which historians may turn. So pervasive has the decades-long dearth of PLATO's exposure been to popular culture and the media that published sources are few and far between. The academic literature has considerable relevant material, to be sure, but PLATO's human story largely eludes the refereed journals. For an author seeking out the PLATO story this meant conducting hundreds of personal interviews and corresponding with an even larger set of people over more than twenty years. But whom to contact? Where to turn? An untenable situation was avoided by setting up a website, running since 1996, announcing the book project, describing its scope, listing questions for which the author was seeking answers, and—breaking no doubt some journalistic rules—listing out names of all who had been interviewed or had corresponded through the mail or online. That ever-growing list prompted ever more people to contact me: a virtuous cycle that has never stopped—it continues right up to the present day. The result is a book largely based on oral history, capturing, before they are forever lost, the stories of the people who participated in the late, great online community known as PLATO.

Abbreviations

BYU, Brigham Young University

CBI-UM, Charles Babbage Institute—University of Minnesota

CERL, Computer-Based Education Research Laboratory, University of Illinois at Urbana-Champaign

CHM, Computer History Museum, Mountain View, California

CSL, Control Systems Laboratory (1952–1957), Coordinated Science Laboratory (1958–)

DI, *The Daily Illini*, the official newspaper of University of Illinois at Urbana-Champaign

ECE, Electrical and Computer Engineering Department, University of Illinois at Urbana-Champaign

NG, Champaign-Urbana, Illinois, *News-Gazette*
PCP, PLATO Corrections Project
UIC, University of Illinois at Chicago (formerly University of Illinois Chicago Circle)
UIUC, University of Illinois at Urbana-Champaign

Interviews

Author's Interviews
If no location specified, interview was conducted via telephone.
Alesandrini, Kathryn. 2002-04-24.
Allen, Brock. 2010-08-03 (San Diego, CA).
Allen, Michael. 1997-02-14.
Allen, Michael. 2016-07-11 (St. Paul, MN).
Alpert, Daniel. 1986-05-01 (Urbana, IL).
Alpert, Daniel. (2003a) 2003-06-14 (Snowmass Village, CO).
Alpert, Daniel. (2003b) 2003-06-20.
Altenbernd, Nick. 2014-11-29.
Anastasio, Ernest. 1997-02-06.
Andersen, David. 2003-11-20.
Anderson, John. 2003-08-23.
Anderson, Richard C. 2002-11-25.
Anderson, Thomas H. 2002-11-19.
Anonymous, 1997-03-01.
Appleman, Don. 2002-07-10.
Aspnes, Chip. 1997-02-21.
Atkinson, Richard. 1987-08-04 (La Jolla, CA).
Avner, Allen. 1986-05-02 (Urbana, IL).
Avner, Allen. 2013-08-01 (Champaign, IL).
Banks, Fred. 2002-12-21.
Beddini, Ann. (2014a) 2014-07-21.
Beddini, Ann. (2014b) 2014-07-23.
Benford, Gregory. 1987-09-02.
Berdahl, Robert. 1997-05-20.
Bereiter, Marilyn (Beckman). 1997-01-28.
Bereiter, Marilyn (Beckman). 2013-08-01 (Urbana, IL).
Bitzer, David. 1997-01-17.
Bitzer, Donald. 1986-05-01 (Urbana, IL).
Bitzer, Donald. 1987-09-21.
Bitzer, Donald. 1996-12-20 (Raleigh, NC).
Bitzer, Donald. 1997-01-20.
Bitzer, Donald. (2003a) 2003-07-31.
Bitzer, Donald. (2003b) 2003-09-29.
Bitzer, Donald. 2012-10-17.
Bitzer, Donald and Maryann. 2002-07-10 (Vandalia, IL).
Bitzer, Donald and Maryann. 2016-07-06 (Vandalia, IL).

Bitzer, Earle, Sandy Bitzer, Carol Lampe, and John Lampe. 2003-09-17 (Collinsville, IL).
Bitzer, Maryann. 1997-01-27.
Bitzer, Maryann. 2003-09-29.
Bitzer, Ruth. 2003-08-29.
Blackmore, Brian. 1997-02-19.
Blahut, Richard. 2003-09-25.
Blair, Lane. 2003-08-19.
Blomme, Rick. 1986-05-01 (Urbana, IL).
Blomme, Rick. 2013-08-01 (Urbana, IL).
Bloomfield, Louis. 2010-04-17.
Bork, Alfred. 1987-02-26 (Irvine, CA).
Borton, Lyle, and Ted Martz. 1997-01-31.
Bowery, Jim. 1987-10-16 (La Jolla, CA).
Braunfeld, Peter. 1997-03-12 (Urbana, IL).
Brown, Doug. 2010-06-08.
Brown, John Seely. 1987-09-16 (Palo Alto, CA).
Brown, Richard M. 2003-09-23.
Bruce, Chip. 2002-12-15.
Bunderson, Victor. 1986-08-25 (Princeton, NJ).
Burr, Garrie. 2002-11-10.
Burson, Craig. 1997-03-07 (Oklahoma City, OK).
Campbell, Lynn. 2015-01-26.
Chapple, Cully. 1997-01-24.
Charp, Sylvia. 2002-12-06.
Chung, Jeff. 2013-12-12.
Cohen, Barry. 2013-10-14.
Cole, Bill. 2002-11-16.
Compton, Dale. 2003-09-11.
Corielle, Priscilla. 2015-10-13.
Cuban, Larry. 2012-08-15 (Palo Alto, CA).
Cundiff, John. 2003-01-29.
Daleske, John. 2002-03-07.
Daleske, John. 2010-05-27.
Daniel, David. 2003-09-23.
Denenberg, Stuart. 1997-02-12.
Desmond, Jack. 2003-09-17.
Deutsch, William. 1987-03-18 (San Antonio, TX).
DiSessa, Andrew. 1987-08-11.
Dugdale, Sharon. 2003-04-30 (Davis, CA).
Duncombe, Kevet. 1997-01-15.
Dutcher, Jim. 2002-12-17.
Eades, David. 2013-08-01.
Eisenberg, John David. 2004-01-22.
Eliot, Mary. 2013-06-02.
Evans, Rupert. 2009-12-29.

Faust, Gerry. 1987-02-27 (San Diego, CA).
Fillman, Lezlie. (1997a) 1997-01-27.
Fillman, Lezlie. (1997b) 1997-01-30.
Fleming, Declan. 2009-12-18.
Fletcher, Dexter. 1987-02-07.
Forsyth, Bill. 2003-09-29.
Fortner, Brand. 2009-10-28.
Fortner, Brand and Monica. 1997-01-12.
Frampton, George. 2013-07-29 (Washington, DC).
Frankel, David. 1997-01-22.
Frankel, Marvin and Matilda. 2002-01-30.
Freyder, Steve. 1997-07-02.
Friedman, H. George. 2003-08-21 (Champaign, IL).
Frishberg, Mort and Sue. (1997a) 1997-02-08.
Frishberg, Mort and Sue. (1997b) 1997-04-09 (West Hollywood, CA).
Fritz, Gary. 1997-01-10.
Frye, G. David. 1997-01-09.
Fugitt, Chris. 2013-02-03.
Fuller, Dave. 2004-01-28.
Galcher, Bill. 1997-01-25.
Gardner, Chester. 2015-07-09 (Santa Fe, NM).
Gardner, Kirt. 2002-12-05.
Gardner, Kirt. 2015-09-25.
Geiger, Andy. 2015-06-23.
Ghesquiere, Jim. 1997-01-08.
Gibbons, Andrew. 1986-11-17 (Orem, UT).
Gilpin, John. 1997-02-02.
Gjerde, Craig. 2013-11-04.
Glish, Jim. 1997-02-03.
Goldberg, Adele. 1987-09-15 (Palo Alto, CA).
Goldberg, Adele. 2003-07-28.
Golden, William. 2003-09-19 (Champaign, IL).
Goldman, Jack. 2003-07-31.
Goldwasser, Ned. 2002-11-19.
Gooch, Sherwin. 1997-01-04.
Green, Douglas. 2013-08-04.
Green, Jay. 1997-02-06.
Grimes, George. 2003-09-06.
Gunsalus, Tina. (1997a) 1997-02-03.
Gunsalus, Tina. (1997b) 1997-05-13.
Hagstrom, Eric. 1997-02-20 (San Diego, CA).
Haken, Lippold. (1997a) 1997-01-27.
Haken, Lippold. (1997b) 1997-03-10 (Champaign, IL).
Haltiwanger, John. 2003-09-24.
Halvorsen, Tim. 1997-01-20.
Hannan, Cecil. 1987-09-09 (San Diego, CA).

Hansen, Barbara. 2016-07-11 (Minneapolis, MN).

Hanson, Andrew. 2013-08-28.

Hanson, Andrew, and Gary Gladding. 2016-07-08 (Urbana, IL).

Harkrader, Al. 1997-01-26.

Hart, Michael. 1997-02-05.

Hendricks, Harold. 1986-11-17 (Provo, UT).

Hill, Jock. 1997-07-07.

Hiltzik, Michael. 2003-09-15.

Hofstetter, Fred. 1986-08-26 (Newark, DE).

Hofstetter, Laura. 1997-03-03.

Hoover, Lenny. 2003-09-23 (Show Low, AZ).

Horne, Jim. 2016-10-22.

House, Ernest. 2012-05-18.

Hunter, Beverly. 1986-01-12 (Amissville, VA).

Hunter, John. 2014-06-25.

Janusz, Greg. (1997a) 1997-02-24.

Janusz, Greg. (1997b) 1997-03-10 (Champaign, IL).

Jarecki, Eugene. 2003-01-12.

Jarecki, Harry and Gloria. 2002-11-17.

Jeffers, Chris. 1997-02-21.

Johansen, Bob. 1997-02-19.

Johnson, Jeff. 2002-08-04.

Johnson, Mark. 2014-10-04.

Johnson, Mike. 2003 03 14.

Johnson, Roger. 2003-04-05 (Encinitas, CA).

Johnson, Roger. (2009a) 2009-11-18 (Del Mar, CA).

Johnson, Roger. (2009b) 2009-12-19.

Johnson, Roger. 2010-06-22.

Johnson, Roger. 2014-06-23 (Encinitas, CA).

Kapor, Mitch. 2003-01-23.

Karsh, Aaron. 2012-12-11.

Kaven, Luke. 2003-10-08.

Kay, Alan. 1987-10-20 (Los Angeles, CA).

Kay, Alan. 1988-02-05.

Kearsley, Greg. 1987-11-28 (Pacific Beach, CA).

Kibbey, David. 2013-03-03.

Kirkpatrick, Greg. 1986-06-17.

Knoke, Jim. 2002-12-08.

Koller, Daphne. 2014-04-18 (Monterey, CA).

Kolstad, Rob. 1997-01-29.

Kraatz, Jim and Celia. 1997-03-11 (Champaign, IL).

Krueger, Scott. 2010-05-02.

Lacey, John. 2015-05-10.

Lamont, Valarie. 1997-02-18.

Layman, Tom. 1997-01-28.

Leeb, Carolyn. 2014-11-23 (San Diego, CA).

Leichner, Gene and Jeannine. 2012-08-15 (Saratoga, CA).
Leonard, George. 1987-09-14 (Big Sur, CA).
LePage, David. 2003-09-25.
Leskera, John. 2002-12-23.
Liddle, Dave. 2003-04-29 (Palo Alto, CA).
Lipsitz, Lawrence and Howard. 1987-04-23 (Englewood Cliffs, NJ).
Lockard, Brodie. 1997-01-23.
Lockard, Brodie. 2013-12-06 (Kailua, HI).
Long, Harvey. 1986-03-25 (Rockville, MD).
Lyman, Tebby. 1986-05-02 (Urbana, IL).
Lynch, Bill. 1997-05-08.
Maggs, Bruce. 1997-01-14.
Maggs, Peter. 1997-03-12 (Urbana, IL).
Manzi, Jim. 2003-01-24.
Mapother, Dillon. 1997-01-26.
Mast, Kim, 2003-11-24.
Mast, Phil. 2003-11-25.
Matheny, John. 1997-04-19.
McGinty, Brendan. 1997-01-14.
McWilliams, Erik. 1987-05-25 (Rockville, MD).
Melmed, Arthur. 1986-10-24 (Washington, DC).
Merrill, David. (1986a) 1986-02-04 (New Orleans, LA).
Merrill, David. (1986b) 1986-11-13 (Crystal City, VA).
Merrill, David. 1987-07-31.
Miller, Chuck. 1997-01-21.
Miller, Harold. 1987-06-04 (Orem, UT).
Molnar, Andrew. 1987-02-18 (Washington, DC).
Morgan, Catherine. 1986-11-04 (Wheaton, MD).
Morris, Bob. 1997-05-19.
Morris, Bob. 2007-10-11.
Mudrick, David. 1986-07-02 (Reston, VA).
Munitz, Barry. 2003-06-06 (Los Angeles, CA).
Nelson, Ted. 1987-02-01.
Noble, Douglas. 2012-05-04.
Nordberg, Steve. 1997-01-28.
Norris, William C. 1986-06-25 (Washington, DC).
Norris, William C. 1997-03-07 (Bloomington, MN).
North, Bob. 2013-03-25 (San Diego, CA).
Nuzzo, Sal. 1986-07-10 (Reston, VA).
Oettinger, Anthony. 1987-04-24 (Cambridge, MA).
Ozzie, Ray, and Len Kawell. 1997-02-27.
Pake, George. 2003-08-01.
Paley, Josh. 2003-01-04.
Papert, Seymour. 1987-04-23.
Parrello, Bruce. 2004-01-21.
Payne, Jim. 2003-02-23.

Peltason, Jack. 2002-01-15.
Peltz, Steve. 1997-03-1 (Champaign, IL).
Peters, G. David. 2017-01-26.
Pondy, Dorothy. 2002-12-06.
Poulos, Cindy. 1997-06-09.
Powers, Richard. 2002-12-07.
Powers, Richard. 2003-08-18 (Urbana, IL).
Price, Robert. 1997-02-21.
Propst, Frank. (1997a) 1997-01-21.
Propst, Frank. (1997b) 1997-01-24.
Propst, Frank. 2013-03-02.
Rader, Bob. 2001-11-24 (Sunnyvale, CA).
Rankaitis, Susan. 2015-03-13 (Claremont, CA).
Risken, John. 2013-10-10.
Risser, Nancy. 2003-06-29.
Rodby, Mike. 2002-08-29.
Roper, Bill. 1997-01-18.
Ross, Clarence J. 1987-02-27 (Scripps Ranch, CA).
Russell, George. 2003-09-12.
Rutherford, Reginald III. 2014-05-10.
Saar, Donna. 2002-12-21.
Safier, David. 2010-08-02.
Scarpelli, Nate. (2014a) 2014-06-03 (San Diego, CA).
Scarpelli, Nate. (2014b) 2014-06-07.
Schaeffer, Bill. 1997-04-09 (Santa Monica, CA).
Schampel, Grant. 1997-01-20.
Schermerhorn, Jerry. 2013-02-19.
Schoch, John. 1997-02-20.
Schultz, Jan. 2012-12-05.
Seidel, Robert. 2015-05-20.
Seiler, Bonnie. 1998-07-18 (Redmond, WA).
Shapira, Andrew. 1997-02-11.
Shapira, Andrew. 1998-08-29 (Bellevue, WA).
Sherwin, Chalmers. 1987-02-07.
Sherwood, Bruce. 1997-01-10.
Sherwood, Bruce, and Ruth Chabay. 1997-01-12.
Sherwood, Bruce, and Ruth Chabay. 2002-06-30 (Raleigh, NC).
Sherwood, Judy. 2003-01-02.
Sides, David. 1997-02-10.
Siegel, Marty. (1997a) 1997-01-15.
Siegel, Marty. (1997b) 1997-01-22.
Skaperdas, Dominic. 2002-11-11.
Skinner, B. F. 1987-04-24 (Cambridge, MA).
Slottow, Irene. 2003-08-20 (Urbana, IL).
Smarr, Larry. 2011-10-24 (La Jolla, CA).
Smith, Kenneth. 2013-05-30.

Smith, Stan. 1997-03-11. (Urbana, IL).
Stake, Bernadine. 2013-10-14.
Steinberg, Esther. 1997-01-30.
Stetten, Ken. 1997-02-13.
Stifle, Jack. 1997-02-06.
Stolurow, Larry. 1997-02-17.
Stolurow, Mara. 2013-09-19.
Stone, Maureen. 2003-12-01.
Stubbs, Patrick. 2002-10-19.
Suhre, Maurice. 2002-10-27.
Suppes, Patrick. 1987-11-08.
Taub, Jack. 1987-08-09.
Taylor, Bob. 2010-05-13 (Mountain View, CA).
Tenczar, Paul. 1986-05-01 (Urbana, IL).
Tenczar, Paul. 1988-02-08.
Tenczar, Paul. 1997-03-11.
Tenczar, Paul. 2003-08-20.
Thompson, Charlene. (2013a) 2013-07-16.
Thompson, Charlene. (2013b) 2013-08-04 (Boulder, CO).
Thrun, Sebastian. 2013-04-18 (Monterey, CA).
Trick, Timothy. 2003-09-25.
Tripp, Dan. 1997-01-21.
Trollip, Stanley. 2009-06-14 (San Diego, CA).
Tucker, Paul. 2002-11-25.
Umpleby, Stuart. 1997-01-03.
Unger, James. 2002-11-21.
Vernon, Nancy. 2016-10-22.
Volpp, Louis. 2003-08-31.
Walker, Mike. 2003-10-21.
Warner, Silas. 1997-01-09.
Watanabe, Nan. 2003-03-23.
Weber, Larry. 2013-01-28.
Weber, Larry. 2014-06-03.
White, Larry. 1997-01-07.
Williams, Mike. 2003-03-13.
Willson, Robert. 2002-11-09.
Woolley, David. 1987-08-25.
Woolley, David. 1997-01-24.
Woolley, David. 2013-07-22.
Worley, James. 2010-07-05.
Wright, Pam. 1988-04-11.
Yeager, Bob. (2013a) 2013-10-10.
Yeager, Bob. (2013b) 2013-10-11.
Zaffron, Andrew. 1997-07-14.
Zelchenko, Peter. 2003-08-23 (Chicago, IL).
Zinn, Karl. 2003-09-26.

Other Interviews and Oral Histories

Alfille, Paul. 2000-05-04. Interview by Denny Cronin on Freecell website. Retrieved 2016-11-04 from http://www.freecell.net/f/c/alfille.html.

Alpert, Daniel. 2002-12-04. Interview by James Hutchinson, UIUC ECE. Unpublished.

Arora, Brij. April 2003. Email interview by James Hutchinson, UIUC ECE. Unpublished.

Bitzer, Donald. 1982-08-17. CBI-UM Oral History Interview OH 283 by Mollie Price.

————. 1984. Unpublished interview by William Taylor for Nader and Taylor, *Big Boys*.

————. (2002a) 2002-11-26. Interview by James Hutchinson, UIUC ECE. Unpublished.

————. (2002b) 2002-12-13. Follow-up interview by James Hutchinson, UIUC ECE. Unpublished.

————. 1988-02-19. CBI-UM Oral History Interview OH 141 with Sheldon Hochheiser.

————. April 1975. "Interview with Don Bitzer, PLATO," Joyce Statz, ed., *SIGCUE Bulletin* 9(2): 2–13.

Bork, Alfred. 1999. "The Future of Learning: An Interview with Alfred Bork," *EDUCOM Review* 34(4), July/August 1999.

Engelbart, Doug. 1986-12-19. Stanford University, Interview 1 of 4. Retrieved 2016-09-15 from http://stanford.edu/dept/SUL/sites/engelbart/engfmst1-ntb.html.

Gallie, Thomas. 1990-07-11. CBI-UM Oral History OH 222 by William Aspray.

Goldberg, Adele. 2002-07-03. IEEE History Center (New Brunswick, NJ) Oral History Interview by Janet Abbate.

Golub, Gene. 1979-06-08. CBI-UM Oral History Interview OH 105 by Pamela McCorduck.

Johnson, Roger. 2002-11-29. Interview by James Hutchinson, UIUC ECE. Unpublished.

Jones, Steve, and Guillaume Latzko-Toth. 2017. "Out from the PLATO Cave: Uncovering the Pre-Internet History of Social Computing." Internet Histories 1, nos. 1-2 (2017). Retrieved 2017-05-17 from https://protect-us.mimecast.com/s/mmYrB6c7KDzwhY?domain=tandfonline.com

Kay, Alan. 2016-06-21. "Alan Kay Has Agreed to Do an AMA Today." *Hacker News*. Retrieved 2016-06-21 from https://news.ycombinator.com/item?id=11939851.

————. 2012-07-10. "Interview with Alan Kay," *Dr. Dobbs Journal*, by A. L. Binstock. Retrieved 2015-05-26 from http://www.drdobbs.com/architecture-and-design/interview-with-alan-kay/240003442.

————. 2013. "An Interview with Computing Pioneer Alan Kay," D. Greelish, *Time*. Retrieved 2013-05-04 from http://techland.time.com/2013/04/02/an-interview-with-computing-pioneer-alan-kay/.

Lafrenz, Dale. 1995-04-13. CBI-UM Oral History Interview OH 315 by Judy E. O'Neill.

Lassner, D. 2000. Untitled transcript of video interview of David Lassner in Hawaii, by Sherwin Gooch.

Licklider, J. C. R. 1988-10-28. CBI-UM Oral History Interview OH 150 by William Aspray and Arthur Nordberg.

Linsenman, Robert. 1984-08-20. Unpublished interview by William Taylor for Nader and Taylor, *Big Boys*.

Merrill, M. David. 2010. "Interview with M. David Merrill, Half a Century of Experience in the Field of Educational Technology and Instructional Design," conducted by Ali Simsek and published in *Contemporary Educational Technology* 1(2): 186–95.

Molnar, Andrew. 1984-09-25. Unpublished interview by William Taylor for Nader and Taylor, *Big Boys*.

———. 1972. "Conversations with Doers," Interview Tape #29 from Educational Technology Publications (Englewood Cliffs, NJ).

———. 1991-09-25. CBI-UM Oral History Interview OH 234 by William Aspray.

Norris, William C. 1986. CBI-UM Oral History Interview OH 116 by Arthur L. Nordberg, recorded on 1986-07-28 and 1986-10-01.

Oettinger, A. 2006. Oral History by Atsushi Akera, January 2006. Association for Computing Machinery.

Patton, Peter. 2001-08-30. CBI-UM Oral History Interview by Philip L. Frana.

Price, Robert. 2012. A collection of interviews in Thomas K. Misa, ed., *Building the Control Data Legacy: The Career of Robert M. Price*, CBI-UM.

———. 2006-06-12. CHM Oral History Interview X3594.2006 by Gardner Hendrie.

Scarpelli, Nate. 2002-12-05. Unpublished interview by James Hutchinson, UIUC ECE.

Schwaiger, Jim, and John Gaby. 2010. "An Interview with the Creators of Oubliette for iPhone." Retrieved 2014-05-05 from http://gabysoft.com/Oubliette/AuthorInterview.aspx.

Sherwin, Chalmers. 1991-06-12. IEEE History Center Oral History Interview by John Bryant. Retrieved 2015-07-17 from http://ethw.org/Oral-History:Chalmers_Sherwin.

Whisenhunt, G., and R. Wood. 2012-05-01. "Interview with DND Creators Gary Whisenhunt and Ray Wood, *RPG(ology) Blog* video, by Carey Martell. Retrieved 2013-09-09 from https://www.youtube.com/watch?v=ijWUQ6ri41Y.

———. 2013. "Additional Notes from Gary Whisenhunt Interview RPG(ology) Not Contained in Video," *RPG(ology)* blog, by Carey Martell. Retrieved 2016-03-15 from http://www.rpgfanatic.net/advanced_game_wiki_database.html?p=news&nrid=5049&game=dnd.

Zraket, C. 1990-05-03. CBI-UM Oral History Interview OH 198 by Arthur L. Nordberg.

Bibliography

Alessi, S. M., and S. R. Trollip. *Computer-Based Instruction: Methods and Development.* Englewood Cliffs, NJ: Prentice Hall, 1985.

Allen, M., ed. *Michael Allen's e-Learning Annual 2008.* San Francisco: Pfeiffer, 2008.

Allison, Nancy. "Need Fast, Error Proof Teacher? Call PLATO." *NG*, September 29, 1963, 51.

Alpert, D., and D. Bitzer. "Advances in Computer-Based Education." *Science* 167, March 1970, 1582–90.

———. "The World After PLATO IV: The Implications of Computer-Based Systems on Education of the Future." Unpublished monograph. UIUC.

Ashby, W. R. *Design for a Brain: The Origin of Adaptive Behavior.* New York: John Wiley & Sons, 1960.

———. *An Introduction to Cybernetics.* London: Chapman & Hall, 1957.

Asimov, I. "The Fun They Had." *Magazine of Fantasy and Science Fiction*, February 1954.

Atkinson, R. C., and H. A. Wilson, eds. *Computer-Assisted Instruction: A Book of Readings.* New York: Academic Press, 1969.

Avner, A. "A Quarter Century of Computer-Based Education." Unpublished draft of paper presented at Nordic CAI Conference, Stockholm, Sweden, 1985.

Baker, F. "Computer-Managed Instruction: A Context for Computer-Based Instruction," in *Computer-Based Instruction: A State-of-the-Art Assessment*, H. F. O'Neill, Jr., ed. New York: Academic Press, 1981, 23–64.

Baldwin, L. V., and K. S. Down, eds. *Educational Technology in Engineering.* Washington, DC: National Academy Press, 1981.

Barber, J. A. "Data Collection as an Improvement Technique for PLATO Lessons." MS degree, UIUC, 1975.

Barlow, A. T. "Computer to Teach Mercy Students 'TLC' for Patients." *NG*, July 1966.

Becker, J. W. "It Can't Replace the Teacher—Yet." *Phi Delta Kappan* 58(5), January 1967, 237–39.

Bell, J. N. "Will Robots Teach Your Children?" *Popular Mechanics*, October 1961, 152.

Bencivenga, J. "Computer Links Prison Inmates to New Career." *Christian Science Monitor*, August 12, 1982.

Benjamin, L. "A History of Teaching Machines." *American Psychologist* 43(9): 703–12.

Bennahum, D. S. *Extra Life: Coming of Age in Cyberspace*. New York: Basic Books, 1998.

Berggreen, A. *RFC #600: Interfacing an Illinois Plasma Terminal to the ARPA-NET*. Computer Systems Laboratory, University of California, Santa Barbara, 1973.

Berners-Lee, T. *Weaving the Web: The Original Design and Ultimate Destiny of the World Wide Web by Its Inventor*. New York: HarperCollins, 1999.

Bernstein, N. "On Minnesota Prison Computer, Files to Make Parents Shiver." *New York Times*, November 18, 1986.

Bielawski, L., and D. Metcalf. *Blended eLearning: Integrating Knowledge, Performance, Support, and Online Learning*. Amherst, MA: HRD Press, 2003.

Bitzer, D. "The PLATO Project at the University of Illinois." *Engineering Education*, December 1986, 175–80.

———. *Signal Amplitude Limiting and Phase Quantization in Antenna Systems*. PhD diss., UIUC Electrical Engineering Department, 1960.

Bitzer, D., and P. Braunfeld, "Computer Teaching Machine Project: PLATO on ILLIAC." *Computers and Automation* 11(2), 1962, 16–18.

Bitzer, D., and F. Propst. Addendum to testimony presented before the House Science and Technology Subcommittee on Domestic and International Planning, Analysis, and Cooperation for the oversight review, "Computers and the Learning Society," October 6, 1977, Washington, DC.

Bitzer, D., P. Braunfeld, and W. Lichtenberger. "PLATO: An Automatic Teaching Machine." *IRE Transactions on Education* E-4(4), December 1961, 157–61.

———. "PLATO II: A Multiple-Student, Computer-Controlled, Automatic Teaching Device. Programmed Learning and Computer-Based Instruction." *Proceedings of the Conference on Applications of Digital Computers to Automated Instruction*, October 10–12, 1961. J. E. Coulson. New York: John Wiley & Sons, 1962, 205–16.

Bjork, D. W. *B. F. Skinner*. New York: Basic Books, 1993.

Blomeyer, R. L. J. "The Use of Computer-Based Instruction in Foreign Language Teaching: An Ethnographically-Oriented Study." PhD diss., UIUC, 1985.

Bloom, A. *The Closing of the American Mind*. New York: Simon & Schuster, 1987.

Bonn, G. S., and S. A. Faibisoff, eds. *Changing Times: Changing Libraries*. Allerton Park Institute, UIUC Graduate School of Library Science, 1978.

Bork, A. "The Future of Learning: An Interview with Alfred Bork." EDU-COM, 1999.

Brickman, W. W., and S. Lehrer, eds. *Automation, Education, and Human Values*. New York: School and Society Books, 1966.

Brong, G. R. "Response. Changing Times: Changing Libraries." G. S. Bonn and S. A. Faibisoff. Urbana-Champaign, IL, University of Illinois Graduate School of Library Science 22, 1978, 91–95.

Brown, J. S. *Learning in and for the 21st Century: CJ Koh Professorial Lecture Series No. 4.* E. L. Low, ed. Singapore: National Institute of Education, 2012.

Bruner, J. *Actual Minds, Possible Worlds.* Cambridge: Harvard University Press, 1986.

———. *The Culture of Education.* Cambridge: Harvard University Press, 1996.

———. "Introduction: The New Educational Technology," in *Revolution in Teaching: New Theory, Technology, and Curricula,* A. de Grazia and D. A. Sohn, eds. New York: Bantam, 1964, 1–7.

———. *Towards a Theory of Instruction.* New York: Belknap Press, 1966.

Brzezinski, M. *Red Moon Rising: Sputnik and the Hidden Rivalries That Ignited the Space Age.* New York: Times Books, 2007.

Burniske, R. W., and L. Monke. *Breaking Down the Digital Walls: Learning to Teach in a Post-Modem World.* Albany: State University of New York Press, 2001.

Bushnell, D. D. "The Computer-Assisted School System." *Computers and Automation* 11(2), 1962, 6–9.

Bushnell, D. D., and D. W. Allen, eds. *The Computer in American Education.* New York: John Wiley & Sons, 1967.

Callahan, R. E. *Education and the Cult of Efficiency: A Study of the Social Forces That Have Shaped the Administration of the Public Schools.* Chicago: University of Chicago Press, 1962.

Carpenter, F. *The Skinner Primer: Beyond Freedom and Dignity: What the B. F. Skinner Debate Is All About.* New York: The Free Press, 1974.

Carroll, M. "The Rumor Chronicle." Unpublished essay, 2011.

Cherian, M. *From PLATO to Podcasts: Fifty Years of Federal Involvement in Educational Technology.* Center on Education Policy, 2009.

Chesnel, M. J. "A PLATO Lesson on Information Structures," UIUC, 1974.

Christensen, C. M., M. B. Horn, and C. W. Johnson. *Disrupting Class: How Disruptive Innovation Will Change the Way the World Learns.* New York: McGraw-Hill, 2008.

Ciraolo, M. Adventures On-line. *ANTIC magazine,* March 1984, 19.

Coulson, J. E. "A Computer-Based Laboratory for Research and Development in Education," in *Programmed Learning and Computer-Based Instruction: Proceedings of the Conference on Applications of Digital Computers to Automated Instruction, October 10–12, 1961,* J. E. Coulson, ed. New York: John Wiley and Sons, 1962, 191–204.

Coulson, J. E., ed. *Programmed Learning and Computer-Based Instruction: Proceedings of the Conference on Applications of Digital Computers to Automated Instruction, October 10–12, 1961.* New York: John Wiley & Sons, 1962.

Crecente, B. "Maria Montessori: The 138-Year-Old Inspiration Behind Spore." *Kotaku,* 2014. Retrieved 2016-05-17 from http://www.kotaku.com

.au/2009/03/maria_montessori_the_138yearold_inspiration_behind _spore-2/.

Crowder, N. A. "Intrinsic and Extrinsic Programming," in *Programmed Learning and Computer-Based Instruction: Proceedings of the Conference on Applications of Digital Computers to Automated Instruction, October 10–12, 1961*, J. E. Coulson, ed. New York: John Wiley & Sons, 58–66.

Crowder, N. A., and G. A. Martin. *Adventures in Algebra*. Garden City, NY: Doubleday, 1960.

Dammeyer, J. W. "Computer-Assisted Learning—or Financial Disaster." *Educational Leadership* 40(5), 1983, 7–9.

Decherney, P., N. Ensmenger, and C. S. Yoo. "Are Those Who Ignore History Doomed to Repeat It?" *University of Chicago Law Review* 78, 2011, 1627–85.

Degim, I. A., J. Johnson, and T. Fu eds. *Theory On Demand #16: Online Courtship: Interpersonal Interactions Across Borders*. Theory On Demand. Amsterdam: Institute of Network Cultures, 2015.

de Grazia, A., and D. A. Sohn, eds. *Programs, Teachers, and Machines*. New York: Bantam, 1964.

——. *Revolution in Teaching: New Theory, Technology, and Curricula*. New York: Bantam, 1964.

Deken, J. *The Electronic Cottage*. New York: Bantam, 1981.

Denenberg, S. A. "A Personal Evaluation of the PLATO System." *ACM SIGCUE Bulletin* 12(2), 1978, 3–10.

Dickson, P. *Sputnik: The Shock of the Century*. New York: Walker, 2001.

Diebold, J., ed. *The World of the Computer*. New York: Random House, 1973.

Digital Equipment Corporation. *Introduction to Computer-Based Education*. Marlborough, MA: Digital Equipment Corporation, 1984.

Driscoll, F. D. "The PLATO System: A Study in the Diffusion of an Innovation." PhD. diss., University of Massachusetts, 1987.

Drucker, Peter. *Adventures of a Bystander*. Piscataway, NJ: Transaction Publishers, 1994.

Duggan, M. A., E. F. McCartan, and M. R. Irwin, eds. *The Computer Utility: Implications for Higher Education*. Lexington, MA: D. C. Heath, 1969.

Dwyer, D. "We're in This Together." *Educational Leadership* 54(3), 1996, 24–26.

Ellul, J. *The Technological Society*. New York: Alfred A. Knopf, 1964.

Engelbart, D. *Augmenting Human Intellect: A Conceptual Framework*. Menlo Park, CA: Stanford Research Institute, 1962.

Evans, Rupert. *Another String to My Bow*. Urbana, IL: Prairie Publications, 2001.

Fields, C., and J. Paris. "Hardware-Software," in *Computer-Based Instruction: A State-of-the-Art Assessment*, H. F. O'Neill, Jr., ed. New York: Academic Press, 1981, 65–90.

Finley, K. "NoSQL: The Love Child of Google, Amazon and . . . Lotus Notes." *Wired*, December 5, 2012. Retrieved from https://www.wired .com/2012/12/couchdb/.

Fleschsig, A. J., and D. A. Seamans. "Determining the Value of PLATO Computer-Based Education for a Freshman Engineering Course." *Engineering Education*, January 1987, 240–42.

Flynn, C. "UI Professor Expresses the Unknown in Computers." *NG*, November 9, 1980.

Friedman, E. A. "Information Technology: Case Studies of Effective Applications: Introduction." *Engineering Education*, 1986, 143–45.

Fuller, R. B. *Education Automation: Freeing the Scholar to Return to His Studies.* Carbondale: Southern Illinois University Press, 1962.

Gaede, O. F. "From PLATO to Tomorrow: The Birth of Virtual Communities." *Interactive Teacher*, 1997.

Galanter, E., ed. *Automatic Teaching: The State of the Art.* New York: John Wiley & Sons, 1959.

Ghesquiere, J., C. Davis, and C. Thompson. *Introduction to TUTOR.* UIUC CERL, 1974.

Glaser, R., ed. *Advances in Instructional Psychology.* Hillsdale, NJ: Lawrence Erlbaum Associates, 1987.

Gleick, J. *The Information: A History, A Theory, A Flood.* New York: Pantheon, 2011.

Goldstein, R. *Desperate Hours: The Epic Rescue of the* Andrea Doria. New York: John Wiley & Sons, 2001.

Goodlad, J. L. "Computers and the Schools in Modern Society," Symposium on Computer-Assisted Learning. *Proceedings of the National Academy of Sciences* 63, 1969, 573–603.

Hafner, K. *The Well: The Story of Love, Death, and Real Life in the Seminal Online Community.* New York: Carroll & Graf, 2001.

Haga, E., ed. *Automated Educational Systems.* Elmhurst, NY: Business Press, 1967.

Hansen, W. J., R. Doring, and L. R. Whitlock. (1976). "A Videotape Analysis of Student Performance on an Interactive Examination." UIUC, 1976.

Harms, V. "PLATO Learning Phone: Hands-on Review by an 8th Grader." *Antic* 4, 1986, 44.

Hart, R. S. "The Illinois PLATO Foreign Languages Project." *CALICO Journal* 12(4). 1995, 15–37.

Heidbreder, E. *Seven Psychologies.* Englewood Cliffs, NJ: Prentice Hall, 1933.

Hernandez, D. "Facebook!? Twitter!? Instagram!? We Did That 40 Years Ago." *Wired*, 2012. Retrieved 2012-12-14 from https://www.wired.com/2012/12/social-media-history/.

Herrick, C. J. *The Thinking Machine.* Chicago: University of Chicago Press, 1929.

Heuston, D. H., and J. W. Parkinson. *The Third Source: A Message of Hope for Education.* Sandy, UT: The Waterford Institute, 2010.

Hillis, W. D. *The Pattern on the Stone.* New York: Basic Books, 1998.

Himwich, H. A., L. Miller, M. F. Kilian, W. B. Goldring, and R. G. Votaw. "A PLATO Simulation for Developing and Assessing Clinical Competence in Childbirth Management." Washington, DC: ADCIS, 1980, 107–09.

Holtzman, W. H., ed. *Computer Assisted Instruction, Testing, and Guidance.* New York: Harper & Row, 1970.

Hughes, J. L., ed. *Programed Learning: A Critical Evaluation.* Chicago: Educational Methods, Inc, 1970.

Hunter, B., C. S. Kastner, M. L. Rubin, and R. J. Seidel, eds. *Learning Alternatives in U.S. Education: Where Student and Computer Meet.* Englewood Cliffs, NJ: Educational Technology Publications, 1975.

Izquierdo, F. J. "A Generator/Grader of Problems About Syntax of Programming Languages to Be Used in an Automated Exam System." UIUC, 1975.

James, L. "Course-Integrated Electronic Socializing on PLATO." 1991. Retrieved 2016-05-03 from http://www.soc.hawaii.edu/leonj/leonj/leonpsy/instructor/leonplato1.html.

Johnson, R. L., D. L. Bitzer, and H. G. Slottow. "The Device Characteristics of the Plasma Display Element." *IEEE Transactions on Electron Devices* ED-18(9): 642–49.

Johnson-Laird, P. N. *The Computer and the Mind: An Introduction to Cognitive Science.* Cambridge: Harvard University Press, 1988.

Johnstone, B. *Never Mind the Laptops: Kids, Computers, and the Transformation of Learning.* New York: iUniverse, 2003.

Jones, D. W. "Run Time Support for the TUTOR Language on a Small Computer System." UIUC, 1977.

Jones, S. "Afterword: Building, Buying, or Being There: Imagining Online Community," in *Building Virtual Communities: Learning and Change in Cyberspace*, K. A. Renninger and W. Shumar, eds. Cambridge: Cambridge University Press, 2002, 368–76.

———. "Everything I Know About the Internet I Learned from PLATO." Presentation delivered at Augustana College, April 19, 2001.

Kearsley, G. "Authoring Systems in Computer-Based Education." *Communications of the ACM* 25(7), 1982, 429–37.

———. *Training for Tomorrow: Distributed Learning Through Computer and Communications Technology.* Reading, MA: Addison-Wesley, 1985.

Kearsley, G., ed. *Online Learning: Personal Reflections on the Transformation of Education.* Englewood Cliffs, NJ: Educational Technology Publications, 2005.

Kent, S. L. *The Ultimate History of Video Games.* Roseville, CA: Prima Publishing, 2001.

Kidder, T. *The Soul of a New Machine.* New York: Avon, 1981.

Kingery, R. A., R. D. Berg, and E. H. Schillinger. *Men and Ideas in Engineering: Twelve Histories from Illinois.* Urbana: University of Illinois Press, 1967.

Koster, R. "The Online World Timeline—Ralph's Website." Retrieved 2016-06-23 from http://www.raphkoster.com/games/the-online-world-timeline/.

Kushner, D. *Masters of DOOM: How Two Guys Created an Empire and Transformed Pop Culture.* New York: Random House, 2003.

Kypta, L. S., and H. Bobotek. "Automatic Air Traffic Control, Part II, an Experimental Control Logic." CSL UIUC, 1963.

Lagemann, E. C. "The Plural Worlds of Educational Research." *History of Education Quarterly.* New York: John Wiley & Sons, 1989, 29, 183–214.

Lammers, S. *Programmers at Work.* Redmond, WA: Microsoft Press, 1987.

Land, E. "Generation of Greatness: The Idea of a University in an Age of Science." Ninth Annual Arthur Denon Little Memorial Lecture at the Massachusetts Institute of Technology, 1957-05-22. Retrieved 2016-03-04 from http://groups.csail.mit.edu/mac/users/hal/misc/generation-of -greatness.html.

Latzko-Toth, G., and S. Jones. "Sharing Digital Resources: PLATO and the Emerging Ethics of Social Computing." Paper presented at the *ETHI-COMP 2014* conference, Paris, June 25–27, 2014.

Lawrie, J. "The History of the Internet in South Africa: How It Began." Unpublished essay, 1997.

Leach, J. "PLATO Computer System Speeds Up the Game of Learning." The Sunday Camera's FOCUS, Boulder, CO, *Daily Camera,* November 20, 1977, 3–6.

Levy, S. *Hackers: Heroes of the Computer Revolution.* New York: Anchor, 1984.

Licklider, J. C. R. "Preliminary Experiment in Computer-Aided Teaching." *Programmed Learning and Computer-Based Instruction.* Proceedings of the Conference on Applications of Digital Computers to Automated Instruction, October 10 12, 1961, J. E. Coulson, ed. New York: John Wiley & Sons, 1962, 217–39.

Loeper, J. J. *Going to School in 1876.* New York: Atheneum, 1984.

Luce, G. G. "Can Machines Replace Teachers?" *Saturday Evening Post* 233(13), September 24, 1960.

Lumsdaine, A. A. "Some Theoretical and Practical Problems in Programmed Instruction." *Programmed Learning and Computer-Based Instruction.* Proceedings of the Conference on Applications of Digital Computers to Automated Instruction, October 10–12, 1961, J. E. Coulson, ed. New York: John Wiley & Sons, 1962, 134–51.

Lyman, E. R. *A Descriptive List of PLATO Programs.* Report R-296, July 1967. CSL UIUC.

Martin, J. S., and M. Szabo (1986). "The Dialogue and Computer-Assisted Instruction in Chemistry at the University of Alberta." 27th International Meeting of the Association for the Development of Computer-Based Instructional Systems, New Orleans, LA, 1986.

McCorduck, P. *The Universal Machine: Confessions of a Technological Optimist.* New York: Harcourt Brace Jovanovich, 1985.

McDonald, C. "The PLATO Computer-Based Education System: Teacher's Tool or Teacher?" SIGCIS Workshop 2010. Tacoma, WA: SIGCIS, 2010.

McGovern, P. J. "Teaching Machines and Programmed Learning—Roster of Organizations and What They Are Doing." *Computers and Automation* 11(2), 1962, 33–40.

Merrill, M. D., E. W. Schneider, and K. A. Fletcher. *TICCIT.* Englewood Cliffs, NJ: Educational Technology Publications, 1980.

Merrill, M. D., and D. G. Twitchell, eds. *Instructional Design Theory.* Englewood Cliffs, NJ: Educational Technology Publications, 1994.

Miller, J. W. "Whatever Happened to New Math?" *American Heritage.* Rockville, MD: American Heritage Publishing Company, 1990, 41.

Molnar, A. "Computers in Education: A Historical Perspective of the Unfinished Task." *T.H.E. Journal* 6(3), August 1990, 80–83.

———. "Computers in eEducation: A Brief History." *T.H.E. Journal,* 1997.

———. "Critical Issues in Computer-Based Learning," in *The Computer and Education.* Englewood Cliffs, NJ: Educational Technology Publications, 1973, 14–18.

———. "ZapMe! Linking Schoolhouse and Marketplace in a Seamless Web." *Phi Delta Kappan,* 2000, 601–3.

Montanelli, R. G. "CS 103 Plato Experiment, Fall 1974." UIUC.

Morris, B. "The Commercialization of PLATO." Control Data Corporation. Unpublished memo, 1982.

Morris, B., and L. Lewytzkyj. "PLATO: A Humanized Technology." Unpublished memo.

Murphy, R. T., and L. R. Appel. *Evaluation of the PLATO IV Computer-Based Education System in the Community College.* Princeton: Educational Testing Service, 1977.

Murray, C. J. *The Supermen.* New York: John Wiley & Sons, 1997.

Nader, R., and W. Taylor. *The Big Boys: Power and Position in American Business.* New York: Pantheon, 1986.

Nair, K. *Blossoms in the Dust: The Human Factor in Indian Development.* New York: Frederick A. Praeger, 1962.

Nelson, C. C., B. J. Hughes, and R. E. Virgo. "CAI Applications in Statics." *Engineering Education,* November 1986, 96–100.

Nelson, T. H. *Computer Lib/Dream Machines.* Redmond, WA: Microsoft Press, 1987.

Noble, D. D. "Mad Rushes into the Future: The Overselling of Educational Technology." *Educational Leadership* 54(3), 1996, 18–23.

———. "Military Research and the Development of Computer-Based Education." PhD diss., University of Rochester, 1990.

Oettinger, A. *Run, Computer, Run: The Mythology of Educational Innovation.* Cambridge: Harvard University Press, 1969.

O'Neal, F. "Waterford School and the WICAT Education Institute: An Alternative Model for CAI and Development Research." *CALICO Journal* 1(1), 1983, 19–24.

O'Neill, H. F., Jr., ed. *Computer-Based Instruction: A State-of-the-Art Assessment.* New York: Academic Press, 1981.

Owens, F. N., A. D. Sherer, and G. Grimes. "Computer-Based Instruction in Nutrition." *NACTA Journal* 18(4), 1974, 71–74.

Paceley, C. "Dan Alpert Continues Shaping Technology and Education." *Physics Illinois News,* Number 1, 2006, UIUC.

———. "The Innovation of PLATO Homework." *Physics Illinois News* 1, 2006, 5, UIUC.

Papert, S. "Computers and Learning," in *The Computer Age: A Twenty-Year Review*, M. L. Dertousoz and J. Mosesm, eds. Cambridge: MIT Press, 1979, 73–86.

———. *The Children's Machine: Rethinking School in the Age of the Computer.* New York: Basic Books, 1993.

———. "Computer as Mudpie," in *Intelligent Schoolhouse: Readings on Computers and Learning*, D. Peterson, ed. Reston, VA: Reston Publishing Company, 1984.

———. *The Connected Family: Bridging the Digital Generation Gap.* Atlanta: Longstreet Press, 1996.

Parry, J. "PLATO." Symposium sponsored by U.S. Office of Education and the National Institute of Education, held at State University of New York at Stony Brook, September 1973. SUNY Stony Brook/Educational Technology Publications, 1975.

Pask, G., and S. Curran. *Micro Man: Computers and the Evolution of Consciousness.* New York: Macmillan, 1982.

Payne, K. "Hocus Pocus: PLATO Inventor Has a Few Tricks up His Sleeve." *Urbana Courier,* 1978.

Peterson, D., ed. *Intelligent Schoolhouse: Readings on Computers and Learning.* Reston, VA: Reston Publishing Company, 1984.

Petroski, H. *To Engineer Is Human: The Role of Failure in Successful Design.* New York: Vintage, 1992.

Phillips, J. L. *Piaget's Theory: A Primer.* San Francisco: W. H. Freeman, 1981.

Piumarta, I., and Kimberly Rose, eds. *Points of View: A Tribute to Alan Kay.* Glendale, CA: Viewpoints Research Institute, 2010.

Postman, N. *The End of Education: Redefining the Value of School.* New York: Alfred A. Knopf, 1995.

———. *Technopoly: The Surrender of Culture to Technology.* New York: Alfred A. Knopf, 1992.

Powers, R. *The Gold Bug Variations.* New York: William Morrow, 1991.

Pradels, J. L. "The Guide: An Information System." UIUC, 1974.

Pressey, S. L., and F. P. Robinson. *Psychology and the New Education.* New York: Harper & Brothers, 1933.

Price, R. M. *The Eye for Innovation: Recognizing Possibilities and Managing the Creative Enterprise.* New Haven: Yale University Press, 2006.

Rahmlow, H. F., R. C. Fratini, and J. Ghesquiere. *PLATO.* Englewood Cliffs, NJ: Educational Technology Publications, 1980.

Rankin, J. "Toward a History of Social Computing: Children, Classrooms, Campuses, and Communities." *IEEE Annals of the History of Computing* 36(2), April–June 2014, 86–88.

Rheingold, H. *Tools for Thought: The History and Future of Mind-Expanding Technology.* New York: Simon & Schuster, 1985.

———. *The Virtual Community: Homesteading on the Electronic Frontier.* Reading, MA: Addison-Wesley, 1993.

Rushkoff, D. *Present Shock: When Everything Happens Now*. New York: Current, 2013.

Rutherford, A. *Beyond the Box: B. F. Skinner's Technology of Behavior from Laboratory to Life, 1950s–1970s*. Toronto: University of Toronto Press, 2009.

Schank, R. C., and P. Childers. *The Cognitive Computer: On Language, Learning, and Artificial Intelligence*. Reading, MA: Addison-Wesley, 1984.

Schank, R. C., and C. K. Riesback. *Inside Computer Understanding: Five Programs Plus Miniatures*. Hillsdale, NJ: Lawrence Erlbaum Associates, 1981.

Seidel, R. J., R. E. Anderson, and B. Hunter. *Computer Literacy: Issues and Directions for 1985*. New York: Academic Press, 1982.

Seidel, R. J., and M. Rubin, eds. *Computers and Communication: Implications for Education*. New York: Academic Press, 1977.

Seil, W. "How the Real HAL Computers Will Change Your Life Before 2001." *Science Digest*, January 1973, 17–21.

Sherwood, B. "Interactive Electronic Media," in *Changing Times: Changing Libraries*, G. S. Bonn and S. A. Faibisoff, eds. University of Illinois Graduate School of Library Science, 1976, 22: 81–91.

———. *The TUTOR Language*. Minneapolis: Control Data Education Company, 1977.

Shlechter, T. M. "Promises, Promises, Promises: History and Foundations of Computer-Based Training," in *Problems and Promises of Computer-Based Training*, T. M. Shlechter, ed. New York: Praeger, 1991, 1–19.

Shlechter, T. M., ed. *Problems and Promises of Computer-Based Training*. New York: Praeger, 1991.

Simpson, P. D. *Alive on the* Andrea Doria!*: The Greatest Sea Rescue in History*. Fleischmanns, NY: Purple Mountain Press, 2006.

Skinner, B. F. *About Behaviorism*. New York: Vintage, 1974.

———. *Beyond Freedom and Dignity*. New York: Alfred A. Knopf, 1971.

———. *A Matter of Consequences*. New York: Alfred A. Knopf, 1983.

———. *Particulars of My Life*. New York: Alfred A. Knopf, 1976.

———. "Programmed Instruction Revisited." *Phi Delta Kappan*, October 1986, 103–10.

———. *The Shaping of a Behaviorist*. New York: Alfred A. Knopf, 1979.

———. *The Technology of Teaching*. New York: Meredith Corporation, 1968.

———. *Walden Two* (reissued ed.). New York: Macmillan, 1976.

Smallwood, R. D. "A Decision Structure for Computer-Based Teaching Machines." *Computers and Automation* 11(2), 1962, 9–14.

Smith, B. *Portrait of India*. New York: J. B. Lippincott, 1962.

Stake, B. "Clinical Studies of Counting Problems with Primary School Children." PhD diss., UIUC School of Education, 1980.

Stifle, J. *The PLATO IV Architecture*. CERL Report X-20, 1972, UIUC Archives.

———. *The PLATO IV Terminal: Description of Operation*. 1974, UIUC Archives.

———. "A Preliminary Report on the PLATO V Terminal." May 19, 1975, CERL UIUC, unpublished.

Stone, A. R. *The War of Desire and Technology at the Close of the Mechanical Age.* Cambridge: MIT Press, 1996.

Sunburst Communications. *How The West Was One + Three X Four.* Designed by Bonnie Seiler. User manual. Pleasantville, NY: Sunburst Communications, 1987.

Suppes, P. "Current Trends in Computer-Assisted Instruction," in *Advances in Computers,* M. C. Yovits, ed. New York: Academic Press, 1979, 18: 173–229.

Thomas, D. *Hacker Culture.* Minneapolis: University of Minnesota Press, 2002.

Thompson, T. J. "The Pascal Series: A PLATO Lesson Sequence." UIUC, 1977.

Thorndike, E. L. *Education: A First Book.* New York: Macmillan, 1912.

Thornton, J. E. *The Design of a Computer: The Control Data 6600.* New York: Scott Foresman, 1970.

Timmins, M. "In the Time of PLATO: How Students at Illinois in the 70s Created the Computer Technology of the Future." *UIUC Alumni News,* 2010. Retrieved 2016-04-18 from https://illinoisalumni.org/2010/09/10/in-the-time-of-plato/.

Truman, J. H., and J. K. Burns. "The Electric Power Plant at the University of Illinois." Bachelor of Science thesis, Electrical Engineering, 1911, UIUC Archives.

Turner, J. A. "Origins of PLATO: A $10 TV Set and the Vision of a Researcher." *Chronicle of Higher Education* 24(14), November 28, 1984, 72.

Tuttle, J. G. "The Historical Development of the Computer Capabilities Which Permitted the Use of the Computer as an Educational System in the United States from 1958 to 1968, with Implications of Trends." PhD diss., New York University, 1960.

UIUC. *Transactions of the Board of Trustees,* July 1, 1958, to June 30, 1960; July 1, 1960, to June 30, 1962; July 1, 1962, to June 30, 1964; July 1, 1964, to June 30, 1966; July 1, 1966, to June 30, 1968; July 1, 1968, to June 30, 1970; July 1, 1970, to June 30, 1972; July 1, 1972, to June 30, 1974; July 1, 1974, to June 30, 1976; July 1, 1976, to June 30, 1978; July 1, 1978, to June 30, 1980; July 1, 1980, to June 30, 1982; July 1, 1982, to June 30, 1984; July 1, 1984, to June 30, 1986; July 1, 1986, to June 30, 1988; July 1, 1988, to June 30, 1990; July 1, 1990, to June 30, 1992; July 1, 1992, to June 30, 1994; July 1, 1994, to June 30, 1996.

Ullman, E. *Close to the Machine: Technophilia and Its Discontents.* San Francisco: City Lights Books, 2001.

U.S. Congress, House of Representatives, Committee on Government Operations. "Federal Information Systems and Plans—Federal Use and Development of Advanced Information Technology." 93rd Congress, 1st Session, April 10 and 17, 1973. Washington, DC: U.S. Government Printing Office.

U.S. Congress, House of Representatives, Committee on Science and Technology. "Computers and the Learning Society: Hearings Before the Subcommittee on Domestic and International Scientific Planning, Analysis, and Cooperation of the Committee on Science and Technology." 95th Congress, October 4, 6, 12, 13, 18, and 27, 1977. Washington, DC: U.S. Government Printing Office.

U.S. Congress, Office of Technology Assessment. *Information Technology and Its Impact on American Education*. Washington, DC: U.S. Government Printing Office, 1983.

Useem, E. L. *Low Tech Education in a High Tech World*. New York: The Free Press, 1986.

Vallee, J. *The Heart of the Internet: An Insider's View of the Origin and Promise of the On-Line Revolution*. Charlottesville, VA: Hampton Roads Publishing, 2003.

Van Meer, E. "PLATO: From Computer-Based Education to Corporate Social Responsibility." *Iterations: An Interdisciplinary Journal of Software History* 2, 2003.

Vinge, V. *True Names . . . and Other Dangers*. New York: Baen Books, 1987.

Watanabe, N. "PLATO Days: Remembering Those Strange Orange Dots." Unpublished essay, 2003.

Watson, J. B. *Behaviorism*. New York: People's Institute Publishing Company, 1924.

Watters, A. "And So, Without Ed-Tech Criticism." *Hack Education*, August 15, 2015. Retrieved 2015-08-17 from http://hackeducation.com/2015/08/15/criticism.

———. "Is It Time to Give Up on Computers in Schools?" *Hack Education*, June 29, 2015. Retrieved 2016-06-30 from http://hackeducation.com/2015/06/29/is-it-time-to-give-up-on-computers.

———. *The Monsters of Education Technology*. Self-published, 2014.

———. *The Revenge of the Monsters of Education Technology*. Self-published, 2015.

———. "Teaching Machines and Turning Machines: The History of the Future of Labor and Learning." *Hack Education*, August 10, 2015. Retrieved 2015-08-10 from http://hackeducation.com/August 10, 2015/digpedlab.

———. "Top Ed Tech Trends of 2015: The Politics of Education Technology." *Hack Education*, December 2, 2015. Retrieved 2015-12-02 from http://hackeducation.com/2015/12/02/trends-politics.

Weber, L. "Blind Student Power." *Illinois Technograph* 84(1), October 1968, 17–20.

Wolfe, T. *The Right Stuff*. New York: Bantam, 1980.

Woolley, D. "You Don't Have to Stay Here." *Just Think of It* blog, 2015, http://just.thinkofit.com/you-dont-have-to-stay-here/.

Worthy, J. C. *William C. Norris: Portrait of a Maverick*. Cambridge, MA: Ballinger, 1987.

Yeomans, E. *The Shady Hill School: The First Fifty Years*. Cambridge, MA: Windflower Press, 1979.

High School and College Yearbooks

Collinsville (IL) High School: 1951–1955.
ILLIO, UIUC: 1911–1996.
Maroon, Champaign (IL) Central High School: 1960–1975.
Rosemary, Urbana (IL) High School: 1957–1986.
U and I, University High School (UIUC): 1955–1985.

Source Notes

Preface

Published Sources
Asimov, Isaac (1951). "The Fun They Had." Originally published in an unnamed 1951 children's newspaper, then reprinted in the February 1954 issue of *The Magazine of Fantasy and Science Fiction*. Retrieved 2014-06-19 from http://visual-memory.co.uk/daniel/funtheyhad.html.

PART I   THE AUTOMATIC TEACHER

Epigraph

Drucker. *Adventures*, 255.
Musk, E. "Elon Musk Interview at World Energy Innovation Forum," May 4, 2016. Retrieved 2016-05-05 from https://www.youtube.com/watch?v=LIIF6WsmzFU.
Papert, "Computer as Mudpie," in *Intelligent Schoolhouse*.

1. *Praeceptor ex Machina*

Interview Sources
Author interviews: Eliot (2013), Gilpin (1997), Papert (1987), Skinner (1987).

Published Sources
Galanter. *Automatic Teaching*.
Petrina, S. "Sidney Pressey and the Automation of Education, 1924–1934." *Technology and Culture* 45(2), 2004, 305–30.
Skinner, B. F. "Baby in a Box." *Ladies' Home Journal*, October 1945, 30–31, 135–36, 138.
———. *A Matter of Consequences*.
———. *Particulars of My Life*.
———. "Programmed Instruction Revisited." *Phi Delta Kappan*, 1986, 103–10.
———. *Shaping of a Behaviorist*.
———. *Technology of Teaching*.

————. *Walden Two.*
Thorndike. *Education*, 165.
Yeomans. *Shady Hill School.*

Unpublished Sources
Email correspondence with author: Buzan, D. (Skinner) (2012); Deutsch, L. P. (2002, 2013); Vargas, J. (Skinner) (1997).

2. An Educational Emergency

Interview Sources
Author interviews: Alpert (1986, 2003); Donald Bitzer (1986, 1987, 1997, 2002); R. Brown (2003); Desmond (2003); G. and J. Leichner (2012); Sherwin (1987); Skaperdas (2002); L. Stolurow (1997).
Other interviews: Alpert (2002); Sherwin (1991).

Published Sources
Associated Press. "Several Countries Tune In on 'Moon': Red Satellite Circles Globe in 1½ Hours." *Urbana Courier*, October 5, 1957, 1.
"Crisis in Education." *Life* 44(12), March 24, 1958, 26.
Dickson. *Sputnik*, 112.
"Approve New University Positions for Dodson, Health Center Heads." *DI*, August 1, 1957.
"Trustees Appoint D. Alpert Control Systems Director." *DI*, October 25, 1958.
"Donald B. Gilles." Wikipedia. Retrieved 2015-01-16 from https://en.wikipedia.org/wiki/Donald_B._Gillies.
CSL quarterly progress report, September 1960. UIUC Archives.
Jorden, W. "Soviet Fires Satellite into Space." *New York Times*, October 5, 1957, 1.
Kosar, K. R. "National Defense Education Act of 1958." *Federal Education Policy History* blog, June 3, 2011. Retrieved 2013-07-14 from https://federaleducationpolicy.wordpress.com/2011/06/03/national-defense-education-act-of-1958-2/.
"List of Nuclear Weapons Tests of the Soviet Union." Wikipedia. Retrieved 2015-03-27, from https://en.wikipedia.org/wiki/List_of_nuclear_weapons_tests_of_the_Soviet_Union.
"List of Nuclear Weapons Tests of the United States." Wikipedia. Retrieved 2015-03-27 from https://en.wikipedia.org/wiki/List_of_nuclear_weapons_tests_of_the_United_States.
Skinner. *A Matter of Consequences*, 132–33.
Wolfe. *Right Stuff*, 57.

Unpublished Sources
Brown, R., letter to Daniel Alpert, UIUC, May 3, 1960.

3. The Super-Achiever

Interview Sources
Author interviews: Alpert (1986, 2003a, 2003b); Donald Bitzer (1987, 1996, 1997, 2002, 2003); E. Bitzer, S. Bitzer, C. Lampe and J. Lampe (2003); R. Bitzer (2003); Braunfeld (1997); R. Brown (2003); Desmond (2003); Dutcher (2002); Forsyth (2003); Skaperdas (2002).

Published Sources
Collinsville High School yearbook, 1955.
Bitzer, Donald. "PLATO@50: PLATO Computer Learning System 50th Anniversary." Presentation at the *PLATO@50 Conference*, June 2, 2010, Computer History Museum, Mountain View, CA. Retrieved 2010-10-30 from https://www.youtube.com/watch?v=THoxsBw-UmM.
"Eta Kappa Nu Holds Smoker," *DI*, October 28.1953.
Graf, R. W. "Case Study of the Development of the Naval Tactical Data System." Prepared for the National Academy of Sciences, Committee on the Utilization of Scientific and Engineering Manpower," January 29, 1964. Cambridge, MA: United Research, Inc., 17.

Unpublished Sources
Email correspondence with author. Bitzer, Donald (2003; confirming the sit-ups story, Bitzer had this to say: "The story is true. The slash was across my nose where I got hit with a baseball bat. I still have that scar. The worst part of the sit-ups was that I wore a blister on my butt and I had to take a final exam in descriptive geometry which required I sit on a stool for three hours."); Faster, W. (2004); Forsyth, B. (2003).
Other documents: D. Alpert letter to Dean Everitt, June 3, 1960.

4. The Diagram

Interview Sources
Author interviews: Avner (1986); David Bitzer (1997); Donald Bitzer (1986, 1987, 1996, 1997, 2002, 2003); R. Bitzer (2003); Blomme (1986, 2013); Braunfeld (1997); Burr (2002); J. Sherwood (2003).

Published Sources
Adderly, S. "ECE Alumnus Wayne Lichtenberger Donates a Piece of Computing History to the University." *ECE Illinois*, 2010, UIUC. Retrieved 2010-12-15 from https://www.ece.illinois.edu/newsroom/article/1179.
Bitzer, D, and P. Braunfeld. "Computer Teaching Machine Project: PLATO on ILLIAC." *Computers and Automation*, 11(2), February 1962, 16–18.
Bitzer, D., P. Braunfeld, and W. Lichtenberger. "PLATO: An Automatic Teaching Device." CSL Report 1-130, CSL, UIUC, June 1961.

————. "PLATO II: A Multiple Student, Computer-Controlled Automatic Teaching Device." CSL Report 1-109, CSL, UIUC, October 1961.

————. "PLATO II." *Proceedings*, 1962, 205–16.

CSL. Quarterly Report, June, July, August 1960. UIUC.

————Quarterly Report, September, October, November 1960. UIUC.

————Quarterly Report, December 1960–February 1961. UIUC.

Kingery et al. *Men and Ideas*, 146–64.

Minow, M. "Dead Media: Paper Tape." Web article, December 26, 2000. Retrieved 2014-10-19 from https://www.merrymeet.com/minow/paper-tape/paper tape.html.

Nash, J. P., ed. "ILLIAC Programming: A Guide to the Preparation of Problems for Solution by the University of Illinois Digital Computer," Digital Computer Laboratory, Graduate College, 1956, UIUC.

"New UI Computer Begins Operations," *DI*, November 11, 1961, 10.

"To Demonstrate Machine in Electronic Learning." *NG*, March 10, 1961.

UIUC Public Information Office. Press release on upcoming first remote demonstration of PLATO at Allerton Park, March 9, 1961.

Ullman. *Close to the Machine*.

Unpublished Sources

Avner, A. "Bitzer I." Note in *PLATO Past* notesfile, CERL PLATO system, Note 7, March 17, 1977.

Bitzer, D., and E. Jordan. "PLATO as a Teacher of Electrical Engineering, Part II." Undated, c. 1962, UIUC Archives.

Lyman, E. Untitled memo to Professor Edward Jordan, September 7, 1980. UIUC Archives.

5. Soldering Irons, Not Switchblades

Interview Sources

Author interviews: Avner (1986); Donald Bitzer (1986, 1987, 1996, 1997, 2002); D. and M. Bitzer (2002, 2016); Fillman (1997a, 1997b); Frampton (2013); Hanson (2013); Hanson and Gladding (2016); Leeb (2014); Volpp (2003); Walker (2003).

Published Sources

Black, J. W. "PLATO—A Teaching Machine Proving Its Worth." *NG*, March 18, 1962, 32.

Goldstein. *Desperate Hours*.

Simpson. *Alive*.

Unpublished Sources

Email correspondence with author: Altenbernd, N. (2014); Hanson, A. (1996–2016); Jorgensen, R. (2014); Kohn, N. (2013–2014); Leeb, C. (2014–2015); Singer, S. (2013); Sokolove, P. (2014); Walker, M. (1996–2015).

6. Gas and Glass

Interview Sources

Author interviews: Donald Bitzer (1986, 1987, 2003a, 2003b, 2012); D. and M. Bitzer (2016); Cole (2002); Desmond (2003); Golden (2003); Roger Johnson (2003, 2009, 2010, 2014); Liddle (2003); Propst (1997a, 1997b, 2013); Scarpelli (2014); Schermerhorn (2013); Skaperdas (2002); Slottow (2003); Weber (2013, 2014); Willson (2002).

Other interviews: Alpert (2002); Arora (2003); Donald Bitzer (2002a, 2002b); Engelbart (1986); Jeff Johnson (2002); Scarpelli (2002).

Published Sources

Arora, B. M., D. L. Bitzer, H. G. Slottow, and R. H. Willson, "The Plasma Display Panel: A New Device for Information Display and Storage." Society for Information Display, Symposium Digest, May 1967.

Bitzer, D. L. "Inventing the AC Plasma Panel." *Information Display*, February 1999, 22–27.

Bitzer, D. L., et al., "Gaseous Display and Memory Apparatus." US Patent 3,559,190, filed 1966-12-22, issued 1971-01-26.

Bitzer, D. L., and Slottow, H. G. "The Plasma Display Panel: A Digitally Addressable Display with Inherent Memory." Presented at the 1966 Fall Joint Computer Conference, Washington, D.C., *AFIPS Conference Proceedings* 29, 1966, 541–47.

Boyd, J. E., and J. Rucker. "A Blaze of Crimson Light: The Story of Neon." *Distillations*, Chemical Heritage Foundation, Summer 2012. Retrieved 2013-03-11 from https://www.chemheritage.org/distillations/magazine/a-blaze-of-crimson-light-the-story-of-neon.

Chatterji, B. N. "From My Memory," Web page. Retrieved 2016-09-22 from http://debashis1.tripod.com/kgphistory/kgphistory8.htm.

CSL. Quarterly Report, December 1962, January 1963, February 1963, CSL, UIUC.

"A Decade of Achievement in India." *Illinois Technograph* 79(1), October, 37–38.

Fett, G. *Final Report: University of Illinois Program in India at the Indian Institute of Technology, Kharagpur, West Bengal.* 1964. UIUC Archives.

Hokamp, H. "Bitzer Pioneers Computer Instruction for India." *NG*, August 28, 1964, 4.

Hood, H. C. "Large Displays: Military Market Now, Civilian Next." *Electronics.* New York: McGraw-Hill, 1963, 24–27.

Hutchinson, J. "Plasma Display Panels: The Colorful History of an Illinois Technology." ECE Illinois Alumni Association Newsletter, 36(1), Winter 2002–2003. Retrieved 2013-03-07 from http://web.archive.org/web/20051001030137/http://www.ece.uiuc.edu/alumni/w02-03/plasma_history.html.

Joseph, F. S. "Versatile PLATO May Join Ranks of 'Old Masters.'" Associated Press, 1967-04-25.

Lewis, R. "Introducing the U of I Computer, a Teaching Assistant That Turns Science Fiction into Reality." *Chicago Sun-Times*, undated, 1967.

———. "Plasma Panel: New Image for TV!" *Chicago Sun-Times*, January 29, 1967, Section 2, 6.

Moore, G. "Cramming More Components onto Integrated Circuits." *Electronics*, April 19, 1965, 38(8), 114–17.

Slottow, H. G. "Plasma Displays." *IEEE Transactions on Electron Devices*, July 1976, ED-23(7): 760–72.

"'Vision Plate' Is Invented at Illinois U. Chicago, IL." *Chicago Sun-Times*, 1967.

Weber, L. "The First Plasma Display Product." *Information Display* 28 (7&8), July–August, 25–26.

———. "History of the Plasma Display Panel." *IEEE Transactions on Plasma Science* 34(2), April 2006, 268–77.

Willson, R. H. "A Capacitively-Coupled Bistable Gas Discharge Cell for Computer Controlled Displays." PhD diss. and CSL Report R-303, June, UIUC.

Unpublished Sources

Email correspondence with author: Arora, B. (September–October 2003, October 2009); Ebeling, Fred (2013–2015); Holz, G. (March 2010); Hutchinson, J. (April 2003); Klotz, J. (February 2016); Walker, M. (1997-01-01); Wedding, D. (February 2013).

Other Documents

Bitzer, D. Author's transcript of unpublished acceptance speech video, Emmy Award, National Academy for Television Arts and Sciences, National Academy for Television Arts and Sciences, New York, NY, 2002.

Ebeling, Fred. "PLATO IV History." Unpublished essay. September 2014.

———. Correspondence with Donald L. Bitzer, 1963–1964. UIUC Archives.

Fett, G. H. "A Digital Computer for IIT, Kharagpur: A Detailed Analysis and Study Report," April 8, 1963. UIUC Archives.

Miscellaneous India documents, 24/2/12, Boxes 1–9. UIUC Archives.

Walker, M. "Alphabat." Note in *PLATO Past* notesfile, CERL PLATO, August 16, 1977.

7. Two's a Crowd

Interview Sources

Author's Interviews: Alpert (1987, 2003); R. C. Anderson (2002); Donald Bitzer (1986, 1987, 1996, 1997, 2002, 2003a, 2003b, 2012); Blomme (1986, 2013); Compton (2003); Desmond (2003); Evans (2009); Faust (1987); Gilpin (1997); Krueger (2010); Lyman (1986); Schultz (2012); Sherwood (1997); K. Smith (2013); L. Stolurow (1997); Tenczar (1986, 1987, 2003); Volpp (2003).

Published Sources

Bitzer, D. L. "PLATO, Plasma Screens, and Computer-Based Education," in Allen, *e-Learning Annual 2008*, 43–57.

Bitzer, D. L., E. R. Lyman, and J. A. Easley. "The Uses of PLATO: A Computer Controlled Teaching System." *Audio-Visual Instruction*, January 1966.

Ghesquiere et al. *Introduction to TUTOR*.

Lyman, E. R. *A Descriptive List of PLATO Programs, 1960–1968*. CERL Report X-2, May 1968.

Sherwood. *TUTOR Language*.

Stolurow, L. M. "Systems Approach to Instruction." Technical Report No. 7, July. Training Research Laboratory, Department of Psychology, UIUC.

Stolurow, L. M., and D. J. Davis. "Teaching Machines and Computer-Based Systems." Technical Report No. 1, August 1963. Training Research Laboratory, Department of Psychology, UIUC.

"TV of Future: Wire-Brained Looking Glass." *Chicago Tribune*, April 26, 1967.

UIUC Public Information Office. Press release announcing Daniel Alpert becoming dean of UIUC graduate college effective September 1. Release date January 20, 1965.

8. Knocking on the Same Doors

Interview Sources

Author interviews: M. Allen (1997, 2016); Alpert (2003a); Atkinson (1987); Bunderson (1986); Faust (1987); Fletcher (1987); Hendricks (1986); Knoke (2002); Leichner (2012); McWilliams (1987); Melmed (1987); Merrill (1986a, 1986b, 1987); Molnar (1987); Risken (2013); Ross (1987); Sherwood (1997); Sherwood and Chabay (1997, 2002); Stetten (1997); Suppes (1987).

Other interviews: Molnar (1991).

Published Sources

Anderson, P. "Dissolving Graduate College Likely." *DI*, July 25, 1972, 5.

———. "Report Suggests UI Reorganization." *DI*, July 18, 1972, 1.

Bowen, E. "The Computer as a Tutor." *Life*, January 27, 1967, 62: 68-70.

Bunderson, C. V. "Reflections on TICCIT," in Allen, *e-Learning Annual*, 1–30.

———. "The TICCIT Project: Design Strategy for Educational Innovation," in *Productivity in Higher Education*, S. A. Harrison and L. M. Stolurow, eds. Washington, DC: National Institutes of Education, 1973.

Bunderson, C. V., and Faust, G. W. "Programmed and Computer-Assisted Instruction," in *The Psychology of Teaching Methods* (75th yearbook of the National Society for the Study of Education), Part I, 44–90. Chicago: University of Chicago Press, 1976.

"Display Panel Forerunner of Flat TV." *Chemical and Engineering News* 47(26): 52–53.

Jones, R. L. (1995). "TICCIT and Clips: The Early Years." *CALICO Journal* 12(4): 84–96.

National Science Foundation. *MOSAIC* 3(3), Summer 1972.

Saltz, I. "Computers Scary? PLATO Friendly." *NG*, September 27, 1971.

———. "PLATO Can Lower Educational Costs." *NG*, September 29, 1971.

———. "PLATO Lets Teachers Be Sympathetic—Not Clerks." *NG*, September 26, 1971.

Silberman, C. "Technology Is Knocking at the Schoolhouse Door." Fortune 74, August 1966, 120–25.

Stetten, K. J. "A Comparison of TICCET and PLATO IV." MITRE Corporation white paper, November 18, 1970.

Suppes, P. "Computer Technology and the Future of Education," in Atkinson and Wilson, *Computer-Assisted Instruction*.

———. "How Far Have We Come? What's Just Ahead?" *Nation's Schools*, 82, 52–53, 86.

———. "On Using Computers to Individualize Instruction," in D. D. Bushnell and D. W. Allen, eds., *The Computer in American Education*. New York: John Wiley & Sons, 1967.

———. "Teacher and Computer-Assisted Instruction." *National Education Association Journal* 56, February, 15–17.

———. "Tomorrow's Education? Computer-Based Instruction in the Elementary School." *Education Age* 2(3), January–February 1966.

———. "The Uses of Computers in Education." *Scientific American* 215, 1967, 206–20.

Tuttle, "Historical Development."

Unpublished Sources

Email correspondence with author: Avner, A. (1997); Luherman, A. (1997); Stetten, K. (1997).

Other Documents

Bitzer, D., Letter to C. Victor Bunderson, 1971-08-31. CERL UIUC Archives.

Black, D. "The Teacher and the TICCIT System." Occasional Paper No. 3, February 1974, Institute for Computer Uses in Education, BYU.

Bunderson, C. V. Letters to Ken Stetten, MITRE Corporation, April 22, 1971; July 19, 1971; October 20, 1971; October 25, 1971; October 27, 1971.

———. "The TICCIT Learner Control Language." Occasional Paper No. 5, December 1975, Institute for Computer Uses in Education, BYU.

Hendricks, H. R. "The TICCIT System at Brigham Young University." CTS Technical Report #10, July 7, 1980, BYU Division of Learning Services.

TICCIT Project. "Minutes of the TICCIT Management Team Meeting," Boessenroth, T., Bunderson, C. V., Fine, S., O'Neal, F. 1974-08-15.

9. A Fork in the Road

Interview Sources

Author's Interviews: Alpert (2003a, 2003b); Donald Bitzer (1986, 1987, 2003); Blomme (2013); Frankel (1997); Goldberg (1987); Goldman (2003); Hiltzik (2003); Kay (1987, 1988); Liddle (2003); Pake (2003); Papert (1987); Propst (1997a, 1997b); Tenczar (1986, 1987, 2003).
Other Interviews: Kay (2012, 2013, 2016).

Published Sources

Doug Engelbart Institute, "Doug's 1968 Demo." Retrieved 2014-09-16 from http://www.dougengelbart.org/firsts/dougs-1968-demo.html.
Engelbart, D., and B. English. "A Research Center for Augmenting Human Intellect," in *Proceedings of the 1968 Fall Joint Computer Conference*, Vol. 33. San Francisco, December 9, 1968, 395–410.
Hiltzik, M. *Dealers of Lightning: Xerox PARC and the Dawn of the Computer Age*. San Francisco: HarperBusiness, 1999.
Isaacson, W. *Steve Jobs*. New York: Simon & Schuster, 2011.
Kay, A. "An Early History of Smalltalk. HOPL-II." *The Second ACM SIG-PLAN Conference on History of Programming Languages*, Association for Computing Machinery, 1993, 69–95.
———. "The Reactive Engine." PhD diss., University of Utah, 1969.
Kay, A., and A. Goldberg. "Personal Dynamic Media." *Computer* 10(3), 1977, 31–41.
Johnstone. *Never Mind*.
Smith, D. K., and R. C. Alexander. *Fumbling the Future: How Xerox Invented, Then Ignored, the First Personal Computer*. New York: William Morrow, 1988.

Unpublished Sources

Email correspondence with author: Bitzer, Donald (2003); Propst, F. (June 2003); Sproull, R. L. (2003).
Other documents: Koning, P. Response posted in Cyber1 notesfile, 2016.

10. Lessons Learned

Interview Sources

Author's Interviews: R. C. Anderson (2002); Donald Bitzer (1986, 1987, 2003a, 2003b); Bloomfield (2010); J. S. Brown (1987); Burr (2002); Cohen (2013); Corielle (2013); Desmond (2003); Dugdale (2003); Eades (2013); Fillman (1997a, 1997b); Ghesquiere (1997); Goldberg (1987); Gooch (1997); Halvorsen (1997); Kibbey (2013); Knoke (2002); Peters (2017);

Risken (2013); Russell (2003); Siegel (1997); S. Smith (1997); Steinberg (1997); Tenczar (1986, 1988, 2003); Yeager (2013a, 2013b).

Published Sources
"Alpert Named to New Posts." *DI*, May 25, 1972, 11.

Bitzer, D., and D. Skaperda. "PLATO IV—An Economically Viable Large-Scale Computer-Based Education System." CERL Report X-5, UIUC, February 1969.

Chabay, R., S. Dugdale, B. A. Seiler, B. Sherwood, and R. Pea. "PLATO@50: Online Education & Courseware: Lessons Learned, Insights Gleaned." Panel at the *PLATO@50* conference, Computer History Museum, Mountain View, CA, June 3, 2010. Retrieved 2010-08-19 from https://www.youtube.com/watch?v=rdDwoUk4ojY.

"Computers: No Longer a Classroom Novelty." *Chemical & Engineering News* 47(25), 1969, 48–49.

Davis, R. B. "What Classroom Role Should the PLATO Computer System Play?" *AFIPS '74 National Computer Conference and Exposition*, Association for Computing Machinery, 1974, 169–73.

Gilpin, M. O. *PLATO at Graham Correctional Center: Starting an Innovative Classroom.* CERL Report E-23, June 1982.

National Clearinghouse for Criminal Justice Planning and Architecture. "PLATO: Computer-Based Education." *Transfer* 11, November 1976, 1–4.

O'Connor, P. "Opinions Vary on Report on Graduate College." *DI*, 1972, July 14, 7.

PLATO Corrections Project Staff. "Computer-Based Education Has Been Introduced in Three Illinois Prisons." *American Journal of Correction*, January–February, 6, 7, 34–37.

Pringle, R. "Center Encourages Scholarship." *DI*, March 2, 1973, 19.

Slattery, J. "PLATO: 'Automated Page Turner'?" *DI* 102(20), October 10, 1972, 7.

UIUC Office of Public Information. Press release regarding appointment of Daniel Alpert to Director of Center for Advanced Study, May 23, 1972. UIUC Archives.

Unpublished Sources
Email correspondence with author: Dugdale, S. (2003); Kibbey, D. (1997, 2013); Siegel, M. (1997, 2010, 2015); Yeager, R. (2013–2015).

Other Documents
Esarey, J. C., letter to Dorothy Pondy, UIUC, November 14, 1977.

PLATO Corrections Project, descriptive handout. UIUC CERL, May 1976.

Siegel, M. "Computer-Based Education in Prison Schools." Undated monograph, UIUC.

———. "Field trip to Menard." CERL PCP memorandum to Don Bitzer, Frank Propst, Paul Tenczar, Bill Golden, November 22, 1976.

———. Letters to J. Clark Esarey, Superintendent, Department of Corrections, September 21, 1977; March 27, 1978; September 15, 1978; January 5, 1979; April 6, 1979.

PART II   The Fun They Had

Epigraph

Tempest, K. *Hold Your Own*. New York: Bloomsbury USA, 2015, 95.

Springer, P. G. "PLATO's Retreatists: The Sordid World of the Computer Addict." *High Times*, January 1984, 44.

Nelson, T. H. *Computer Lib/Dream Machines*. Redmond, WA: Microsoft Press, 1987, 13.

11. Impeachment

Interview Sources
Author's Interviews: Donald Bitzer (1996); Blomme (1986); B. and M. Fortner (1997); Golden (2003); R. Johnson (2003); J. and C. Kruutz (1997); Lamont (1997); Risser (2003); Sherwood and Chabay (1997); J. Sherwood (2003); Tenczar (1997); Umpleby (1997); Weber (2013).

Published Sources
Alternative Futures Project at the University of Illinois, Newsletter 1, January 1971. UIUC.
Alternative Futures Project at the University of Illinois, Newsletter 2, January 1972. UIUC.
Alternative Futures Project at the University of Illinois, Newsletter 3, March 1973. UIUC.
"Citizen Involvement: Can We Develop the Full Peoplepower of Democracy?" *New Ways*, C. F. Kettering Foundation, Summer 1973.
"Community Use of PLATO." *Prairie Dispatch* 1(1), April 5, 1973, 7.
"Electronic Electorate." *Engineering Outlook*, UIUC College of Engineering, 14(1), September 1972.
Lurie, L. *The Impeachment of Richard Nixon*. New York: Berkley, 1973.
Osgood, C., and S. Umpleby. "A Computer-Based System for Exploration of Possible Futures for Mankind 2000: A Progress Report." August 1967, UIUC.
"PLATO: One Teaching Computer for Two Games." *Gaming Newsletter* 1(3), September–October 1969, National Gaming Council.
UIUC Public Information Office. Press release, March 28, 1969.
———. Press release, June 8, 1972.
Umpleby, S. "The Alternative Futures Project at the University of Illinois." Retrieved 2012-12-02 from https://www2.gwu.edu/~umpleby/afp.html.
Wade, P. "City Agrees to Plug Pipe Near Boneyard." *NG*, April 13, 2011. Retrieved June 19, 2012 from http://www.news-gazette.com/news/local/2011-04-13/city-agrees-plug-pipe-near-boneyard.html.

———. "Why Is It Called the Boneyard Creek?" *NG*, April 2011. Retrieved June 19, 2012 from http://www.news-gazette.com/blogs/voice
-vote/2011-04/why-it-called-boneyard-creek.html.

Unpublished Sources
Email correspondence with author: Carter, G. (1997); Gilfillan, J. (May
2003); Robeson, Eric (2013); Pour, Ivan (April 2003); Woolley, D. (1997);
Umpleby, S. (1997).

Other Documents
Andreessen, M. Twitter correspondence with author, January 21, 2014.
PLATO Past notesfile, CERL PLATO system. Notes "afp" posted by carter/
comm on March 9, 1977 with three responses, and "creek" posted by
blomme/s on May 6, 1977 with twenty-four responses.
Umpleby's impeachment discussion in lesson "Discuss," printouts, October
1973, UIUC Archives.

12. The New Wave

Interview Sources
Author's Interviews: Andersen (2003); Beddini (2014a, 2014b); Bereiter
(2013); Bitzer, D. (1986, 1987, 1996, 1997, 2002); Blomme (1986, 2013); D.
Brown (2010); Frankel (1997); Fuller (2004); Mike Johnson (2003); J. and
C. Kraatz (1997); K. Mast (1997); P. Mast (1997); Rader (2001); Risser
(2003); Woolley (1987, 1997).

Published Sources
"High Tech Heroes #3." Mark Rustad interviewed by Sherwin Gooch,
1988. Retrieved 2013-01-25 from https://www.youtube.com/watch?v=cb
JB4PSG5nU.

Unpublished Sources
Email correspondence with author: Brown, D. (December 1996, June 2016);
Green, D. (2001, 2002, 2013); Kopf, D. (January 2002, April 2010); Mid-
den, M. (December 1996); Rustad, M. (July 2014, June 2016); Sherwood,
B. (1996, 1997); Woolley, D. (June 2016).

13. The Big Board

Interview Sources
Author's Interviews: Andersen (2003); Anonymous (1997); Banks (2002);
Blomme (1986, 2013); Bloomfield (2010); Bowery (1987); Duncombe
(1997); Galcher (1997); D. Green (2013); Harkrader (1997); J. Johnson
(2002); McGinty (1997); Rutherford (2014); Warner (1997); White (1997);
Woolley (1987).

Published Sources
Brandom, R. "'Spacewar!' The Story of the World's First Digital Video Game." *Verge*, February 4, 2013. Retrieved 2014-04-19 from http://www .theverge.com/2013/2/4/3949524/the-story-of-the-worlds-first-digital -video-game.

Unpublished Sources
Email correspondence with author: Bloomfield, L. (2010); Green, D. (2001, 2002, 2010, 2013); Joyner, R. (2004); Moss, R. (2016); Shankman, B. (1997); Witz, E. (2017); Woolley, D. (2010); Zweig, D. (2004, 2012).

14. The Killer App

Interview Sources
Author's Interviews: Andersen (2003); D. Brown (2010); Eisenberg (2004); Galcher (1997); Gooch (1997); K. Mast (2003); B. Sherwood (1997); Tenczar (1987, 1988); Warner (1997); Woolley (1987, 1997); Yeager (2013a, 2013b); Zinn (2003).

Published Sources
Hafner. *The WELL*.
Latzko-Toth, G. *La co-construction d'un dispositif sociotechnique de communication: le cas de l'internet relay chat*. PhD diss., Université du Québec à Montréal, 2010.
"On the Internet, Nobody Knows You're a Dog." Wikipedia. Retrieved 2014-05-02 from https://en.wikipedia.org/wiki/On_the_Internet,_nobody _knows_you're_a_dog.
Rheingold, *Virtual Community*.
Woolley, D. "Between PLATO and the Social Media Revolution." 1983. Retrieved 2013-01-04 from http://just.thinkofit.com/between-plato-and -the-social-media-revolution/.
———. "PLATO: The Emergence of Online Community," 1994. Retrieved 1995-12-14 from http://thinkofit.com/plato/dwplato.htm.

Unpublished Sources
Email correspondence with author: Anonymous (1997); Banks, F. (2002); Brinati, T. (2003); Brown, C. (2004); Brown, D. (1996, 1997); Cole, C. (2003); Daleske, J. (2002); Hobbs, E. (2005); Lionel, S. (1997); Marley, R. (2002); McNeil, A. (2004); Ozzie, R. (1997, 2014–2016); Parnes, B. (2002); Parry, J. (2007); Reichmann, L. (2015); Woolley, D. (1996, 1997, 2014, 2016).

15. Empire

Interview Sources
Author's Interviews: D. and M. Bitzer (2002); Blackmore (1997); Bowery (1987); Brown (2010); Daleske (2002); B. and M. Fortner (1997); Fritz

(1997); Galcher (1997); Golden (2003); Halvorsen (1997); Miller (1997); Munitz (2003); Rodby (2002); Shapira (1997, 1998); Warner (1997).

Published Sources

Daleske, J. "How Empire Came to Be." Daleske.com personal website essay. Retrieved 2016-01-19 from http://daleske.com/plato/empire.php.

"The Doomsday Machine (episode)." Wikia. Retrieved 2014-07-07 from http://memory-alpha.wikia.com/wiki/The_Doomsday_Machine _(episode).

Kriz, M. "Mr. Spock: Nimoy Faces the Challenges of Theatre." *DI*, May 9, 1974.

Nimoy, L. *I Am Spock*. New York: Hyperion, 1995.

———. *I Am Not Spock*. Berkeley, CA: Celestial Arts, 1975.

Viodova, B. "Roddenberry Raps." *DI*, November 8, 1974.

Woolley, D. "The Day I Met Leonard Nimoy." *Just Think of It* blog post. Retrieved 2015-02-28 from http://just.thinkofit.com/the-day-i-met -leonard-nimoy/.

Unpublished Sources

Email correspondence with author: Anonymous (1997); Bordelon, E. (1997); Brown, C. (2004); Churches, D. (2010); Clifford, P. (1997); Daleske (2002–2009); Frankel, D. (2002); Frye, G. (2001); Gilomen, B. (2002); Harbaugh, L. (2003); Kriz Hobson, M. (2010); Little, G. (2003); Midden, M. (2001); Rankaitis, S. (2003); Woolley, D. (1997).

16. Into the Dungeon

Interview Sources

Author's Interviews: Banks (2002); Blackmore (1997); Borton and Martz (1997); Daleske (1997); Duncombe (1997); Fritz (1997); Frye (1997, 2004); Galcher (1997); Glish (1997); Gooch (1997); Hagstrom (1997); Harkrader (1997); Jarecki (1997); J. Johnson (2002); Mark Johnson (2014); Kaven (2003); Lockard (1997, 2013); Maggs (1997); McGinty (1997); Miller (1997); Nordberg (1997); Paley (2003); Ozzie and Kawell (1997); Powers (2002); Roper (1997); Rutherford (2014); Shapira (1997, 1998); Sherwood and Chabay (1997); Sides (1997); Woolley (1997).

Published Sources

Barton, M. *Dungeons and Desktops: The History of Computer Role-Playing Games*. Wellesley, MA: A. K. Peters, 2008.

Peterson, J. *Playing at the World: A History of Simulating Wars, People, and Fantastic Adventures from Chess to Role-Playing Games*. San Diego: Unreason Press, 2012, 608–26.

Wright, W. "Lessons in Game Design." Lecture given at CHM on November 20, 2003. Retrieved 2016-01-29 from https://www.youtube.com/ watch?v=CdgQyq3hEPo.

——. "Sculpting Possibility Space." *Accelerating Change* podcast, November 7, 2004. Retrieved 2015-10-01 from http://web.archive.org/web/20130729231215id_/http://itc.conversationsnetwork.org/shows/detail376.html.

Unpublished Sources
Email correspondence with author: Bartle, R. (2003); Boland, N. (2001); Logg, E. (2017); Pellett, D. (1997); Pellett, F. (1997); Rubin, O. (2017); Walczak, D. (2001); Wei, M. (2003, 2013); Whisenhunt, G. (1997); Wood, R. (1997); Witz, E. (1997, 2002, 2017); Zelchenko, P. (1997, 2002).

17. The Zoo

Interview Sources
Author's Interviews: Anonymous (1997); Avner (1986); Blackmore (1997); Blomme (2013); B. and M. Fortner (1997); Galcher (1997); Glish (1997); Golden (2003); Gooch (1997); Harkrader (1997); Johnson, J. (2002); Kaven (2003); Kibbey (2013); Matheny (1997); McGinty (1997); Paley (2003); Powers (2002, 2003); Shapira (1997); Woolley (1987).

Published Sources
Powers, R. *Plowing the Dark*. New York: Farrar, Straus & Giroux, 2000, 1.
Richard Powers—American Novelist. Official website for R. Powers. Retrieved 2014-01-16 from http://www.richardpowers.net/.
Wood, P. "Addiction! Late-Night Computer Craziness on Campus." *NG*, December 13, 1986.

Unpublished Sources
Email correspondence with author: Anonymous (1997–2009); Clifford, P. (1997); Darling, D. (2006); Dewey, J. (2002); Esker, S. (2009); Gilomen, B. (2002); Giourdas, J. (2003); Harkrader (2010); Johnson, L. (2002); Krause, P. (2002); Kutzko, M. (2002); Leach, L. (2003); Manton, J. (1997, 2006); Paley, J. (2002); Powers (2002, 2003, 2010); Reichmann, L. (2015).

18. Red Sweater

Interview Sources
Author's Interviews: Aspnes (1997); Banks (2002); Bereiter (1997); Daleske (2002); Duncombe (1997); J. D. Eisenberg (2004); B. and M. Fortner (1997); Friedman (2003); Frye (1997); Fuller (2004); Galcher (1997); Ghesquiere (1997); Gilpin (1977); Glish (1997); Jay Green (1997); Halvorsen (1997); Kaven (2003); K. Mast (2003); McGinty (1997); Parrello (2004); Shapira (1997); Sherwood and Chabay (1997); Siegel (1997a, 1997b); Warner (1997); White (1997), Woolley (1997).

Published Sources
Asteroff, J. F. "Paralanguage in Electronic Mail: A Case Study." PhD diss., Teachers College, Columbia University, 1987.
Watkins, F. L. "NewsReport Paper Printed into Computer." *DI*, March 19, 1975, 18–20.

Unpublished Sources
Email correspondence with author: Anonymous (1997); Banks, F. (2002); Christenson, J. (2002); Fumento, D. (1997, 2010); Gold, S. (2005); Hobbs, E. (2005); Huben, M. (2003, 2005); Jones, D. W. (2006); Koning, P. (1997); Marley, R. (2002); McNeil, A. (2004); Midden, M. (October 2012); Parrello, B. (1996, 1997, 2002); Roper, B. (2010); Sebby, P. (1997); Simkin, R. (2003); Vojak, B. (2006); Ware, D. (2006); Weast, G. (2002, 2004); Winberg, D. (2005); Woolley, D. (1996, 1997); Zweig, D. (2004).

Other Documents
Groupe, Al. Printouts of NewsReport articles, undated.
Parrello, B. Message posted to Discuss forum on PLATO, October 21, 1973. UIUC Archives.

19. The Supreme Being and the Master of Reality

Interview Sources
Author's Interviews: Aspnes (1997); Daleske (2002); Eisenberg (2004); B. and M. Fortner (1997); Frye (1997); Galcher (1997); Gooch (1997); Graper (1997); Green (1997); Harkrader (1997); Kaven (2003); Lynch (1997); Parrello (1997); Roper (1997); Shapira (1998); White (1997); Woolley (1997).

Published Sources
Grapenotes website. Retrieved 1997-09-10 from http://www.grapenotes.com.
Turkle, S. *Life on the Screen*. New York: Simon & Schuster, 1997.
————. *The Second Self*. New York: Simon & Schuster, 1984.

Unpublished Sources
Email correspondence with author: Graper, D. (1996, 1997); Lynch (1997); Neuman, M. (1997); Parrello, B. (1997); Schwartz, R. (2005); Wacaser, T. (2007); Warfield, A. (2003); Widitz, R. (1997).

20. Climbing the Ziggurat

Interview Sources
Author's Interviews: Anonymous (1997); Bloomfield (2003); Donald Bitzer (1997); D. and M. Bitzer (2002); B. and M. Fortner (1997); Golden (2003); D. Green (2013); Harkrader (1997); E. Jarecki (2003), H. and G. Jarecki (2002); McGinty (1997); Ozzie and Kawell (1997); J. Paley (2003); Powers

(2002); Rader (2001); Risken (2013); Shapira (1997, 1998); Woolley (1997); Yeager (2013a, 2013b).

Published Sources
Wolfe. *The Right Stuff*, 18–19.

Unpublished Sources
Email correspondence with author: Green, D. (2001, 2010, 2013); Jarecki, G. (2002); Jarecki, H. (2002); Johnson, L. (2002); Lipper, A. (1996, 1997, 2002, 2003); Lipper, G. (1996, 1997); Redman, B. (1997); Soussan, D. (2005).

21. Coming of Age

Interview Sources
Author's Interviews: Anonymous (1997); Blackmore (1997); Bowery (1987); Corielle (2013); Daleske (1997); Duncombe (1997); Frye (1997); Fuller (2004); Galcher (1997); Golden (2003); D. Green (2013); Harkrader (1997); J. Johnson (2002); Kaven (2003); Kirkpatrick (1986); B. Maggs (1997); Matheny (1997); McGinty (1997); Nelson (1987); Ozzie and Kawell (1997); Sides (1997); Woolley (1997).
Other interviews: Bitzer (1975).

Published Sources
Barnett, D. C. "A Suicide Prevention Incident Involving Use of the Computer." *Professional Psychology* 13(4), August 1982, 565–70.
DalSanto, Bob. "The Wonderful World of PLATO." *DI*, September 20, 1975, S-4, S-5.
Metcalfe, Bob. "Metcalfe's Law: A Network Becomes More Valuable as It Reaches More Users." *Infoworld* 17(40), October 2, 1995, 53.
Nelson. *Computer Lib/Dream Machines*, 93–95.
Pesman, C. "PLATO Games People Play." *DI*, September 20, 1975, S-2, S-3.
Rheingold. *Virtual Community*.
Springer, P. G. "PLATO's Retreatists: The Sordid World of the Computer Addict." *High Times*, January 1982, 44–59.
UIUC Office of Public Information. Press release on NSF funding and 4,000-terminal PLATO IV, July 15, 1970. UIUC Archives.
———. Press release on upcoming PLATO IV system, April 20, 1971. UIUC Archives.
Wood, P. "Addiction! Late-Night Computer Craziness on Campus." *NG*, December 13, 1986.

Unpublished Sources
Email correspondence with author: Barnett (1997, 2017); Carroll, M. (2010); McGee, A. (2005); Miller, N. H. (2004); Otto, A. (2001); Rankaitis (2003); Reichmann (2015); Rose, S. (2001); Schwager, M. (2003); Schwartz, R. (2005, 2007); Stawicki, R. (2001); Woodhead (1996, 1997, 2002, 2004).

Other Documents:
PLATO Past notesfile. CERL PLATO system. "Press Help," June 9, 1977.
Public Notes notesfile. CERL PLATO system. "marriage" May 1, 1977; "Wedding," January 9, 1981.
Public Notes notesfile, NovaNET system. "Another Couple?," November 30, 1991, "NovaNET plays Cupid," April 12, 1995.

PART III   Getting to Scale

Epigraph

Roosevelt, T. Speech entitled "The Strenuous Life," April 10, 1899. Reprinted on *Voices of Democracy* website. Retrieved 2016-09-09 from http://voicesofdemocracy.umd.edu/roosevelt-strenuous-life-1899 -speech-text/.
Kahneman, D. "How to Be Amazing" podcast, hosted by Michael Ian Black, Episode 42. Retrieved 2016-12-29 from http://howtobeamazingshow .com/episodes/2016/10/12/episode-42-daniel-kahneman.
Frazier, C. *Cold Mountain.* New York: Atlantic Monthly Press, 1997, 380.

22. The Business Opportunity

Interivew Sources
Author interviews: M. Allen (1997, 2016); Alpert (2003a, 2003b); Donald Bitzer (1986, 1987, 2002, 2003); Borton and Martz (1997); Cole (2002); Fillman (1997); M. and S. Frishberg (1997); Glish (1997); Golden (2003); Hill (1997); Mark Johnson (2014); P. Maggs (1997); Morris (1997, 2007); Nordberg (1997); Norris (1986, 1997); Poulos (1997); Seidel (2015); Sherwood and Chabay (1997); Sides (1997); S. Smith (1997); Stifle (1997); Stubbs (2002); Tenczar (1986, 1987, 1997, 2003); Tripp (1997).
Other interviews: Bitzer (1975, 1982, 1984, 1988, 2002a, 2002b), Gallie (1990), Golub (1979); Molnar (1991); Norris (1986); Patton (2001); Price (2006, 2012).

Published Sources
Bitzer, D. "The Million Terminal System of 1985," in R. J. Seidel and M. L. Rubin, eds., *Computers and Communication: Implications for Education,* proceedings of the Conference Computer Technology in Education for 1985 at Airlie House, Warrenton, VA, on September 15–18, 1975, New York: Academic Press, 1977, 59–70.
"Frustrating Warfare of Business," *Life, 62(10),* May 5, 40–52.
Melton, M. PLATO Co-inventor Seeks Sidewalk Computers." *Urbana Courier,* February 21, 1977.
Nader and Taylor. *Big Boys,* 449–503.
Nelson. *Computer Lib/Dream Machines,* 93–95.

Pantages, A. "Control Data's Education Offering: 'Plato Would Have Enjoyed PLATO.'" *Datamation*, May 1976, 183, 186–87.

"PLATO to Go Commercial." *DI*, April 22, 1971, 8.

"Russia Interested in UI Computer." *NG*, January 24, 1974.

"Russians Interested in PLATO Hook-up." *DI*, January 24, 1974.

UIUC Office of Public Information. Press release describing CERL PLATO demo in Soviet Union, November 21, 1973. UIUC Archives.

U.S. Congress, House of Representatives, 95th Congress, First Session. Committee on Science and Technology. *Computers and the Learning Society*, No. 47. October 4, 6, 12, 13, 18, and 27, 1977, 154–83.

Unpublished Sources

Email correspondence with author: Cole, B. (1998, 2011); Dallas, A. (2002); Moe, R. (1997); Rizza, P. (2016); Robertson, P. (2002); Smith, M. (2002); Sherwood, B. (2016); Stawicki, R. (2001); Vernon, N. (2016); Warren, G. (1998, 2016, 2017).

Other Documents

Bitzer, D. Speech and PLATO demonstration given in Perth, Australia, 1976. Author's transcript of Control Data Corporation video.

Dammeyer, J. "Some Notes on the History of CDC PLATO." Handwritten memo, 1980. CBI-UM Series 33, Box 2, Folder 5.

Drygulch Help Lesson, screens captured by author, 1987–88.

Morris, R. "The Commercialization of PLATO: Random Thoughts and Notes." Unpublished monograph, March 1982.

Rizza, P. "Peter Rizza and PLATO." Unpublished essay, February 2016.

Stawicki, R. "John Dammeyer and PLATO." Unpublished essay, May 2004.

Warren, G. "CDC PLATO—John Dammeyer." Unpublished essay, December 2016.

23. The Whim of Iron

Interview Sources

Author's Interviews: Borton and Martz (1997); Burson (2002); Campbell (2015); M. and S. Frishberg (1997); Ghesquiere (1997); Glish (1997); Harkrader (1997); Hansen (2016); Kaven (2003); Lacey (2015); C. Miller (1997); Morris (1997); Nordberg (1997); Price (1997); Rader (2001); Risken (2013); Sherwood and Chabay (1997); Siegel (1997); Stubbs (2002); Woolley (1997).

Other interviews: Bitzer (1984); Linsenman (1984); Price (2006, 2012).

Published Sources

Jensen, M. *HR Pioneers: A History of Human Resource Innovations at Control Data Corporation*. St. Cloud, MN: North Star Press, 2013.

Nader and Taylor. *Big Boys*, 449–503.

PLATO Homelink, software package. Control Data Publishing Company, 1984.

Unpublished Sources
Email correspondence with author: Allard, J. (2006); Harris, M. A. (2002); Kaven (2017); Kohlmiller, P. (2002); Peper, E. (2003); Porter, E. C. (2002, 2009, 2013); Quade, S. (2007); Rizza, P. (2016); Rothenberg, R. (2002); Samarias J. (2010); Sebby, P. (1997); Vernon, N. (2016).

Other Documents
Adkins, S. "PLATO experiences." Personal communication, 2011.
Dammeyer, J. "Some Notes on the History of CDC PLATO." Handwritten memo, 1980. CBI-UM Series 33, Box 2, Folder 5.
Rizza, P. "Peter Rizza and PLATO." Unpublished essay, February 2016.

24. Diaspora

Interview Sources
Author's Interviews: Gooch (1997); F. Hofstetter (1987); L. Hofstetter (1997); Lynch (1997).
Other interviews: Price, R. (2006, 2012).

Published Sources
Hofstetter, F. "Back to the Future of Educational Technology," in Allen, *e-Learning Annual*, 59–82.
———. *Computer Literacy for Musicians*. Englewood Cliffs, NJ: Prentice Hall, 1988.
Price. *Eye for Innovation*, 190–208.
Worthy. *William C. Norris*, 83–106.

Unpublished Sources
Email correspondence with author: Graper (1997); Schucker, F. (2016); Scriven, M. (2003).

Other Documents
University of Delaware PLATO System Public Notes (=pbnotes=) notesfile, 1978–1986.
University of Delaware PLATO System UDNotes (=udnotes=) notesfile, 1982–86.

25. The Crash Pit

Interview Sources
Author's interviews: Chung (2013); Geiger (2015); Lockard (1997, 2013); Poulos (1997).

Published Sources

Fregger, B. *Lucky That Way: Stories of Seizing the Moment While Creating Games Millions Play*. Austin, TX: Groundbreaking Press, 2009.

"Mahjong." Wikipedia. Retrieved 2013-12-15 from https://en.wikipedia.org/wiki/Mahjong.

Reifkind, M. "The Best Pommel Horse Gymnast of All Time: Ted Marcy," *Rifs* blog, 2010. Retrieved 2013-12-12 from http://rifsblog.blogspot.com/2010/01/best-pommell-horse-gymnast-of-all-time.html.

Rep, J. *The Great Mahjong Book*. Rutland, VT: Tuttle Publishing, 2007.

Stone, A. "Gymnast Remains Paralyzed After December Accident." *Stanford Daily*, 1980, 3.

Unpublished Sources

Email correspondence with author: Chung (2014); Fregger (2013, 2014); Lockard (1997, 2013, 2014); Marcy, T. (2013); Wright, G. (2014).

Other Documents

Lockard, B. *Mah-Jongg* PLATO lesson, 1983. Retrieved 2009-10-20 on Cyber1.org system.

Lockard, B., and W. Lockard. *Edges* lesson on CERL PLATO system, 1979. Retrieved 2013-12-09 on NovaNET system.

26. A Changing of the Guard

Interview Sources

Author's Interviews: Berdahl (1997); Donald Bitzer (1986); Fillman (1997a, 1997b); Fleming (2009); Friedman (2003); C. Gardner (2015); K. Gardner (2002, 2015); Goldwasser (2002); Haken (1997a, 1997b); Harkrader (1997); J. and C. Kraatz (1997); Lyman (1986); Matheny (1997a, 1997b); Nordberg (1997); Norris (1986); Peltason (2002); Pondy (2002); Propst (1997a, 1997b, 2013); Siegel (1997a, 1997b); Smarr (2011); Tenczar (1986); Walker (2003). Several additional interviews were off the record.

Other interviews: Bitzer (1988).

Published Sources

Nader and Taylor. *Big Boys*, 449–503.

"Plato Learning, Inc.—Company Profile, Information, Business Description, History, Background Information." Reference for Business. Retrieved 2016-12-16 from http://www.referenceforbusiness.com/history2/62/Plato-Learning-Inc.html.

United States Court of Appeals, Seventh Circuit (1994). *Propst, B. and Propst, M. v. Bitzer, D. L., Weir, M. W., Leibman, J. S., and Berdahl, R. M.*, 39 F.3d 148, 95 Ed. Law Rep. 505. Nos. 93-3825, 93-3826. Retrieved 2016-11-19 from http://law.justia.com/cases/federal/appellate-courts/F3/39/148/512186/.

Unpublished Sources

Email correspondence with the author: Burr (2002); Gunsalus (1997); Kannan, N. (2016); Michael, Mary (2003); Propst (1997, 2002, 2004, 2005); Scaletti, C. (2016).

Other Documents

Avner, A. "nova > cerl sys," note posted to CERL Public Notes, PLATO system, Note 94, 1991-02-08.

———. "PLATO Leaves Hawaii." A note posted in Public Notes announcing the shutdown of the PLATO system at University of Hawaii. Note 125, June 26, 1995.

Bitzer, D. Letter to Chancellor Thomas E. Everhart, March 8, 1987.

Frye, D. "Up," note posted to CERL Public Notes, PLATO system, January 8, 1987.

Gormley, J. "A Fond Farewell." A note posted amidst the final years of CERL when many were leaving the organization voluntarily or involuntarily, in Public Notes on NovaNET announcing her leaving her job of fifteen years at CERL. Several weeks' worth of well-wishing responses followed. Then, on February 11, 1993, a month after her farewell note, a user replied announcing she had passed away suddenly from a burst aneurysm.

Jones, L. Posted a reply to a question raised in Public Notes about whether NovaNET was closing down. Jones said it was not. Note 58, response 1, March 29, 1993. She followed up with additional facts in Note 58, response 4, March 30, 1993.

———. Posted a response to a note in Public Notes amid continued swirling rumors about NovaNET's demise announcing that "UIUC's Vice Chancellor for Research, Chester S. Gardner, had decided to phase out the CERL lab, and have University Communications (UCI) assume NovaNET operations . . ." Note 99, response 4, July 2, 1993.

Kraatz, J. "JMK leaves," a farewell note posted by Jim Kraatz after being at CERL since 1969. Note 115, August 5, 1993.

McGinty, B. "Transition," a note posted in Public Notes on NovaNET announcing transition of NovaNET system operations and development from CERL to UCI, Note 104, July 20, 1993.

Milosevich, S. "Bitzer retires," note posted in *PLATO Past* notesfile on CERL PLATO system, August 11, 1989, archiving a note posted by "jill/cerl" on August 13, 1989 publicizing a memo sent to "PLATO/NovaNET community" with the text of Bitzer's retirement announcement.

Propst, F. Draft of an essay on PLATO as a new communications medium, 1983.

———. Letter to author, November 14, 002.

Zvilius, M., et al. "novanet," Note posted in Pad notesfile, CERL PLATO system March 26, 1987.

Zvilius, M., L. Haken, and M. Walker. "Zephyr computer," *PLATO Past* notesfile note and responses, CERL PLATO system, February 28, 1990 through April 8, 1991.

27. Leaving the Nest

Interview Sources
Author interviews: M. Allen (1997, 2016); B. and M. Fortner (1997); Halvorsen (1997); Horne (2016); Kapor (2003); Manzi (2003); Ozzie and Kawell (1997); J. Sherwood (2003); Tenczar (1986, 1987, 1997, 2003); Warner (1997); Woolley (1997); Yeager (2013a, 2013b).
Other interviews: Alfille (2000).

Published Sources
"Company News: Lotus Buying Developer of Its Notes Software." *New York Times*, May 24, 1994. Retrieved 2016 from http://www.nytimes .com/1994/05/24/business/company-news-lotus-buying-developer-of -its-notes-software.html.
IBM Corporation. "The History of Notes and Domino." Retrieved 2017-01-15 from http://www.ibm.com/developerworks/lotus/library/ ls-NDHistory/.
Lammers. *Programmers at Work*, 175–89.
Maher, J. "Castle Wolfenstein." *The Digital Antiquarian* blog, 2012. Retrieved 2016-10-10 from http://www.filfre.net/2012/04/castle-wolfenstein/.
———. "Making Wizardry." *The Digital Antiquarian* blog, 2012. Retrieved 2016-10-20 from http://www.filfre.net/2012/03/making-wizardry/.
———. "The Wizardry and Ultima Sequels." *The Digital Antiquarian* blog, 2012. Retrieved 2016-10-20 from http://www.filfre.net/2012/03/making -wizardry/.
Ozzie, Ray. "Speaking Mind to Mind." Op-ed, *New York Times*, December 1, 2002.
UIUC Computer Science Department. "Bruce Artwick Is Still Flying." *Alumni News* 1(6), Spring 1996, 10–11.

Unpublished Sources
Email correspondence with author: Alfille (2002); Baumann, K. (2003); Greenberg, A. (1996–1997); Gorback, M. (2003); Huben, M. (2005); Klass (2017); Maher, J. (2016); Morris (1997); O'Hara, M. J. (1997); Ozzie, R. (1997–2015); Platt, R. C. (2003); Sachs, J. (2003); Sides, D. (1997); Tenczar (2017); Woodhead, R. (1996, 1997, 2002, 2004); Zarnowski, P. (1997).

Other Documents
Ozzie, R. "First All-Hands September 1998 Groove Meeting PowerPoint Presentation." 1998.
———. "Life Beyond Lotus." Presentation at *PC Forum 1991*. New York: EDventure Holdings, 1991.

Epilogue

Interview Sources
Author's interviews: Appleman (2002); Koller (2014); Thrun (2014).

Published Sources
Kingery et al. *Men and Ideas*, 164.
Nader and Taylor. *Big Boys*, 449–503.

Unpublished Sources
Email correspondence with author: Appleman, D. (2015, 2016).

Index

Page numbers in *italics* refer to photos and illustrations.

Illustration Credits

201: Photo provided by Stu Umpleby

222: Used with permission of the *Champaign-Urbana News-Gazette*, 1974.

279: Photo courtesy of Chuck Miller and John Daleske

308: Photo by Brian K. Johnson, *News-Gazette*, Champaign, Illinois, 1986

388: Drawing by Ted Nelson. Used with permission.

406: Image courtesy of the Charles Babbage Institute, University of Minnesota Libraries, Minneapolis, Minnesota

412: Photo provided by Bob Morris

414: Photo provided by Bob Morris

416: Photo by Gary Brown, 2007

419: Photo provided by Bob Morris

420: Photo provided by Bob Morris

422: Photo provided by Bob Morris

427: Image courtesy of the Charles Babbage Institute, University of Minnesota Libraries, Minneapolis, Minnesota

460: Screen image provided by Bill Lynch

A Note About the Author

Brian Dear is a longtime tech start-up entrepreneur and founder of companies, including Coconut Computing, FlatWorks, Eventful, and Nettle. He has also worked at a variety of dot-com companies, including MP3 and eBay. He worked in the field of computer-based education for eight years, including five on the PLATO system. He has written for *Educational Technology, BYTE, IEEE Expert,* and the *San Diego Reader.* He lives with his wife in Santa Fe, New Mexico.

A Note on the Type

This book was set in Janson, a typeface named for the Dutchman Anton Janson, but is actually the work of Nicholas Kis (1650–1702). The type is an excellent example of the influential and sturdy Dutch types that prevailed in England up to the time William Caslon (1692–1766) developed his own incomparable designs from them.

Composed by North Market Street Graphics,
Lancaster, Pennsylvania

Printed and bound by Berryville Graphics,
Berryville, Virginia

Design by Michael Collica